Monte Carlo Methods and Models in Finance and Insurance

CHAPMAN & HALL/CRC
Financial Mathematics Series

Aims and scope:
The field of financial mathematics forms an ever-expanding slice of the financial sector. This series aims to capture new developments and summarize what is known over the whole spectrum of this field. It will include a broad range of textbooks, reference works and handbooks that are meant to appeal to both academics and practitioners. The inclusion of numerical code and concrete real-world examples is highly encouraged.

Published Titles

American-Style Derivatives; Valuation and Computation, *Jerome Detemple*
Analysis, Geometry, and Modeling in Finance: Advanced Methods in Option Pricing, *Pierre Henry-Labordère*
Credit Risk: Models, Derivatives, and Management, *Niklas Wagner*
Engineering BGM, *Alan Brace*
Financial Modelling with Jump Processes, *Rama Cont and Peter Tankov*
Interest Rate Modeling: Theory and Practice, *Lixin Wu*
An Introduction to Credit Risk Modeling, *Christian Bluhm, Ludger Overbeck, and Christoph Wagner*
Introduction to Stochastic Calculus Applied to Finance, Second Edition, *Damien Lamberton and Bernard Lapeyre*
Monte Carlo Methods and Models in Finance and Insurance, *Ralf Korn, Elke Korn, and Gerald Kroisandt*
Numerical Methods for Finance, *John A. D. Appleby, David C. Edelman, and John J. H. Miller*
Portfolio Optimization and Performance Analysis, *Jean-Luc Prigent*
Quantitative Fund Management, *M. A. H. Dempster, Georg Pflug, and Gautam Mitra*
Robust Libor Modelling and Pricing of Derivative Products, *John Schoenmakers*
Stochastic Financial Models, *Douglas Kennedy*
Structured Credit Portfolio Analysis, Baskets & CDOs, *Christian Bluhm and Ludger Overbeck*
Understanding Risk: The Theory and Practice of Financial Risk Management, *David Murphy*
Unravelling the Credit Crunch, *David Murphy*

Proposals for the series should be submitted to one of the series editors above or directly to:
CRC Press, Taylor & Francis Group
4th, Floor, Albert House
1-4 Singer Street
London EC2A 4BQ
UK

Chapman & Hall/CRC FINANCIAL MATHEMATICS SERIES

Monte Carlo Methods and Models in Finance and Insurance

Ralf Korn
Elke Korn
Gerald Kroisandt

CRC Press is an imprint of the
Taylor & Francis Group, an **informa** business
A CHAPMAN & HALL BOOK

CRC Press
Taylor & Francis Group
6000 Broken Sound Parkway NW, Suite 300
Boca Raton, FL 33487-2742

Printed in the United States of America on acid-free paper
10 9 8 7 6 5 4 3 2 1

International Standard Book Number: 978-1-4200-7618-9 (Hardback)

Library of Congress Cataloging-in-Publication Data

Korn, Ralf.
Monte Carlo methods and models in finance and insurance / Ralf Korn, Elke Korn, Gerald Kroisandt.
p. cm. -- (Financial mathematics series)
Includes bibliographical references and index.
ISBN 978-1-4200-7618-9 (hardcover : alk. paper)
1. Business mathematics. 2. Insurance--Mathematics. 3. Monte Carlo method. I. Korn, Elke, 1962- II. Kroisandt, Gerald. III. Title. IV. Series.

HF5691.K713 2010
518'.282--dc22 2009045581

Visit the Taylor & Francis Web site at
http://www.taylorandfrancis.com

and the CRC Press Web site at
http://www.crcpress.com

Contents

List of Algorithms

Chapter 1

Introduction and User Guide

1.1 Introduction and concept

Monte Carlo methods are ubiquitous in applications in the finance and insurance industry. They are often the only accessible tool for financial engineers and actuaries when it comes to complicated price or risk computations, in particular for those that are based on many underlyings. However, as they tend to be slow, it is very important to have a big tool box for speeding them up or – equivalently – for increasing their accuracy. Further, recent years have seen a lot of developments in Monte Carlo methods with a high potential for success in applications. Some of them are highly specified (such as the Andersen algorithm in the Heston setting), others are general algorithmic principles (such as the multilevel Monte Carlo approach). However, they are often only available in working papers or as technical mathematical publications.

On the other hand, there is still a lack of understanding of the theory of finance and insurance mathematics when new numerical methods are applied to those areas. Even in very recent papers one sees presentations of big breakthroughs when indeed the methods are applied to already solved problems or to problems that do not make sense when viewed from the financial or insurance mathematics side.

We therefore have chosen an approach that combines the presentation of the application background in finance and insurance together with the theory and application of Monte Carlo methods in one book. To do this and still keep a reasonable page limit, compromises in the presentation are unavoidable. In particular, we will not to give strict formal proofs of results. However, one can often use the arguments given in the book to construct a rigorous proof in a straightforward way. If short and nontechnical arguments are not easy to provide, then the related references are given. This will keep the book at a reasonable length and allow for fluent reading and for getting fast to the point while avoiding burdening the reader with too many technicalities. Also, our intention is to give the reader a feeling for the methods and the topics via simple pedagogical examples and numerical and graphical illustrations. On this basis we try to be as rigorous and detailed as possible. This in particular means that we introduce the financial and actuarial models in great detail and also comment on the necessity of technicalities.

In our approach, we have chosen to separately present the Monte Carlo techniques, the stochastic process basics, and the theoretical background and intuition behind financial and actuarial mathematics. This has the advantage that the standard Monte Carlo tools can easily be identified and do not need to be separated from the application by the reader. Also, it allows the reader to concentrate on the main principles of financial and insurance mathematics in a compact way. Mostly, the chapters are as self-contained as possible, although the later ones are often building up on the earlier ones. Of course, all ingredients come together when the applications of Monte Carlo methods are presented within the areas of finance and insurance.

1.2 Contents

We have chosen to start the book with a survey on random number generation and the use of random number generators as the basis for all Monte Carlo methods. It is indeed important to know whether one is using a good random number generator. Of course, one should not implement a new one as there are many excellent ones freely available. But the user should be able to identify the type of generator that his preferred programming package is offering or the internal system of the company is using. Modern aspects such as parallelization of the generation of random numbers are also touched and are important for speeding up Monte Carlo methods.

This chapter is followed by an introduction to the Monte Carlo method, its theoretical background, and the presentation of various methods to speed them up by what is called variance reduction. The application of this method to stochastic processes of diffusion type is the next step. For this, basics of diffusion processes and the Itô calculus are provided together with numerical methods for solving stochastic differential equations, a tool that is essential for simulating paths of e.g. stock prices or interest rates. Here, we already present some very recent methods such as the statistical Romberg or the multilevel Monte Carlo method.

The fifth chapter contains an introduction to both classical stock option pricing, more recent stock price models in the diffusion context, and interest rate models. Here, many nonstandard models (such as stochastic volatility models) are presented. Further, we give a lot of applications of Monte Carlo methods in option pricing and interest rate product pricing. Some of the methods are standard applications of what have been presented in the preceding chapters, some are tailored to the financial applications, and some have been developed more recently.

In the sixth and seventh chapters we leave the diffusion framework. Stochastic processes that contain jumps such as jump-diffusions or Lévy processes enter the scene as the building blocks for modelling the uncertainty inherent in financial markets. Developing efficient Monte Carlo methods for Lévy models is a very active area that still is at its beginning. We therefore only present the basics, but also include some spectacular examples of specially tailored algorithms such as the variance gamma bridge sampling method.

Finally, in Chapter 8, some applications of Monte Carlo methods in actuarial mathematics are presented. As actuarial (or insurance) mathematics has many aspects that we do not touch, we have only chosen to present some main areas such as premium priniciples, life insurance, nonlife insurance, and asset-liability management.

1.3 How to use this book

This book is intended as an introduction to both Monte Carlo methods and financial and actuarial models. Although we often avoid technicalities, it is our aim to go for more than just standard models and methods. We believe that the book can be used as an introductory text to finance and insurance for the numerical analyst, as an introduction to Monte Carlo methods for the practitioners working in banks and insurance companies, as an introduction to both areas for students, but also as a source for new ideas of presentation that can be used in lectures, as a source for new models and Monte Carlo methods even for specialists in the areas. And finally the book can be used as a cooking book via the collection of the different algorithms that are explicitly stated where they are developed.

There are different ways to read the contents of this book. Although we recommend to have a look at Chapter 2 and the aspects of generation of random numbers, one can directly start with the presentation of the Monte Carlo method and its variants in Chapters 3 and 4. If one is interested in the applications in finance then Chapter 5 is a must. Also, it contains methods and ideas that will again be used in Chapters 6 through 8.

If one is interested in special models then they are usually dealt with in a self-contained way. If one is only interested in a special algorithm for a particular problem then this one can be found via the table of algorithms.

1.4 Further literature

All topics covered in this book are popular subjects of applied mathematics. Therefore, there is a huge amount of monographs, survey papers, and lecture notes. We have tried to incorporate the main contributions to these areas. Of course, there exist related monographs on Monte Carlo methods. The book that comes closest to ours in terms of applications in finance is the excellent monograph by Glasserman (2004) which is a standard reference for financial engineers and researchers. We have also benefitted a lot from reading it, but have of course tried to avoid copying it. In particular, we have chosen to present financial and actuarial models in greater detail. A further recent and excellent reference is the book by Asmussen and Glynn (2007) that has a broader scope than the applications in finance and insurance and that needs a higher level of preknowledge as our book. In comparison to both these references, we concentrate more on the presentation of the models.

More classic and recent texts dealing with Monte Carlo simulation are Rubinstein (1981), Hammersley and Handscomb (1964), or Ugur (2009) who also considers numerical methods in finance different from Monte Carlo.

1.5 Acknowledgments

As with all books written there are many major and minor contributors besides the authors. We have benefitted from many discussions, lectures, and results from friends and colleagues. Also, our experiences from recent years gained from lecture series, student comments, and in particular industry projects at the Fraunhofer Institute for Industrial Mathematics ITWM at Kaiserslautern entered the presentations in the book.

We are happy to thank Christina Andersson, Roman Horsky and Henning Marxen for careful proof reading and suggestions. Assistance in providing numbers and code from Georgi Dimitroff, Nora Imkeller, and Susanne Wendel is gratefully acknowledged. Further, the authors thank Hansjörg Albrecher and participants of the 22nd Summer School of the SWISS Association of Actuaries at the University of Lausanne for many useful comments, good questions, and discussions.

Finally, the staff at Taylor & Francis/CRC Press has been very friendly and supportive. In particular, we thank (in alphabetical order) Rob Calver, Kevin Craig, Shashi Kumar, Linda Leggio, Sarah Morris, Katy Smith, and Jessica Vakili for their great help and encouragement.

Chapter 2

Generating Random Numbers

2.1 Introduction

2.1.1 How do we get random numbers?

Stochastic simulations and especially the Monte Carlo method use random variables (**RV**s). So the ability to provide random numbers (**RN**s) with a specified distribution becomes necessary. The main problem is to find numbers that are really random and unpredictable. Of course, throwing dice is much too slow for most applications as usually a lot of RNs are needed. An alternative is to use physical phenomena like radioactive decay, which is often considered as a synonym for randomness, and then to transform measurements into RNs. With the right equipment this works much faster. But how can one ensure that the required distribution is mimicked? Indeed, modern research makes it possible to use physical devices for random number generation by transforming the measurements so that they deliver a good approximation to a fixed distribution, but those devices still are too slow for extensive simulations. Another disadvantage is that the sequence of RNs cannot be reproduced unless they have been stored. Reproducibility is an important feature for Monte Carlo methods, e.g. for variance reduction techniques, or simply for debugging reasons. However, physical random number generators (**RNG**s) are useful for applications in cryptography or gambling machines, when you have to make sure that the numbers are absolutely unpredictable.

The RNs for Monte Carlo simulations are usually generated by a numerical algorithm. This makes the sequence of numbers deterministic which is the reason why those RNs are often called **pseudorandom numbers**. But if we look at them without knowing the algorithm they appear to be random. And in many statistical tests they behave like true random numbers. Of course, as they are deterministic, they will fail some tests. When choosing a RNG we should make sure that the RNs are thoroughly tested and we should be aware which tests they fail.

First of all, we concentrate on generating uniformly distributed RNs, especially uniformly distributed real RNs on the interval from zero to one, $u \sim U[0,1]$, $U(0,1)$, $U(0,1]$ or $U[0,1)$. Theoretically, whether 1 or 0 are included or not does not matter, as the probability for a single number is zero. But the user should be aware whether the chosen RNG delivers zeroes and ones as this might lead to **programme crashes**, as e.g. $\ln(0)$.

In a second step these uniformly distributed RNs will be transformed into RNs of a desired distribution (such as e.g. the normal or the gamma one).

Nowadays there exist a lot of different methods to produce RNs and the research in this area is still very active. The reader should be aware that the top algorithms of today might be old-fashioned tomorrow. Also the subject of writing a good RNG is so involved that it cannot be covered here in all its details. So, we will not describe the perfect RNG in this chapter (in fact it does not exist!). Here we will give the basics for understanding a RNG and being able to judge its quality. Also, we will try to give you some orientation to choose the one that suits your problem. After reading this chapter you should **not** go and write your own RNG, except for research reasons and to collect experience. It is better to search for a good, well-programmed and suitable up-to-date RNG, or to check the built-in one in your computer package, accept it as usable, and be happy with it (for some time) or otherwise, if it is no good, mistrust the simulation results.

Although implementing RNGs is such an enormous and complicated field, we will present some simple algorithms here so that you can experiment a little bit for yourself. But do not see those algorithms as recommendations, the only aim is to get familiar with the technical terms.

2.1.2 Quality criteria for RNGs

For a reliable simulation good random numbers are imperative. A bad RNG could cause totally stupid simulation results which can lead to false conclusions. Before using an built-in RNG of a software programme, it should be checked first. Here are some quality criteria the user should be aware of.

• Of course, uniformly distributed RNs $\sim U[0,1]$ should be **evenly distributed** in the interval $[0,1]$, but the structure should not be too regular. Otherwise they will not be considered as random. However, in some simulation situations we are able to take advantage of regular structures, especially when working with smooth functions. In these cases we can use **quasirandom sequences**, which do not look random at all, but are very evenly distributed.

• As in Monte Carlo simulations, where we often need lots of RNs, it should be possible to produce them very fast and efficiently without using too much memory. So, **speed** and **memory requirements** matter.

• RN algorithms work with a finite set of data due to the construction principle of the algorithm and due to the finite representation of real numbers in the computer. So the RNG will eventually repeat its sequence of numbers. The maximal length of this sequence before it is repeated is called the **period** of the RNG. For the extensive simulations of today huge amounts of RNs are needed. Therefore the period must be long enough to avoid using the same

RNs again. A rule of thumb says that the period length of the RNG should be at least as long as the square of the quantity of RNs needed. Otherwise the deterministic aspect of the RNG comes into play and makes the RNs correlated. In former times RNGs with a period of about 10^{19} were considered as good, but nowadays those algorithms are considered as insufficient. Even worse, there are still RNGs with a period of only 10^5 in use, especially built-in RNGs, which are often the cause for severely bad Monte Carlo simulations.

• In cryptography the unpredictability of the RNs is essential. This is not really necessary for Monte Carlo methods. It is more important that RNs appear to be random and independent, and **pass statistical tests** about being independent, identically distributed (i.i.d.) RNs $\sim U[0,1]$.

• Within Monte Carlo methods the sequence of RNs should be **reproducible**. First because of debugging reasons. If we get strange simulation results, we will be able to examine the sequence of RNs. Also it is quite common in some applications, e.g. in sensitivity analysis, to use the same sequence again. Another advantage is the possibility to compare different calculation methods or similar financial products in a more efficient way by using the same RNs.

• The algorithm should be programmed so that it delivers the same RNs in every computer, also called **portability**. It should be possible to repeat calculations in other machines and always obtain the same result. The same seed for initialization should always return the same RN sequence.

• To make calculations very fast it is often desirable to use a computer with parallel processors. So the possibility of **parallelization** is another important point of consideration. One option is e.g. to jump ahead quickly many steps in the sequence and produce disjoint substreams of the stream of RNs. Another option is to find a family of RNGs which work equally well for a large set of different parameters.

• The **structure** of the random points is very important. A typical feature of some RNGs is that if d-dimensional vectors are constructed out of consecutive RNs, then those points lie on hyperplanes. Bad RNGs have only a few hyperplanes in some dimensions which are far apart, leaving huge gaps without any random vectors in space. Further, one should be aware if there is a severe correlation between some RNs. For example, if the algorithm constructs a new RN out of two previous RNs, groups of three RNs could be linked too closely. This correlation can lead to false simulation results if the model has a structure that is similar.

• Sometimes it is necessary to choose random number generators which are **easy to implement**, i.e. the code is simple and short. This is a more general point and applies to all parts of the simulation. Very often it is useful to have a second simulation routine to check important simulation results. In this case, code which is easy to read increases the possibility that it is bug-free.

2.1.3 Technical terms

All RNGs that are based on deterministic recurrences can be described as a quintuple $(\mathcal{S}, \mu, f, \mathcal{U}, g)$ (see L'Ecuyer [1994]), where

• $\mathcal{S}$ is the finite set of states, the so-called **state space**.

• $\mathbf{s}_n \in \mathcal{S}$ is a particular **state** at iteration step n.

• μ is the probability measure for selecting $\mathbf{s}_0$, the initial state, from $\mathcal{S}$. $\mathbf{s}_0$ is called the **seed** of the RNG.

• The function f, also called the **transition function**, describes the algorithm, $\mathbf{s}_{n+1} = f(\mathbf{s}_n)$.

• $\mathcal{U}$ is the **output space**. We will mainly consider $[0,1]$, $[0,1)$, $(0,1]$ or $(0,1)$.

• The **output function** $g : \mathcal{S} \to \mathcal{U}$ maps the state $\mathbf{s}_n \in \mathcal{S}$ into a number $u_n \in \mathcal{U}$, the final random number that we are interested in.

REMARK 2.1 1. The importance of the seed that invokes the RNG should not be underestimated. One cause of strange simulation results is often that the RNG has not been seeded or has been started with a zero vector. Another reason is that some seeds have to be avoided as they lead to bad RN sequences, e.g., a seed that contains mainly zeroes might lead to RNs that do not differ much. Further, it is advisable to save the seed for debugging.

2. As real numbers can only be presented with finite accuracy in computers, the set of states $\mathcal{S}$ must be finite and so the range $g(\mathcal{S}) \subset \mathcal{U}$ is also finite. As we are interested in an even distribution, we want the gaps in $[0,1)$ not to be too large. Therefore, $\mathcal{S}$ has to be a large set.

3. Another consequence of $\mathcal{S}$ being finite is that $\mathbf{s}_{n+\rho} = \mathbf{s}_n$ for some integer ρ, meaning the sequence will repeat itself eventually. The smallest number ρ for which this happens is called the **period** of the RNG. It cannot be larger than $|\mathcal{S}|$, another reason for the set of states being huge. In some generators the period might be different for different seeds and the RN cycles are then disjoint. So we have to be careful with which seed we start. □

2.2 Examples of random number generators

2.2.1 Linear congruential generators

The linear congruential method was one of the first RNGs. Linear congruential generators (**LCGs**) were first introduced by Lehmer (1949) and were very popular for many years.

Algorithm 2.1 LCG

$$s_{n+1} = (as_n + c) \mod m, \quad n \in \mathbb{N},$$

where

- $m \in \mathbb{N} \setminus \{0\}$ is called the **modulus**,
- $a \in \mathbb{N}$ is called the **multiplier**, $a < m$,
- $c \in \mathbb{N}$ is called the **increment**, $c < m$,
- $s_0 \in \mathbb{N}$ is called the **seed**, $s_0 < m$.

Numbers in $[0, 1)$ are obtained via

$$u_n = \frac{s_n}{m}.$$

These generators are often denoted by LCG(m, a, c). The state space can be described as

$$\mathcal{S} \subseteq \mathbb{N},\ \mathcal{S} = \{0, 1, \ldots, m-1\} \text{ or } \{1, 2, \ldots, m-1\}.$$

For the increment $c = 0$ the generator cannot be started with $s_0 = 0$. Then 0 is excluded from the set of states.

Choosing coefficients

Often the modulus m is chosen as a prime. Then all calculations are done in the finite field $\mathbb{Z}_m$. Preferred moduli are Mersenne Primes, primes of the form $2^k - 1$, e.g. $2^{31} - 1 = 2147483647$, the largest 32-bit signed integer. Sometimes a power of 2 is chosen as the modulus, because calculations can be made faster by exploiting the binary structure of computers. But then the lower order bits of the RNs are highly correlated, especially the last bit of the states s_n is either constant or strictly alternates between 0 and 1. This means that all our integer RNs s_n are either even or odd or we have a regular even/odd pattern.

If the period of such a generator is m or $m-1$ (in the case of $c = 0$) we call this property **full period** as this is the maximum possible period length. In order to achieve a long period it is advisable to choose a very large number as the modulus. Further, the increment and the multiplier have to be selected to create a generator with full period. If $c = 0$, m is a prime, a is a primitive root modulo m, then the generator has full period. For $c \neq 0$ the criteria for full period are more complicated (see e.g. Knuth [1998]): the only common divisor of m and c should be 1, every prime number that divides m should also divide $a - 1$; if m is divisible by 4, then $a - 1$ should also have the divisor 4.

Implementation issues

Our next task is the implementation of the algorithm. If the modulus m is near the maximum integer and also the multiplier a is quite large, then the calculation step done with integer arithmetics can lead to an overflow. We then get errors or, in the better case, our code will be machine dependent and is no longer portable. To circumvent this difficulty there are several possibilities:

• Instead of integer arithmetic we can work with floating-point arithmetic. For example, in 64-bit floating-point arithmetic integers up to 2^{53} are represented exactly. So we just take care that $a\,(m-1) \leq 2^{53}$.

• As we work in the ring $\mathbb{Z}_m$, we could consider multiplying with $-\bar{a}$ instead of multiplying with a, where $\bar{a} = m - a$ is the additive inverse of a.

• Another technique is called **approximate factoring**. If m is large and a not too large, e.g. $a < \sqrt{m}$, then we decompose $m = aq + r$, i.e. $r = m \mod a$, $q = \lfloor m/a \rfloor$. The complicated part of the iteration step is done by

$$(as_n) \mod m = \begin{cases} a\,(s_n \mod q) - r\lfloor z/q \rfloor & \text{if it is } \geq 0 \\ a\,(s_n \mod q) - r\lfloor z/q \rfloor + m & \text{otherwise} \end{cases} \tag{2.1}$$

As the parameter r is small this method avoids overflows. This trick also works when $a = \lfloor m/k \rfloor$ with an integer $k < \sqrt{m}$.

• The **powers-of-two decomposition** can be applied when a can be written as a sum or a difference of a small number of powers of 2, e.g. $a = 2^3 + 2^8$. Then the modulus is decomposed as $m = 2^k - h$ with integers $k, h \in \mathbb{N}$. Now $(as_n) \mod m$ can be efficiently calculated using shifts, masks, additions, subtractions, and a single multiplication by h.

Disadvantages of LCGs

LCGs should no longer be used for important simulations. First, much better RNGs exist, and then they also have a lot of disadvantages:

• There exists a high sequential correlation between the RNs, so they do not mimic randomness very well. For example, if the multiplier a is small compared to the modulus m, then very small RNs are **always** followed by another small RN. In a simulation this means that rare events could be too close together or happen too often.

• The modulus m cannot be larger than the largest integer available in the programming language and so the period, which is maximally m, is usually not very long. In 32-bit computers the period is often only $2^{31} - 1$ and this is simply too short nowadays.

• We take a closer look at the set of all possible t-dimensional vectors filled with consecutive RNs, the finite set $\Psi_t := \{(u_1, u_2, \ldots, u_t)\,|\,s_0 \in \mathcal{S}\} \subseteq [0,1)^t$

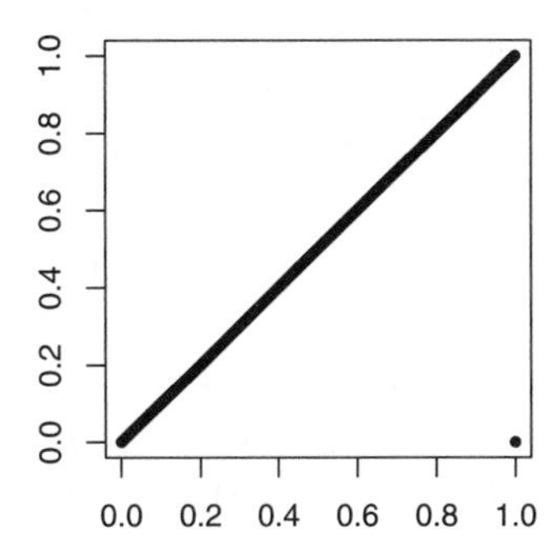

FIGURE 2.1: Hyperplanes in a simple LCG.

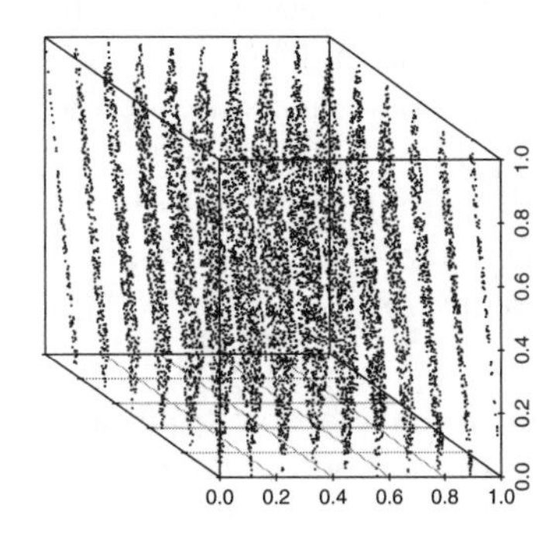

FIGURE 2.2: Hyperplanes in RANDU.

(see Definition 2.2). If our pseudorandom numbers were indeed random and independent, the t-dimensional hypercube would be filled evenly, without any recognizable structure. But the pattern structure of Ψ_t from an LCG is very regular – all points lie on equidistant, parallel hyperplanes. This is called the **lattice structure** of the LCG.

Example 1: As a toy example, we take a look at the LCG $s_{n+1} = (1 \cdot s_n + 1) \mod 3331$. This "RNG" has full period, i.e. $\rho = 3331$. But Ψ_2 consists of only one "hyperplane" and one single point (see Figure 2.1).

Example 2: The LCG RANDU – given by $a = 65539, c = 0$, $m = 2^{31}$ – was very popular for many years. But a look at the three-dimensional vectors that are constructed out of consecutive RNs reveals that all the points from Ψ_3 lie on just 15 "hyperplanes"(see Figure 2.2).

Improving LCGs

The advantages of LCGs are that they are very fast, do not need much memory, are easy to implement, and easy to understand. Still there are applications where good representatives of this type of generator are sufficient.

• One cannot avoid the lattice structure, so when working with an LCG, a variant with many hyperplanes should be chosen. The key to judge this is the **spectral test** (see Knuth [1998]). It computes the largest distance $1/l_t$ between two successive hyperplanes for a family of hyperplanes. Often the value l_t is computed for several Ψ_t, $t \in \mathbb{N}$. It gives us a rough impression of the number of hyperplanes in Ψ_t. If Ψ_t has many hyperplanes or $1/l_t$ is small for many t, the better is the generator. A rule says that $l_t \leq 1 + a^2$ for the case $c = 0$, so the multiplier a should not be too small.

• There is the possibility to shuffle the sequence of RNs, eventually with the help of another LCG (or any RNG) (see Knuth [1998]). The j-th value of the original sequence is not the j-th output, instead it is kept in waiting position. The j-th value of the sequence or the value of the other RNG decides which waiting position is freed, and this position is again filled up with the momentary RN. This method removes parts of the serial correlation in the RN sequence.

• Generally, we can combine the output of two (or even more) different LCGs. Then we have a new type of RNG, called **combined LCG**. There are two different methods to combine the generators (see also the next section):

Method 1: If $u_n^{(1)} \in [0,1)$ is the output of LCG1, $u_n^{(2)}$ the output of LCG2, then $u_n = \left(u_n^{(1)} + u_n^{(2)}\right) \mod 1$ is the output of the combined generator.

Method 2: If $s_n^{(1)} \in \mathcal{S}$ is the n-th integer value in the sequence of LCG1, $s_n^{(2)}$ is that from LCG2, then we combine $s_n = \left(s_n^{(1)} + s_n^{(2)}\right) \mod m_1$, where m_1 is the modulus from LCG1 with $m_1 > m_2$.

Recommended LCGs

LCGs of good quality can e.g. be used for shuffling sequences or seeding other RNGs. A combined generator is good enough for small-scale simulations. In Table 2.1, we list some recommended choices for the parameters (see Entacher [1997], Press et al. [2002]).

Known as	LCG(m, a, c)	Period
Park and Miller	LCG$(2^{31}-1, 16807, 0)$	$2^{31}-2$
Fishman and Moore	LCG$(2^{31}-1, 950706376, 0)$	$2^{31}-2$
Fishman	LCG$(2^{31}-1, 48271, 0)$	$2^{31}-2$
L'Ecuyer	LCG$(2^{31}-249, 40692, 0)$	$2^{31}-250$
ran2 (combined LCG)	LCG1$(2147483563, 40014, 0)$	
	LCG2$(2147483399, 40692, 0)$	2.3×10^{18}

Table 2.1: Linear Congruential Generators

2.2.2 Multiple recursive generators

Multiple recursive generators (**MRGs**) are a generalization of linear generators and are as easy to implement. With the same modulus their periods are much longer and their structure is improved. We are considering only homogeneous recurrences with $c = 0$ as every inhomogeneous recurrence can be replaced by a homogeneous one of higher order with suitable initial values.

Algorithm 2.2 MRG

$$s_n = (a_1 s_{n-1} + \ldots + a_k s_{n-k}) \mod m, \quad n \in \mathbb{N}, n \geq k,$$

where

- $m \in \mathbb{N} \setminus \{0\}$ is the **modulus**,
- $a_i \in \mathbb{N}$ are the **multipliers**, $i = 1, \ldots, k$, $a_i < m$,
- k with $a_k \neq 0$ is the **order** of the recursion, $k \geq 2$,
- $\mathbf{s}_0 = (s_0, s_1, \ldots, s_{k-1}) \in \mathbb{N}^k$ is the **seed**, $s_i < m$, $i = 0, \ldots, k-1$.

Numbers in $[0, 1)$ are obtained by

$$u_n = \frac{s_{n+k-1}}{m}, \; n > 0.$$

The state space of an MRG can be described as

$$\mathcal{S} \subseteq \mathbb{N}^k, \; \mathcal{S} = \{0, 1, \ldots, m-1\}^k \setminus \{\mathbf{0}\}.$$

The initial vector $\mathbf{s}_0 \in \mathcal{S}$ can be chosen arbitrarily, only the vector $\mathbf{0}$ has to be excluded from the set of states. The n-th vector in the sequence is $\mathbf{s}_n = (s_{n+k-1}, \ldots, s_{n+1}, s_n) \in \mathcal{S}$. Besides $a_k \neq 0$ there has to be at least one other coefficient $a_i \neq 0$. This algorithm can be rewritten with a matrix $\mathbf{A} \in \mathbb{N}^{k,k}$ via

$$\mathbf{A} = \begin{pmatrix} 0 & 1 & & 0 \\ \vdots & 0 & & 0 \\ 0 & 0 & & \ldots 1 \\ a_k & a_{k-1} & \ldots & a_1 \end{pmatrix}, \tag{2.2}$$

$$\mathbf{s}_{n+1} = (\mathbf{A}\mathbf{s}_n) \mod m.$$

Choosing coefficients

Nearly the same considerations as for LCGs apply here. With modulus m the maximal possible period length is $m^k - 1$, with k the order of the recurrence. Hence, for k large, very long periods are possible, even with a smaller modulus.

- If m is a prime number, it is possible to choose the coefficients a_i, so that the maximal period length can be achieved. The MRG has maximal period if and only if the characteristic polynomial of the recurrence is primitive over the field $\mathbb{Z}_m$ (see L'Ecuyer [1999a]).

$$P(z) = det(z\mathbf{I} - \mathbf{A}) = z^k - a_1 z^{k-1} - \ldots - a_k. \tag{2.3}$$

- If nearly all a_i are zero (consider MRGs with $a_k \neq 0$, $a_r \neq 0$ for $0 < r < k$, $a_i = 0$ for $i \neq k, r$) then the algorithm is very fast, but the structure of

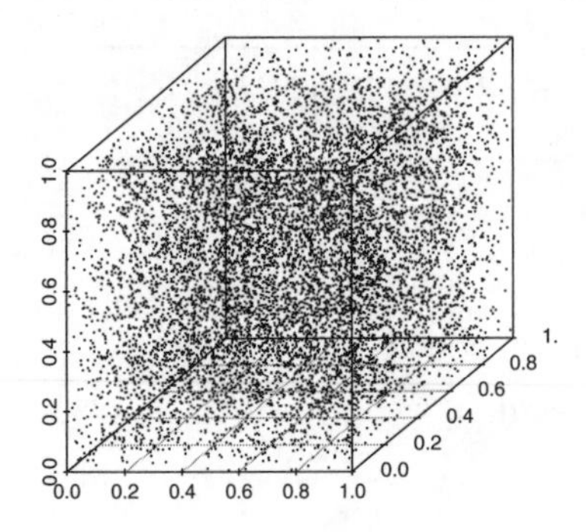

FIGURE 2.3: Vectors from consecutive RNs.

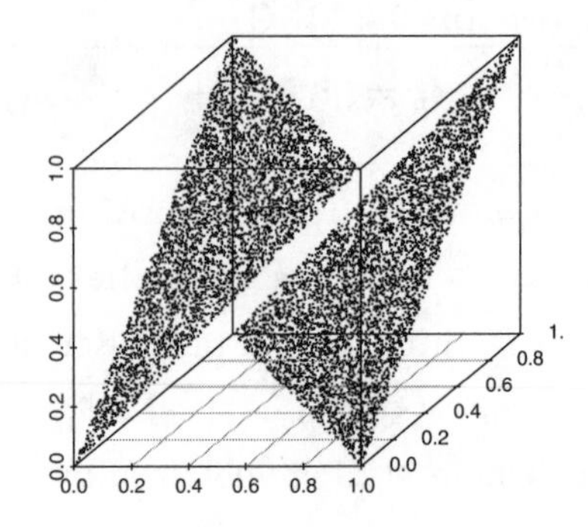

FIGURE 2.4: Vectors from selected RNs.

the RNs is too coarse. There are too many gaps in space when vectors are filled with RNs. If m and k are large with many coefficients $a_i \neq 0$, then the structure is usually excellent, but the algorithm becomes rather slow.

• We take a closer look at the finite set Ψ_t of t-dimensional random vectors, for which we will give a formal definition here as we will encounter it more often in the following text.

DEFINITION 2.2

The set $\Psi_t := \{(u_1, u_2, \ldots, u_t) \,|\mathbf{s}_0 \in \mathcal{S}\} \subseteq [0,1)^t$ is the set of all t-dimensional vectors produced by consecutive RNs from a distinctive RNG. s_0 symbolizes the seed, u_1 is the first real RN produced with this seed, u_2 is the second one, and so on. This set is seen as a multiset and so $|\Psi_t| = |\mathcal{S}|$. We will call it **"the set of all t-dimensional output vectors"**.

We further consider a generalized set $\Psi_I := \{(u_{i_1}, u_{i_2}, \ldots, u_{i_t}) \,|\mathbf{s}_0 \in \mathcal{S}\} \subseteq [0,1)^t$ where $I = \{i_1, \ldots, i_t\}, i_r \in \mathbb{N}$, is a finite index set.

With MRGs, these point sets have lattice structure and consist of equidistant parallel hyperplanes. The largest distance between two successive hyperplanes $1/l_t$ can be computed with the spectral test (see Section 2.3.1). The coefficients a_i should be chosen in such a way that $1/l_t$ is small for as many t as possible. It can be shown that for $t > k$ the distance is $1/l_t \geq (1+a_1^2+\ldots+a_k^2)^{-1/2}$ (see L'Ecuyer [1997]). If the sum of squares of the coefficients is small there will be only a few hyperplanes. The same relation is true for sets I containing all indices corresponding to nonzero coefficients, i.e. $I = \{1, k+1, \text{ all } j+1 \text{ with } a_{k-j} \neq 0\}$.

Example: We look at the generator proposed by Mitchell and Moore in 1958 (unpublished by the authors; see Knuth [1998]):

$$s_n = (s_{n-24} + s_{n-55}) \mod 2^{31} \tag{2.4}$$

This is a RNG with a nonprime modulus, so for the initialization not all numbers in the seed vector should be even. Here we have the enormous period of $2^{30}(2^{55} - 1)$. Further, the method is extremely fast, because there is no multiplication. In Figure 2.3 we look at a subset of 10,000 points of the set $\Psi_3 = (u_1, u_2, u_3)$ filled with successive RNs. The points are evenly distributed in space, and there are no huge gaps to be seen. In Figure 2.4 we see 10,000 points of the set $\Psi_I = (u_1, u_{32}, u_{56})$. This is exactly the set of RNs that corresponds to the set of nonzero coefficients. This is indeed an extreme example, because there are only two planes (and one single point). RNGs with only two nonzero coefficients, both being one, have once been quite popular, but they always have a low-dimensional index set with only three hyperplanes. They may seem to be good in most applications, but if a special simulation just depends on the most critical RNs in the sequence, which happens in practice, the results will be unreliable.

2.2.3 Combined generators

MRGs with a good lattice structure have many nonzero coefficients which are large as well. But these generators are no longer fast as a lot of multiplications have to be done. One way to improve fast multiple recursive generators with poor lattice structure is to combine them. Typically, the component generators of the combined RNGs have different moduli m.

Consider J distinct MRGs with

$$s_{n,j} = \left(a_{1,j}s_{n-1,j} + \ldots + a_{k_j,j}s_{n-k_j,j}\right) \mod m_j$$
$$n \in \mathbb{N},\ k_j \in \mathbb{N},\ n \geq k_j,\ j = 1, \ldots, J, \tag{2.5}$$

where $m_1 = \max\{m_j | j = 1, \ldots, J\}$, $a_{k_j,j} \neq 0$, and at least one other coefficient $a_{i,j} \neq 0$ for each j. Now there are two possibilities to combine these generators to obtain a new one:

$$\begin{aligned} v_n &= (d_1 s_{n+k-1,1} + \ldots + d_J s_{n+k-1,J}) \mod m_1 \\ u_n &= \frac{v_n}{m_1} \in [0,1), \end{aligned} \tag{2.6}$$

or

$$u_n = \left(\frac{d_1 s_{n+k-1,1}}{m_1} + \ldots + \frac{d_J s_{n+k-1,J}}{m_J}\right) \mod 1 \in [0,1) \tag{2.7}$$

with $k = \max\{k_1, \ldots, k_J\}$ and $n > 0$. The weight factors $d_j, j = 1, \ldots, J$ are integers $-m_j < d_j < m_j$. The state space can be described as

$$\mathcal{S} \subseteq \mathbb{N}^{k_1,k_2,\ldots,k_J}, \quad \mathcal{S} \subseteq \{0,1,\ldots,m_1-1\}^{k_1} \times \ldots \times \{0,1,\ldots,m_J-1\}^{k_J}.$$

Known as	Coefficients	Period
combMRG96a	$m_1 = 2^{31} - 1, m_2 = 2145483479,$ $a_{11} = 0, a_{12} = 63308, a_{13} = -183326,$ $a_{21} = 86098, a_{22} = 0, a_{23} = -539608$	$\approx 2^{185}$
MRG32k3a	$m_1 = 2^{32} - 209, m_2 = 2^{32} - 22853,$ $a_{11} = 0, a_{12} = 1403580, a_{13} = -810728,$ $a_{21} = 527612, a_{22} = 0, a_{23} = -1370589$	$\approx 2^{191}$
MRG32k5a	$m_1 = 2^{32} - 18269, m_2 = 2^{32} - 32969,$ $a_{11} = a_{13} = 0, a_{12} = 1154721, a_{14} = 1739991,$ $a_{15} = -1108499, a_{21} = 1776413, a_{22} = a_{24} = 0,$ $a_{23} = 865203, a_{25} = -1641052$	$\approx 2^{319}$

Table 2.2: Combined Multiple Recursive 32-Bit Generators

The vector $\mathbf{0}$ should be excluded from the set of states, also vectors where a complete section for one $j \in \{1, \dots, J\}$ is zero. After that the seed vector $\mathbf{s}_0 = (s_{0,1}, \dots, s_{k_1-1,1}, \dots, s_{0,J}, \dots, s_{k_J-1,J}) \in \mathcal{S}$ can be chosen arbitrarily.

If all coefficients are carefully chosen (e.g. $m_1, \dots, m_J$ distinct primes, $d_j < m_j$, $k = k_1 = \dots = k_J$, each recurrence with maximal period $\rho_j = (m_j^k - 1))$ (see L'Ecuyer [1996a]), then it can be shown that the combination in Equation (2.7) is equivalent to a MRG with modulus $m = m_1 \cdots m_J$ and period ρ, where ρ is the least common multiple of $\rho_1, \dots, \rho_J$. Further, both variants of the combined generator deliver nearly the same RNs if the moduli are close to each other. So a combined RNG can be seen as a clever way to work with a RNG with a huge modulus m, long period ρ, and a lot of nonzero coefficients, but which is however rather fast when the components have a lot of zero coefficients.

One of the first combined MRGs, combMRG96a (see Table 2.2), can be found in L'Ecuyer (1996a), where also an implementation in C with q-r decomposition is given. However, an implementation with floating-point arithmetic is much faster (see L'Ecuyer [1999a], where also 64-bit generators are described). The other generators in Table 2.2 have also been described by L'Ecuyer (1999a). Negative coefficients $a < 0$ are regarded as an additive inverse to $-a$ in the field $\mathbb{Z}_m$, i.e. $\bar{a} = a + m > 0$. All combinations are constructed as $v_n = (s_{n+k-1,1} - s_{n+k-1,2}) \mod m_1$, $u_n = v_n / m_1$ (see Equation (2.6) with $d_1 = 1, d_2 = -1$)).

2.2.4 Lagged Fibonacci generators

The lagged Fibonacci RNGs (**LFGs**) are a generalization of the Fibonacci series

$$s_n = s_{n-1} + s_{n-2}. \tag{2.8}$$

Note that this is obviously not a good RNG. However, some representatives of the generalized versions create usable RNGs:

$$s_n = s_{n-q} \odot s_{n-p}. \tag{2.9}$$

The operator $\odot$ can be addition, subtraction, or multiplication modulo m, or the bitwise exclusive XOR-function. The lagged Fibonacci generators will be noted as LFG(p,q,$\odot$), where p and q are called the **lags**, $p, q \in \mathbb{N} \setminus \{0\}$, $p > q$. Multiplication must be done on the set of odd integers. The modulus m can be any integer, 1 included. In the latter case, s_n are floating-point numbers in the interval $(0, 1)$.

These generators can be generalized by using three or more lags. LFGs with addition or subtraction are special cases of MRGs, e.g. the Mitchell and Moore RNG in Equation (2.4). As we have seen above these RNGs are not good, some point sets in low dimensions consist of only a few hyperplanes. The formerly very popular shift register generators, e.g. the infamous R250, are another special case of LFGs, using the XOR-operation. Although very fast, they are no longer recommended as they fail important statistical tests.

Multiplicative LFGs seem to belong to the group of good RNGs although they are slower than the additive versions. If the modulus is a power of two, 2^b, and the lags are chosen so that they are the exponents of a primitive polynomial, then the period is $2^{b-2}(2^{p-1})$. Multiplicative LFGs have the speciality to own several independent, full-period RN cycles. Seed tables can be created for seeds that start disjoint cycles.

2.2.5 $\mathbb{F}_2$-linear generators

Another idea to speed up implementation is to exploit the binary representation of numbers in the computer. So we are looking for algorithms using only 0's and 1's. Therefore, we work in the field $\mathbb{F}_2$ with elements $\{0, 1\}$, i.e. all operations are performed modulo 2. Addition in $\mathbb{F}_2$ is just the binary exclusive-or (XOR) operation, $\oplus$:

$$(x + y) \mod 2 \quad \text{is equivalent to } x \oplus y,\ x, y \in \mathbb{F}_2. \tag{2.10}$$

To start the sequence, $\mathbb{F}_2$-linear generators have to be initialized with a seed vector $\mathbf{x}_0 = (x_0, x_1, \ldots, x_{k-1}) \in \mathcal{S}$. The state space can be described as

$$\mathcal{S} = \{0, 1\}^k \setminus \{\mathbf{0}\}.$$

With this in mind, the period of the generator cannot be larger than $2^k - 1$ as this is the maximum of possible values for $\mathbf{x}_n$, zero vector excluded. The length of the period can be determined by analyzing the characteristic polynomial of the matrix $\mathbf{A}$,

$$P(z) = \det(z\mathbf{I} - \mathbf{A}) = z^k - \alpha_1 z^{k-1} - \ldots - \alpha_{k-1} z^1 - \alpha_k, \tag{2.11}$$

Algorithm 2.3 $\mathbb{F}_2$-linear generators

$$\begin{aligned} \mathbf{x}_{n+1} &= \mathbf{A}\mathbf{x}_n, \quad \mathbf{x}_n \in \mathbb{F}_2^k, \ \mathbf{A} \in \mathbb{F}_2^{k,k} \\ \mathbf{y}_{n+1} &= \mathbf{B}\mathbf{x}_{n+1}, \quad \mathbf{y}_{n+1} \in \mathbb{F}_2^w, \ \mathbf{B} \in \mathbb{F}_2^{w,k} \end{aligned}$$

- $\mathbf{x}_n$ is called the k-bit **state vector** at step n, $n \in \mathbb{N}, k \in \mathbb{N} \setminus \{0\}$,
- $\mathbf{y}_n$ is called the w-bit **output vector**, $w \in \mathbb{N} \setminus \{0\}$,
- $\mathbf{A}$ is called the **transition matrix**,
- $\mathbf{B}$ is called the **output matrix** or **tempering**.

The final output $u_n \in [0,1)$ is then generated via

$$u_n = \sum_{i=1}^{w} y_{n,i-1} 2^{-i} = 0.y_{n,0} y_{n,1} \ldots y_{n,w-1} \text{ (in binary representation).}$$

where $\alpha_j \in \mathbb{F}_2, j = 1, \ldots, k$, $\alpha_1 = trace\,(\mathbf{A})\,, \alpha_k = \det \mathbf{A}$. With this polynomial a linear recurrence in $\mathbb{F}_2$ can be defined: (see also L'Ecuyer [1994])

$$v_n = \alpha_1 v_{n-1} + \ldots - \alpha_{k-1} v_{n-k+1} + \alpha_k v_{n-k} \tag{2.12}$$

If $\alpha_k = 1$, meaning the rank of $\mathbf{A}$ is maximal, then this recurrence is of order k and every $x_{n,i}$, $n > 0$, $i = 0, \ldots, k-1$, from the Algorithm 2.3 follows this recursion. If the polynomial $P(z)$ is primitive over $\mathbb{F}_2$, the recurrence (2.12) has maximal period $2^k - 1$ and consequently the first recurrence in Algorithm 2.3 also. The matrix $\mathbf{B}$ is often used to improve the distribution of the output, but in many cases it is just the $w \times w$-identity matrix with $k - w$ zero columns added.

Linear feedback shift register generators

Special cases of $\mathbb{F}_2$-linear generators are the **Tausworthe generators**, also known as **linear feedback shift register (LFSR)** generators.

Algorithm 2.4 Linear feedback shift register generators

$$x_n = (a_1 x_{n-1} + \ldots + a_k x_{n-k}) \mod 2, \ n \geq k, \ a_i \in \{0,1\}$$

with output

$$u_n = \sum_{i=1}^{w} x_{ns+i-1} 2^{-i}, \ w, s \in \mathbb{N}.$$

The generation of uniformly distributed real numbers can be seen as taking a block of w bits (**word length**) every s steps (**stepsize**). To start the algorithm, a seed vector $\mathbf{x}_0 = (x_0, x_1, \ldots, x_{k-1}) \in \mathcal{S}$ has to be chosen. The characteristic polynomial of the first recurrence in Algorithm (2.4) is

$$P(z) = z^k - a_1 z^{k-1} - \ldots - a_k. \tag{2.13}$$

The recurrence has period length $2^k - 1$ if and only if P is a primitive polynomial. If the stepsize s is relatively prime to $2^k - 1$, then the sequence of real RNs also has period $2^k - 1$. The algorithm can be rewritten with the matrix $\mathbf{A} = \tilde{\mathbf{A}}^s$, where

$$\tilde{\mathbf{A}} = \begin{pmatrix} 0 & 1 & & & 0 \\ 0 & 0 & & & 0 \\ & & \ddots & & \\ 0 & & & 1 & 0 \\ 0 & & & 0 & 1 \\ a_k & a_{k-1} & \ldots & a_2 & a_1 \end{pmatrix} \in \mathbb{F}_2^{k,k}, \tag{2.14}$$

$$\mathbf{x}_n = (x_{ns}, x_{ns+1}, \ldots, x_{ns+k-1})^T,$$

$$\mathbf{x}_{n+1} = \tilde{\mathbf{A}}^s \mathbf{x}_n.$$

An example of a popular LFSR generator is the formerly popular R250 $x_n = x_{n-103} \oplus x_{n-250}$ with period $2^{250} - 1$, which was already mentioned in Section 2.2.4. The R250 belongs to the class of **trinomial-based LFSR** generators, because the characteristic polynomial has only three nonzero coefficients. The generators in this class are very fast but they fail some important statistical tests. LFSR generators do not have good equidistribution properties (see Section 2.3.2). Nevertheless, LFSR generators are useful for producing random signs or ideal for the Monte Carlo exploration of a binary tree for decisions whether to branch left or right. And most importantly, they are a good basis for combined generators, which finally have good equidistribution properties.

Combined $\mathbb{F}_2$-linear generators

We can combine J different mod-2-RNGs with parameters $(k_j, w, \mathbf{A}_j, \mathbf{B}_j)$, $j = 1, \ldots, J$. All the $\mathbf{B}$-matrices must have the same number w of rows. Then we get J different $\mathbf{x}_{j,n} = \mathbf{A}_j \mathbf{x}_{j,n-1}$-sequences, $j = 1, \ldots, J$, which are combined into a single $\mathbf{y}_n$-sequence

$$\mathbf{y}_n = \mathbf{B}_1 \mathbf{x}_{1,n} \oplus \mathbf{B}_2 \mathbf{x}_{2,n} \oplus \ldots \oplus \mathbf{B}_J \mathbf{x}_{J,n}. \tag{2.15}$$

The random number in $[0, 1)$ is then constructed as

$$u_n = \sum_{i=1}^{w} y_{n,i} 2^{-i} = 0.y_{n,1} \ldots y_{n,w} \tag{2.16}$$

It can easily be seen that the combined generator is just a normal $\mathbb{F}_2$-linear generator with $k = k_1 + \ldots + k_J$, the matrix $\mathbf{A}$ is a big block-diagonal matrix $\mathbf{A} = diag(\mathbf{A}_1, \ldots, \mathbf{A}_J)$, and the columns of $\mathbf{B}$ consist of the columns of all $\mathbf{B}_j$, i.e. $\mathbf{B} = (\mathbf{B}_1, \ldots, \mathbf{B}_J)$.

The combined generator cannot have full period. The period is as large as the least common multiple of its component generators. But with a good choice of parameters it is possible to get very close to the theoretical full period of $2^k - 1$ (see L'Ecuyer [1999b]).

Interesting combined generators are the combination of three or four trinomial-based LFSR generators. With their few nonzero coefficients each single generator works very fast. If the combined generators are chosen cleverly, the combined generators can have very large periods with characteristic polynomials that have a lot of nonzero coefficients, which take care that the RNs are well distributed. So we can achieve fast RNGs with excellent distribution properties. L'Ecuyer (1999b) presents several tables of combinations of three or four fast LFSR generators with excellent equidistribution properties. Those combinations are maximally equidistributed and further collision-free (see Section 2.3.2). Also, implementations in C++ are given in L'Ecuyer's paper.

As an example we reformulate L'Ecuyer's entry no. 62 from table 1 as an algorithm (see Algorithm 2.5). This RNG has a period of about 2^{113} and is a combination of four different 32-bit LFSR generators.

Algorithm 2.5 Example of a combined LFSR generator

This RNG consists of four trinomial-based LFSR generators of degrees 31, 29, 28, and 25.

$$\begin{aligned} x_{1,n} &= x_{1,n-25} + x_{1,n-31} \mod 2, \\ x_{2,n} &= x_{2,n-27} + x_{2,n-29} \mod 2, \\ x_{3,n} &= x_{3,n-15} + x_{3,n-28} \mod 2, \\ x_{4,n} &= x_{4,n-22} + x_{4,n-25} \mod 2. \end{aligned}$$

Numbers in $[0, 1)$ for each generator are generated by

$$u(j, n) = \sum_{i=1}^{w} x_{j,ns_j+i-1} 2^{-i}, \quad j = 1, \ldots, 4.$$

The stepsizes are $s_1 = 16, s_2 = 24, s_3 = 11, s_4 = 12$, and the word length is $w = 32$, the final RN in $[0, 1)$ is the XOR-combination

$$u_n = u_{1,n} \oplus u_{2,n} \oplus u_{3,n} \oplus u_{4,n}.$$

Generalized feedback shift register generator and Mersenne Twister

The most famous RNG in the $\mathbb{F}_2$-category is the Mersenne Twister MT19937 (see Matsumoto and Nishimura [1998]) with the tremendously large period length of $2^{19937} - 1$. The Mersenne Twister is a variant of a **generalized feedback shift register (GFSR)** generator.

Algorithm 2.6 GFSR

The n-th iterate $\tilde{\mathbf{x}}_n \in \{0,1\}^{pq}$ is calculated as $\tilde{\mathbf{x}}_n = \mathbf{A}\tilde{\mathbf{x}}_{n-1}$, where

$$\mathbf{A} = \begin{pmatrix} \mathbf{S}_1 & \mathbf{S}_2 & & \mathbf{S}_{q-1} & \mathbf{S}_q \\ \mathbf{I}_p & & & & \mathbf{0} \\ & \mathbf{I}_p & & & \\ & & \ddots & & \vdots \\ \mathbf{0} & & & \mathbf{I}_p & \mathbf{0} \end{pmatrix}$$

with $\mathbf{S}_i \in \{0,1\}^{p,p}$, $\mathbf{I}_p$ the $p \times p$-identity matrix.

In the case of $w = p$, $\mathbf{B}$ consisting of the first rows of an identity matrix, $\mathbf{S}_r = \mathbf{S}_q = \mathbf{I}_p$ for some $1 \leq r < q$, $\mathbf{S}_i = 0$ for $i \notin \{r, q\}$, we have got a trinomial-based generalized feedback shift register (GFSR) generator. By denoting $\mathbf{x}_n := (\tilde{x}_{n,1}, \ldots, \tilde{x}_{n,p}) \in \{0,1\}^p$ we have

$$\mathbf{x}_n = \mathbf{x}_{n-r} \oplus \mathbf{x}_{n-q}.$$

The lower bits of the vector $\tilde{\mathbf{x}}_n$ are just shifted.

Sometimes GFSR generators have more than two matrices $\mathbf{S}_i = \mathbf{I}_p$. This means that $\mathbf{x}_n$ is constructed with several p-dimensional parts from previous vectors combined by bitwise XOR. GFSR generators can be effectively implemented and are therefore very fast, but the maximal period of a GFSR generator is just $2^q - 1$, no matter how many matrices are nonzero. This is a bit disappointing, because with $\tilde{\mathbf{x}} \in \{0,1\}^{pq}$ we would expect a much longer period. This motivated the construction of the Mersenne Twister.

In the Mersenne Twister framework we work with matrix $\mathbf{S}_q$ in the form

$$\mathbf{S}_q = \begin{pmatrix} 0 & 0 & & 0 & s_1 \\ 1 & 0 & & 0 & s_2 \\ 0 & 1 & & & \vdots \\ 0 & & \ddots & 0 & s_{p-1} \\ 0 & 0 & & 1 & s_p \end{pmatrix}, \quad \mathbf{S_q} \in \{0,1\}^{p,p}, \; s_i \in \{0,1\}. \tag{2.17}$$

If further $\mathbf{S}_r = \mathbf{I}_p$ for some $r \in \{1, \ldots, q\}$, $\mathbf{S}_i = \mathbf{0}$ for $i \notin \{r, q\}$, then this

RNG is called the **twisted GFSR** (**TGFSR**) generator. Instead of the last equation in Algorithm 2.6 we then have

$$\mathbf{x}_n = \mathbf{x}_{n-r} \oplus \mathbf{S}_q \mathbf{x}_{n-q}. \tag{2.18}$$

If the parameters are carefully chosen, then the period length is 2^{pq}. But this generator needs a special matrix $\mathbf{B}$, which improves the uniformity of the RNs. The operation implemented by matrix $\mathbf{B}$ is called **tempering**.

Niederreiter (1995) examined a generalization of the TGFSR, the **multiple recursive matrix methods** (**MRMMs**), with various matrices $\mathbf{S}_i$

$$\mathbf{x}_n = \mathbf{S}_1 \mathbf{x}_{n-1} \oplus \ldots \oplus \mathbf{S}_q \mathbf{x}_{n-q}. \tag{2.19}$$

The Mersenne Twister type also belongs to this class, i.e. it alters recursion (2.18) by

$$\mathbf{x}_n = \mathbf{x}_{n-r} \oplus \mathbf{S}_q \left(\mathbf{x}_{n-q}^u, \mathbf{x}_{n-q}^l \right) \tag{2.20}$$

where $u = w - l$ denote the upper bits of vector $\mathbf{x}_{n-q}$ and l the lower (i.e. the last) bits of $\mathbf{x}_{n-q}$, $0 \leq l < p$. Usually, in Mersenne Twister types we have $p = w$.

The most famous RNG of the Mersenne Twister type is MT19937, meanwhile implemented in many computer programmes and freely downloadable from various sources. Although this RNG is still the state of the art at the time of writing, we would recommend that the reader also tries other good generators. Mersenne Twisters often have good equidistribution properties but all these generators have one severe weakness. Once we are in a state with only a few 1's and many 0's (or we accidently start with such a seed!), we stay in such a situation for a long time, which means that the states do not differ much for some time. This problem is called lack of **diffusion capacity** (see Section 2.3.3).

2.2.6 Nonlinear RNGs

Linear RNGs typically have quite a regular structure such as the lattice structure of t-tuples of RNs from MRGs. To get away from this regularity, one can either transform the RNs, discard or skip some of the RNs – or use a nonlinear RNG. Until today not many nonlinear RNGs have been discussed in the literature as those RNGs are difficult to analyze theoretically. That they perform well is often just shown with a battery of statistical tests. There are inversive congruential generators (ICGs), explicit inversive congruential generators (EICGs), digital inversive congruential generators (DICGs), and combinations. They may have quadratic or cubic functions (see e.g. Eichenauer-Herrmann [1995], Knuth [1998]).

Characteristic for nonlinear RNGs is that there are no lattice structures. But the operation inversion in computers is more time-consuming than addition, bit shifting, subtraction, or even multiplication. So those generators

ICG(m, a, c)	Period
ICG(1039, 173, 1)	1039
ICG(2027, 579, 1)	2027
ICG($2^{31}-595$, 858993221, 1)	$2^{31}-595$
ICG($2^{31}-1$, 1288490188, 1)	$2^{31}-1$
ICG($2^{31}-1$, 9102, 36884165)	$2^{31}-1$

Table 2.3: Inversive Congruential Generators

are usually significantly slower than linear RNGs. But nevertheless, they are useful to verify important simulation results. Another idea is to add nonlinearity to a RNG by combining an excellent fast linear RNG with a nonlinear RNG (see L'Ecuyer and Granger-Piché [2003]).

Here we present two examples of nonlinear RNGs. Let $m \in \mathbb{Z}$ be a prime number to ensure that $\mathbb{Z}_m$ is a field. We define $\bar{c} := c^{-1}$ in $\mathbb{Z}_m$ if $c \neq 0$ and $\bar{c} := 0$ if $c = 0$, i.e. $c\bar{c} = 1$ for all $c \neq 0$. The first type, **ICG**(m, a, c), looks like a LCG but includes the inverse of the RN, see Algorithm 2.7.

Algorithm 2.7 Inversive congruential generators

$$s_{n+1} = (a\bar{s}_n + c) \mod m, \quad n \in \mathbb{N},$$

with $m \in \mathbb{N}$ prime, $c \in \mathbb{N}$, $c < m$, $a \in \mathbb{N} \setminus \{0\}$, $a < m$, seed $s_0 \in \mathcal{S} \subseteq \mathbb{N}$, $\mathcal{S} = \mathbb{Z}_m$. Numbers in $[0, 1)$ are obtained via

$$u_n = \frac{s_n}{m}.$$

ICG(m, a, c) has a maximal period length of m. A sufficient condition for maximum period is that the polynomial $x^2 - cx - a$ is primitive over $\mathbb{Z}_m$ (see Eichenauer-Herrmann and Lehn [1986]). If ICG$(m, a, 1)$ has maximal period, then so has ICG(m, t^2a, t) for $0 < t < m$. Examples for ICGs with maximal period are given in Table 2.3 (see also Hellekalek [1995]).

The second type, **EICG**(m, a, c), has the enormous advantage that one can easily produce disjoint substreams, which is particularly useful for parallelization techniques.

$$s_n = \overline{(a\,(n + s_0) + c)} \mod m \tag{2.21}$$

Selecting the parameters for a maximal period is easy – m must be prime and $a \neq 0$. Unfortunately, we do not have many different RNGs as most of them are equivalent. The sequence EICG$(m, a, 0)$ is obtained from EICG$(m, 1, 0)$ by choosing every a-th element.

2.2.7 More random number generators

We will mention some more random number generators with good properties only briefly:

• The recent WELL RNGs by L'Ecuyer et al. (2006) have excellent equidistribution properties. They belong to the class of $\mathbb{F}_2$-linear generators. Their period lengths are comparable to the Mersenne Twister types. Matrix $\mathbf{A}$ has a block structure and consists mainly of zero blocks. The nonzero blocks describe fast operations that are easy to implement as shifting, bitwise XOR, bitwise AND, or they are identity matrices. $\mathbf{A}$ is composed in such a way that the bit mixing is improved which results in a better diffusion capacity.

• Marsaglia (2003) has described XORshift generators, which are extremely fast RNGs that mainly work with binary shifts and bitwise XORs. They are a special case of the multiple recursive matrix method.

• Instead of working in the field $\mathbb{F}_2$ we can use the ring $\mathbb{F}_{2^m}$. This turns these generators into mod 2^m-generators. If m is chosen according to the number of bits needed to represent a real number or integer in the computer, we can take advantage of the binary structure of computers and implement fast calculations.

• Recently, there has been research on generalized Mersenne Twisters which use 128-bit arithmetic or a combined 32-bit arithmetic that adds up to 128 bits, and take advantage of special processor features to speed up calculations (see Matsumoto and Saito [2008]). However, such RNGs are no longer machine independent.

2.2.8 Improving RNGs

• As mentioned with LCGs, the output can be shuffled with the help of another generator or with the same generator. However, this removes only part of existing serial correlations and the RNs stay the same. Another drawback is that this kind of RNG cannot be used for parallelization techniques because the n-th output is no longer foreseeable.

• Some generators that have deficencies in their structure can be improved by dumping some of the produced RNs. An example is the RANLUX generator of Lüscher (1994) in which one "takes the luxury" to discard a certain number. It is based on a subtract-with-borrow RNG:

$$\begin{aligned} s_n &= (s_{n-10} - s_{n-24} - c_n) \mod 2^{24} \\ c_{n+1} &= [s_{n-10} < s_{n-24} + c_n] \end{aligned} \tag{2.22}$$

If the luxury level is LUX $= 0$, no numbers are ignored; with LUX $= 1$, 24 points are skipped, LUX $= 2$ skips 73 numbers, and so on. Although the RNG becomes slower with higher luxury levels, the structure improves significantly.

• It is possible to split the sequence of an RNG into several substreams and then to alternate between these streams.

• Two or more RNGs can be combined. Usually, combined generators show a much better performance, but this cannot be guaranteed for all combinations.

2.3 Testing and analyzing RNGs

There are two ways of analyzing the quality of a sequence of RNs: one is to examine the mathematical properties of the RNG analytically, the other is to submit the RNG to a battery of statistical tests, e.g. the test suite TestU01 by L'Ecuyer and Simard (2002) or the Diehard test battery by Marsaglia (1996).

2.3.1 Analyzing the lattice structure

As t-dimensional vectors formed by consecutive RNs from LCGs, MRGs, and other generators lie on a fixed number of hyperplanes, this number can be calculated for a range of dimensions. A good RNG should have a lot of hyperplanes in as many dimensions as possible, or alternatively, the distance between the parallel hyperplanes should be small, so that the RNG does not leave big gaps in the t-dimensional space.

The **spectral test** (see Knuth [1998]) analyzes the lattice structure in the set Ψ_t of all t-dimensional vectors constructed from t consecutive RNs from a special RNG (see Definition 2.2), started with every possible seed from state space $\mathcal{S}$:

$$\Psi_t := \{(u_1, \ldots, u_t) \,|s_0 \in \mathcal{S}\} \subseteq [0,1)^t \qquad (2.23)$$

The traditional spectral test is only applicable if this point set has indeed a lattice structure. Then it measures the maximum distance l_t between two successive hyperplanes. The value $1/l_t$ is called the **accuracy**, a value that is closely related to the minimum number of parallel hyperplanes. This number depends on the slope of the hyperplanes and their position to the coordinate axes of the t-dimensional cube.

The accuracy has to be calculated for each dimension. It may happen that the quality of some RNGs is totally different in selected dimensions. So sometimes the spectral test gives no clear ranking among RNGs, then it only shows what RNGs to avoid.

The search for the maximum distance between hyperplanes must be done efficiently as one cannot check all sequence points of the RNG separately when the period of the generator is very long. There also exist variants of the spectral test for general point sets.

2.3.2 Equidistribution

Here we consider the equidistribution of uniformly distributed real RNs $u \sim U[0,1)$. There also exists a similar concept for uniformly distributed integers $n = 1, \ldots, N \in \mathbb{N}$, called the k-distribution test (see e.g. Matsumoto and Kurita [1992]). We take a close look at the set Ψ_t, defined in Definition 2.2. In every simulation, there should be a chance to get close to each vector in the t-dimensional unit hypercube, so a good RNG should cover the unit hypercube $[0,1)^t$ with Ψ_t as even and as dense as possible. So, firstly, $\mathcal{S}$ should be a large set. Then we have to check if the RNs are well distributed, which we are going to make more precise now. Indeed, there exist several uniformity measures. Here we present the concept of equidistribution because values can be computed efficiently in many cases (see L'Ecuyer [1996b]).

DEFINITION 2.3 Equidissection
The partition of $[0,1)$ into 2^l equal segments defines a partition of the t-dimensional unit hypercube $[0,1)^t$ into 2^{tl} equal cells. This partition is called a (t,l)-equidissection in base 2.

DEFINITION 2.4 Equidistribution
The set Ψ_t is said to be (t,l)-equidistributed if each cell of a (t,l)-equidissection in base 2 of the unit hypercube contains the same number of points of the set Ψ_t.

A theoretical problem is that perfect equidistribution is only possible when the number of cells divides $|\Psi_t|$. Practically, the problem is neglible, also because $|\Psi_t| = |\mathcal{S}| = 2^k$ for a $k \in \mathbb{N}$ in most cases (this time zero vectors are included in the state space in order to make counting easier). For every RNG there exist special limitations up to which degree equidistribution is possible. The number of cells has to be smaller than the number of points in the state space $2^{tl} \leq |\Psi_t|$, and the partition should not be finer than the bit representation of numbers in the computer, $l \leq w$, w word length. For example, in the case of LFSR generators this means $l \leq w$ and $tl \leq k$.

DEFINITION 2.5 Maximal equidistribution
Set $k := \log_{base2}(|\Psi_t|)$, $l_t^* := min\{w, \lfloor k/t \rfloor\}$ the maximal number for which the multiset Ψ_t of an $\mathbb{F}_2$-linear generator can be (t,l)-equidistributed. If Ψ_t is (t,l_t^*)-equidistributed for all $0 \leq t \leq k$, then it is called maximally equidistributed (ME).

L'Ecuyer (1999b) described several combined Tausworthe generators which are maximally equidistributed and further collision-free.

DEFINITION 2.6 Collision-free

The set Ψ_t is said to be collision-free, if each nonempty cell of a (t, l)-equi-dissection in base 2 of the unit hypercube contains exactly one point for $l_t^* < l \leq w$.

The criterion equidistribution gives more importance to the most significant bits of the RN, while the lower order bits may have quite a poor distribution. One can multiply the RNs by a power of two, modulo 1, and then analyze the equidistribution again. RNGs that have the same good equidistribution properties for all bits are called **resolution-stationary**. This property is not very important as it suffices to know that sometimes lower order bits do not have a good distribution and to avoid placing importance on them in simulations. For example, a die is best simulated as $\lceil 6u_n \rceil$ and not as $(s_n \mod 6) + 1$, where s_n is the random integer in the iteration before it has been converted to a real number.

2.3.3 Diffusion capacity

A RNG is said to have a good diffusion capacity if started with two different seeds that are in some sense close to each other, though the sequences of RNs still differ significantly. If a state s_n contains many 0's and just a few 1's, then some RNGs have trouble getting away from that point and the following states s_{n+k} also contain many 0's for a long time. This is called **low diffusion capacity**. Some variants of the Mersenne Twister are known to have such a disadvantage.

One cause for the bad diffusion capacity is that often the characteristic polynomial of the recurrence contains too many zeroes. This means that during the recursion too many bits remain unchanged. Once we are in a state with many zeroes, we are stuck in it for a while as the algorithm only changes a few bits in every step, e.g. in trinomial-based generators.

In this context often the **Hamming weight** $H(\mathbf{s})$ is evaluated, where $H(\mathbf{s})$ is the number of bits set to 1 in the bit vector $\mathbf{s}$. One idea is to calculate a moving average of the fraction of bits that are 1 of a selection of successive output vectors $\mathbf{s}_n$. If the RNG is started in a state with many 0's, then this average should reach approximately 0.5 after some, but not too many, iteration steps. An alternative is to examine moving averages of the RN sequences $u_i \in [0, 1)$, when started close to 0 or 1, which gives more weight to the more significant bits.

2.3.4 Statistical tests

The RN sequences should be uniformly distributed on $[0, 1)$ and appear to be independent. These are properties that can be tested with several statistical tests. Therefore, the RNs u_n are tested against the null hypothesis:

$$H_0\text{: The numbers } u_n,\ n = 1, \ldots, N, \text{ are realizations of i.i.d. } U(0,1)\text{-distributed random variables.} \tag{2.24}$$

A **test statistic** for a RN sequence u_n, $n = 1, \ldots, N$ is a random variable $X : [0,1)^N \to \mathbb{R}$ for which a good approximation of the distribution under the null hypothesis H_0 is known. We define the left and right p-values of the test by

$$p_r := \mathbb{P}\left(X \geq \hat{x} | H_0\right), \quad p_l := \mathbb{P}\left(X \leq \hat{x} | H_0\right), \tag{2.25}$$

where $\hat{x}$ is the observed value of the RV X. Usually, the observations are divided into J subsets, then the theoretical probability p_j, $j = 1, \ldots, J$, of each subset under H_0 is calculated and compared with the empirical frequency, usually by a weighted difference.

What does passing a test in the RNG context mean? If the p-value is not too small, e.g. more than 0.05, we can say that this test has been passed. But it is not like in medicine that e.g. a p-value of less than 0.01 leads to the conclusion that the null hypothesis should be rejected. Here things are not so easy to decide. We do not fix significance levels here. We even accept smaller p-values, because we want the less likely combinations also to appear with our RNG. But if repeated tests always give small p-values, we get the suspicion that the RNG might not work well. Also very small p-values like 0.00000001 or smaller give a hint that the RNG might be bad.

Passing those tests does not prove that the RNG is suitable for all kinds of simulations. This does not even show that we have found a good RNG. But with every test that is passed, our confidence in the generator is improved. As the RNs are not truly random but deterministic, there will always be a statistical test that they will fail. Due to this, we only require that good RNGs should not fail simple statistical tests. Better generators should pass more tests. We also should make sure that the RNs are suitable for our special application. So at least one test should be designed in such a way to check the generator in our special simulation situation.

Different tests are needed to test different departures from the null hypothesis. It should be possible that any test can be applied to any subsequence, extracted at random. Further the behaviour of the RNG must be equal for each seed, so that not only one seed should be tested.

Well-known test collections for RNGs are Marsaglia's Diehard test battery (see Marsaglia [1996]), the library TestU01 from L'Ecuyer and Simard (2002) or the NIST (National Institute of Standards and Technology) test suite from Rukhin et al. (2001), which test i.i.d. $U(0,1)$- sequences and strings of random bits.

To get a feeling for testing we are going to present some examples of statistical tests here. By no means can the following tests be considered as the most important tests, they are more of a kind of instructional list, although some of them are indeed very important.

0-1-test

Here, we want to test whether a RNG, which produces random bits, actually generates zeroes and ones with equal probability. Suppose we have generated the **0-1**-sequence $s_n, n = 1, \ldots, N$. We look at the sum $S_N = \sum_{i=1}^{N} s_i$. If the sequence is random with independent RNs and equal probability, then the sum is $B(N, 1/2)$-distributed. We use the test statistic

$$X := \frac{2}{\sqrt{N}} \left(\sum_{i=1}^{N} s_i - \frac{N}{2} \right). \tag{2.26}$$

If N is large (i.e. $N > 35$), X is approximately $N(0,1)$-distributed under H_0.

χ^2-test

This is a generalization of the **0-1**-test. Suppose that we have to produce r RNs to get a result for one special observation, and that every observation falls into one of k categories. For example, we observe three random bits to get an integer $0 \leq i < 8$, so $r = 3$ and $k = 8$. Let p_i, $0 \leq i < k$, be the probability under H_0 that the observation falls into category i. Now we produce $N \cdot r$ independent RNs. Let Y_i be the number of observations in category i. This value is compared with the theoretical value $\mathbb{E}Y_i = Np_i$, and the following statistic is formed:

$$X := \sum_{i=0}^{k-1} \frac{(Y_i - Np_i)^2}{Np_i} = \frac{1}{N} \sum_{i=0}^{k-1} \frac{Y_i^2}{p_i} - N. \tag{2.27}$$

If the null hypothesis is true, then this value is approximately χ^2-distributed with $(k-1)$ degrees of freedom for sufficiently large N. A common rule says that N should be at least so large that every theoretical value is $Np_i \geq 5$.

Frequency test

This test is an application of the χ^2-test and checks in a simple way if RNs are evenly distributed over the interval $[0, 1)$. We therefore split a sample of N RNs u_n, $n = 1, \ldots, N$ into k subgroups via the rule $v_n = \lfloor ku_n \rfloor$. The number k can be chosen according to your needs, e.g. the choice $k = 128 = 2^7$ tests the leading seven bits of the real RNs. Let Y_i be the number of RNs in category i, $i = 0, \ldots, k-1$. Then under H_0 the test statistic

$$X := \sum_{i=0}^{k-1} \frac{(Y_i - N/k)^2}{N/k} = \frac{k}{N} \sum_{i=0}^{k-1} Y_i^2 - N \tag{2.28}$$

is approximately χ^2-distributed with $k-1$ degrees of freedom, provided that $N/k \geq 5$.

Serial test or m-tuple test

The frequency test can be generalized in examining pairs of successive RNs. Again we have partitioned the single RNs into k equally possible subgroups, so every possible pair $(v_{2n-1}, v_{2n}) = (r, s)$, $0 \leq r, s \leq k-1$, $n = 1, \ldots, N$ should appear with the same frequency. For this test we apply the χ^2-test to k^2 categories, each with probability $1/k^2$. k should not be too large, but N should be large, at least $5k^2$. Let Y_i be the number of RNs in category i, $i = 0, \ldots, k^2 - 1$. Under the null hypothesis the statistic

$$X := \sum_{i=0}^{k^2-1} \frac{\left(Y_i - N/k^2\right)^2}{N/k^2} = \frac{k^2}{N} \sum_{i=0}^{k^2-1} Y_i^2 - N. \tag{2.29}$$

is approximately χ^2-distributed with $k^2 - 1$ degrees of freedom. This test can further be generalized to triples or quadruples or more.

This test is one of the **important tests** and can be seen as the "book-writing test." If you work with 26 categories which symbolize letters of the alphabet, this test checks if every two, or three-letter word appears with equal probability, when you choose the letters randomly with the help of your RNG. Sometimes this test reveals astonishing facts, as some RNGs would **never** write the word "cat" in some cases (see Marsaglia [1996]).

There are many tests based on this scheme, e.g. the poker test, where the RNs are tested whether they would be a usable basis for a computer poker simulation.

Kolmogorov-Smirnov test

In contrast to the χ^2-test, the Kolmogorov-Smirnov-type tests also check probability distributions with infinitely many values, especially with continuous cumulative distribution functions $F(x)$. This test compares the empirical with the desired theoretical distribution function. With the sample $u_1, \ldots, u_N \in (0, 1)$ the empirical distribution function is defined by

$$F_N(x) := \frac{|\{u_i | \, u_i \leq x, i = 1, \ldots, N\}|}{N}, \qquad x \in \mathbb{R}. \tag{2.30}$$

The Kolmogorov-Smirnov test measures the maximal distance between the theoretical and empirical distributions:

$$K_N^+ = \sqrt{N} \sup_{-\infty < x < +\infty} \left(F_N(x) - F(x)\right) \tag{2.31}$$

$$K_N^- = \sqrt{N} \sup_{-\infty < x < +\infty} \left(F(x) - F_N(x)\right) \tag{2.32}$$

The advantage of this test is that it is also applicable for small sample sizes as the distributions of the two test statistics are exactly known and can be found in tables or computer routines. As we want the RNs uniformly distributed, we can test for uniform distribution, or we could test a series of independent χ^2-test results if they are really χ^2-distributed.

Maximum-of-t-test

This is an application of the Kolmogorov-Smirnov test. In a series of RNs $\in [0, 1)$ we determine the maximum in independent groups of t RNs $v_i := max\{u_{ti+1}, \ldots, u_{ti+t}\}$. Thus, we have the theoretical distribution function $F(x) = x^t$, $0 \leq x < 1$, which can be compared to the empirical one. Alternatively, the sequence $v_1^t, \ldots, v_N^t$, which should be uniformly distributed under H_0, can be analyzed with the Kolmogorov-Smirnov test.

Application-based tests

One of the most important tests is the test run of the RNG with a similar problem to the one that should be solved but from which the exact solution is already known. If the Monte Carlo method performs well with this task, the RNG seems to be suitable for the more complicated problem.

A famous allround application-based test is the two-dimensional Ising model test with the Wolff algorithm or Metropolis algorithm (see Knuth [1998]), which is an application in physics, where the solution is well known. This test is good in discovering long-term correlations. Many generators in the past that on first glance appeared to be excellent, failed in this test. This fuelled further research and the disadvantages of these generators came to light.

2.4 Generating random numbers with general distributions

Let us assume that we have found a good RNG which generates RNs that behave like i.i.d. RV$\sim U[0, 1)$. Our next step is to transform these RNs into nonuniform RNs, e.g. normally distributed, χ^2-distributed, or Poisson-distributed ones. Often the desired distribution can only be approximated. Of course, the approximation should be as exact as possible. Further robustness is important. If the distribution depends on parameters, then the approximation should also be good for nearby parameters. The transformation method should be efficient, it should be fast, and should not use too much memory. Another important point is the compatibility with variance reduction techniques in Monte Carlo simulations. Then often the only acceptable method for transformation is the inversion method.

2.4.1 Inversion method

The best way to transform RNs is the inversion method because it preserves structures. If the distribution structure of the uniformly distributed RNs is good, so will the structure of the transformed RNs. Also it is compatible with variance reduction techniques like e.g. antithetic variates. But, if speed matters, this method might not be the best, as often complicated functions are involved or one has to work with approximation algorithms.

Assume that the RV X has the cumulative distribution function (c.d.f.) F where F is strictly increasing and continuous. Then there exists the inverse F^{-1}. For the uniformly distributed RV $U \sim U[0,1)$ the RV $F^{-1}(U)$ has the same distribution as X, i.e. it has c.d.f. F due to

$$P\left(F^{-1}(U) \leq x\right) = P\left(U \leq F(x)\right) = F(x). \tag{2.33}$$

Note that there is a monotone relationship between the uniformly distributed RV and the transformed variable. If the c.d.f. F is not strictly increasing or not continuous, we can define a general inverse by

$$F^{\leftarrow}(u) := \min\{x \mid F(x) \geq u\}. \tag{2.34}$$

With this generalized inverse, we are able to formulate the inversion method.

Algorithm 2.8 Inversion method

Let F be a univariate c.d.f.

1. Sample a uniformly distributed RN u on $[0,1)$.
2. Obtain a RN with c.d.f. F via $x = F^{\leftarrow}(u)$.

If it is not possible to invert F analytically, it can be inverted numerically (by e.g. the Newton-Raphson method) or with an explicit approximation formula. Explicit approximation formulae can be improved by a Newton-Raphson, a regula falsi, or an interpolation step.

Discrete RVs

We now want to simulate a **die**, i.e. an equidistribution on $\{1,2,3,4,5,6\}$. As the c.d.f. is

$$F(x) = \sum_{i=1}^{min(6,\lfloor x \rfloor)} \frac{1}{6}, \quad x \geq 0, \tag{2.35}$$

the generalized inverse is given by

$$F^{\leftarrow}(u) = \lceil 6u \rceil, u \in (0,1] \tag{2.36}$$

Hence, we simply multiply a uniformly distributed RN $u \sim (0,1]$ by 6 and use the computer built-in rounding mechanism.

For an **equidistribution** on the set $1, \ldots, N$ and $u \sim (0, 1]$, we obtain the desired distribution via

$$x := \lceil Nu \rceil . \tag{2.37}$$

Finally, for the integer-valued RV X with $P(X = i) = p_i$, $i \in \mathbb{N}$,

$$x := \min \left\{ k \in \mathbb{N} \,\middle|\, \sum_{i=0}^{k} p_i \geq u \right\} \tag{2.38}$$

is the desired RN given by the inversion method. If this is coded in a straightforward way, the expected number of steps in search of the minimum is $E(X+1)$. The search can be accelerated by using tables (see Devroye [1986]).

Exponential distribution

The exponential distribution appears as waiting time between independent Poisson events and is suitable for measuring life spans or modelling radioactive decay. When the RV X is exponentially distributed with rate $\lambda > 0$, it has the c.d.f.

$$F(x) = 1 - e^{-\lambda x} \quad \text{for } x \geq 0, \tag{2.39}$$

and the inverse of F is given by

$$F^{-1}(u) = -\frac{\ln(1-u)}{\lambda} \quad \text{for } 0 \leq u < 1. \tag{2.40}$$

Then $y = -\ln(1-u)/\lambda$ with $u \sim U[0, 1)$ is exponentially distributed with rate λ. In practice, we rather use $-\ln(u)/\lambda$ as the RV $(1-U)$ has the same distribution as U. In software packages other algorithms for generating RNs with the exponential distribution are often implemented, as the evaluation of the logarithm function is rather time-consuming.

2.4.2 Acceptance-rejection method

Some distributions are so complicated, that the inversion of the c.d.f. F is much too difficult or only approximations exist. Then generating a RN with the acceptance-rejection method could be much faster and even easier. An acceptance-rejection algorithm can be constructed whenever we have got a density.

Suppose we want to simulate the RV X with density $f(x)$. Then we look for another RV Y with density $g(y)$, from which samples can be easily drawn by transforming uniformly distributed RNs, and which has the property

$$f(x) \leq C g(x), \quad x \in \mathbb{R} \text{ or } x \in \mathbb{R}^d, \tag{2.41}$$

with a constant $1 \leq C < \infty$. g is often called the comparison density or the majorizing function. Indeed, this can always be achieved with some simple,

Algorithm 2.9 Acceptance-rejection method

Let the densities f and g satisfying (2.41) be given.

1. Generate a uniformly distributed RN u on $[0,1)$.
2. Generate a RN y with the distribution given by the density g.
3. If $u \leq f(y)/(Cg(y))$ then accept y as the new RN x with density f. Otherwise reject it and go back to Step 1.

suitable density and a large constant C. In the case of a bounded density with compact support, the majorizing function can be constructed by choosing a uniform density on the support.

To produce one RN with distribution f we need more than one uniformly distributed RN. The speed of this RNG is determined by the time to generate the sample y, the time to calculate $f(y)$, and the constant C, because

$$\mathbb{P}\left(U \leq \frac{f(Y)}{Cg(Y)}\right) = \mathbb{E}\left[\mathbb{P}\left(U \leq \frac{f(Y)}{Cg(Y)} \,\Big|\, Y\right)\right]$$
$$= \mathbb{E}\left[\frac{f(Y)}{Cg(Y)}\right] = \int \frac{f(y)}{Cg(y)} g(y)\, dy = \frac{1}{C}. \quad (2.42)$$

So $1/C$ gives us the acceptance probability. For a fast algorithm the constant C should be as close to 1 as possible. In that case, nearly all RNs y are accepted and not too many RNs u and y have to be generated in vain.

If the function f is too time-consuming to evaluate, we can use squeeze functions q_1, q_2 with

$$q_1(x) \leq f(x) \leq q_2(x) \leq Cg(x), \quad (2.43)$$

which can be computed much faster. If $u \leq q_1(y)/Cg(y)$, then y can immediately be accepted. Otherwise, if $u > q_2(y)/Cg(y)$ then y can immediately be rejected. Only if both cases do not apply, should the function f be evaluated.

To illustrate in detail how the acceptance-rejection method works, we take a look at a special area.

DEFINITION 2.7 Body of a function

Let $f : \mathbb{R}^d \to \mathbb{R}$ be a nonnegative, integrable function. Then

$$B_f := \left\{(x, z) \in \mathbb{R}^d \times \mathbb{R} \,|\, 0 \leq z \leq f(x)\right\} \quad (2.44)$$

is called the body of f.

The acceptance-rejection method is based upon the following theorem (see Devroye [1986]).

THEOREM 2.8

Let X be a multivariate RV with density f on $\mathbb{R}^d$, $U \sim U[0,1)$ independent from X, and $C > 0$. Then $(X, UCf(X))$ is uniformly distributed on B_{Cf}.

Vice versa, if the multivariate RV $(X,Z) \in \mathbb{R}^{d+1}$ is uniformly distributed on B_f, then X has density f on $\mathbb{R}^d$.

So, the aim is to find a way to pick points randomly from the area B_f, where f is the desired density, with equal distribution. Because we are able to simulate the RV Y with density g, uniformly distributed points on B_{Cg} can easily be determined by generating one RN $y \sim Y$ and another independent RN $u \sim U[0,1)$. Then $(y, uCg(y))$ is uniformly distributed on B_{Cg}. If $uCg(y) \leq f(y)$ we have also found a point in B_f. Otherwise we have to do a new trial. The assumption $Cg(x) \geq f(x)$ makes sure that we do not cut out points from B_f. This is the core of the acceptance-rejection method.

Application: Standard normal distribution

First, we look at the distribution of a RV X given by the absolute value of a standard normal distribution, $X = |Z|$, $Z \sim N(0,1)$. So X has the density

$$f(x) = \sqrt{\frac{2}{\pi}} e^{-x^2/2} \text{ for } x \geq 0. \tag{2.45}$$

The exponential function in the density reminds us of the exponential distribution with density $g(y) = e^{-y}$, from which we already know how to draw samples with the inversion method: $u \sim U[0,1) \Rightarrow y = -ln(u) \sim exp(1)$. We now try to find the constant C so that

$$\frac{f(x)}{g(x)} \cdot \frac{1}{C} = \sqrt{\frac{2}{\pi}}\, e^{x-x^2/2} \cdot \frac{1}{C} \leq 1. \tag{2.46}$$

This term is maximized for $x = 1$, so if we choose $C = \sqrt{\frac{2e}{\pi}}$, the acceptance probability is about 0.76. In a final step we can transform our accepted RN into a normal RN by assigning a random sign to it, determined by another independent uniformly distributed RN.

As three and more RNs are needed for just one normal RN, the acceptance probability is not very close to 1. Further, the exponential function has to be exploited, so this method is very slow and therefore not the standard method for generating normal RNs.

For discrete probabilities the acceptance-rejection method works as well, just substitute the density with the mass probability function.

2.5 Selected distributions

2.5.1 Generating normally distributed random numbers

The task we encounter most often in Monte Carlo simulations is to generate normally distributed RNs. Here we concentrate on standard normal RNs. If we need RNs Z with distribution $N(\mu, \sigma^2)$, we first generate RNs $X \sim N(0,1)$, then we obtain $Z \sim N(\mu, \sigma^2)$ via

$$Z = \sigma X + \mu. \tag{2.47}$$

The first problem with a standard normal RV X with density

$$\phi(x) = \frac{1}{\sqrt{2\pi}} e^{-x^2/2}$$

and c.d.f.

$$\Phi(x) = \int_{-\infty}^{x} \frac{1}{\sqrt{2\pi}} e^{-x^2/2} dx$$

is that we cannot calculate values of the c.d.f. Φ by a simple formula. Instead we have to evaluate the integral numerically. A fast approximation avoiding the time-consuming exponential function is given e.g. by the rational function (see Abramowitz and Stegun [1972]) in Algorithm 2.10.

Algorithm 2.10 Approximation of the standard normal c.d.f.

This function gives approximate values for the standard normal distribution:

$$\begin{aligned} d := (&0.0498673470, 0.0211410061, 0.0032776263, \\ &0.0000380036, 0.0000488906, 0.0000053830) \end{aligned}$$

For $x \in [0, \infty)$ approximate

$$\Phi(x) \approx 1 - 0.5\left(1 + d_1 x + d_2 x^2 + d_3 x^3 + d_4 x^4 + d_5 x^5 + d_6 x^6\right)^{-16}$$

This approximation has a maximum relative error of $1.5 \cdot 10^{-7}$.

For many applications it suffices to work with approximations on the interval $[0, \infty)$ as due to the symmetry of Φ we have

$$\Phi(x) = 1 - \Phi(-x). \tag{2.48}$$

When more precision is needed, then it is better to have a special approximation on the interval $(-\infty, 0)$, which takes care that events with very small possibilities are not neglected (see e.g. Marsaglia [2004]).

When trying to generate normally distributed random numbers with the inverse method we encounter a similar problem: we do not have an explicit formula for the inverse. Again, we have to work with a numerical approximation. There exist several good approximations as e.g. the Acklam inverse, the Moro inverse, the Beasley-Springer approximation. As an example we present the Beasley-Springer-Moro algorithm (see Glasserman [2004]). We only have to work out an approximation on the interval $[0.5, 1)$ as due to the symmetry of Φ we have

$$\Phi^{-1}(u) = -\Phi^{-1}(1-u), 0 < u < 1. \tag{2.49}$$

Algorithm 2.11 Beasley-Springer-Moro algorithm for the inverse standard normal

---

Let

$$\begin{aligned} a : &= (2.50662823884, -18.61500062529, 41.39119773534, -25.44106049637) \\ b : &= (-8.47351093090, 23.08336743743, -21.06224101826, 3.13082909833) \\ c : &= (0.3374754822726147, 0.9761690190917186, 0.1607979714918209, \\ &\quad 0.0276438810333863, 0.0038405729373609, 0.0003951896511919, \\ &\quad 0.0000321767881768, 0.0000002888167364, 0.0000003960315187) \end{aligned}$$

1. For $u \in [0.5, 0.92]$ approximate

$$\Phi^{-1}(u) \approx \frac{\sum_{n=0}^{3} a[n]\left(u - \frac{1}{2}\right)^{2n+1}}{1 + \sum_{n=0}^{3} b[n]\left(u - \frac{1}{2}\right)^{2n}}$$

2. For $u \in (0.92, 1)$ approximate

$$\Phi^{-1}(u) \approx \sum_{n=0}^{8} c[n]\left(ln\left(-ln\left(1-u\right)\right)\right)^{n}$$

This approximation has a maximum absolute error of 3×10^{-9} over the range $[\Phi(-7), \Phi(7)]$.

It is also possible to search for the root of the equation $\Phi(x) = u$ with the Newton algorithm

$$x_{n+1} = x_n - \frac{\Phi(x_n) - u}{\phi(x_n)}. \tag{2.50}$$

Adding just one Newton step to the Beasley-Springer-Moro algorithm improves the accuracy even more.

The classical method for generating standard normal RNs is the Box-Muller method which samples from the two-dimensional standard normal distribution (see Algorithm 2.12). Due to the functions ln, cos, and sin, this method is rather slow. But it is very easy to implement, so it is a good start for normal RNs.

Algorithm 2.12 The Box-Muller method

---

1. Generate two independent RNs $u_1, u_2 \sim U(0, 1]$.

2. Obtain two independent standard normal RVs via

$$y_1 = \sqrt{-2\ln(u_1)}\sin(2\pi u_2), \quad y_2 = \sqrt{-2\ln(u_1)}\cos(2\pi u_2).$$

2.5.2 Generating beta-distributed RNs

The just cited Box-Muller method belongs to the group of polar methods as the functions sin and cos are involved. For the beta distribution a similar transformation exists. A RV $X \sim Beta(a, b)$, $a, b > 0$ has the density

$$f(x) = \frac{x^{a-1}(1-x)^{b-1}}{B(a.b)}, \quad 0 \leq x \leq 1, \tag{2.51}$$

with

$$B(a,b) = \frac{\Gamma(a)\Gamma(b)}{\Gamma(a+b)}, \quad \Gamma(a) = \int_0^\infty t^{a-1}e^{-t}dt. \tag{2.52}$$

A symmetric RV $X \sim Beta(a, a)$ with $a \geq 0.5$ can be generated as

$$x = \frac{1}{2}\left(1 + \sqrt{1 - u_1^{2/(2a-1)}}\cos(2\pi u_2)\right), u_1, u_2 \sim U(0, 1] \text{ independent.} \tag{2.53}$$

Devroye (1996) describes a more general method valid for all $a > 0$. For $0 < a, b < 1$ we can use the acceptance-rejection algorithm called Jöhnk's beta generator. This method is not recommended for $a, b > 1$ as it is too slow in this case. It requires on average $\Gamma(a+b+1)/(\Gamma(a+1)\Gamma(b+1))$ trials, which increases rapidly with a and b. Approximation methods for the symmetrical beta distribution and its inverse have been described by L'Ecuyer and Simard (2006).

2.5.3 Generating Weibull-distributed RNs

The Weibull distribution appears in insurance mathematics. It is also a good basis for an acceptance-rejection algorithm for the gamma distribution

Algorithm 2.13 Jöhnk's beta generator

1. Generate two independent RNs $u_1, u_2 \sim U(0,1]$.
2. Transform them into $x = u_1^{1/a}, y = u_2^{1/b}$.
3. If $x + y \leq 1$ then return $z := \dfrac{x}{x+y} \sim Beta(a,b)$, else go back to step one.

(see Section 2.5.4). The *Weibull*(a) distribution has density and c.d.f.

$$f(x) = ax^{a-1}/\exp(-x^a), \quad F(x) = 1 - \exp(-x^a), \quad a, x > 0. \tag{2.54}$$

Random numbers can be generated with the inversion method

$$x := (-\ln(u))^{1/a}, \quad u \sim U(0,1]. \tag{2.55}$$

2.5.4 Generating gamma-distributed RNs

The gamma distribution appears naturally when discrete events are Poisson-distributed. Consider the waiting time D_a until the a-th event happens, where the events are distributed according to $Poisson(\lambda)$. Then D_a is gamma-distributed with parameters a and $\theta = 1/\lambda$. More general: A RV X is gamma-distributed with shape parameter a and scale parameter θ, $X \sim Gamma(a,\theta)$, if it has the density

$$f_{a,\theta}(x) = x^{a-1}\frac{\exp(-x/\theta)}{\Gamma(a)\theta^a}, \quad x > 0,\ a, \theta > 0. \tag{2.56}$$

Properties of the gamma distribution:

- $\mathbb{E}X = a\theta$, $\mathbb{V}arX = a\theta^2$, $mode = (a-1)\theta$ for $a \geq 1$.
- **Scaling:** If X is $\sim Gamma(a,\theta)$, then $cX \sim \Gamma(a, c\theta)$ for any $c > 0$.
- **Summation:** If $X_i \sim Gamma(a_i,\theta), i = 1,\ldots,N$, independent, then $\sum_{i=1}^N X_i \sim Gamma(\sum_{i=1}^N a_i, \theta)$.
- **Limiting behaviour:** For $a < 1$ the density $f_{a,\theta}$ is monotone decreasing with $lim_{x \searrow 0} f_{a,\theta}(x) = \infty$.
- **Exponential distribution:** $Gamma(1,\theta)$ is simply the exponential distribution $\exp(1/\theta)$.

The scaling property tells us that it is only necessary to know how to generate $Gamma(a,1)$-distributed RNs. The last property and the summation property give us an easy method to generate $Gamma(a,\theta)$ random numbers when a is a small integer. Then the gamma distribution can be generated as the sum of exponentially distributed RVs. In general, the gamma distribution could be split up into an integer part and a part with $a < 1$:

$$Gamma(a, 1) \sim Gamma(\lfloor a \rfloor, 1) + Gamma(a - \lfloor a \rfloor, 1). \tag{2.57}$$

So, it would suffice to design a method for $a < 1$ only, but if a is large then we have to generate a lot of uniform RNs which makes this method slow. Further, the product of many numbers less than 1 is extremely small, so instability problems appear. This is the reason why this method normally should not be chosen.

There exist different acceptance-rejection methods for $a < 1$ and $a > 1$ as the comparison functions must follow the different limiting behaviour. Well-known algorithms for $a > 1$ are the ones by Ahrens and Dieter, Best, and Cheng (for details see Devroye [1986]). The last one will be described here as it is an interesting example for constructing an acceptance-rejection algorithm. Cheng's algorithm is based on the Burr XII density $g(x)$ with c.d.f. $G(x)$

$$g(x) = \lambda_a \mu_a \frac{x^{\lambda_a - 1}}{(\mu_a + x^{\lambda_a})^2}, \quad G(x) = \frac{x^{\lambda_a}}{\mu_a + x^{\lambda_a}}, \quad x \geq 0, \tag{2.58}$$

where λ_a and μ_a are parameters chosen according to a. RNs with this distribution can be generated with the inversion method $G^{-1}(u) = [(\mu_a u)/(1-u)]^{1/\lambda_a}$, $u \sim [0, 1]$. Cheng chooses $\mu_a = a^{\lambda_a}$ and $\lambda_a = \sqrt{2a-1}$. Thus, the rejection constant is

$$C = \frac{4a^a e^{-a}}{\lambda_a \Gamma(a)}, \tag{2.59}$$

which asymptotically tends to 1.13 for large a, which is quite good. To speed up calculations a squeeze step is added.

Algorithm 2.14 Cheng's algorithm for gamma-distributed RNs with $a > 1$

---

1. Generate two independent RNs $u_1, u_2 \sim U(0, 1)$.
2. Calculate
   $y := \frac{1}{\sqrt{2a-1}} \ln\left(\frac{u_1}{1-u_1}\right)$, $x := ae^y$, $z := u_1^2 u_2$,
   $r := a - \ln(4) + \left(a + \sqrt{2a-1}\right) y - x$.
3. Squeeze step: If $r \geq \frac{9}{2} z - 1 - \ln\left(\frac{9}{2}\right)$, then accept x.
4. Else: If $r \geq \ln(z)$, then accept x.
   Otherwise reject x and go back to the first step.

We still need an algorithm for $0 < a < 1$. The example we present here is constructed with the help of the Weibull distribution (see Devroye [1986]) and was chosen because of its elegant form. There are algorithms by Ahrens, Ahrens/Dieter, and Ahrens/Best (see Knuth [1998] or Devroye [1986]), which seem to work better. We have the rejection constant

$$C = \frac{\exp\left((1-a)\left(a^{a/(1-a)}\right)\right)}{\Gamma(a+1)} \leq 3.07 \text{ for all } a \in (0,1). \tag{2.60}$$

The rejection constant tends to 1 as $a \nearrow 1$ or $a \searrow 0$.

Algorithm 2.15 Algorithm for gamma-distributed RNs with $0 < a < 1$

---

1. Generate two independent RNs $u_1, u_2 \sim U(0,1]$ and transform them to exponentially distributed RNs $e_1 = -\ln(u_1)$, $\quad e_2 = -\ln(u_2)$.
2. Generate a RN with *Weibull*(a) distribution $x := e_1^{1/a}$.
3. If $e_1 + e_2 - (1-a)\, a^{a/(1-a)} \geq x$, then accept x, else go back to the first step.

More properties of the gamma distribution

- If $X \sim Gamma(a, \theta)$ then $1/X$ is inverse gamma-distributed with parameters a and θ^{-1}.
- If $X \sim Gamma(a, \theta)$ and $Y \sim Gamma(b, \theta)$, X, Y independent, then $X/(X+Y)$ is beta-distributed with parameters a, b.
- If $X \sim Gamma(a, 2)$ then X is chi-square-distributed with $2a$ degrees of freedom, i.e. $X \sim \chi^2_{2a}$.
- If $Y \sim Gamma(b, 1)$, $Z \sim Beta(a, b-a), b > a > 0$, independent, then $X_1 = YZ \sim Gamma(a, 1)$ and $X_2 = Y(1-Z) \sim Gamma(b-a, 1)$, also independent.

The second property is very important as it shows an interesting relation between the gamma and the beta distributions. It can be used when working with conditional gamma-distributed RV and is a basis for the bridge-sampling of the gamma process. With the help of the last property we can formulate Jöhnk's method to generate gamma-distributed RNs with shape parameter $a < 1$. Here, the rejection constant satisfies $C \leq 4/\pi \approx 1.27$ for $0 < a < 1$, which is quite good, but the calculation of two powers and one logarithm is time-consuming.

Algorithm 2.16 Jöhnk's generator for gamma-distributed RNs with $0 < a < 1$

1. Generate a $Beta(a, 1-a)$-distributed RN z with Jöhnk's beta generator.
2. Generate another independent uniformly distributed RN $u \sim U(0,1]$ and transform it into an exponentially distributed RN $y = -\ln(u)$.
3. Then $x := yz$ is $Gamma(a, 1)$-distributed for $a < 1$.

2.5.5 Generating chi-square-distributed RNs

The chi-square distribution is a special case of the gamma distribution as indicated in the properties of the gamma distribution. A RV X is chi-square-distributed, $X \sim \chi_k^2$, with k degrees of freedom, if it has the density

$$f_k(x) = \frac{x^{k/2-1} e^{-x/2}}{2^{k/2}\Gamma(k/2)}, \quad x, k > 0. \tag{2.61}$$

We have $\mathbb{E}X = k$ and $\mathbb{V}arX = 2k$. Chi-square-distributed RNs can be generated with the help of gamma-distributed RNs.

Algorithm 2.17 Chi-square-distributed RNs

Let $k > 0$ be given.

1. Simulate a RN $y \sim Gamma(k/2, 1)$
2. Obtain a RN $x \sim \chi_k^2$ via $x := 2y$

Chi-square-distributed RVs with degree $k \in \mathbb{N} \setminus \{0\}$ can also be described with the help of normally distributed RVs:

$$X_1, \ldots, X_k \sim N(0,1) \Rightarrow Z := \sum_{i=1}^{k} X_i^2 \sim \chi_k^2. \tag{2.62}$$

If $X_i \sim N(\mu_i, 1)$, $i = 1, \ldots, k$, we obtain the noncentral chi-square distribution $\chi_k^2(\delta)$ with noncentrality parameter $\delta = \sum_{i=1}^k \mu_i^2$ as the distribution of $Z := \sum_{i=1}^k X_i^2 \sim \chi_k^2(\delta)$. These considerations lead us to an Algorithm 2.18 for sampling RNs with chi-square distribution and even noncentral chi-square distribution. If k is large, this method might be rather slow. Generally for all $k < 0$, we can decompose a noncentral RV into a noncentral chi-square part and a standard chi-square part with one degree of freedom less:

$$\chi_k^2(\delta) = \chi_1^2(\delta) + \chi_{k-1}^2, \quad k > 0. \tag{2.63}$$

Algorithm 2.18 Chi-square-distributed RNs with the help of normally distributed RNs

Let $\delta > 0$, $k \in \mathbb{N} \setminus \{0\}$ be given.

1. Sample k independent RNs $n_1, \ldots, n_k$ with $n_i \sim N(0,1)$.
2. Set $\hat{n}_1 := n_1 + \sqrt{\delta}$ so that $\hat{n}_1 \sim N(\sqrt{\delta}, 1)$.
3. Then $x := \hat{n}_1^2 + n_2^2 + \ldots + n_k^2$ is $\chi_k^2(\delta)$-distributed.

This in particular means that we have to sample a normally distributed RN and a chi-square-distributed RN.

2.6 Multivariate random variables

The multivariate generation problem can be reduced to the univariate generation problem. But then we have to know a lot about the distributions, especially the conditional ones. When working with densities we should know the decomposition of the density f of the multivariate RV X into

$$f(x_1, \ldots, x_d) = f_1(x_1) f_2(x_2|x_1) \cdots f_d(x_d|x_1, \ldots, x_{d-1}) \tag{2.64}$$

where the f_i's are conditional densities, i.e. f_1 is the marginal density of X_1, f_2 is the conditional density of X_2 given X_1, and so on. So we have to know a whole series of conditional densities. The same applies when working with distribution functions, we have to know all the conditional distributions.

2.6.1 Multivariate normals

A d-dimensional multivariate normal RV $Z \sim N(\mu, \mathbf{\Sigma})$ is described by its mean vector μ and its covariance matrix $\mathbf{\Sigma}$. If we know a decomposition of the matrix $\mathbf{\Sigma} = \mathbf{A}^T \mathbf{A}$, we can concentrate on generating normal random vectors $\mathbf{X} \sim N(\mathbf{0}, \mathbf{I})$, $\mathbf{I}$ the identity matrix, as with the transformation

$$\mathbf{Z} = \mathbf{A}\mathbf{X} + \mu \tag{2.65}$$

we obtain $Z \sim N(\mu, \mathbf{\Sigma})$. So we can simply work with d independent standard normal RVs, $X_i, i = 1, \ldots, d$.

There are several possibilities to obtain a decomposition of $\mathbf{\Sigma}$, it is not uniquely determined. One method is the **Cholesky factorization**, which gives us a lower triangular matrix. This half-filled matrix has the advantage that the transformation of the standard normal vector is done nearly twice as fast as with a full matrix.

Algorithm 2.19 Cholesky factorization

Let a positive-definite matrix $\mathbf{\Sigma}$ be given. We find a lower triangular matrix $\mathbf{A}$ with $\mathbf{A}^T\mathbf{A} = \mathbf{\Sigma}$, $a_{i,j} = 0$ for $j > i$ by:

1. $a_{11} = \sqrt{\sigma_{11}}$,
2. $a_{ij} = \left(\sigma_{ij} - \sum_{k=1}^{j-1} a_{ik}a_{jk}\right)/a_{jj}$ for $1 \leq j < i \leq d$,
3. $a_{ii} = \sqrt{\sigma_{ii} - \sum_{k=1}^{i-1} a_{ik}^2}$, $1 < i \leq d$.

Other factorizations: Another possibility to get a suitable matrix $\mathbf{A}$ is the Eigenvector factorization. When the covariance matrix is symmetric and positive definite it has d real eigenvalues $\lambda_1, \ldots \lambda_d > 0$. Corresponding to it we can find a set of orthonormal eigenvectors $\mathbf{v}_1, \ldots, \mathbf{v}_d$, $\mathbf{v}_i^T\mathbf{v}_i = 1$, $\mathbf{v}_i^T\mathbf{v}_j = 0$ for $i \neq j$, $\mathbf{v}_i = \mathbf{\Sigma}\lambda_i\mathbf{v}_i$. Then $\mathbf{\Sigma} = \mathbf{V\Lambda V}^T$, where $\mathbf{V} = (v_1 \ \ldots \ v_d)$, $\mathbf{\Lambda} = diag\,(\lambda_i, \ldots, \lambda_i)$, and we can choose $\mathbf{A} := \mathbf{V\Lambda}^{1/2}$. Usually this matrix is dense, so there is no computational advantage. If the eigenvalues are ordered $\lambda_1 \geq \lambda_2 \geq \ldots \geq \lambda_d$, then this factorization can be useful for variance reduction techniques in Monte Carlo simulations, as one can focus these techniques on the first k components of $\mathbf{X} \sim N(\mathbf{0}, \mathbf{I})$, which explain the fraction of

$$\frac{\lambda_1 + \ldots + \lambda_k}{\lambda_1 + \ldots + \lambda_d} \tag{2.66}$$

of the variance.

2.6.2 Remark: Copulas

Sometimes we want to generate multivariate RVs with a certain dependence structure. This can be achieved with copula functions. Copulas are a way to extend one-dimensional distributions to multidimensions. They are practical if one has a knowledge of the marginal distributions but one is not so sure about the dependencies, especially when working with multivariate RVs which are not multivariate normal. For more on copulas see Chapter 8, Section 8.4.

2.6.3 Sampling from conditional distributions

If the RV X has c.d.f. F, the RV X given $X \in [a, b)$, $-\infty < a < b < \infty$, has c.d.f.

$$H(x) = \frac{F(x) - F(a)}{F(b) - F(a)} \text{ for } x \in [a, b], \tag{2.67}$$

provided that $F(b) - F(a) > 0$. With RNs $u \sim U[0, 1)$ the RNs $v := F(a) + (F(b) - F(a))\,u$ are uniformly distributed between $F(a)$ and $F(b)$. If the inverse of F on the interval $[a, b)$ has been given, RNs with the conditional distribution can easily be generated as

$$H^{-1}(u) = F^{-1}(F(a) + [F(b) - F(a)]\, u). \tag{2.68}$$

Assume that we have got a density f that is log-concave and differentiable in a. We want to generate a RV conditioned to $[a, b)$, $-\infty < a < b < \infty$. As the density is log-concave, it is bounded by

$$\frac{f(a)}{F(b) - F(a)} e^{\beta(x-a)} \tag{2.69}$$

where $\beta = f'(a)/f(a)$. This function can be used as a comparison function in the acceptance-rejection method, as it is easy to sample from it by inversion.

More general: If we want to sample X, conditioned on $X \in A$ for some set A, we can use the crude procedure: "Sample X until $X \in A$, return X."This is some kind of acceptance-rejection method with

$$f(x)/g(x) = 1/P(X \in A), \quad x \in A, \tag{2.70}$$

so the acceptance factor depends on $P(A)$. If this value is rather small, then a lot of RNs will be rejected with this method. In that case this might not be the method of choice.

Conditioning with multivariate normals: Suppose we have got a multivariate normal variable and we are interested in the distribution of one random vector given the other. We can write

$$\begin{pmatrix} \mathbf{X}_1 \\ \mathbf{X}_2 \end{pmatrix} \sim N\left(\begin{pmatrix} \mu_1 \\ \mu_2 \end{pmatrix}, \begin{pmatrix} \mathbf{\Sigma}_{11} & \mathbf{\Sigma}_{12} \\ \mathbf{\Sigma}_{21} & \mathbf{\Sigma}_{22} \end{pmatrix}\right) \tag{2.71}$$

with the matrix $\mathbf{\Sigma}_{22}$ having full rank. The conditioned distribution can be described as

$$(\mathbf{X}_1 | \mathbf{X}_2 = \mathbf{x}) \sim N\left(\mu_1 + \mathbf{\Sigma}_{12}\mathbf{\Sigma}_{22}^{-1}(\mathbf{x} - \mu_2), \mathbf{\Sigma}_{11} - \mathbf{\Sigma}_{12}\mathbf{\Sigma}_{22}^{-1}\mathbf{\Sigma}_{21}\right), \tag{2.72}$$

and so it is still a normal distribution.

2.7 Quasirandom sequences as a substitute for random sequences

Quasirandom sequences are not random at all. They do not look random and they do not pass statistical tests. But they are nearly perfectly evenly distributed. Quasirandom sequences give vectors for a special dimension, which is comparable to a grid. But it is usually not necessary to decide in advance how many points are needed. The sequences build up a pattern which becomes finer from point to point. Also the collection of points is optimized in such a way that there are no double values, no clusters, and no big gaps. Those sequences of t-dimensional vectors are also called low-discrepancy sequences as the discrepancy is used as a measurement for evenness.

DEFINITION 2.9 Discrepancy
The **discrepancy** *of a finite set* $\Phi_t = \left\{x_i \in [0,1)^t, i = 1, \ldots, n\right\}$ *of t-dimensional points is defined as*

$$D(x_1, \ldots, x_n) := \sup_{A \in \mathcal{A}} \left| \frac{|\Phi_t \cap A|}{n} - volume(A) \right| \tag{2.73}$$

where $\mathcal{A} = \left\{\prod_{j=1}^{t} [a_j, b_j) \,|\, 0 \leq a_j < b_j \leq 1\right\}$.

The **star discrepancy** $D^*(x_1, \ldots, x_n)$ *restricts the set of rectangles* $\mathcal{A}$ *to the ones with vertices* $a_j = 0,\ j = 1, \ldots, t$.

It is widely believed that the best rate that can be achieved is $\left((\ln n)^{t-1}/n\right)$ for point sets of fixed size $n > 1$. Methods for constructing t-dimensional sequences are usually considered as low-discrepancy methods if the star discrepancy of the first n points is about $O\left((\ln n)^t/n\right)$. Because low-discrepancy point sets are no longer random, the error estimates of the Monte Carlo method, which are based on stochastic arguments, are no longer applicable. But as all points are evenly distributed, quasirandom sequences often deliver much better results in some Monte Carlo simulations. The Koksma-Hlawka inequality makes this plausible for small dimensions (see Niederreiter [1995]):

THEOREM 2.10 Koksma-Hlawka inequality
For smooth functions f with finite variation $V(f)$ *in the sense of Hardy and Krause we have*

$$\left| \int_{[0,1)^t} f(\mathbf{x})\, d\mathbf{x} - \frac{1}{n} \sum_{i=1}^{n} f(\mathbf{x}_i) \right| \leq V(f)\, D^*(x_1, \ldots, x_n). \tag{2.74}$$

This means that the quality of the quasi-Monte Carlo approximation of the integral of the function f depends on the variation and the star discrepancy of the integration points. As the variation of the function is fixed we have to choose a suitable low-discrepancy set to improve the approximation.

This error bound has advantages and disadvantages. Using quasirandom sequences in Monte Carlo simulations gives us a strict error bound, in contrast to the probabilistic error bounds of usual Monte Carlo methods, which only produce confidence intervals. But experience has shown that the error bound often overestimates the integration error. Convergence is usually much better. Further this bound is difficult to compute and often the variation is unknown or sometimes even infinite. Indeed, convergence with quasirandom numbers is often very good. One reason is that some of the functions that are integrated have special smoothness properties which can be effectively exploited by the regular grids of quasirandom numbers. Another reason is that in financial applications nominal dimensions are often very large, but the functions can be approximated well by functions of low dimensions. In this case, grids, which have projections with a good uniform distribution in lower dimensions, especially in two dimensions, seem to work very effectively.

Well-known quasirandom sequences include: the Halton, the Faure, the Niederreiter, and the Sobol sequences.

2.7.1 Halton sequences

Halton sequences use the representation of integers in another base to produce numbers between 0 and 1. For each dimension another base is used. As

---

Algorithm 2.20 One-dimensional Halton sequences (Van-der-Corput sequences)

---

indexVan-der-Corput sequences

1. Select a prime number b as base.
2. For the j-th number in the Halton sequence write $j \in \mathbb{N}$ in base b, $j = \sum_{i=0}^{\infty} a_i b^i$, where $a_i \in \{0, \ldots, b-1\}$.
3. Obtain the Halton grid number as $H_j = \sum_{i=0}^{\infty} a_i b^{-i-1} \in [0, 1)$.

---

an example consider $j = 17$ in base $b = 5$ which is 32. Then, the output would be $2 \cdot 1/5 + 3 \cdot 1/25 = 0.52$. Van-der-Corput sequences are low-discrepancy sequences.

For generating grid point number j it is not necessary to use exactly the number j, there exist choices that work as well, e.g. the Gray code $G(j)$. With the help of the Gray code, Halton grids can be calculated efficiently. But when the dimensions get larger the Halton sequences begin to deteriorate. The two-dimensional projections of higher dimensions reveal gaps and clusters. So Halton sequences are only useful for numerical integration in lower dimensions.

Algorithm 2.21 t-dimensional Halton sequences

1. Select t different prime numbers b_k, $k = 1, \ldots, t$ as bases, typically the first t primes.
2. For the j-th point in the Halton sequence write $j \in \mathbb{N}$ in each base b_k, $j = \sum_{i=0}^{\infty} a_{k,i} b_k^i$, where $a_{k,i} \in \{0, \ldots, b_k - 1\}$.
3. Obtain the t-dimensional Halton grid point as $H_j = (h_1, \ldots, h_t)$ with $h_k = \sum_{i=0}^{\infty} a_{k,i} b_k^{-i-1}$, $k = 1, \ldots, t$.

2.7.2 Sobol sequences

In contrast to the Halton sequences, the Sobol ones use only base 2 for the expansion of integers. This makes the Sobol grids more regular in higher dimensions, as the large prime numbers in the Halton grids are responsible for the big gaps in the projections of higher dimensions. Working in base 2 also has the advantage that the binary structure of computers can be exploited.

Each coordinate of the t-dimensional vectors of a Sobol sequences follows the same construction principle, but each one has its own generator matrix $\mathbf{C} \in \{0, 1\}^{w,w}$. The columns of C consist of the binary expansions of so-called **direction numbers** c_j, $1, \ldots, w$.

In the k-th iteration step the number k is represented in base 2,

$$k = \sum_{i=0}^{w-1} a_{k,i} 2^i, \quad \mathbf{a}_k := (a_{k,0}, \ldots, a_{k,w-1})^T \in \{0, 1\}^w, \tag{2.75}$$

where the number w is chosen large enough for all necessary iteration steps. The integer w often corresponds to the number of bits used to represent an integer in the computer. After that, a binary vector $\mathbf{y}_k$ is computed

$$\mathbf{y}_k = \mathbf{C}\mathbf{a}_k, \tag{2.76}$$

which is the basis for one coordinate of the k-th grid point

$$x = \sum_{i=0}^{w-1} y_{k,i} 2^{-i-1}. \tag{2.77}$$

This algorithm has similarities with the algorithm for the Halton sequences, but the generator matrix $\mathbf{C}$, which mixes the bits, is an important difference. For a t-dimensional Sobol sequence we need t matrices $\mathbf{C}_i$, $i = 1, \ldots, t$, each of it consisting of the binary expansion of w direction numbers. For these Sobol chooses a primitive polynomial in the field $\mathbb{F}_2$ of degree q, $q \in \mathbb{N}$,

$$P(z) = z^q + \alpha_1 z^{q-1} + \ldots + \alpha_{q-1} z^1 + 1, \quad \alpha_q := 1. \tag{2.78}$$

This polynomial is the basis for a recurrence relation

$$\mathbf{m}_j = \alpha_1 2^1 \mathbf{m}_{j-1} \oplus \ldots \oplus \alpha_q 2^q \mathbf{m}_{j-q} \oplus \mathbf{m}_{j-q}, \tag{2.79}$$

where $\mathbf{m}_j$ is a binary vector, representing the binary expansion of an integer, and $\oplus$ the bitwise exclusive XOR. The direction numbers can be described as

$$c_j = \frac{\mathbf{m}_j}{2^j}, \quad j = 1, \ldots, w. \tag{2.80}$$

The direction numbers, initialized with odd integers $0 < \mathbf{m}_j < 2^j$, $j = 1, \ldots, q$, consequently lie in $(0, 1)$. The columns of matrix $\mathbf{C}$ are then filled with the binary expansions of those w different direction numbers. The special polynomial of degree zero, $P(z) \equiv 1$, defines the identity matrix as generator matrix.

One important aspect is the initialization of the Sobol sequences, which has an enormous impact on the structure of the distribution and uniformity properties. But we will not present it here as it goes too far. Large tables of primitive polynomials in $\mathbb{F}_2$ can be found in the Internet or in several books. Again, as in the case of Halton sequences, you can work in step k either with the integer k or a variant of the Gray code $G(k)$. An efficient implementation has been described by Bratley and Fox (1988).

2.7.3 Randomized quasi-Monte Carlo methods

Because the points of the quasirandom sequences do not even try to mimic randomness, we can no longer use the stochastic error bounds in the usual Monte Carlo method. We have been given deterministic worst-case error bounds by the Koksma-Hlawka inequality, but they overestimate the real error in such a way that some good stochastic bounds would be helpful. The idea is now to reintroduce a little bit of randomness without destroying the excellent structure too much. Then we are able to design methods that deliver confidence intervals and still take advantage of the fast convergence with quasirandom sequences (see L'Ecuyer [2004]). Here we take a look at the t-dimensional deterministic point set $\Phi_t = \{\mathbf{u}_1, \ldots, \mathbf{u}_N\} \subseteq [0, 1)^t$.

• One possibility to randomize the point set Φ_t is to add a random shift. Let $\mathbf{x}$ be a random vector uniformly distributed in $[0, 1)^t$, then the randomized set is calculated as $\Phi_t(\mathbf{x}) = \{(\mathbf{u}_i + \mathbf{x}) \mod 1 | \, i = 1 \ldots, N\}$. The effect is that each point of the set is now uniformly distributed over the hypercube, but nevertheless, the points are not independent.

• Another idea is to mix up the digits of the components of each vector randomly. Let $u_{k,j}$ be the j-th coordinate of point k, $k = 1, \ldots, N$, $j = 1, \ldots, t$. The expansion in base b then is $u_{k,j} = \sum_{i=0}^{r-1} v_i b^{-i-1}$, where r is chosen large enough. Then r independent permutations π_i on $\{0, \ldots, b-1\}$ are selected randomly and applied to the representation in base b, $\tilde{u}_{k,j} = \sum_{i=0}^{r-1} \pi_i(v_i) b^{-i-1}$.

The same digit permutations are applied to the same coordinate of all vectors in Φ_t. For the other coordinates independent permutations are randomly drawn.

• The digits can be scrambled in many other ways. Another idea is the linear permutation of digits: $\tilde{u}_{k,j} = \sum_{i=0}^{r-1} \tilde{v}_i b^{-i-1}$, with $\tilde{v}_i = \sum_{l=1}^{i} h_{l,i} v_l + g_i \mod b$. The integers $h_{l,i}$ and g_i are drawn randomly from $\{0, \ldots, b-1\}$, $h_{i,i} \neq 0$.

2.7.4 Hybrid Monte Carlo methods

Quasirandom sequences and especially Sobol sequences usually work very well in Monte Carlo simulations. The initial coordinates of these sequences – for most methods only a few dimensions (Sobol usually offers much more, but it depends on the implementation) – have a better distribution than those in the higher dimensions. So, it is recommended to assign the more important variables of the simulation to the lower dimensions. This can be done e.g. by a change of variables, Brownian bridge construction, or a principal component construction (see Chapter 4 or Section 2.6.1). Often only a few variables influence the value that should be estimated, so we talk about the **effective dimension** $\hat{d}$, which is often much smaller than the real dimension d (see Equation 2.66) where a few components explain most of the variance). This idea can be spun even further. One uses a low-dimensional quasirandom sequence, $q_k \in (0,1)^{\hat{d}}$, for the first components, and fills the rest of the vector with independent, uniformly distributed pseudorandom numbers $p_k \in (0,1)^{d-\hat{d}}$. This method is called the **hybrid Monte Carlo method** (see Asmussen and Glynn [2007]). It has the advantage that projections of the components in higher dimensions have a better structure compared to quasirandom structures in higher dimensions. Another application of this method is the possibility to use quasirandom sequences in situations where the dimensionality of the problem is not clear in advance (as e.g. valuing of path-dependent exotic options).

2.7.5 Quasirandom sequences and transformations into other random distributions

Quasirandom sequences give us very evenly distributed grids in many dimensions, which are often a good substitute for uniformly distributed vectors and even speed up convergence when working with smooth functions due to the evenness. But if other distributions are needed, e.g. the normal distribution, then care must be taken not to destroy the excellent grid structure. Then the only suitable method to transform the numbers is the inversion method.

2.8 Parallelization techniques

It is often possible to work with more than one computer processor, especially when doing a large-scale simulation. Then we will need RNGs that can be split up to be distributed on the different processors. There are several ideas for such so-called parallel RNGs:

1. The sequence of RNs is split up into several very long disjoint substreams and then each processor is fed with the seed of the appropriate substream.

2. With the leap-frog technique the RN $u_{n \cdot L+j}$ is assigned to the j-th processor, $n \in \mathbb{N}$, $1 \leq j \leq L$. $L > 1$ is called the lag and is usually the same as the number of processors.

3. We could supply different processors with different RNGs.

4. Some RNGs have several independent, disjoint subcycles with large periods. Then we can feed each processor with the seed of a different subcycle.

Parallel RNGs should have the following properties:

- The RNG should work for any number of processors up to an upper bound.
- The RNG should deliver the same RNs for any number of processors.
- The quality of the simulation should not depend on the number of processors.
- The RN sequences for each processor should satisfy the criteria for a good RNG.
- The sequences on different processors should not be correlated.
- Communication between the processors should not be necessary; once started the simulation on each processor runs by itself.

Finding a good RNG for parallelization is still a challenge. We require independence within streams and between the streams of different processors. Small correlations can be inflated by distributing a previously good RN sequence on several processors. Also the method to initialize a RNG needs more attendance, as seeding several similar RNGs with a seed that is also very similar may lead to catastrophically correlated RNs which would have otherwise been good.

2.8.1 Leap-frog method

The advantage of the leap-frog method is that we get the same stream of RNs in every calculation independent of the number of processors. So, these simulations are reproducible without problems. Leap frogging is especially easy with LCGs, combined LCGs, or shift-register generators. We only need to know how to jump ahead a few steps. Jumping ahead for L steps in the case of a LCG becomes

$$s_{n+L} = \left[\left(a^L \mod m \right) s_n + \frac{a^L - 1}{a - 1} c \right] \mod m \tag{2.81}$$

One problem becomes obvious here: Although the multiplier a may have good properties in the spectral test, the new multiplier $(a^L \mod m)$ might have bad properties, so the individual RN substream on each processor might be quite unsuitable. Also, when using a modulus which is a power of 2, and the number of processors is also a power of 2 (which is often the case), there are serious correlations between the RN sequences on different processors. This problem can be avoided by using a prime modulus.

The problem with MRGs or other recursions with matrices is that the new matrix $\mathbf{A}^L$ has to be computed in advance, which can be rather time-consuming. But this step is only done once. The more serious drawback is that a formerly sparse matrix can become full, and so the calculations in every iteration step will no longer be fast.

2.8.2 Sequence splitting

For the sequence splitting method we divide the sequence of RNs into several substreams. If we have got a RNG with period length ρ, we could split up the whole sequence into L subsequences of equal length, where L is the number of processors. If the RNG has an extremely long period, an alternative method is to cut off several very long pieces with a length of $\tilde{\rho}$. In most cases the jump-ahead step, with a jump of ρ/L or $\tilde{\rho}$, has to be done only once, in the initialization, then the RNG runs as usual on the different processors. So, just the setting-up time becomes longer. But here the jump is rather big and precalculations can be very time-consuming. Efficient algorithms for jumping ahead have to be found.

With jumping ahead a huge step, long-range correlations may become apparent that would not have otherwise been significant. Another disadvantage here is that with a different number of processors the RN sequences will be different, the simulation becomes machine-dependent. One solution could be to assign some streams to nonexisting virtual processors. But this only works up to a certain limit, which is not very high.

The possibility of splitting the RN sequence into substreams is a very useful feature for some simulation situations, not only for parallel processors. For example consider the application to compare two different parameter settings. Here it is an advantage to have mainly the same RNs. If the number of RNs needed changes with every parameter setting, then with several RN substreams you can make sure that every main simulation part starts with the same RNs.

2.8.3 Several RNGs

For the third method we need several reliable RNGs of nearly equal quality, which must further deliver uncorrelated sequences. Usually, RNGs of the same type but with different parameters are chosen. The task now is to find enough good RNGs for all processors. The problem remains that unknown correlations may exist.

A first approach is to use nonlinear RNGs, as the EICGs are so far considered to be uncorrelated. However, not much research has been done in this direction yet.

A promising idea is to use a collection of Mersenne Twister type generators. Matsumoto and Nishimura (2000) describe how to create them efficiently. These generators seem to be independent, have long periods, good equidistribution properties, are fast, and many of them exist. The disadvantage is the long setting-up time before the simulation can be run.

2.8.4 Independent sequences

Multiplicative lagged Fibonacci generators are not usable in the two previous methods because jumping ahead in this case is much too time-consuming. But they have lots of disjoint full-period RN cycles. Those cycles can be distributed on several processors by careful initialization. The seeds for the different cycles can be tabulated or it is even possible to choose a random seed for each processor, as there are generators that have such a huge number of disjoint cycles, so that the possibility of choosing overlapping cycles is very small.

2.8.5 Testing parallel RNGs

If sequence splitting is the method of choice, the RNG must be submitted to much larger tests as in the single processor case. In general, one has to test for correlations within a RN stream the way it is used in a single processor and for correlations between different streams. For the latter test the different RN streams can be interweaved and can be submitted to the usual statistical tests.

Chapter 3

The Monte Carlo Method: Basic Principles

3.1 Introduction

The main idea of the Monte Carlo method is to approximate an expected value $\mathbb{E}(X)$ by an arithmetic average of the results of a big number of independent experiments which all have the same distribution as X. The basis of this method is one of the most celebrated results of probability theory, the strong law of large numbers. As expected values play a central role in various areas of applications of probabilistic modelling, the Monte Carlo method has a widespread use. Examples of such areas of application are the analysis and design of queueing systems (such as in supermarkets or in large factories), the design of evacuation schemes for buildings, the analysis of the reliability of technical systems, the design of telecommunication networks, the estimation of risks of investments or of insurance portfolios, just to name a few.

Historically, the Monte Carlo method dates back to 1949 when the article "The Monte Carlo Method" by Metropolis and Ulam appeared in the *Journal of the American Statistical Association.* However, it was already developed during World War II. J. von Neumann and S. Ulam are commonly regarded as the founders of the Monte Carlo method. The name **Monte Carlo method** should indicate that one uses a sort of gambling to obtain an approximation procedure.

Nowadays one performs no physical gambling in the Monte Carlo method. The outcomes of the independent experiments, needed to perform the method, are replaced by suitable random numbers that are generated by a computer. As the amount of random numbers has to be very high to ensure that the Monte Carlo estimate is close to the exact expected value, the method tends to be quite slow when applied in its **crude** form. As the Monte Carlo estimator is a random variable, each run of it typically produces new values. As the estimator is unbiased, the variance of the estimator is a measure for its accuracy. Reducing this variance by suitable methods is therefore the usual way of speeding up the Monte Carlo method.

In this chapter we introduce the crude Monte Carlo method, give some first simple applications, and then concentrate on various methods for obtaining variance reductions. While due to our focus on finance most our examples and considerations are done in the context of probability distributions with densities, nearly all ideas carry over to the discrete distribution setting in the obvious way: Simply switch from the density function to the probability function of the discrete distribution.

3.2 The strong law of large numbers and the Monte Carlo method

3.2.1 The strong law of large numbers

The strong law of large numbers (together with its various variants) is one of the most powerful theorems of probability theory and has been a central object of research during the history of probability theory. It is the basis of the Monte Carlo method and states that the arithmetic mean of a sequence of independent, identically distributed random variables $(X_n)_{n\in\mathbb{N}}$ converges almost surely to the expected value $\mu = \mathbb{E}(X_1)$ which of course is the same for all X_n. We will state it here in its simplest form which is also referred to as **Kolmogorov's version** of the strong law:

THEOREM 3.1 Strong law of large numbers
Let $(X_n)_{n\in\mathbb{N}}$ be a sequence of integrable, real-valued random variables that are independent, identically distributed (i.i.d.), and defined on a probability space $(\Omega, \mathcal{F}, \mathbb{P})$. Let further

$$\mu = \mathbb{E}(X_1). \tag{3.1}$$

Then, we have for $\mathbb{P}$-almost all $\omega \in \Omega$

$$\frac{1}{n}\sum_{i=1}^{n} X_i(\omega) \overset{n\to\infty}{\to} \mu, \tag{3.2}$$

i.e. the arithmetic mean of the (realizations of) X_i tends to the theoretical mean of every X_i, its expectation μ.

Note that the convergence in the strong law is the almost sure convergence. This is an ω-wise convergence, i.e. the convergence of the arithmetic mean in the theorem can be reduced to the convergence of sequences of **real numbers** $\frac{1}{n}\sum_{i=1}^{n} X_i(\omega)$.

REMARK 3.2 1. The above form of the strong law is not the most general one. One can, for example, relax the independence assumption and replace it by pairwise independence (i.e. all pairs X_i and X_j have to be independent for $i \neq j$). For this version of the strong law see Etemadi (1981).

2. One can also relax the assumption of having an identical distribution for the different random variables X_n. In this case, one needs the independence of the X_n, $\sigma_j^2 = \mathbb{V}ar\left(X_j\right) < \infty$ and that we have

$$\sum_{j=1}^{\infty} \frac{\sigma_j^2}{j^2} < \infty. \tag{3.3}$$

Then one obtains:

$$\frac{1}{n} \sum_{j=1}^{n} \left(X_j - \mathbb{E}\left(X_j\right)\right) \xrightarrow{a.s} 0 \text{ for } n \to \infty. \tag{3.4}$$

3. There are also versions for stochastic processes such as the martingale strong law which is not presented here. For nearly all our purposes with regard to the Monte Carlo method, Kolmogorov's strong law is sufficient. □

3.2.2 The crude Monte Carlo method

Let X be a real-valued random variable with a finite expectation $\mathbb{E}(X)$. One popular method to compute this expectation (approximately) is given in Algorithm 3.1.

Algorithm 3.1 The (crude) Monte Carlo method

Approximate $\mathbb{E}(X)$ by the arithmetic mean $\frac{1}{N}\sum_{i=1}^{N} X_i(\omega)$ for some $N \in \mathbb{N}$.

Here, the $X_i(\omega)$ are the results of N independent experiments that have the same probability distribution as X.

The method in this pure form is called the **crude** Monte Carlo method to distinguish it from all its variants that are presented in the following.

For considering the accuracy of the Monte Carlo method, we have to point out that as a stochastic method, different runs of the Monte Carlo method typically lead to different results (although they might be quite close to each other!) when approximating a certain expression. We therefore have to deal with a stochastic error. Let us first state that the Monte Carlo method approximates the relevant expectation correctly in the mean.

THEOREM 3.3 Unbiasedness of the Monte Carlo estimator
Let $(X_n)_{n\in\mathbb{N}}$ be a sequence of integrable real-valued random variables that are

independent and identically distributed as X. All random variables are defined on a probability space $(\Omega, \mathcal{F}, \mathbb{P})$.

Then, the Monte Carlo estimator

$$\bar{X}_N := \frac{1}{N} \sum_{i=1}^{N} X_i, \; N \in \mathbb{N}, \tag{3.5}$$

is an unbiased estimator for $\mu = \mathbb{E}(X)$, i.e. we have

$$\mathbb{E}\left(\bar{X}_N\right) = \mu. \tag{3.6}$$

Although this already ensures that a Monte Carlo estimator is correct in the mean, it does not help us to get a feeling for the absolute value of the error. Therefore, we look at the standard deviation of the difference between $\bar{X}_N$ and μ. As we have

$$\mathbb{V}ar\left(\bar{X}_N - \mu\right) = \mathbb{V}ar\left(\bar{X}_N\right) = \frac{1}{N^2} \sum_{i=1}^{N} \mathbb{V}ar\left(X_i\right) = \frac{\sigma^2}{N}, \tag{3.7}$$

the standard deviation of the error is of order $O(1/\sqrt{N})$. As the standard deviation is a measure for the (mean) accuracy of the crude Monte Carlo method, this calculation has the following important consequence:

Increasing the accuracy of the Monte Carlo estimator

Increasing the (mean) accuracy of the crude Monte Carlo estimate by one digit (i.e. reducing its standard deviation by a factor 0.1) requires increasing the number of Monte Carlo runs by a factor of 100.

This, in particular, means that simply repeating a Monte Carlo run a certain number of times does not significantly improve the accuracy of the estimator. In the sense of the above insight, if we want to achieve a higher accuracy, it needs a significant effort. Methods to speed up this slow rate of convergence will be one of the main subjects of this chapter.

One can justify the use of the standard deviation of the error as a measure for the accuracy of the Monte Carlo estimator by the central limit theorem.

THEOREM 3.4 Central limit theorem (i.i.d. case)

Let $(X_n)_{n \in \mathbb{N}}$ be a sequence of independent real-valued random variables that are identically distributed and are defined on a probability space $(\Omega, \mathcal{F}, P)$. Assume also that they all have a finite variance $\sigma^2 = \mathbb{V}ar(X)$. Then, the normalized and centralized sum of these random variables converges in distribution towards the standard normal distribution, i.e. we have

$$\frac{\sum_{i=1}^{N} X_i - N\mu}{\sqrt{N}\sigma} \xrightarrow{D} \mathcal{N}(0,1) \text{ as } N \to \infty. \tag{3.8}$$

From the central limit theorem one can infer that for large values of N the crude Monte Carlo estimator is approximately $\mathcal{N}\left(\mu, \sigma^2/N\right)$-distributed. As the standard deviation σ uniquely characterizes the spread of the values of a normal distribution around its mean μ, using the standard deviation as a measure for the accuracy of the Monte Carlo estimator is justified. As we know that the asymptotic distribution of the Monte Carlo estimator is approximately normal, we obtain

An approximate $(1-\alpha)$-confidence interval for the expectation μ

$$\left[\frac{1}{N}\sum_{i=1}^{N} X_i - z_{1-\alpha/2}\frac{\sigma}{\sqrt{N}}, \frac{1}{N}\sum_{i=1}^{N} X_i + z_{1-\alpha/2}\frac{\sigma}{\sqrt{N}}\right]. \tag{3.9}$$

Here, $z_{1-\alpha/2}$ is the $1-\alpha/2$-quantile of the standard normal distribution. As the 97.5%-quantile of the standard normal distribution is about 1.96, a popular choice for an approximative symmetric 95%-quantile for the expectation estimated by the Monte Carlo method in applications is given by the

2σ-rule for an approximate 95%-confidence interval for μ

$$\left[\frac{1}{N}\sum_{i=1}^{N} X_i - 2\frac{\sigma}{\sqrt{N}}, \frac{1}{N}\sum_{i=1}^{N} X_i + 2\frac{\sigma}{\sqrt{N}}\right]. \tag{3.10}$$

REMARK 3.5 1. As the length of the confidence interval is proportional to $1/\sqrt{N}$, one has to increase the number N of simulation runs by a factor of 100 to reduce this length by a factor of 0.1. Again, this underlines the slow convergence of the crude Monte Carlo method.

2. Typically, the standard deviation σ needed for setting up the confidence intervals is unknown. So, to use them as approximate confidence intervals, one has to estimate σ^2 by the sample variance

$$\bar{\sigma}_N = \sqrt{\frac{1}{N-1}\sum_{i=1}^{N}\left(X_i - \bar{X}_N\right)^2} = \sqrt{\frac{N}{N-1}\left(\frac{1}{N}\sum_{i=1}^{N} X_i^2 - \bar{X}_N^2\right)} \tag{3.11}$$

and then obtains a usable 2σ-rule for an approximate 95%-confidence interval for μ

$$\left[\frac{1}{N}\sum_{i=1}^{N} X_i - 2\frac{\bar{\sigma}_N}{\sqrt{N}}, \frac{1}{N}\sum_{i=1}^{N} X_i + 2\frac{\bar{\sigma}_N}{\sqrt{N}}\right]. \tag{3.12}$$

Of course, one can in the same way establish a *usable* version for the general confidence interval for μ. We will however in the following always use 1.96 instead of 2 when we compute an approximate 95%-confidence interval.

3. It has to be emphasized that the above confidence interval is an approximate, asymptotic confidence interval. As on one hand, we are only estimating

the variance and on the other hand, we do not *a priori* know if N is sufficiently large to make the central limit theorem work, we should still have a careful look at the whole situation if the confidence interval seems to indicate something surprising. In particular, there might occur situations (such as the Monte Carlo estimation of expectations of functions that are only nonzero far away from the center of the underlying distribution) where the approximation for the 95%-confidence interval might not be valid. □

3.2.3 The Monte Carlo method: Some first applications

The basis of applying the Monte Carlo method is to calculate or approximate certain expressions via *guessing* them with the help of drawing a usually big amount of suitable random numbers. As we need the strong law of large numbers for proving that this method works, it is necessary that the expression under study can be related to an expectation which is then itself approximated by the arithmetic mean of the above sequence of random numbers. We will illustrate this idea by some easy examples.

Example 3.6 Monte Carlo calculation of π
An experimental way to calculate π approximately is to consider the part of the unit circle C with center in the origin that intersects with the positive unit square $[0,1]^2$ (see Figure 3.1).

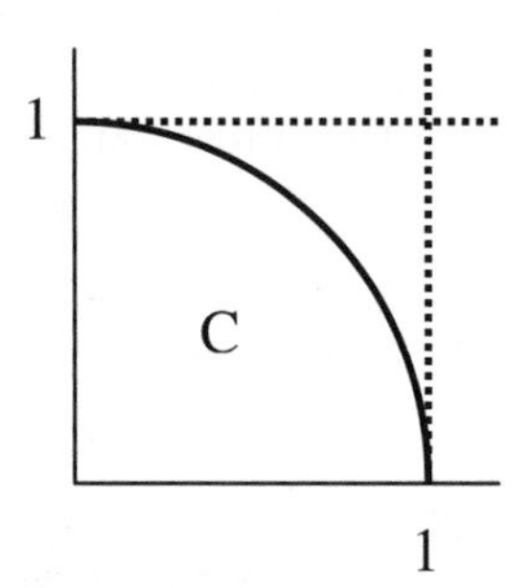

FIGURE 3.1: Estimating π by Monte Carlo.

Our experiment consists of choosing randomly points $P_1, ..., P_N$ of the unit square and consider

$$X_i = 1_{P_i \in C}, \tag{3.13}$$

the functions indicating if P_i is inside the unit circle or not. By this, we implicitly assume that the chosen points are uniformly distributed on $[0,1]^2$. We then have

$$\mathbb{P}(P_i \in C) = \pi/4 \tag{3.14}$$

as the hitting probability of C equals its area (note that the area of $[0,1]^2$ equals 1). Because the indicator function 1_{P_i} satisfies

$$\mathbb{E}\left(1_{P_i}\right) = \mathbb{P}\left(P_i \in C\right) = \pi/4, \tag{3.15}$$

we can thus estimate π by the corresponding arithmetic mean over all P_i to obtain the Monte Carlo estimate

$$\hat{\pi}\left(\omega\right) = \frac{4}{N}\sum_{i=1}^{N} 1_{P_i \in C}\left(\omega\right) . \tag{3.16}$$

The speed of the convergence of the method is illustrated by the results in Table 3.1 where we have chosen the values of 100, 10,000, and 100,000 for N.

N	100	10,000	100,000
$\hat{\pi}$	2.84	3.1268	3.14144

Table 3.1: Crude Monte Carlo Estimates for π

Note that even for a seemingly high number of trials such as N=10,000 one does not necessarily obtain the first three digits right!

This should be regarded as an impressive indicator that the Monte Carlo method converges indeed quite slowly. However, one should also note that the relative error of this estimate is below 0.5%. Compared to this, the estimate for N=100,000 is then extraordinarily precise!

Even more, it might happen that due to the randomness of the Monte Carlo estimator, one has a better performance for $N = 100$ than for $N = 100,000$. It is therefore absolutely necessary that also confidence bounds $[\hat{\pi}_{low}, \hat{\pi}_{up}]$ for the Monte Carlo estimate are computed.

N	100	10,000	100,000
$\hat{\pi}_{low}$	2.477	3.0938	3.13105
$\hat{\pi}_{up}$	3.203	3.1598	3.15183

Table 3.2: Monte Carlo 95%-Confidence Bounds for π

The bounds are shown in Table 3.2 for a 95%-confidence level. And indeed, they all contain the value of π, but their lengths differ significantly.

As in the case of an estimator that is mainly based on estimating the probability of a specific event, we have a simpler formula as in the general case for estimating the variance needed for the confidence interval. We will explain how to obtain it in the next example.

Example 3.7 Estimating the probability of an event
This example has already been hinted at in the preceeding one. We only formalize it here as estimating the probability of an event is an important application of the Monte Carlo method. Therefore, let A be a certain event. We want to estimate the probability of its occurrence $\mathbb{P}(A)$. By using the relation between the expectation of the indicator function 1_A of A,

$$1_A(\omega) = \begin{cases} 1, & \text{if } \omega \in A \\ 0, & \text{if } \omega \notin A \end{cases} \tag{3.17}$$

and the probability of A,

$$\mathbb{E}(1_A) = \mathbb{P}(A), \tag{3.18}$$

the Monte Carlo estimate for $\mathbb{P}(A)$ simply is the relative frequency of the occurrence of A in N independent experiments.

Formally, let A_i denote the occurrence of A in experiment i. We then define the Monte Carlo estimator for $\mathbb{P}(A)$ as

$$rf_N(A) = \frac{1}{N}\sum_{i=1}^{N} 1_{A_i}. \tag{3.19}$$

As one also has

$$\mathbb{V}ar(1_A) = \mathbb{P}(A)(1 - \mathbb{P}(A)), \tag{3.20}$$

we introduce

$$\hat{\sigma}_N^2 = rf_N(A)(1 - rf_N(A)) \tag{3.21}$$

and obtain an approximate 95%-confidence interval for $\mathbb{P}(A)$ as

$$\left[rf_N(A) - \frac{1.96}{\sqrt{N}}\hat{\sigma}_N, rf_N(A) + \frac{1.96}{\sqrt{N}}\hat{\sigma}_N\right]. \tag{3.22}$$

REMARK 3.8 By noting that the Monte Carlo estimator for π in Example 3.6 is just the relative frequency of the event C multiplied by 4, we obtain an estimator for the variance of $4 * 1_C$ by simply multiplying $rf_n(C)(1 - rf_n(C))$ by 16. □

Example 3.9 Monte Carlo integration
A very simple, but often efficient application of the Monte Carlo approach is to calculate the value of deterministic integrals of the form

$$\int_{[0,1]^d} g(x)\,dx \tag{3.23}$$

with g a real-valued, bounded function. By introducing the density function $f(x)$ of the d-dimensional uniform distribution on $[0,1]^d$ via

$$f(x) = 1_{[0,1]^d}(x), \quad x \in \mathbb{R}^d, \tag{3.24}$$

we can rewrite the above integral artificially as an expected value of $g(X)$ where X is a random variable which is uniformly distributed on $[0,1]^d$, i.e.

$$I = \int_{[0,1]^d} g(x)\,dx = \int g(x) f(x)\,dx = \mathbb{E}(g(X)). \tag{3.25}$$

Again, this allows us to compute a Monte Carlo estimator $\hat{I}$ for the above integral via simulating N random variables $X_1, \ldots, X_N$ which are all independent and uniformly distributed on $[0,1]^d$, and then to obtain

$$\hat{I}_n(\omega) = \frac{1}{N}\sum_{i=1}^{N} g(X_i(\omega)). \tag{3.26}$$

REMARK 3.10 1. Although the integral has to be calculated over a subset of $\mathbb{R}^d$, the real-valued random variables $Z_i = g(X_i)$ allow the application of the strong law of large numbers. The rate of convergence of $O(N^{-\frac{1}{2}})$ of the Monte Carlo estimator stays valid **independent** of the dimension d.

As deterministic quadrature formulae typically have a rate of convergence of $O(N^{-\frac{2}{d}})$, we expect the Monte Carlo method to outperform these formulae (at least in the mean) for dimensions $d > 4$. This is often paraphrased as **Monte Carlo methods beat the curse of dimensionality**.

2. Application of this Monte Carlo integration method is not limited to the unit interval. Exactly the same method can be carried out for general, bounded d-dimensional rectangles. Of course, then the random variables X_i have to be uniformly distributed on the corresponding rectangle.

In the general case of an unbounded domain, one needs a suitable transformation h^{-1} that maps this domain to the unit interval. Then, the integral equals $\mathbb{E}(g(h(X))h'(X)$ with X uniformly distributed on the unit interval.

3. The method of Monte Carlo integration can be imitated for calculating discrete sums of a function $g(x)$ over a countable set A. Indeed, by

$$\sum_{x\in A} g(x) = \sum_{x\in A} \frac{g(x)}{p(x)} p(x) = \mathbb{E}\left(\frac{g(X)}{p(X)}\right) \tag{3.27}$$

where $\mathbb{P}$ is a discrete probability distribution on A with

$$\mathbb{P}(X = x) = p(x) > 0 \;\forall\; x \in A, \tag{3.28}$$

every discrete sum can be interpreted as an expectation.

Again, the Monte Carlo method can be applied to sample a large number N of random variables from this distribution $\mathbb{P}$ and then to calculate the crude

Monte Carlo estimator for the sum

$$\hat{S}_{N,\mathbb{P}} = \frac{1}{N} \sum_{i=1}^{N} \frac{g(X_i)}{p(X_i)} \tag{3.29}$$

to approximate the expectation in Equation (3.27). Of course, the choice of a suitable probability distribution in the infinite sum case is not straightforward as there is no uniform distribution on a set with infinitely many elements. □

For general, unbounded domains of integration $D \subset \mathbb{R}^d$ one can make use of a probability distribution which ideally has a support exactly equal to D. As, depending on the form of D, it might be quite hard to find such a distribution, a very crude method would be to use a multidimensional normal distribution with the identity matrix I as the variance-covariance matrix and an expectation vector μ lying inside D (ideally in some central position). With $\varphi_{\mu,I}(x)$ being the corresponding density function, we then obtain

$$\int_D g(x)\,dx = \int_{\mathbb{R}^d} 1_D(x) \frac{g(x)}{\varphi_{\mu,I}(x)} \varphi_{\mu,I}(x)\,dx = \mathbb{E}(\tilde{g}(X)) \tag{3.30}$$

from which the construction of a Monte Carlo estimator works in the usual way. However, note that because we now have to require that this last expectation exists and is finite,

$$\tilde{g}(x) := 1_D(x) \frac{g(x)}{\varphi_{\mu,I}(x)} \tag{3.31}$$

is no longer bounded. For more on Monte Carlo integration we refer to the monograph by Evans and Swartz (2000).

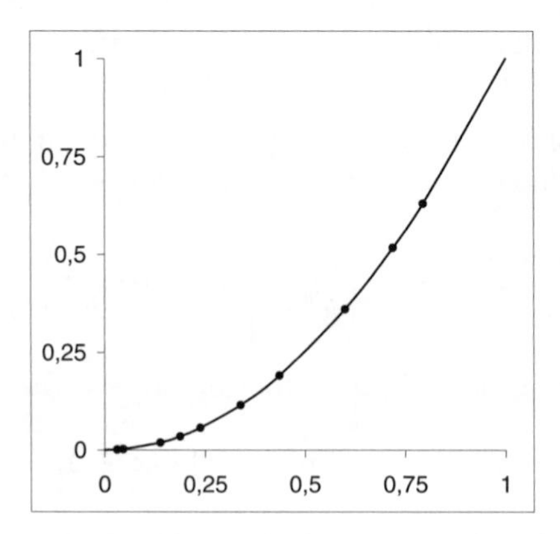

FIGURE 3.2: Monte Carlo integration of $g(x) = x^2$ on $[0, 1]$, dots represent the sampled values, $N = 10$.

In Figure 3.2 we show an illustration of how the integral of $g(x) = x^2$ on $[0,1]$ is approximated by

$$\frac{1}{10}\sum_{i=1}^{10} g(X_i(\omega)) = \frac{1}{10}\sum_{i=1}^{10} X_i(\omega)^2, \tag{3.32}$$

the mean over all the values calculated at 10 randomly chosen points. Note that the dots denote the function values at the randomly generated points. In particular, there are two points simultaneously close to the origin.

We also integrate this function on $[0,1]$ with the help of the Monte Carlo method for further values of N. The corresponding values together with upper and lower 95%-confidence bounds are given in Table 3.3. Note the very bad performance for $N = 10$ which results from the fact that there is no sampling point close to 1. However, even in this case, the 95%-confidence interval contains the correct value of 1/3. Still, only in the case of N=10,000 we see a satisfying behaviour, again a hint on the slow convergence of the crude Monte Carlo method. Further, this very simple example clearly highlights one problem of Monte Carlo integration: If there is a small dominating region (such as the neighbourhood of 1 in our example) then the crude Monte Carlo method needs an enormous amount of sample points to deliver a satisfying performance. From this, it should be clear that one should usually not use crude Monte Carlo methods for one-dimensional integration.

N	10	100	10,000
$\hat{I}_{low}$	0.047	0.297	0.325
$\hat{I}_N$	0.192	0.360	0.331
$\hat{I}_{up}$	0.338	0.423	0.337

Table 3.3: Monte Carlo Integration with 95%-Confidence Bounds for x^2 on $[0,1]$ (Exact Value = 1/3)

3.3 Improving the speed of convergence of the Monte Carlo method: Variance reduction methods

The main disadvantage of the crude Monte Carlo method is its slow convergence. In probabilistic terms, this is expressed by the fact that the standard deviation of the error only decreases as a square root in terms of the required number of simulations. Thus, if one is able to modify the method resulting in a faster decrease of the variance, one could speed up the computations in the sense that achieving a desired accuracy requires less simulation runs. Any such modification of the crude Monte Carlo method is called a **variance reduction** method. In the following sections we will introduce some popular variance reduction methods. As most of them are standard material in theory and application of Monte Carlo methods we will often not explicitly refer to the literature. Their treatment can also be found in the monographs by Asmussen and Glynn (2007), Glasserman (2004), Hammersley and Handscomb (1964), Ripley (1987), or Rubinstein (1981), just to name a few.

Before we will go into details we should keep some aspects in mind:

- Simulation procedures cannot only be speeded up by reducing the variance of the estimator. Careful implementation and storage management should also be optimized to save computing time.

- Implementing and adapting some of the following variance reduction methods requires quite some effort in programming and mathematical considerations. The gain in variance reduction should also be judged against this additional effort. To put it clearly, is it really worth using a variance reduction method in a specific situation?

- If the computational work per sample under sampling method A denotes W_A and under method B denotes W_B, then we speak of a **more effective reduction of variance** by using method A if we have

$$W_A \cdot \mathbb{V}ar\left(\bar{X}_N^A\right) < W_B \cdot \mathbb{V}ar\left(\bar{X}_N^B\right) \tag{3.33}$$

with $\bar{X}_N^Y$ denoting the Monte Carlo estimator used in method Y and based on N samples. Indeed, this relation is meaningful as we have already seen that the variance of a Monte Carlo estimator is proportional to the number of samples needed to obtain an approximate confidence interval of a given length for the expectation of interest (with the same constant of proportionality for both methods!). For this, just recall that the form of an approximate confidence interval is

$$\left[\bar{X}_N - z_{1-\alpha/2}\frac{\sigma}{\sqrt{N}}, \bar{X}_N + z_{1-\alpha/2}\frac{\sigma}{\sqrt{N}}\right]. \tag{3.34}$$

So when talking about variance reduction, we should also keep the amount of computational work per sample in mind. As the variance reduction methods typically require a higher amount of work per sample (sometimes negligible, sometimes significant), it is only worthwhile using them if they predict a substantial reduction of variance.

3.3.1 Antithetic variates

The method of antithetic variables is the easiest variance reduction method. It is based on the idea to combine a random choice of points with a systematic one. Its main principle is **variance reduction by introducing symmetry**. Assume we want to compute $\mathbb{E}(f(X))$ with X a random variable uniformly distributed on $[0,1]$. While the crude Monte Carlo estimate would be

$$\bar{f}(X) = \frac{1}{N}\sum_{i=1}^{N} f(X_i), \tag{3.35}$$

with X_i being independent copies of X, in the method of antithetic variates we would also use the numbers $1-X_1$, ..., $1-X_N$ and introduce the **antithetic Monte Carlo estimator**

$$\bar{f}_{anti}(X) = \frac{1}{2}\left(\frac{1}{N}\sum_{i=1}^{N} f(X_i) + \frac{1}{N}\sum_{i=1}^{N} f(1-X_i)\right). \tag{3.36}$$

Note that as X and $1-X$ have the same distribution, both sums on the right-hand side of Equation (3.36) are unbiased estimators for $\mathbb{E}(f(X))$. Therefore, the antithetic estimator is also unbiased. Let $\sigma^2 = \mathbb{V}ar(f(X))$. Then the variance of the antithetic estimator is given by

$$\mathbb{V}ar\left(\bar{f}_{anti}(X)\right) = \frac{\sigma^2}{2N} + \frac{1}{2N}\mathbb{C}ov(f(X), f(1-X)), \tag{3.37}$$

i.e. we have a reduction of the variance compared to the crude Monte Carlo estimator based on $2N$ random numbers if $f(X)$ and $f(1-X)$ are negatively correlated. Further, we have to notice that we also save computational effort as we only have to generate N random numbers instead of $2N$.

A theoretical result that can often be used to justify the method of antithetic variates is the following proposition (see Asmussen and Glynn [2007]):

PROPOSITION 3.11 Chebyschev's covariance inequality
Let X be a real-valued random variable. Let f, g be nondecreasing functions with $\mathbb{C}ov(f(X), g(X))$ being finite. Then we have:

$$\mathbb{E}(f(X)g(X)) \geq \mathbb{E}(f(X))\mathbb{E}(g(X)). \tag{3.38}$$

Indeed, by choosing $g(x) = -f(1-x)$, this proposition directly implies:

PROPOSITION 3.12 Variance reduction in the uniform case
Let f be a nondecreasing or a nonincreasing function, let X be uniformly distributed on $[0,1]$ with $\mathbb{C}ov(f(X), f(1-X))$ being finite. Then we have:

$$\mathbb{C}ov(f(X), f(1-X)) \leq 0. \tag{3.39}$$

In particular, the antithetic Monte Carlo estimator based on N random numbers has a smaller variance than the crude Monte Carlo estimator based on $2N$ random numbers.

We demonstrate the usefulness of the method by looking at the integration example from Figure 3.2. Its bad performance in the case of $N = 10$ mainly was due to the lack of random numbers close to 1. If we add the points $1 - X_i$, we obtain an antithetic Monte Carlo estimate of 0.340 which is even better than the crude Monte Carlo estimate for $N = 100$. The reason for this is that due to the symmetrized set of random numbers, now also the critical neighbourhood of 1 is covered by the sample. Figure 3.3 illustrates this fact.

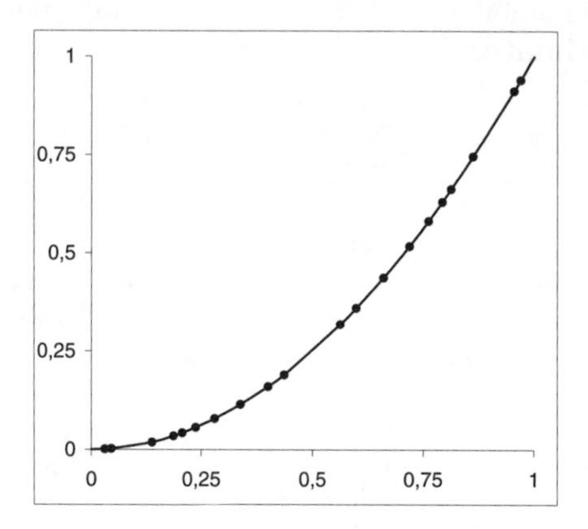

FIGURE 3.3: Antithetic Monte Carlo integration of $g(x) = x^2$ on $[0, 1]$, dots represent the sampled values, $N = 10$.

REMARK 3.13 Whenever we are generating random variables Y via the inverse transformation method out of uniformly distributed random variables X, the foregoing result can be used. To see this, note that the distribution function $F(.)$ of Y is a nondecreasing function and so is its inverse $F^{-1}(.)$. Thus, if we want to estimate an expectation of the form $\mathbb{E}(h(Y))$ with a nondecreasing (a nonincreasing function) h, we can simply apply the proposition to the nondecreasing (nonincreasing) function $f(x) := h\left(F^{-1}(x)\right)$. □

The introduction of antithetic variables is of course not only limited to uniformly distributed random variables. Whenever one can easily generate variables $\tilde{X}_i$ out of X_i such that

- $\tilde{X}_i$ has the same distribution as X_i
- $\mathbb{C}ov\left(f\left(\tilde{X}_i\right), f(X_i)\right) \leq 0$

we can use the above method. Examples for this are symmetric distributions such as the normal distribution: For $X_i \sim \mathcal{N}\left(\mu, \sigma^2\right)$ the suitable antithetic variable is given by

$$\tilde{X}_i = 2\mu - X_i. \tag{3.40}$$

In particular, for $\mu = 0$ the antithetic variate is simply $-X_i$. Again, we can use Chebyschev's covariance inequality to prove a variance reduction if f is a monotonic function.

PROPOSITION 3.14 Variance reduction in the normal case
Let f be a nondecreasing or a nonincreasing function, let X be $\mathcal{N}(\mu, \sigma^2)$-distributed with $\mathbb{C}ov(f(X), f(2\mu - X))$ being finite. Then we have:

$$\mathbb{C}ov(f(X), f(2\mu - X)) \leq 0. \tag{3.41}$$

REMARK 3.15 Here are some further aspects of applying the method of antithetic variates.

1. Note that in both our examples the sample mean of the random variables equals the population mean, i.e. we have

$$\frac{1}{2N} \sum_{i=1}^{N} \left(X_i + \tilde{X}_i \right) = \mathbb{E}(X_1). \tag{3.42}$$

We thus have an **automatic moment matching** for the first moment in the sample of the used random numbers.

2. As can be seen in the above numerical example, the effect of using antithetic variates normally does not dramatically improve the speed of convergence of the Monte Carlo method. This behaviour is often observed in practical techniques (see also Section 3.4).

3. If we want to compute an expectation of the form

$$\mathbb{E}(Y) = \mathbb{E}(h(X_1, ..., X_k)) \tag{3.43}$$

where the X_i are independent on $[0, 1]$ uniformly distributed random variables and h is a real-valued function, then one can use component-wise antithetic variates. For example, for each simulated k-dimensional vector $X^j = (X_1^j, ..., X_k^j)$ one can also use

$$\tilde{X}^j = \left(1 - X_1^j, ..., 1 - X_k^j \right)$$

for constructing an antithetic variate Monte Carlo estimator as in the one-dimensional case. It can be shown that if h is nondecreasing in each component then this method yields a variance reduction.

4. **Confidence intervals for the antithetic variate Monte Carlo estimator**. To obtain a confidence interval for the antithetic variate Monte

Carlo estimator, one has to consider the confidence interval for the crude Monte Carlo estimator for $\frac{1}{2}(f(X)+f(\tilde{X}))$ where $\tilde{X}$ is the antithetic variate for X. This is an estimator based on only N observations

$$h\left(X_i\right)=\frac{1}{2}\left(f\left(X_i\right)+f\left(\tilde{X}_i\right)\right). \tag{3.44}$$

The variance σ^2 for the confidence interval in the antithetic method therefore has to be estimated by

$$\bar{\sigma}_{anti}^2=\frac{1}{N-1}\sum_{i=1}^{N}\left(\frac{1}{2}\left(f\left(X_i\right)+f\left(\tilde{X}_i\right)\right)-\bar{f}_{anti}\left(X\right)\right)^2 \tag{3.45}$$

leading to the approximate 95%-confidence interval for $\mathbb{E}\left(f\left(X\right)\right)$ of

$$\left[\bar{f}_{anti}\left(X\right)-1.96\frac{\bar{\sigma}_{anti}}{\sqrt{N}},\bar{f}_{anti}\left(X\right)+1.96\frac{\bar{\sigma}_{anti}}{\sqrt{N}}\right]. \tag{3.46}$$

□

3.3.2 Control variates

The principle of control variates is based on the idea that if we want to compute $\mathbb{E}\left(X\right)$, we **should try to compute as much as possible exactly** and should only compute that part by Monte Carlo simulation that we cannot avoid. More precisely, if we know a random variable Y which is (in some sense) close to X and for which we can compute $\mathbb{E}\left(Y\right)$ exactly, then this random variable can be chosen as a **control variate**, i.e. we use the relation

$$\mathbb{E}\left(X\right)=\mathbb{E}\left(X-Y\right)+\mathbb{E}\left(Y\right), \tag{3.47}$$

which motivates the following **control variate Monte Carlo estimator**

$$\bar{X}_Y=\frac{1}{N}\sum_{i=1}^{N}\left(X_i-Y_i\right)+\mathbb{E}\left(Y\right) \tag{3.48}$$

for $\mathbb{E}\left(X\right)$ with X_i, Y_i being independent copies of X and Y. By the relation

$$\begin{aligned}\mathbb{V}ar\left(\bar{X}_Y\right)&=\frac{1}{N}\mathbb{V}ar\left(X-Y\right)\\&=\frac{1}{N}\left(\mathbb{V}ar\left(X\right)+\mathbb{V}ar\left(Y\right)-2\mathbb{C}ov\left(X,Y\right)\right)\end{aligned} \tag{3.49}$$

we obtain a reduction of the variance for the control variate estimator compared to the crude one if we have

$$\mathbb{V}ar\left(X\right)\geq\mathbb{V}ar\left(X-Y\right). \tag{3.50}$$

The amount of reduction of $\mathbb{V}ar(X)$ is given by

$$2\mathbb{C}ov(X,Y) - \mathbb{V}ar(Y). \tag{3.51}$$

If Y is very close to X this can lead to the elimination of nearly all the variance of the crude Monte Carlo estimator by using its control variate variant. However, the efficiency of the control variate also hinges on the fact

- that we are able to directly simulate the difference $X_i - Y_i$ as one random variable (i.e. we know its exact distribution and it is enough to simulate just one random number and use a suitable transformation)
- or that we have to use the inverse transformation method for both X_i and Y_i separately (but still use the same random number for both or can at least draw X_i and Y_i from their joint distribution) which of course needs a higher amount of work than in the previous case.

In the worst case the introduction of Y can nearly double the computing time (if we ignore the time for calculating the exact value $\mathbb{E}(Y)$). Therefore, in the worst case, the introduction of the control variate Y only increases the efficiency of the Monte Carlo method if the variance of $X - Y$ is at most half as big as $\mathbb{V}ar(X)$. Also, finding a suitable covariate often needs some intuition and is not always based on a systematic search algorithm.

REMARK 3.16 To obtain a confidence interval for the control variate Monte Carlo estimator, one has to use the confidence interval for the crude Monte Carlo estimator for $\mathbb{E}(X - Y)$ and then simply has to add $\mathbb{E}(Y)$ to this interval, i.e. we obtain an approximate 95%-confidence interval by

$$\left[\bar{X}_Y - 1.96\frac{\hat{\sigma}_{X-Y}}{\sqrt{N}}, \bar{X}_Y + 1.96\frac{\hat{\sigma}_{X-Y}}{\sqrt{N}}\right] \tag{3.52}$$

with

$$\hat{\sigma}^2_{X-Y} = \frac{1}{N-1}\sum_{i=1}^{N}\left(X_i - Y_i - \frac{1}{N}\sum_{j=1}^{N}(X_j - Y_j)\right)^2. \tag{3.53}$$

□

We first illustrate the method by our standard integration example. For this, we simply choose Y to be uniformly distributed on $[0,1]$, hence we have $X_i = Y_i^2$. As we know $\mathbb{E}(Y) = 0.5$ we only have to simulate

$$X_i - Y_i = Y_i^2 - Y_i \tag{3.54}$$

Looking at the graph of this difference and comparing it to the integrand, we can clearly expect a variance reduction (see Figure 3.4). This is simply due to the fact that $g(x) = x^2$ has a much bigger variation than $\hat{g}(x) = x^2 - x$.

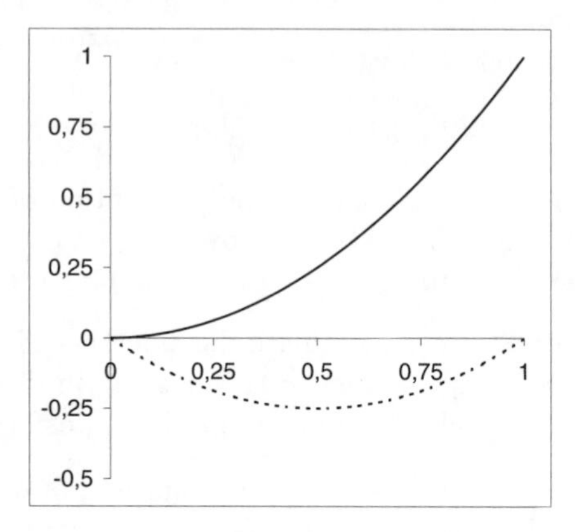

FIGURE 3.4: Function $g(x) = x^2$ and control variate version $\hat{g}(x) = x^2 - x$ (dotted line) on $[0, 1]$.

Indeed, the numbers shown in Table 3.4 impressively indicate the reduction of variance compared to the performance of the crude Monte Carlo estimator (see Table 3.3). The length of the confidence bounds have been greatly reduced which underlines the power of the control variate method when a suitable control is at hand. Note that due to rounding, the confidence intervals are not exactly symmetric. Further, the (point) estimate for $N = 10$ is slightly better than the one for $N = 100$. However, the confidence interval for $N = 100$ is much shorter than the one for $N = 10$, which underlines that we have a more reliable estimate in the case of $N = 100$.

N	10	100	10,000
$\hat{I}_{low}$	0.292	0.327	0.331
$\hat{I}_N$	0.340	0.343	0.333
$\hat{I}_{up}$	0.389	0.359	0.334

Table 3.4: Monte Carlo Integration with Control Variate x with 95%-Confidence Bounds for $\int_0^1 x^2 dx$ on $[0, 1]$ (Exact Value = 1/3)

Further aspects of applying the method of control variates

1. Optimizing the control variate:

If we have found a candidate Y then due to the construction of the control variate estimator also aY can be used as a control variate for $a > 0$. This is due to the fact that the new control variate estimator is still unbiased due to the linearity of the expectation. It also generates a variance reduction for a positive a if Y has already lead to a variance reduction (note that a negative a would have increased (!) the variance of the control variate estimator). So the optimal use of the control variate Y is achieved via introducing a multiplicator a^* that minimizes

$$\begin{aligned} g(a) &= \mathbb{V}ar(X - aY) = \mathbb{V}ar(X) + a^2\mathbb{V}ar(Y) - 2a\mathbb{C}ov(X,Y) \\ &= \sigma^2 + a^2\sigma_Y^2 - 2a\sigma_{X,Y}. \end{aligned} \tag{3.55}$$

From this, one obtains

$$a^* = \frac{\sigma_{XY}}{\sigma_Y^2} \tag{3.56}$$

with an actual variance reduction of

$$2a^*\mathbb{C}ov(X,Y) - (a^*)^2\,\mathbb{V}ar(Y) = \frac{\sigma_{XY}^2}{\sigma_Y^2}. \tag{3.57}$$

By denoting the correlation between X and Y as $\rho_{X,Y}$ and using the relation

$$\sigma_{XY} = \rho_{X,Y}\sigma_X\sigma_Y \tag{3.58}$$

we obtain that the maximum relative variance reduction is given by

$$\frac{2a^*\mathbb{C}ov(X,Y) - (a^*)^2\,\mathbb{V}ar(Y)}{\mathbb{V}ar(X)} = \rho_{X,Y}^2. \tag{3.59}$$

Note that the maximum achievable variance reduction decreases quadratically with the correlation between X and its control variate Y. Thus, only using a control variate with a high value of $\rho_{X,Y}$ will help in making the control variate method effective. If we have a control with, say, $\rho_{X,Y} = 0.4$ then the maximum possible reduction of variance is only 16% of the original σ_X^2!

If both the variance of Y and the covariance between X and Y are known then we can directly use a^*Y as control variate. If this is not the case then one either has the chance to estimate both (or only σ_{XY} if at least σ_Y^2 is known) via a simulation procedure that can be performed before starting the control variate procedure. We could also estimate the unknown parameters during the performance of the control variate procedure as a by-product and then update parameter a^* continuously for the control variate estimator. In our standard integration example with the control variate Y, the optimal choice of a can explicitly be computed as $a^* = 1$. So we have indeed used the **best linear control variate**.

2. Multiple controls

Due to the way we have constructed the control variate estimator, we can add another control variate, say Z, in the form

$$\bar{X}_{Y,Z} = \bar{X}_Y - \frac{1}{N}\sum_{i=1}^{N} Z_i + \mathbb{E}(Z). \tag{3.60}$$

This again yields an unbiased estimator for $\mu = \mathbb{E}(X)$. Further, it leads to a variance reduction if we have

$$\mathbb{V}ar(Z) < 2\mathbb{C}ov(X_Y, Z). \tag{3.61}$$

In the particular case of uncorrelated Z_i and Y_i this reduces to requiring

$$\mathbb{V}ar(Z) < 2\mathbb{C}ov(X, Z). \tag{3.62}$$

One can thus use as many control variates as one would like. A particular application of multiple controls to a multivariate situation as

$$X = f(Y_1, ..., Y_d) \tag{3.63}$$

will be the method of unconditional mean control variates (see below).

3. Control variates and series approximations

It is often not easy to find a good control variate. In our standard integration example we have already seen that the control X is the best linear control variate (as a function of X). One could also imagine best estimators of higher polynomial degree. We thus assume that we would like to estimate

$$\mu = \mathbb{E}(f(X)) \tag{3.64}$$

and that we have a Taylor approximation of order k of the form

$$f_k(x) = \sum_{j=0}^{k} \frac{f^{(j)}(x_0)}{j!}(x - x_0)^j. \tag{3.65}$$

The main question then is whether we are able to determine an optimal value x_0 such that we obtain the highest possible reduction of variance when using $f_k(X)$ as a covariate. Of course, this highly depends on our ability to compute all kinds of moments of X up to order k and in particular to compute all the covariances $\mathbb{C}ov\left(X^j, f(X)\right)$. In our integration example we have implicitly done that as we could replace the control X for $f(X) = X^2$ by the control

$$f_1(X) = f'\left(\frac{1}{2}\right)\left(X - \frac{1}{2}\right) + f\left(\frac{1}{2}\right) = X - \frac{1}{4}, \tag{3.66}$$

because the constant $-1/4$ cancels out in the covariate part. So, actually the best linear estimate that we achieved in this case is the Taylor approximation of order 1 in the point $x_0 = 1/2$ that is directly in the center of the range of X (see Figure 3.5 for an illustration).

We note however that a Taylor approximation is only locally a good one. So x_0 should be in some sense in the center of the distribution of X. Further, the quality of this covariate depends of course on the approximation quality of the Taylor polynomial. We might have good bounds from the explicit form of the Taylor remainder term in special cases such as convex/concave functions f or for functions that have a series representation. On the other hand, we cannot present a general result here.

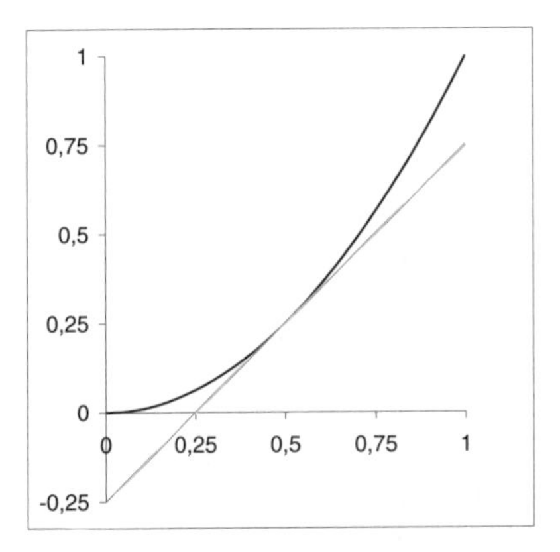

FIGURE 3.5: Integrand x^2 (black) and best linear Taylor approximation of order 1 (grey).

4. Unconditional mean control variates

The method of unconditional control variates is a natural one for approximating expectations of multivariate functions

$$\mathbb{E}\left(g\left(X\right)\right)=\mathbb{E}\left(g\left(X^{(1)},...,X^{(d)}\right)\right) \tag{3.67}$$

by d univariate controls

$$Y^{UM_j}\left(X\right)=g\left(\mu_1,...,\mu_{i-1,}X^{(i)},\mu_{i+1},...,\mu_d\right),\ j=1,...,d, \tag{3.68}$$

with $\mu_j=\mathbb{E}\left(X^{(j)}\right)$, the so-called **unconditional mean control variates**. Thus, one is using the univariate versions of $g\left(X\right)$ where only component i is allowed to vary freely while the other components are set equal to the component mean. For the method to work, we have to be able to calculate all the expected values of the controls. We can then introduce the **unconditional mean control variate estimator** as

$$\bar{X}_N^{UMC}=\frac{1}{N}\sum_{i=1}^{N}\left(g\left(X_i\right)-\sum_{j=1}^{d}Y^{UM_j}\left(X_i\right)\right)+\sum_{j=1}^{d}\mathbb{E}\left(Y^{UM_j}\left(X\right)\right). \tag{3.69}$$

In the context of finance, this method has been introduced by Pellizzari (2001). With regard to confidence intervals and the possible variance reduction all that has already been said above when discussing multiple controls is valid as the method is just a particular example of a multiple control variate strategy. We illustrate it with a simple example where we have a random variable $X=\left(X^{(1)},X^{(2)},X^{(3)}\right)$ with a multivariate normal distribution, i.e.

$$X\sim\mathcal{N}\left(\begin{pmatrix}1\\1\\1\end{pmatrix},\begin{pmatrix}1&0.8&0.8\\0.8&1&0.64\\0.8&0.64&1\end{pmatrix}\right). \tag{3.70}$$

We want to estimate

$$\mathbb{E}\left(g\left(X\right)\right) = \mathbb{E}\left(X^{(1)} \cdot X^{(2)} \cdot X^{(3)}\right). \tag{3.71}$$

This implies that now the control variates are simply the components $X^{(i)}$. Thus, the unconditional mean control variate estimator is given by

$$\bar{X}_N^{UMC} = \frac{1}{N}\sum_{i=1}^{N}\left(X_i^{(1)} \cdot X_i^{(2)} \cdot X_i^{(3)} - X_i^{(1)} - X_i^{(2)} - X_i^{(3)}\right) + 3. \tag{3.72}$$

A simulation with N=10,000 resulted in the numbers reported in Table 3.5. Note that the variance has been reduced by approximately 25% which also resulted in a smaller confidence interval.

Method	Mean	Lower quantile	Upper quantile
CMC	3.240	3.120	3.361
UMCV	3.201	3.096	3.306

Table 3.5: $\mathbb{E}\left(X^{(1)} \cdot X^{(2)} \cdot X^{(3)}\right)$ Estimated with Crude (CMC) and Unconditional Mean Control Variate Method (UMCV), N=10,000

5. Control variates are not always good

It is actually trivial, but there do exist bad control variates. Whenever there is a good control variate Y (measured in terms of a high value of $\rho_{X,Y}$), there is also a bad one, namely $-Y$. To see this simply note that we have

$$\rho_{X,Y} = -\rho_{X,-Y}, \tag{3.73}$$

i.e. the use of $-Y$ as a control variate would result in an increase of variance of the corresponding estimator. The purpose of this artificial example however is to highlight that one should have at least a heuristical argument that there is a positive correlation between the random variable X and the control variate Y if we do not know $\rho_{X,Y}$ explicitly.

If one is indeed unsure about the variance reducing effect of a control variable then one should sample a few realizations of both X and Y, estimate the covariance σ_{XY} and σ_Y^2 from these realizations, and finally use aY with $a = \hat{\sigma}_{XY}/\hat{\sigma}_Y^2$ as control variate. This is indeed the optimization suggested in our first point on further aspects of control variates.

3.3.3 Stratified sampling

The main principle underlying stratified sampling is a natural one: **sample in a small subpopulation that mirrors the properties of the total**

population as much as possible. Of course, for this we need an indicator of these properties of the population (this indicator will often be called Y in the following). Indeed, this is quite a familiar method as it is the base for opinion polls often met in newspapers (such as e.g. on the population's attitude towards the work of the government, or on the forecast of sports results). An opinion poll is always based on a so-called **representative sample** of the population. The reason behind this is that each (sufficiently large) subgroup of the population should be present in the sample in the same fraction as it is present in the full population. Mathematically, this means that one wants to eliminate the variance caused by the sample population differing in its characteristics from the whole population. Only the variance (= the different opinions) inside the different subgroups should remain.

In the method of stratified sampling the distribution of a random variable X is divided into d different parts that are determined by the values $y_1, ..., y_d$ of a second random variable Y. One can typically think about the purpose of Y to indicate from which of the d different parts of the sample space Ω the random variable X should be drawn. If

- the probability distribution of Y is known and easy to calculate and
- $X\,|Y$ can easily be simulated

then one can use this information to calculate $\mu = \mathbb{E}(X)$ via

$$\mathbb{E}(X) = \sum_{i=1}^{d} \mathbb{E}(X\,|Y = y_i)\,P(Y = y_i). \tag{3.74}$$

If now all the probabilities $p_i = P(Y = y_i)$ are known then one only has to simulate the d different conditional expectations with the appropriate crude Monte Carlo method. To show that this indeed leads to a reduction of variance, let us define for $i = 1, ..., d$:

$$\bar{X}_{i,N_i} := \frac{1}{N_i}\sum_{j=1}^{N_i} X_j^{(i)}, \quad \mu_i := \mathbb{E}(X\,|Y = y_i), \quad \sigma_i^2 := \mathbb{V}ar(X\,|Y = y_i).$$

All the random variables $X_j^{(i)}$ must have the same distribution as $X\,|Y = y_i$. We then introduce the **stratified Monte Carlo estimator** for μ by

$$\bar{X}_{strat,N} = \sum_{i=1}^{d} p_i \bar{X}_{i,N_i} \tag{3.75}$$

with $N = N_1 + ... + N_d$. Due to the above representation of the expectation as a weighted sum of the conditional expectations, the stratified Monte Carlo estimator is an unbiased estimator for μ. Even more, one can show that the stratified estimator naturally has a smaller variance than the crude Monte

Carlo estimator. To see this, note that due to the (conditional) independence of the subestimators $\bar{X}_{i,N_i}, i = 1, ..., d$ we have

$$\mathbb{V}ar\left(\bar{X}_{strat,N}\right) = \mathbb{V}ar\left(\sum_{i=1}^{d} p_i \bar{X}_{i,N_i}\right) = \sum_{i=1}^{d} p_i^2 \frac{\sigma_i^2}{N_i} = \sum_{i=1}^{d} \frac{p_i}{N_i} p_i \sigma_i^2. \tag{3.76}$$

By using the relations

$$\mathbb{E}\left(\mathbb{V}ar\left(X\,|Y\right)\right) = \sum_{i=1}^{d} \mathbb{V}ar\left(X\,|Y = y_i\right) P\left(Y = y_i\right) = \sum_{i=1}^{d} p_i \sigma_i^2, \tag{3.77}$$

$$\sigma^2 = \mathbb{V}ar\left(X\right) = \mathbb{E}\left(\mathbb{V}ar\left(X\,|Y\right)\right) + \mathbb{V}ar\left(\mathbb{E}\left(X\,|Y\right)\right) \geq \mathbb{E}\left(\mathbb{V}ar\left(X\,|Y\right)\right) \tag{3.78}$$

with a strict inequality in the second relation if $\mathbb{E}\left(X\,|Y\right)$ is not almost surely constant, we obtain:

PROPOSITION 3.17 Variance reduction for well-chosen weights
(a) With the above notation there exist $N_1, ..., N_d$ *such that the variance of the stratified Monte Carlo estimator is smaller than that of the crude Monte Carlo estimator for* μ.
(b) Let us assume that all values of Np_i *are integers. Then, for the choice of proportional stratification* $N_i = Np_i$ *the variance of the stratified Monte Carlo estimator is strictly smaller than that of the crude estimator if* $\mathbb{E}\left(X\,|Y\right)$ *is not almost surely constant.*
(c) The highest variance reduction is obtained for the choice of

$$N_i^* := N \frac{p_i \sigma_i}{\sum_{j=1}^{d} p_j \sigma_j} \tag{3.79}$$

(where without loss of generality we have assumed that all σ_j *are positive).*

While the first two claims follow from the relations preceding the proposition, the third one can be obtained from minimizing the explict expression for the variance of the stratified Monte Carlo estimator in the variables $N_1, ..., N_d$ under the constraints of $N_1 + ... + N_d = N$ and $N_i \geq 0$. This can be achieved with the help of the Lagrangian method of constrained optimization.

REMARK 3.18 1. Note that the optimal weight N_i^*/N of the subgroup determined by $Y = y_i$ simply measures the contribution of this subgroup to the total weighted variation in terms of $\sum_{j=1}^{d} p_j \sigma_j$.

2. The performance of the stratified sampling method depends heavily on the variation in the (conditional) expectation between the different subgroups determined by $Y = y_i$. If the values of $\mathbb{E}\left(X\,|Y = y_i\right)$ are significantly different then the reduction in variance can be tremendous; if $\mathbb{E}\left(X\,|Y = y_i\right)$ is

close to being constant as a function of Y then there will be nearly no variance reduction by using the method of stratified sampling. For those familiar with statistics, a comparison to the variance decomposition in the analysis of variance is enlightening.

3. Note also that the applicability of the stratified sampling method crucially relies on the availability of both the distribution of Y (or more precisely, the probabilities p_i) and the variances in the subgroups determined by $Y = y_i$. If the σ_i are not available then one could still use part (b) of the proposition to obtain a variance reduction with the choices of $N_i = Np_i$. An alternative would be to estimate the σ_i in some preliminary simulations and then switch over to the choices of the N_i^* where the σ_i have to be replaced by the just obtained estimate. This however introduces additional errors in the method and it is *a priori* not clear if this method is efficient. □

In many applications, the stratifying variable Y is only formally identified as a random variable. To illustrate this, we use again our simple integration example of the preceding sections. Here, we choose Y as a random variable that attains the values $1, 2, 3, 4$ each with the same probability, and $Y = j$ means that one then samples the function $g(x) = x^2$ in the interval $[0.25(j-1), 0.25j]$.

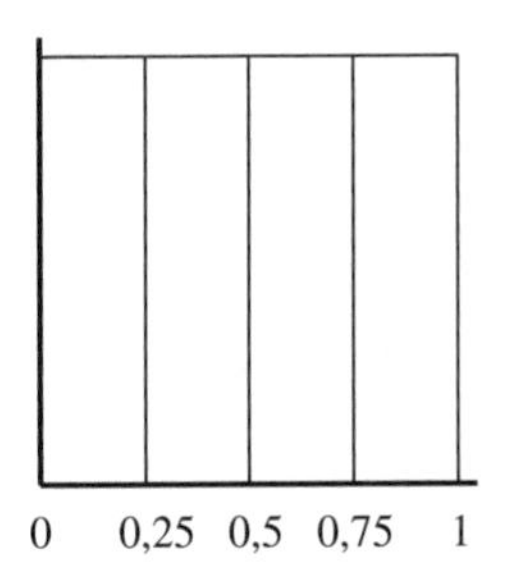

FIGURE 3.6: Stratifying $\mathcal{U}[0,1]$ in four equally probable strata.

Indeed, we take a sample of size $N/4$ for each such interval (see Figure 3.6) and use this to compute the crude Monte Carlo estimate for

$$\int_{0.25(j-1)}^{0.25j} x^2 dx, \; j = 1, ..., 4.$$

For N=100 we already obtain a very precise estimate,

$$\bar{X}_{strat,100} = 0.332 \;\; \text{with} \; \mathbb{V}ar\left(\bar{X}_{strat,100}\right) = 0.008, \tag{3.80}$$

as always rounded to the third digit.

Further aspects of applying stratified sampling

1. Stratifying a general distribution

Let the variable X that we are going to simulate have a general, nondiscrete distribution that can be generated out of the uniform distribution with the help of the inverse transformation method. We can then easily create strata of the distribution of X with given probabilities $p_i, i = 1, ..., d$ (i.e. we can *stratify the distribution*). Therefore, let F be the distribution function of X. Then we obtain the required strata B_i as

$$B_i = \left(F^{-1}\left(\sum_{j=1}^{i-1} p_j\right), F^{-1}\left(\sum_{j=1}^{i} p_j\right)\right] , \quad ,i = 1, ..., d \tag{3.81}$$

where $F^{-1}(.)$ is the inverse of the distribution function F. Note also that we have implicitly used $F^{-1}(0) = -\infty$ and also that for the right-hand side of B_d we are using the open interval if we have $F^{-1}(1) = +\infty$. The variable Y is then again formally introduced with the interpretation that $Y = i$ means that we are (conditionally) sampling from stratum B_i.

2. Multidimensional stratification and the curse of dimensionality

The last example can also be generalized to more than one dimension when we have a vector $X = (X_1, ..., X_d)$ of independent real-valued random variables X_j. For this, we assume that we can generate the distribution of X_j out of the uniform distribution U_j on $[0, 1]$ with the help of the inverse transformation method, i.e. we assume that we have the representation

$$X = (X_1, ..., X_d) = \left(F_1^{-1}(U_1), ..., F_d^{-1}(U_d)\right) . \tag{3.82}$$

We can then directly imitate the method of the foregoing example to generate the strata component-wise. It is thus enough to stratify the uniform distribution U on $[0, 1]^d$. The straightforward choice is to use a product set approach on $[0, 1]^d$. For simplicity, we stratify each coordinate j into n_j equally probable sets and then define product sets

$$A_{i_1,...,i_d} = \prod_{j=1}^{d} \left(\frac{i_j - 1}{n_j}, \frac{i_j}{n_j} \right], \; i_j = 1, ..., n_j, \; j = 1, ..., d \tag{3.83}$$

(see Figure 3.7 for an illustration in the case of $d = 2$ and $n_j = 4$) with

$$\mathbb{P}(U \in A_{i_1,...,i_d}) = \prod_{j=1}^{d} \frac{1}{n_j} =: \bar{p}. \tag{3.84}$$

We then obtain the desired sets $B_{i_1,...,i_d}$ that all have the same probability $\bar{p}$ under the distribution function F by applying the inverse transformation method, i.e. via

$$B_{i_1,\ldots,i_d} = \prod_{j=1}^{d} \left(F_j^{-1}\left(\frac{i_j - 1}{n_j}\right), F_j^{-1}\left(\frac{i_j}{n_j}\right) \right],$$

$$i_j = 1, ..., n_j,\ j = 1, ..., d. \quad (3.85)$$

The main problem with this approach is the number of strata A_i (respectively B_i) that we are obtaining in this way. It equals $n = n_1 \cdot ... \cdot n_d$ which gets extremely large even for small values of n_j if the dimension d is large. As one needs at least one (but typically some more!) random number per strata, one realizes that the number of required random numbers explodes with increasing dimension d. This fact is usually called the *curse of dimensionality.* The same problem has already been mentioned for product quadrature rules of numerical integration.

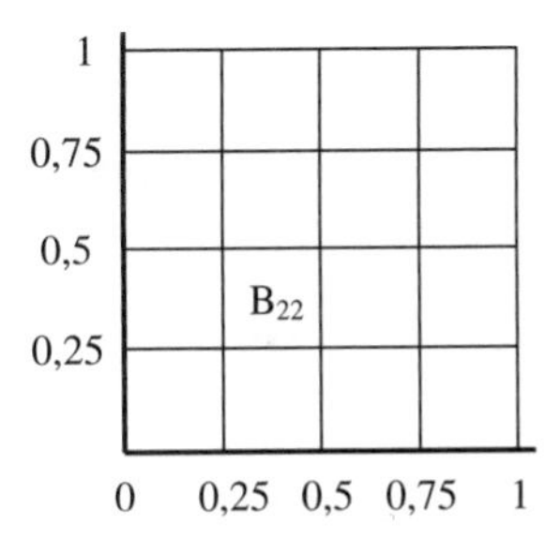

FIGURE 3.7: Stratifying $\mathcal{U}[0,1]^2$ in 16 equally probable strata.

3. Confidence intervals for stratified sampling

To obtain a confidence interval for the expected value $\mu = \mathbb{E}(X)$ with the help of stratified sampling, we take a look at the crude Monte Carlo estimators in the different strata. For this, we assume that we have d strata $A_1, ..., A_d$ all indicated by the random variable Y attaining the values $y_1, ..., y_d$. To avoid trivial cases, we further assume that we have

$$\mathbb{P}(Y = y_i) = p_i > 0,\ i = 1, ..., d, \quad (3.86)$$

and that the sizes N_i of the subsamples in the strata are chosen to guarantee

$$\frac{N_i}{N} \overset{N\to\infty}{\longrightarrow} p_i,\ i = 1, ..., d. \quad (3.87)$$

This in particular ensures that all N_i tend to ∞ if the sample size N does. Thus, by the central limit theorem, for each crude Monte Carlo estimator in the different strata we obtain

$$\frac{1}{\sqrt{N_i}} \sum_{j=1}^{N_i} \left(X_j^{(i)} - \mu_i \right) \xrightarrow{D} N\left(0, \sigma_i^2\right) \text{ for } N_i \to \infty. \tag{3.88}$$

Hence, for large N (and thus large N_i) we obtain that all subestimators $\bar{X}_{i,N_i}$ are approximately $N\left(\mu_i, \frac{\sigma_i^2}{N_i}\right)$-distributed. As further all the subestimators are independent, we have that the stratified Monte Carlo estimator $\bar{X}_{strat,N}$ is approximately $N\left(\mu, \frac{1}{N}\sum_{i=1}^{d} \frac{\sigma_i^2 p_i^2 N}{N_i}\right)$-distributed. If we further use that for large sample size N, the relative frequency N_i/N of strata A_i approximates the probability p_i, we can use a $N\left(\mu, \frac{1}{N}\sum_{i=1}^{d} \sigma_i^2 p_i\right)$ distribution for the construction of the confidence interval. It only remains to estimate the strata variances σ_i^2 by the sample variances

$$\hat{\sigma}_i^2 = \frac{1}{N_i - 1} \sum_{j=1}^{N_i} \left(X_j^{(i)} - \bar{X}_{i,N_i} \right)^2.$$

This then results in an approximate 95%-confidence interval for μ of

$$\left[\bar{X}_{strat,N} - \frac{1.96}{\sqrt{N}} \sqrt{\sum_{i=1}^{d} \hat{\sigma}_i^2 p_i}, \bar{X}_{strat,N} + \frac{1.96}{\sqrt{N}} \sqrt{\sum_{i=1}^{d} \hat{\sigma}_i^2 p_i} \right] \tag{3.89}$$

4. The method of poststratification

It might be possible that the probabilities p_i of the different strata A_i are known, but one is not interested in a conditional sampling on the different strata. A reason for this might be that the transformation required for the stratified sampling is not at hand. Then it is still possible to use a kind of **automatic stratification procedure**. This is based on the fact that the relative frequencies of occurrence of stratum A_i in a large sample tends to its probability p_i. The method of poststratified sampling then works as:

- Generate a sample $X_1, ..., X_N$ in the usual way.
- Classify the different values of this sample in groups related to the different strata A_i, i.e. assign each X_j to the sample A_i that is related to it. Set $N_i = |X_j \in A_j : j = 1, ..., N|$.
- On each A_i calculate the crude Monte Carlo estimator with the assigned X values of the sample to obtain the strata means $\hat{X}_{i,N_i}$.
- Obtain the poststratified Monte Carlo estimator $\hat{X}_{strat,N}$ as

$$\hat{X}_{strat,N} = \sum_{i=1}^{d} p_i \hat{X}_{i,N_i} \tag{3.90}$$

with $p_i = \mathbb{P}(A_i)$.

We first have to emphasize that the absolute frequencies N_i of observations related to stratum A_i are **not chosen before** the sampling. They are indeed random variables and determined **after** (i.e. ex-*post*) $X_1, ..., X_N$ have been generated. The assignment of the X values to the different strata can be formally expressed as we actually simulate pairs (X_j, Y_j) and we assign X_j to A_i if we have $Y_j = y_i$. However, in situations where the distribution of X is already directly stratified, the assignment to the strata is straightforward. For the question of the performance of the poststratified sampling, note that we can represent the crude Monte Carlo estimator of the full sample as

$$\bar{X}_N = \frac{1}{N}\sum_{i=1}^{N} X_i = \frac{1}{N}\sum_{i=1}^{d} N_i \hat{X}_{i,N_i} = \sum_{i=1}^{d} \frac{N_i}{N}\hat{X}_{i,N_i} . \tag{3.91}$$

Thus, the poststratified Monte Carlo estimator differs from the crude one by the choice of the weights of the strata means. Note that the poststratified estimator gives those strata means a higher weight that are underrepresented in the total sample, i.e. for which we have $N_i/N < p_i$. In the same way it assigns a lower weight to the overrepresented strata means. Consequently, as the strong law of large numbers implies the convergence

$$\frac{N_i}{N} \overset{a.s.}{\to} p_i \text{ for } n \to \infty \tag{3.92}$$

poststratification obtains the same variance reduction as proportional stratification asymptotically. However, it is not clear how large the sample size N has to be such that the poststratification method can benefit from this large sample size property. Thus, the advantage of using this method is not clear.

5. Latin hypercube sampling

As already discussed above, if the dimension d of the random variable $X = \left(X^{(1)}, ..., X^{(d)}\right)$ is large then the use of stratified sampling is somewhat restricted. Indeed, assume that all the components of X are independent. Then, stratifying each component with k equally likely strata would require generating k^d samples of X to fill each of the resulting product strata (in the sense of point 2 above) with just one sample X_i. Thus, eliminating all the variance between the different strata via stratified sampling is inefficient in such a situation.

Latin hypercube sampling then is an idea of how to eliminate at least some of this variance between the strata by a specific way of multidimensional sampling if the sample size is already fixed. Assume for this that we are in the situation of the d-dimensional random vector X above that has independent, identically distributed components. Assume further that we will sample N such random vectors $X_1, ..., X_N$. The main steps of Latin hypercube sampling are then:

- Stratify each component $X^{(j)}, j = 1, ..., d$ with N equally likely strata $A_1, ..., A_N$.
- Sample exactly one observation $Y_i^{(j)}$ from stratum A_i for all $i = 1, ..., N$ for each component $j = 1, ..., d$.
- Choose d permutations $\pi_1, ..., \pi_d$ randomly from the set of permutations of $\{1, 2, ..., N\}$.
- Set

$$X_i^{(j)} = Y_{\pi_j(i)}^{(j)} \text{ for } i = 1, ..., N, j = 1, ..., d. \tag{3.93}$$

We have thus perfectly stratified each component $X^{(j)}$, i.e. each A_i is nonempty. Further, we have first sampled inside each stratum component-wise and then we have just randomly paired the components. If one is now arranging the pairs of strata (A_i, A_j) of two diffferent dimensions in a cross table, then we have exactly one entry per row and per column. This resembles a so-called Latin square. As the method just presented is a multidimensional generalization of the Latin square method known from experimental design, this justifies the name Latin hypercube sampling. The **Latin hypercube estimator** of $\mu = \mathbb{E}(f(X))$ is then given by the crude Monte Carlo estimator over the sample $X_1, ..., X_N$ obtained by Latin hypercube sampling:

$$\bar{X}_{LHS,N} = \frac{1}{N} \sum_{i=1}^{N} f(X_i). \tag{3.94}$$

To demonstrate the sampling procedure, we look at the following example where we simulate 4 vectors $X_i \sim \mathcal{U}[0,1]^3$ by the Latin hypercube method:

- Each $X^{(j)}$ is stratified into $A_1 = [0, 0.25], A_2 = (0.25, 0.5], A_3 = (0.5, 0.75], A_4 = (0.75, 1]$.
- We then sample all $Y_i^{(j)}$ by stratified sampling to obtain:

$$\begin{aligned} Y^{(1)} &= (0.095500046, 0.493293558, 0.701216163, 0.866725669) \\ Y^{(2)} &= (0.025170141, 0.349131748, 0.705786309, 0.897030549) \\ Y^{(3)} &= (0.149121067, 0.273186438, 0.546647542, 0.844218268) \end{aligned}$$

- We use the permutations:

$$\pi_1 = (1,\ 2,\ 3,\ 4),\ \pi_2 = (2,\ 1,\ 3,\ 4),\ \pi_3 = (4,\ 3,\ 1,\ 2)\ .$$

- This results in the sample

$$\begin{aligned} X_1 &= (0.095500046,\ 0.349131748,\ 0.546647542) \\ X_2 &= (0.493293558,\ 0.025170141,\ 0.844218268) \\ X_3 &= (0.701216163,\ 0.705786309,\ 0.273186438) \\ X_4 &= (0.866725669,\ 0.897030549,\ 0.149121067) \end{aligned}$$

Latin hypercube sampling was introduced by McKay et al. (1979) and was further developed by Stein (1987) and Owen (1992). A proof that under strong assumptions on the function f the (asymptotic) variance of $\bar{X}_{LHS,N}$ is smaller than that of the crude Monte Carlo estimator is given by Stein (1987). Further, Stein (1987) and Loh (1996) provide work on large sample properties and an examination of the efficiency of the method. We do not go into further details, but refer the interested reader to the above given references.

3.3.4 Variance reduction by conditional sampling

A method that is at the first sight conceptually very similar to stratified sampling is the conditional sampling approach. Again the aim is to estimate $\mu = \mathbb{E}(X)$ by a suitable variant of the Monte Carlo method. Here, variance reduction is obtained with the help of a second variable Y and the use of conditional expectations. However, in the stratified sampling approach, the distribution of the Y variable was already known and the conditional expectations $\mathbb{E}(X|Y)$ had to be estimated by the (crude) Monte Carlo method. In the conditional sampling approach the roles of the conditional expectations and of the distribution of Y are reversed. We here assume that

- $\mathbb{E}(X|Y)$ can be computed exactly by a given analytical formula,
- the distribution of Y is estimated by the (crude) Monte Carlo method.

By using the representation

$$\mu = \mathbb{E}(X) = \mathbb{E}(\mathbb{E}(X|Y)), \tag{3.95}$$

we obtain a Monte Carlo estimator for μ by sampling $\mathbb{E}(X|Y)$. While in the stratified sampling method we fixed $Y = y_i$ and then sampled $\mathbb{E}(X|Y = y_i)$, we here simply sample Y to obtain various values of $\mathbb{E}(X|Y)$. The **conditional Monte Carlo estimator** is thus obtained by

- Sample Y N times to obtain $Y_1, ..., Y_N$
- Compute $\mathbb{E}(X|Y_i)$
- Set

$$\bar{X}_{cond,N} = \frac{1}{N}\sum_{i=1}^{N}\mathbb{E}(X|Y_i) \tag{3.96}$$

By construction, the conditional Monte Carlo estimator is unbiased. By the conditional variance decomposition formula – already used in the stratified sampling method – we obtain

$$\begin{aligned}\sigma^2 = \mathbb{V}ar(X) = \mathbb{E}(\mathbb{V}ar(X|Y)) + \mathbb{V}ar(\mathbb{E}(X|Y))\\ \geq \mathbb{V}ar(\mathbb{E}(X|Y)) =: \sigma^2_{cond}\end{aligned} \tag{3.97}$$

with a strict inequality if $X|Y$ is not almost surely constant. We thus have:

PROPOSITION 3.19 Variance reduction by conditioning
With the above notation, the variance of the conditional Monte Carlo estimator never exceeds the variance of the crude Monte Carlo estimator. If $X|Y$ is not almost surely constant then there is a positive variance reduction by using the conditional Monte Carlo estimator.

REMARK 3.20 1. This time, the Monte Carlo error by estimating the mean in the subgroups characterized by the different values of Y is fully eliminated. We will thus have a big variance reduction if the variation inside the groups is large and the (conditional) group means do not differ much.

2. As the conditional Monte Carlo estimator is a crude Monte Carlo estimator constructed by using the different realizations $\mathbb{E}(X|Y=y_i)$ of the conditional expectation, we can construct confidence intervals for $\mathbb{E}(X)$ in the usual way with the help of the usual estimate for the variance of the observations $\mathbb{E}(X|Y=y_i)$, i.e. as

$$\left[\bar{X}_{cond,N} - 1.96\frac{\bar{\sigma}_{cond}}{\sqrt{N}}, \bar{X}_{cond,N} + 1.96\frac{\bar{\sigma}_{cond}}{\sqrt{N}}\right]. \tag{3.98}$$

□

We highlight the use of the method and its main difference to stratified sampling by the following simple example.

Example 3.21 Conditional versus stratified sampling
Assume that one wants to estimate the average costs μ of a night spent in a hotel in holiday (per person) of people from – say – Germany. Let us further assume that there are two companies. Company A – which is specialized in lifestyle – knows the exact fractions of p_i of the n different countries $C_1, ..., C_n$ where German people spend their holidays. To estimate μ, the company should use the stratified sampling method. More precisely, given suitable weights $\hat{p}_i$ (such as $\hat{p}_i = N_i/N$ assuming that this can be achieved with N_i being an integer [otherwise use a suitable rounding procedure]) the company should ask N_i people who spent their holidays in country C_i. From this, it can estimate the hotel costs in country C_i by the crude Monte Carlo method and then determine the stratified Monte Carlo estimator.

On the other hand, assume that company B – which specializes in tourist offers – knows the average cost per person and night, $\bar{C}_i$, in each of the countries, but has no idea about the preferences, Y, of the German population with respect to their holiday habits. Here, $Y = i$ means that the relevant person spent the holiday in country C_i. Company B should apply the conditional

Monte Carlo method, i.e. it should simply ask N Germans about where they spent their holidays. The answers $C_{i_1}, ..., C_{i_N}$ can then be used to compute the conditional Monte Carlo estimator

$$\bar{X}_{cond,N} = \frac{1}{N} \sum_{j=1}^{N} C_{i_j}. \tag{3.99}$$

3.3.5 Importance sampling

While in the two preceding variance reduction methods, we looked at transformed distributions of X obtained by conditioning or stratifying the distribution, importance sampling however builds upon a direct transformation of the density function of X (or a transformation of the probability function, in case of X being a discrete random variable). The main idea of importance sampling simply is to find a distribution for the underlying random variable that **assigns a high probability to those values that are important** for computing the quantity of our interest, $\mathbb{E}(g(X))$.

To motivate the method, we again have a look at the Monte Carlo integration of Example 3.9. There, with $f(x)$ being the density of the uniform distribution on $\mathcal{U}[0,1]^d$, the relation

$$\int_{[0,1]^d} g(x)\, dx = \int g(x) f(x)\, dx = E(g(X)) \tag{3.100}$$

enabled us to estimate the deterministic integral via the use of N independent, on $[0,1]^d$ uniformly distributed random variables $X_1, ..., X_N$ to obtain a crude Monte Carlo estimator

$$\bar{I}_N(\omega) = \frac{1}{N} \sum_{i=1}^{N} g(X_i(\omega)). \tag{3.101}$$

However, we could imagine that if we closely examine the integrand $g(x)$, then we could improve the accuracy of the above estimate if we generate more random numbers in the areas where $g(.)$ has large values (in absolute terms) while in areas where it is nearly 0, only a few samples are necessary to predict the contribution of the function on this area to the integral. Clearly, to take advantage of our knowledge of $g(.)$ in the above way, we have to change the distribution of the random variable X in a suitable way, as otherwise we would change the value of the expectation.

As an example consider the case of $d = 1$, $g(x) = x \cdot (1-x)$ which obviously is nonnegative and symmetric on $[0,1]$, vanishes for $x \in \{0,1\}$, and attains its maximum of 0.25 in $x = 0.5$. In the spirit of the discussion above, instead of using a uniform distribution on $[0,1]$, it would be better to use a triangular distribution for the random variable X, i.e. it should have the probability density (see Figures 3.8 and 3.9):

$$\tilde{f}(x) = \begin{cases} 0 & , x \leq 0 \text{ or } x \geq 1 \\ 4x & , 0 < x < \frac{1}{2} \\ 4 - 4x, & \frac{1}{2} \leq x < 1 \end{cases} \tag{3.102}$$

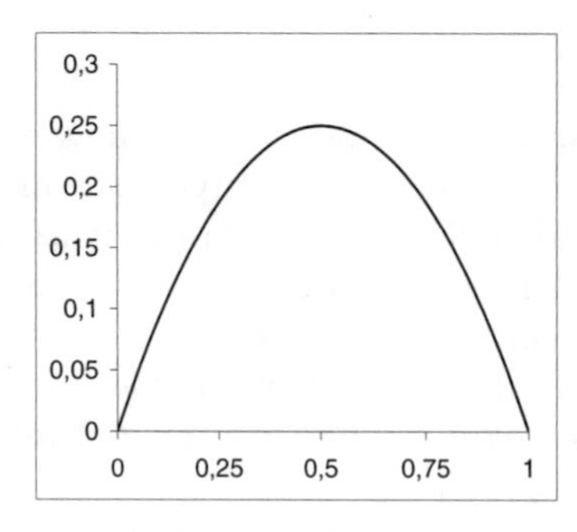

FIGURE 3.8: Integrand $x((1-x))$.

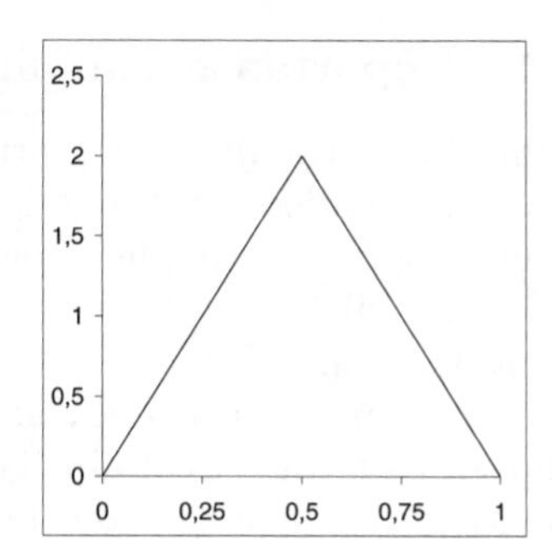

FIGURE 3.9: Sampling density $f^*(x)$.

As the value of the integral should not be changed when we sample according to the new density $\tilde{f}$, we have to divide $g(x)$ by $\tilde{f}(x)$:

$$\int_0^1 x(1-x)\,dx = \int_0^1 \frac{x(1-x)}{\tilde{f}(x)} \tilde{f}(x)\,dx. \tag{3.103}$$

This means that when we use the new distribution we actually have to sample $X(1-X)/\tilde{f}(X)$ to obtain the new Monte Carlo estimate

$$\bar{I}_{imp} = \frac{1}{N} \sum_{i=1}^{N} \frac{X_i(1-X_i)}{\tilde{f}(X_i)}. \tag{3.104}$$

Note that the X_i are now distributed according to the density $\tilde{f}(.)$. A simple comparison of using the Monte Carlo estimate with N=1,000 between the uniform approximation and the triangular approximation shows the superiority of the new method, the new one resulting in 0.168 and a 95%-confidence interval of $[0.166, 0.170]$, the crude one in 0.163 with a 95%-confidence interval of $[0.158, 0.167]$ while the exact value is $1/6$. The new method also leads to a much smaller variance of the estimator as we have

$$\mathbb{V}ar\left(\bar{I}_{crude}\right) = \frac{1}{180N}, \tag{3.105}$$

$$\mathbb{V}ar\left(\bar{I}_{imp}\right) = \frac{1}{1152N} = \frac{1}{64 \cdot 18N}. \tag{3.106}$$

We thus have reduced the variance to less than one sixth of the variance of the original crude Monte Carlo estimator by using the suggested method.

For the general case we try to formalize this approach to apply the idea of importance sampling to the computation of

$$\mathbb{E}\left(g\left(X\right)\right) = \int g\left(x\right) f\left(x\right) dx, \tag{3.107}$$

where the $\mathbb{R}^d$-valued random variable X has the density $f(x)$, and where we assume that the expectation exists for the function $g : \mathbb{R}^d \to \mathbb{R}$. For every density function $\tilde{f}(x)$ on $\mathbb{R}^d$ with the property of

$$\tilde{f}(x) > 0 \text{ for all } x \text{ with } f(x) > 0 \tag{3.108}$$

and its associated probability measure $\tilde{\mathbb{P}}$ we introduce $\tilde{g}(.)$ via the relation

$$\begin{aligned}\mathbb{E}\left(g\left(X\right)\right) &= \int g\left(x\right) f\left(x\right) dx = \int g\left(x\right) \frac{f\left(x\right)}{\tilde{f}\left(x\right)} \tilde{f}\left(x\right) dx \\ &= \tilde{\mathbb{E}}\left(g\left(X\right) \frac{f\left(X\right)}{\tilde{f}\left(X\right)}\right) = \tilde{\mathbb{E}}\left(\tilde{g}\left(X\right)\right).\end{aligned} \tag{3.109}$$

Here, $\tilde{\mathbb{E}}(.)$ denotes the expectation with respect to $\tilde{\mathbb{P}}$. The weight function $f(X)/\tilde{f}(X)$ is called the **likelihood ratio function** of the above change of measure from $\mathbb{P}$ to $\tilde{\mathbb{P}}$. The **importance sampling estimator** (with respect to $\tilde{f}(.)$) for $\mu = \mathbb{E}(g(X))$ is defined as

$$\bar{I}_{imp,\tilde{f},N}\left(g\left(X\right)\right) = \frac{1}{N}\sum_{i=1}^{N} \tilde{g}\left(X_i\right) = \frac{1}{N}\sum_{i=1}^{N} g\left(X_i\right) \frac{f\left(X_i\right)}{\tilde{f}\left(X_i\right)} \tag{3.110}$$

where the X_i are independent and are distributed according to the **importance sampling density function** $\tilde{f}$. So the importance sampling estimator is a weighted crude Monte Carlo estimator where the weights for each observation X_i are determined by the likelihood ratio function. Note that due to the representation (3.109) the importance sampling estimator is unbiased and consistent. Its variance is given by

$$\begin{aligned}\sigma^2_{imp,\tilde{f},N} &= \tilde{\mathbb{V}ar}\left(\bar{I}_{imp,\tilde{f},N}\left(g\left(X\right)\right)\right) \\ &= \frac{1}{N}\tilde{\mathbb{V}ar}\left(\tilde{g}\left(X\right)\right) = \frac{1}{N}\left(\tilde{\mathbb{E}}\left(\tilde{g}\left(X\right)^2\right) - \mu^2\right) \\ &= \frac{1}{N}\left(\int \frac{g\left(x\right)^2 f\left(x\right)}{\tilde{f}\left(x\right)} f\left(x\right) dx - \mu^2\right).\end{aligned} \tag{3.111}$$

As the importance sampling estimator is again an averaging over independent identically distributed random variables, the central limit theorem yields an

Approximate 95%-confidence interval for $\mathbb{E}(g(X))$

$$\left[\bar{I}_{imp,\tilde{f},N}(g(X)) - 1.96\frac{\tilde{\sigma}_{imp,\tilde{f},N}}{\sqrt{N}}, \bar{I}_{imp,\tilde{f},N}(g(X)) + 1.96\frac{\tilde{\sigma}_{imp,\tilde{f},N}}{\sqrt{N}}\right] \quad (3.112)$$

with $\tilde{\sigma}_{imp,\tilde{f},N}$ denoting the sample standard deviation of the importance sampling estimator. If we assume that we have $g(x) \geq 0$ for all $x \in \mathbb{R}^d$, then there is one particularly striking feature in this representation. Namely, if we choose

$$\tilde{f}(x) = c \cdot f(x) \cdot g(x) = \frac{f(x) \cdot g(x)}{\int f(y) \cdot g(y)\, dy} \quad (3.113)$$

then $\tilde{f}(x)$ is a density function on $\mathbb{R}^d$ and we would have $\tilde{g}(X) = 1/c$, i.e.

$$\tilde{\mathbb{V}ar}\left(\bar{I}_{imp,\tilde{f},N}(g(X))\right) = 0. \quad (3.114)$$

However, the drawback of this approach is that the constant c is essentially the value we are trying to calculate by our Monte Carlo approach as we have $\mu = 1/c$. So, if we already know c, we would not be interested in performing importance sampling anyway. On the positive side, the above choice has the following consequence.

***PROPOSITION 3.22* Variance reduction by importance sampling**
Let $g(.)$ be a nonnegative function. Then there exist choices of importance sampling density functions $\tilde{f}$ such that we have

$$\tilde{\mathbb{V}ar}\left(\bar{I}_{imp,\tilde{f},N}(g(X))\right) < \mathbb{V}ar\left(\bar{I}(g(X)_N)\right) \quad (3.115)$$

with $\bar{I}(g(X))_N$ being the crude Monte Carlo estimator for $\mathbb{E}(g(X))$. Moreover, for all functions $\tilde{f}$ with property (3.108) we obtain

$$\begin{aligned}&\mathbb{V}ar\left(\bar{I}(g(X))_N\right) - \tilde{\mathbb{V}ar}\left(\bar{I}_{imp,\tilde{f},N}(g(X))\right)\\ &= \frac{1}{N}\left(\int g(x)^2\left(1 - \frac{f(x)}{\tilde{f}(x)}\right) f(x)\, dx\right).\end{aligned} \quad (3.116)$$

Note that the last relation follows from representation (3.111), the fact that both estimators are unbiased and the usual representation of $\mathbb{E}(g(X)^2)$. Further, this last relation gives a hint on the structure of a good importance sampling density $\tilde{f}$. To obtain a variance reduction, we should have:

- $\tilde{f}(x)$ should be large (in the sense of $\tilde{f}(x) > f(x)$) whenever $g(x)^2 f(x)$ is large.
- $\tilde{f}(x)$ should be small (in the sense of $\tilde{f}(x) < f(x)$) whenever $g(x)^2 f(x)$ is small.

Of course, there have to be values x with $\tilde{f}(x) < f(x)$. With regard to the purpose of variance reduction, they should therefore only appear when the product $g(x)^2 f(x)$ has no large impact. Besides the two requirements above, an importance sampling density should also satisfy practical issues such as:

- $\tilde{f}(x)$ should be easy to evaluate.
- random variables with density $\tilde{f}$ should be easy to simulate.

Thus, a good importance sampling density should be a compromise between being similar to $g^2 \cdot f$ and being tractable.

Some popular methods to obtain an importance sampling density

We present easy, popular, and tractable methods which are based on shifting the original density $f(x)$ (**translation**) and on adjusting its shape by **scaling**.

1. Shifting the density and the maximum principle

The idea of this method is simply to replace $f(x)$ by

$$\tilde{f}(x) = f(x - c) \tag{3.117}$$

for a constant c leading to

$$\tilde{g}(x) = \frac{f(x)}{f(x-c)} g(x). \tag{3.118}$$

The so-called **maximum principle** consists of choosing c in such a way that $\tilde{f}(x)$ and $g(x) f(x)$ attain their maximum at the same point x_{max}. This, of course, requires that we can explicitly determine this maximum point. Also, if it is not unique, it is not always clear how to choose c. In the special case of a multivariate normal density function

$$f(x) = \varphi_{\nu,\Sigma}(x) = \frac{1}{2^{d/2}\left|det(\Sigma)\right|} \exp\left(-\frac{1}{2}(x-\nu)' \Sigma^{-1} (x-\nu)\right) \tag{3.119}$$

one knows that its maximum point is the mean ν. So, one chooses

$$c = \nu^* - \nu \tag{3.120}$$

with

$$\nu^* = arg\ max_x \{g(x) f(x)\}. \tag{3.121}$$

Such an approach is particularly suitable for computing expressions related to extreme events. We illustrate this by the following example.

Example 3.23 Computing costs of extreme events for a normal distribution

Suppose that we have $X \sim \mathcal{N}(0,1)$ and that we face costs of $g(X)$ if we observe values of X bigger than 10. This is extremely unlikely, but one might think of events such as a default of the United States or a serious accident in a nuclear power plant. If we now use a crude Monte Carlo estimator then, even for a very high number of N, we would typically not observe a single value of X_i exceeding 10 and thus estimate the mean costs $\mathbb{E}(g(X))$ to be zero. If we use

$$g(x) = C \cdot x \cdot 1_{[10,\infty)}(x) \tag{3.122}$$

with C a typically very large constant then it is easy to verify that we have

$$10 = arg\ max_x \left\{ C \cdot x \cdot 1_{[10,\infty)}(x) \frac{1}{\sqrt{2\pi}} \exp\left(-x^2/2\right) \right\}. \tag{3.123}$$

Thus, we use

$$\tilde{f}(x) = \varphi_{0,1}(x - 10) = \frac{1}{\sqrt{2\pi}} \exp\left(-(x-10)^2/2\right) \tag{3.124}$$

which leads to the importance sampling estimator of

$$\bar{I}_{imp,\tilde{f},N}(g(X)) = \frac{1}{N} \sum_{i=1}^{N} C \cdot X_i \cdot 1_{X_i \geq 10} \exp(50 - 10X_i) \tag{3.125}$$

with all independent $X_i \sim \mathcal{N}(10,1)$. With N=10,000 and $C = 10^9$ we obtained an estimate of $7.530 \cdot 10^{-14}$ with an approximate 95%-confidence interval of $\left[7.029 \cdot 10^{-14}, 8.031 \cdot 10^{-14}\right]$. Compare this to the exact value of $C \cdot \exp(-50)/\sqrt{2 * \pi} = 7.695 \cdot 10^{-14}$. This is an impressive performance, in particular as the crude Monte Carlo estimator delivers a value of 0 with a zero variance! See Figure 3.10 for an illustration of the shift of the sampling density. The figure and the form of the importance sampling estimator nicely highlight the way importance sampling works: The shift yields sampling values in the area of importance for calculating the expectation while the likelihood ratio functions assign these samples their probability weights. In the crude Monte Carlo method this has already been done **before** the sampling which results in (nearly) no samples in the region of interest.

2. Changing the shape of the density by scaling

Here, the idea is to steepen or flatten the density via replacing $f(x)$ by

$$\tilde{f}(x) = \frac{1}{c} f\left(\frac{x}{c}\right) \tag{3.126}$$

for a constant $c > 0$ leading to

$$\tilde{g}(x) = c \frac{f(x)}{f\left(\frac{x}{c}\right)} g(x). \tag{3.127}$$

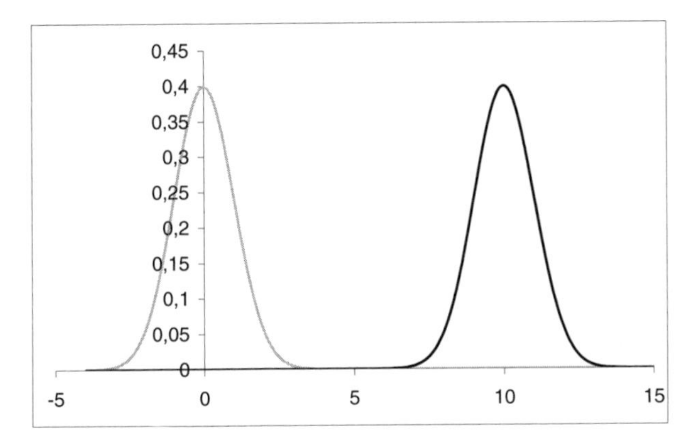

FIGURE 3.10: Original density $f(x)$ (grey) and shifted importance sampling density $\tilde{f}(x)$ (black).

Choosing a big value of $c >> 1$ spreads out the distribution which can be advantageous to estimate costs of extreme events. However, an obvious drawback is that this spreading is symmetric for symmetric distributions. So regions far away from our region of interest obtain more weight, too. Further, the variance of the distribution corresponding to $\tilde{f}$ equals the variance corresponding to the original density multiplied by c^2. The use and the advantages/disadvantages of this approach can again be illustrated by

Example 3.23 (continued)
In the unchanged situation of Example 3.23, we could choose the shape parameter c such that there is a significant probability to enter the region $[10, \infty)$ when we simulate random numbers according the transformed density $\tilde{f}$. As for an $N(0, \sigma^2)$-distributed random variable X_σ we have

$$\mathbb{P}(X_\sigma \geq \sigma) = 1 - \Phi(1) = 0.159 \ ,$$

one could choose $\sigma = 10$. Then in the mean a sixth of the generated random numbers X_i would enter $[10, \infty)$ and thus would make a nonzero contribution to the so-obtained importance sampling estimator (see Figure 3.11).

However, by choosing $\sigma = 10$ we have increased the standard deviation of the sampling distribution by a factor of 10 which is an undesirable feature. This is also underlined by the estimate of $8.259 \cdot 10^{-14}$ with an approximate 95%-confidence interval of $\left[5.956 \cdot 10^{-14}, 1.056 \cdot 10^{-13}\right]$. Although the true value of $7.695 \cdot 10^{-14}$ is inside the 95%-confidence interval, the estimator is still quite unstable. Note also the much bigger confidence interval compared to the one of the mean-shifting technique.

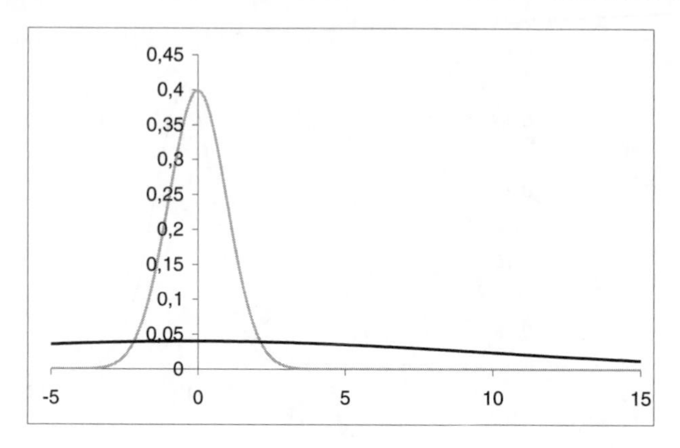

FIGURE 3.11: Original density $f(x)$ (grey) and scaled importance sampling density $\tilde{f}(x)$ (black).

3. Exponential twisting

For a distribution with a density f on $\mathbb{R}^d$, the method of **exponential twisting** with parameter $\theta \in \mathbb{R}^d$ is characterized by the relation

$$f_\theta(x) = \frac{\exp(\theta' x)}{\mathbb{E}(\exp(\theta' X))} f(x) \tag{3.128}$$

where X is distributed according to $\mathbb{P}$. The function $M(\theta) = \mathbb{E}(\exp(\theta' X))$ in the denominator is called the **moment generating function** of X. By defining the **cumulant generating function** $C(\theta) = \ln(M(\theta))$ one can directly verify that we have

$$C'(\theta) = \mathbb{E}(X \exp(\theta' X - C(\theta))) = \mathbb{E}_\theta(X) \tag{3.129}$$

where $\mathbb{E}_\theta$ denotes the expectation under the transformed density f_θ. So, exponential twisting with parameter θ shifts the mean of X to the value $C'(\theta)$. When $C'(x)$ can easily be calculated, we thus know what to do if we want to have an appropriate mean shift.

The same change of measure as above will be applied under the name **Esscher method** in Chapter 7 in the context of option pricing in incomplete financial markets.

4. Conditional sampling restricted to the important area

So far, our methods tried to **bring more probability** into the area that is relevant for the calculation of the expectation of interest. We now make a more radical move and **bring all the probability** into this area. We do this by conditional sampling. The first problem however is that now the importance sampling density $\tilde{f}$ is zero on an area where the original one might be positive. However, we repair this by requiring that

$$\tilde{f}(x) > 0 \quad \text{for all } x \text{ with } \; g(x) f(x) \neq 0 \tag{3.130}$$

instead of requirement (3.108). Then the defining relation for the importance sampling density is still true, and due to the relation between a density and the conditional density on an interval $[a, b]$ (with both a, b possibly equalling $-\infty$ respectively ∞)

$$f_{\{X|X \in [a,b]\}}(x) = \frac{f(x)}{\mathbb{P}(X \in [a, b])} \tag{3.131}$$

we obtain a very simple form of the likelihood ratio function if we choose such a conditional density as our importance sampling density:

$$\frac{f(x)}{\tilde{f}(x)} = \mathbb{P}(X \in [a, b]). \tag{3.132}$$

This also makes the importance sampling estimator easier to calculate as we then have the representation

$$\bar{I}_{imp,\tilde{f},N}(g(X)) = \frac{1}{N} \sum_{i=1}^{N} \tilde{g}(X_i) = \frac{1}{N} \mathbb{P}(X \in [a, b]) \sum_{i=1}^{N} g(X_i) \tag{3.133}$$

where of course the X_i have to be sampled from the conditional density.

If the interval we are conditioning on has a very small probability then calculation of this probability might be a hard numerical problem. One could avoid this by performing a **combined conditioned-shift method**: First move the density (partly) into the area of interest by an appropriate shifting, and then use the shifted density as the basis for conditioning on the area of interest. The price of this gain in numerical stability is that we now have a conventional importance sampling estimator again, i.e.

$$\bar{I}_{imp,\tilde{f}_{cond},N}(g(X)) = \frac{1}{N} \tilde{\mathbb{P}}(X \in [a, b]) \sum_{i=1}^{N} g(X_i) \frac{f(X_i)}{\tilde{f}_{cond}(X_i)} \tag{3.134}$$

with $\tilde{f}_{cond}(x)$ being the density obtained from the shifted density $\tilde{f}(x)$ by conditioning. We highlight the differences between both approaches:

Example 3.23 (continued)
In the well-known situation of Example 3.23, a pure conditioning would lead to an importance sampling estimator of

$$\bar{I}_{imp,\tilde{f},N}(g(X)) = \frac{1}{N} \mathbb{P}(X \in [10, \infty)) \sum_{i=1}^{N} C \cdot X_i. \tag{3.135}$$

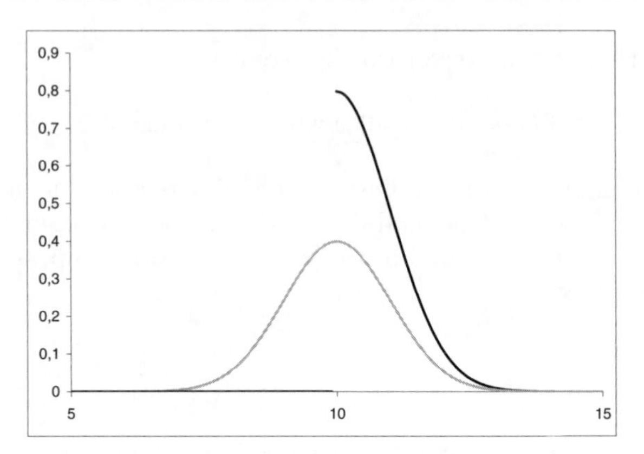

FIGURE 3.12: Shifted density (by the maximum method) $\tilde{f}(x)$ (grey) and conditional shifted density $\tilde{f}_{cond}(x)$ (black).

However, the probability occurring in the estimator is extremely small and very hard to distinguish from zero. For the combined approach, we first perform a shift according to the maximum principle leading us to a normal distribution with unit variance and mean 10. Then, conditioning on only sampling values larger than 10 only results in a multiplying factor of two to obtain the conditional density on $[10, \infty)$. We thus obtain the combined importance sampling estimator of

$$\bar{I}_{imp,\tilde{f}_{cond},N}(g(X)) = \frac{1}{N}\sum_{i=1}^{N} C \cdot X_i \cdot 1_{X_i \geq 10} \cdot \frac{1}{2}\exp(50 - 10X_i) \tag{3.136}$$

where now the X_i are sampled from the conditional distribution. Still, there is a very small multiplying factor inside the importance sampling estimator, but this one is much easier to calculate than the corresponding probability for the direct conditioning. The way this combined estimator is obtained can be seen in Figure 3.12. The result from performing the relevant simulation is the most exact one of the three methods considered, as shown in the comparison in Table 3.6.

Method	Estimator	Lower quantile	Upper quantile
Crude MC	0	0	0
Mean Shift	$7.530 \cdot 10^{-14}$	$7.029 \cdot 10^{-14}$	$8.030 \cdot 10^{-14}$
Scaling	$8.259 \cdot 10^{-14}$	$5.956 \cdot 10^{-14}$	$1.056 \cdot 10^{-13}$
Comb. Conditioning	$7.530 \cdot 10^{-14}$	$7.190 \cdot 10^{-14}$	$7.870 \cdot 10^{-14}$

Table 3.6: Different Importance Sampling Methods with 95%-Confidence Bounds (Exact Value = $7.695 \cdot 10^{-14}$)

5. Importance sampling for discrete random variables

So far, we have restricted ourselves to the case of probability distributions with densities. However, the principle of importance sampling is also valid for discrete distributions. Note that if X has a discrete distribution $\mathbb{P}$ with $p_i = \mathbb{P}(X = x_i)$, we have the following relation (assuming that all expressions appearing are well defined):

$$\begin{aligned} \mathbb{E}(g(X)) &= \sum_{i=1}^{\infty} g(x_i)\, p_i = \\ &= \sum_{i=1}^{\infty} \frac{g(x_i)\, p_i}{\tilde{p}_i} \tilde{p}_i = \tilde{\mathbb{E}}\left(g(X) \frac{p(X)}{\tilde{p}(X)} \right) \end{aligned} \tag{3.137}$$

Here, the probability function $p(x)$ is defined as

$$p(x) = \begin{cases} p_i & \text{if } x = x_i \text{ for some } i \in \mathbb{N} \\ 0 & \text{else} \end{cases}. \tag{3.138}$$

The above discussed methods and suggestions for finding a suitable importance sampling distribution can all easily be adapted to the discrete setting. In particular, discrete exponential families such as the binomial distributions or the Poisson distributions can use exponential changes of measure as a convenient transformation.

REMARK 3.24 If the importance sampling density $\tilde{f}$ becomes smaller much faster in the tails than the original density f then the likelihood function $f(x)/\tilde{f}(x)$ can attain very high values. As this is a tail event, it might happen only very rarely but can totally bias the value of the importance sampling estimator. To avoid this, one should try to use importance sampling densities that do not have lighter tails than the original density f. □

We will meet further applications of importance sampling in both finance and insurance in Chapter 5 and Section 8.2.5. However, there are also numerous examples of applications of importance sampling in other areas of science such as biology, physics, information processing, and many others.

3.4 Further aspects of variance reduction methods

3.4.1 More methods

The last section did not contain all possible methods of variance reduction in our survey. Indeed, there are many more, some of them efficient, some of them with an unclear impact. One such method is **moment matching**.

Moment matching

The idea of moment matching is that the samples generated for Monte Carlo estimation should have the same statistical properties as the underlying distribution. By this we mean that the empirical moments of the sample should coincide with the theoretical moments of the distribution. While this seems to be desirable at first sight, it also poses some important questions. To underline this, let us have a look at a popular adjustment of the crude Monte Carlo method that guarantees that the corrected sample has the same first two moments as the underlying distribution, $\mu = \mathbb{E}(X)$ and $\sigma^2 = \mathbb{V}ar(X)$. To achieve this, replace the sample elements X_i by

$$X_i^c = \left(X_i - \bar{X}_N\right) \frac{\sigma}{\bar{\sigma}_N} + \mu \tag{3.139}$$

with $\bar{\sigma}_N^2$ being the usual sample variance based on N observations. It is easy to verify that we have

$$\bar{X}_N^c = \frac{1}{N} \sum X_i^c = \mu, \quad (\bar{\sigma}_N^c)^2 = \frac{1}{N-1} \sum \left(X_i^c - \bar{X}_N\right)^2 = \sigma^2. \tag{3.140}$$

However, this is paid for by the following two problems:

- The X_i^c are no longer independent.
- The X_i^c no longer have the same distribution as X. Note in particular that dividing by $\bar{\sigma}_N$ is a nonlinear transformation that can lead us to completely different classes of distributions!

Further, in a lot of studies the impact of this method with regard to variance reduction is unclear. Note also that the antithetic variates method is an example of a mean matching method.

Weighted Monte Carlo estimation

A different method to match moments is to assign weights w_i to the elements $X_1, ..., X_N$ of the sample of random variables from which the Monte Carlo estimator for $\mathbb{E}(g(X))$ is computed. Then the **weighted Monte Carlo estimator** for $\mathbb{E}(g(X))$ is computed as

$$\bar{I}_{w,N} = \sum_{i=1}^{N} w_i g(X_i). \tag{3.141}$$

What remains is to determine the weights to match the mean in the underlying sample, i.e. we require

$$\mathbb{E}(X) = \sum_{i=1}^{N} w_i X_i \ . \tag{3.142}$$

To have consistency with the usual weight of $1/N$ in the crude Monte Carlo estimator, it makes sense to ask in addition for

$$\sum_{i=1}^{N} w_i = 1. \tag{3.143}$$

As these two conditions do not uniquely determine the weights w_i, one could

- introduce additional moment constraints (up to $N-2$) similar to those in Equation (3.142) or
- introduce an additional criterion such as the length of the weight vector w and use that w which is optimal for this criterion under the additional constraint (3.143).

In Glasserman (2004) and in Glasserman and Yu (2005) properties of such estimators resulting from e.g. the solution of the least-squares criterion

$$\min_{w \in \mathbb{R}^N} \|w\|_2^2 \tag{3.144}$$

under the additional constraints (3.142) and (3.143) are examined in detail. An interesting connection to the control variate method is presented there. However, this connection also shows that there is no improvement over the control variate method. So we do not go into details of the method.

Common random numbers

This method is not really a variance reduction method like the ones we have presented so far. It is more a principle that if we compare two expressions that are simulated by different Monte Carlo estimators then the variance of the difference attains its minimum when we use the same random numbers for computing the two Monte Carlo estimators. A particular example where this is made more precise is the computation of option price sensitivities with the help of finite differences in Chapter 5.

Combined variance reduction methods

Typically, the variance reduction methods we have presented here work as a two-step procedure:

1. Apply a transformation to the problem (such as stratification, conditioning, transformation of the distribution, or subtraction of a similar random variable).
2. Use crude Monte Carlo estimators for estimating expected values in the transformed problem.

As the second step consists of the use of a crude Monte Carlo estimator, again a variance reduction technique can be applied which reduces the variance in the transformed problem. So, in principle, a second iteration of the above two-step procedure is possible which itself could be followed by a third iteration, and so on. In practice, many situations can be imagined where actually a combination is justifiable such as

- using stratified sampling after the distribution has been transformed to the important part of the distribution via importance sampling,
- using a control variate after the sample has been symmetrized by antithetic sampling.

Other combinations are possible, and there are examples that support them and others where the combination has no advantage over just using one of both methods. Thus, there is no general rule for finding an efficient combination.

3.4.2 Application of the variance reduction methods

We have seen various methods to speed up the convergence of the crude Monte Carlo method in the preceding section. Of course, the most interesting questions are which method to use when, and which method is the best. There is no clear answer to these questions. However, we try to formulate some simple advice to help apply Monte Carlo methods:

- If there is an obvious good control variate at hand, then use it. Often, in this situation using antithetic variates is beneficial before the control variate method is applied.
- For computing expectations where rare events play an important role, importance sampling is usually the preferable method, sometimes the only one that works.
- If there is no obvious argument for a variance reduction method then be careful in applying such a method as it can also lead to a waste of computing time.

To judge if stratified sampling should be introduced depends on the problem. In the one-dimensional setting it is often easy to introduce a stratification that greatly reduces the variance when $\mathbb{E}(g(X))$ has to be estimated. There it is crucial to use both detailed knowledge on the distribution of X and the behaviour of the function $g(x)$. In such a case, it will typically be preferable to a control variate approach. For higher dimensions, however, stratified sampling suffers under the curse of dimensionality.

Another important role in deciding on the level of sophistication of the variance reduction technique is the further use of the developed algorithm. If it will be used very often on parametrically changing problems (such as an option pricing routine in financial software) where furthermore a reliable solution should be computed fast, a detailed analysis of the problem is advisable. There, even a combination of different variance reduction techniques to reduce as much variance as possible should be considered. However, one should again note that variance reduction is not the only aspect of efficiency. As already said in the introducing part to Section 3.2, the computing and the variance reduction factor have to be considered simultaneously. If on the other hand one chooses a Monte Carlo approach just to compute one single expectation, then sometimes the search for a suitable variance reduction method can take longer than using a crude Monte Carlo estimator with a very high number of replications N.

In the following chapters, we will often apply Monte Carlo methods to compute expectations over paths of stochastic processes. There, high dimensions are introduced by the discretization of a path of a stochastic process. We will also comment on variance reduction techniques in these settings. Further, there will be many examples where a particular variance reduction technique will be tailored to the particular application in problems of finance or actuarial problems.

Chapter 4

Continuous-Time Stochastic Processes: Continuous Paths

4.1 Introduction

Stochastic processes are the main modelling tool when dynamically evolving phenomena with a random component are considered. Of course, we will consider such phenomena as stock prices, interest rates, and premium processes, but one can also think about examples from nature such as weather or technical systems such as the flow of interacting particles through some filter. We will therefore introduce the notion of a stochastic process in this chapter.

As the normal distribution plays a popular role in probabilistic modelling, Brownian motion and the Itô integral as the corresponding stochastic process versions of the normal distribution are surveyed. To deal with functionals of Brownian motion and Itô integrals the so-called Itô calculus will be presented in a comprehensive way (see i.e. Karatzas and Shreve [1991] for a rigorous treatment). It is then a natural step to introduce stochastic differential equations as the modelling tool based on these building blocks. After collecting some basic theoretical results about their existence and uniqueness, we then mainly concentrate on discretization methods to simulate the solutions of stochastic differential equations. Here, we mainly follow the standard reference, the monograph by Kloeden and Platen (1999), for presenting the basic principles. However, we also introduce some very recent methods such as the statistical Romberg method (see Kebaier [2005]) and the multilevel Monte Carlo method (see Giles [2008]).

4.2 Stochastic processes and their paths: Basic definitions

A stochastic process is an indexed family of random variables. Typically, this index is interpreted as the running time. Thus, a stochastic process can be seen as a model for describing the behaviour of a family of random experiments performed one after another in time. Alternatively, it can be thought of as the result of a random experiment where the exact value is partly revealed over time, one piece after another. This second interpretation explains that the sets of possible events corresponding to the evolution of the stochastic process changes over time. This flow of information over time leads to the concept of a filtration which is the collection of all scenarios of the evolution of the stochastic process. We put this together in the following definition.

DEFINITION 4.1
Let $(\Omega, F, \mathbb{P})$ be a probability space with sample space Ω, σ-field F, and probability measure $\mathbb{P}$. Let I be an ordered index set.
(a) A family $\{F_t\}_{t\in I}$ of sub-σ-fields of F with $F_s \subset F_t$ for $s < t,\ s,t \in I$, is called a **filtration**.
(b) A family $\{(X_t, F_t)\}_{t\in I}$ consisting of a filtration $\{F_t\}_{t\in I}$ and a family of $\mathbf{R}^n$-valued random variables $\{X_t\}_{t\in I}$ such that X_t is F_t-measurable is called a **stochastic process** *with respect to the filtration $\{F_t\}_{t\in I}$.*
(c) For a fixed $\omega \in \Omega$, the set

$$X_{\cdot}(\omega) := \{X_t(\omega)\}_{t\in I} = \{X(t,\omega)\}_{t\in I} \tag{4.1}$$

can be interpreted as a function of time t and is called a **sample path** *or a* **realization of the stochastic process**.

Standard examples of stochastic processes are temperature curves, the evolution of a stock price index, the sequence of the wealth of a gambler taking part in a series of games of chance, or the size of a population as a function of time, just to present a few.

REMARK 4.2 1. We often simply write X to denote a stochastic process. We either do this if the filtration $\{F_t\}_{t\in I}$ is clearly identifiable from the current context or if the filtration is the so-called **natural filtration**, i.e. if we have

$$F_t := \sigma\{X_s \ :\ s \leq t\}. \tag{4.2}$$

2. If the index set I in the definition of the stochastic process is an interval $I \subset \mathbb{R}$ (or even more typically $I \subset [0,\infty)$) then we speak of a **continuous-time stochastic process**. If the index set I is a discrete subset of $\mathbb{R}$ (e.g. a sequence in $\mathbb{N}$) then we speak of a **discrete-time stochastic process**. □

One of the main differences between a stochastic process describing temperature curves and one showing the evolution of a population size above is that temperature curves change continuously over time while population sizes change by jumps of integer sizes. Such properties of the paths of a stochastic process are very important for both theory and application.

DEFINITION 4.3
If the sample paths of a stochastic process $X_\cdot(\omega)$ are all (but up to a set of P-measure zero) continuous (right-continuous, left-continuous) then we speak of a **continuous** *(right-continuous, left-continuous)* **stochastic process**.

We will in this chapter mainly concentrate on continuous-time stochastic processes with continuous paths, i.e. on continuous stochastic processes. The treatment of (continuous-time) stochastic processes with jumps and their applications in financial and actuarial models will be the subject of Chapters 6, 7, and 8. However, the next definitions are independent of the continuity of the paths of the processes. They generalize the idea of performing one experiment after another in time by introducing the notion of the increments of a stochastic process as the difference between the values at two times.

DEFINITION 4.4
(a) A stochastic process $\{(X_t, F_t)\}_{t\in I}$ is said to have **independent increments** *if for all $r \le u \le s \le t$ with $r, u, s, t \in I$ we have*

$$X_t - X_s \text{ is independent of } X_u - X_r. \tag{4.3}$$

(b) A stochastic process $\{(X_t, F_t)\}_{t\in I}$ is said to have **stationary increments** *if for all $s \le t$ with $s, t \in I$ we have*

$$X_t - X_s \sim X_{t-s}. \tag{4.4}$$

REMARK 4.5 Both these properties will simplify the analysis and in particular the simulation of a stochastic process considerably. If a stochastic process X has independent increments then to describe its future evolution, only the current value X_t and the increments after time t are necessary. Thus, no past values other than that of X_t have to be stored. If the process X has stationary increments then the distributional properties of the process do not change over time. This does not mean that each X_t has the same distribution but that the distribution of the increments $X_t - X_s$ only depends on the time difference $t - s$. We will also introduce two fundamental classes of stochastic processes that generalize these two properties. The first one is the class of **Markov processes** where the distribution of the future values of the process only depends on the past via its present value. The second concept is that of the **martingale** that generalizes the idea of a fair game. □

DEFINITION 4.6
An $\mathbb{R}^d$-valued stochastic process $\{(X_t, F_t)\}_{t\in I}$ on a probability space $(\Omega, F, \mathbb{P})$ is called a **Markov process** *with initial distribution ν if we have*

$$\mathbb{P}(X_0 \in A) = \nu(A) \;\; \forall A \in B\left(\mathbb{R}^d\right), \tag{4.5}$$

$$\mathbb{P}(X_t \in A | F_s) = \mathbb{P}(X_t \in A | X_s) \;\; \forall A \in B\left(\mathbb{R}^d\right), \; t \ge s. \tag{4.6}$$

In particular, the distribution of future values of X only depends on the past via the present value X_t.

A process X_t with independent increments is an example of a Markov process. To see this, note that the Markov property is a consequence of the independence of the increments from the past and the representation

$$X_t = X_s + (X_t - X_s)\,. \tag{4.7}$$

Many processes that we will consider in our applications in finance and insurance are indeed Markov processes.

DEFINITION 4.7
The real-valued process $\{(X_t, F_t)\}_{t\in I}$ with $\mathbb{E}\,|X_t| < \infty$ for all $t \in I$ is called
(a) a **super-martingale**, *if we have*

$$\mathbb{E}\,(X_t\,|F_s\,) \le X_s \text{ for all } s,t \in I \text{ with } s \le t \;\; \mathbb{P}\text{-a.s.}; \tag{4.8}$$

(b) a **sub-martingale**, *if we have*

$$\mathbb{E}\,(X_t\,|F_s\,) \ge X_s \text{ for all } s,t \in I \text{ with } s \le t \;\; \mathbb{P}\text{-a.s.}; \tag{4.9}$$

(c) a **martingale**, *if we have*

$$\mathbb{E}\,(X_t\,|F_s\,) = X_s \text{ for all } s,t \in I \text{ with } s \le t \;\; \mathbb{P}\text{-a.s.}. \tag{4.10}$$

REMARK 4.8 Martingales in discrete time are often used to model games of chance. Indeed, if the sequence X_n, $n \in \mathbb{N}$ denotes the evolution of the wealth of a gambler taking part in a series of fair games then it should satisfy the martingale condition $\mathbb{E}\,(X_{n+1}\,|F_n\,) = X_n$. We show this in a special case. Let therefore F_n be the natural filtration, i.e. the flow of information that is generated by the values of the process X until time n. If the outcomes Y_i of the different games are independent and satisfy the **fairness condition** $\mathbb{E}\,(Y_i) = 0$, then

$$X_n = x + \sum_{i=1}^{n} Y_i \tag{4.11}$$

is a stochastic process with independent increments (here, x is the player's initial wealth). We then obtain for $n, m \in \{0, 1, 2, ...\}$:

$$\begin{aligned}\mathbb{E}\,(X_{n+m}\,|F_n\,) &= \mathbb{E}\left(X_n + \left(\sum_{i=1}^{m} Y_{n+i}\right) |F_n\right)\\ &= \mathbb{E}\,(X_n\,|F_n\,) + \mathbb{E}\left(\sum_{i=1}^{m} Y_{n+i}\,|F_n\right) = X_n + \mathbb{E}\left(\sum_{i=1}^{m} Y_{n+i}\right) = X_n \end{aligned} \tag{4.12}$$

Thus, in the mean, the gambler is as rich after the game as he was before. From his point of view, a sub-martingale is thus a favourable game, while a super-martingale represents an unfavourable one. □

4.3 The Monte Carlo method for stochastic processes

As a stochastic process is just a family of random variables, its simulation seems to be a straightforward task. However, there are some facts and aspects that have to be considered first:

- The elements $X_t,\ t \in I$ of a stochastic process are usually **not** independent.
- The index set I can be noncountable which means that a completely detailed simulation of a corresponding process is simply impossible.
- What is our aim when simulating a stochastic process? Do we want to **imitate** the real process as well as possible or are we only interested in **consequences** of the process such as the mean of a functional of it?

We deal with those facts and questions in some detail in the following sections.

4.3.1 Monte Carlo and stochastic processes

Let us start by answering the last questions. In this book, we are mainly interested in calculating expected values by the Monte Carlo method. So, we first generalize it to the stochastic process situation. Let $X = \{X_t,\ t \in I\}$ be a stochastic process and let $g(X) = g(X_t(\omega),\ t \in I)$ be a functional on the path of this stochastic process. We assume that

$$\mu = \mathbb{E}(g(X)) = \mathbb{E}(g(X_t,\ t \in I)) \tag{4.13}$$

is defined and is finite. If we are then able to simulate independent copies

$$X_i(\omega) = \{X_{t,i}(\omega),\ t \in I\} \tag{4.14}$$

of the path of the stochastic process X, then $g(X)$ is just a real-valued random variable and we can define

The (crude) Monte Carlo method for stochastic processes:

Approximate $\mathbb{E}(g(X))$ *by the arithmetic mean* $\frac{1}{N}\sum_{i=1}^{N} g(X_i(\omega))$.

Note that in contrast to the definition of the crude Monte Carlo estimator for real-valued random variables, we now have to use a functional g as otherwise talking of an expectation would make no sense. We thus only have to be able to simulate independent replications of paths of a stochastic process to apply the crude Monte Carlo method. All the properties of the crude Monte Carlo estimator then follow as in the situation of Section 3.2. These include the unbiasedness and strong consistency of the estimator. Also, the central limit theorem yields asymptotic normality of the estimator and can be used for calculating approximate confidence intervals. Of course, the variance reduction methods such as control variate, importance sampling, or stratification now have to be adapted in a suitable way as we will see later.

One could distinguish the Monte Carlo estimation problems by considering the main different types of functionals:

1. If the functional $g(x)$ only depends on the value of the stochastic process X at a particular time, i.e. if we have

$$g(X) = h(X_T) \tag{4.15}$$

for a fixed time T and a real-valued function $h(.)$, then we only have to know the distribution of the stochastic process at time T. For processes where this distribution is explicitly known, the Monte Carlo simulation reduces to a simple one of ordinary random variables, and there is no additional complexity due to the fact that X_T is the result of a stochastic process. If the distribution of X_T is not known explicitly then simulation of the paths of the stochastic process until time T will require discretization methods. We comment on that in later sections.

2. If the functional $g(x)$ depends on the values of the stochastic process X at a finite set of fixed time points $t_1, ..., t_n$, i.e. if we have

$$g(X) = h(X_{t_1}, ..., Xt_n) \tag{4.16}$$

for a real-valued function $h(.)$, then we are again in the situation of the previous chapter. We now have a multidimensional problem as we have to simulate realizations of the vectors $(X_{t_1}, ..., X_{t_n})$ where the components X_{t_i} are not independent. As in the previous chapter, knowledge of their joint distribution is necessary for simulating them.

3. In the general case, i.e. if the functional $g(x)$ cannot be reduced to one of the two foregoing cases, one is often not able to determine the distribution of $g(X)$. This is new compared to the previous chapter, and we thus have to use suitable approximation methods. They are often tailored to the specific problem related to the functional and there is no general method available.

4.3.2 Simulating paths of stochastic processes: Basics

The fact that the random variables $X_t, t \in I$ that constitute a stochastic process are related to each other has the consequence that we cannot just independently simulate random numbers that have the same distribution as the different X_t. We have to take care for the relation among the X_t which can be extremely strong. Just imagine the requirement of continuity of the paths of a stochastic process. Then, it is clear that $X_{t+\epsilon}$ is nearly completely determined by X_t for a small value of ϵ. We will deal with this in greater detail when we look at particular examples of stochastic processes such as Brownian motion or solutions of stochastic differential equations.

To motivate most of what follows with regard to the simulation of the paths of stochastic processes, we have a look at a simple example in Algorithm 4.1.

Algorithm 4.1 Simulation of a discrete-time stochastic process with independent increments

---

Let $\{X_t, t \in \{1, 2, ..., n\}\}$ be a discrete-time stochastic process with independent increments. Let $\mathbb{P}_k$ be the distribution of the k-th increment $X_k - X_{k-1}$ with X_1 being the first increment by setting $X_0 = 0$. Then, we obtain a path $X_{\cdot}(\omega)$ via:

1. Set $X_0(\omega) = 0$.
2. Simulate random numbers $Y_k(\omega), k = 1, ..., n$ with $Y_k \sim \mathbb{P}_k$.
3. Set $X_k(\omega) = X_{k-1}(\omega) + Y_k(\omega)$, $k = 1, ..., n$.

Here, the independent increment assumption allows us to simulate the increments by independent sampling. The value of the stochastic process at the next time instants is obtained by summing up its current value plus the newly simulated increment. The assumptions of this example can be relaxed. Indeed, the simulation scheme for paths of such finite-step stochastic processes goes through without modification if at time-step $k-1$ we know the **conditional distribution** of $X_k - X_{k-1}$. Then, the random numbers Y_k above are sampled from that conditional distribution to obtain the increments.

Simulation of continuous-time stochastic processes

Now what happens if instead of just a finite set $I = \{1, 2, ..., n\}$ we would have an uncountable index set, such as $I = [0, T]$, in particular, if we would have a continuous-time stochastic process? Such an index set does not in general allow a simulation of a path of the corresponding stochastic process at each time t simultaneously. However, one could imitate the above construction on a sufficiently fine grid $0 = t_0 < t_1 < ... < t_n = T$ in $[0, T]$ and then continue the process in a suitable way in between the grid points. In the special case of a continuous stochastic process, it is reasonable to use a linear interpolation between the simulated grid points. This guarantees that we approximate an underlying continuous path by a continuous path. In the case that we know the conditional distribution of the increments of the stochastic process at t_{k-1} given the value $X_{t_{k-1}}$ at this grid point, we can state the simulation Algorithm

Algorithm 4.2 Simulation of a continuous-time stochastic process with continuous paths

Let $0 = t_0 < t_1 < ... < t_n = T$ be a partition of $[0,T]$. Let $\mathbb{P}_k$ denote the conditional distribution of X_{t_k} given $X_{t_{k-1}}$. We obtain a path $X_.(\omega)$ via:

1. Set $X_0(\omega) = 0$.
2. For $k = 1$ to n do
   (a) Simulate a random number $Y_k(\omega)$ with $Y_k \sim \mathbb{P}_k$.
   (b) Set $X_{t_k}(\omega) = X_{t_{k-1}}(\omega) + Y_k(\omega)$.
   (c) Between t_{k-1} and t_k obtain X_t via linear interpolation, i.e. set

$$X_t(\omega) = X_{t_{k-1}}(\omega) + \frac{t - t_{k-1}}{t_k - t_{k-1}} Y_k(\omega), \ t \in (t_{k-1}, t_k).$$

4.2 that generalizes the above one for the finite-set process.

Of course, there remain the questions of how to discretize and of the convergence of the discretized version towards the stochastic process. These will be dealt with in the next sections.

Exact simulation versus approximate simulation.

If we know the distributions of X_0 and of all the increments $X_{t_k} - X_{t_{k-1}}$ (given the past path of the process) then for every t_k we can simulate random variables with exactly the same distribution as X_{t_k}. To do so, simulate X_0 and all the increments $X_{t_j} - X_{t_{j-1}}$ for $j = 1, ..., k$. Adding all those random numbers leads to a sum that has the distribution of X_{t_k}. This is called an **exact simulation** (at the times t_k). We will also encounter situations where we do not know the above distributions of the increments of the stochastic process at time points different from $t = 0$. However, then the dynamics of the evolution of the process over time are typically given by an equation (such as a stochastic differential equation). In such a situation, discretization methods are available that yield an approximation of the stochastic process. While it is usually advisable to use exact simulation if the distributions are known, there can occur situations when the exact simulation is so inefficient that a sufficiently fine discretization method will be preferable.

4.3.3 Variance reduction for stochastic processes

If we can simulate a path of a stochastic process X then we are back at the question of variance reduction of the crude Monte Carlo estimator of $\mathbb{E}(g(X))$. The methods developed in the preceding chapter can be used here, too. We will not go too much into detail as many modifications are specific to the processes and problems dealt with in later sections and chapters. However, here are some simple comments on the different methods.

Control variate techniques

Applications of control variate techniques that we will see, are often of the type to approximate the functional $g(X)$ by a functional $h(X)$ that is simpler to compute than using a different process as a control variate. A typical situation occurs when an infinite-dimensional functional is approximated by a finite-dimensional one of the form

$$\mathbb{E}(g(X)) \approx \mathbb{E}(h(X_{t_1}, ..., X_{t_n})) \tag{4.17}$$

for a suitable function $h(.)$ and suitable times $t_1, ..., t_n$. For computing the expectation of $h(.)$ the unconditional mean control variate would be a possible further variance reducing approximation method.

Stratified sampling

If the value of the expected value $\mathbb{E}(g(X))$ depends strongly on the distribution of the underlying stochastic process X at some particular times $t_1, ..., t_n$, then it might be a good idea to stratify the joint distribution of $(X_{t_1}, ..., X_{t_n})$ if n is not too big.

Importance sampling

Importance sampling is always a good candidate for variance reduction when the functional $g(X)$ is only nonzero if the process X does not leave a specified area O on a given time interval $[0, T]$. Applications of this method are manifold. Two of them will be given for barrier option pricing in Section 5.6.2 and for estimating extreme events in Section 8.2.5.

4.4 Brownian motion and the Brownian bridge

The stochastic process that is the most important building block of financial modelling and that also has applications in various other fields of science is the so-called *Brownian motion* process:

DEFINITION 4.9

(a) The real-valued stochastic process $\{W_t\}_{t\geq 0}$ with continuous paths and the

properties

$$W_0 = 0 \quad \mathbb{P}\text{-a.s.} \tag{4.18}$$

$$W_t - W_s \sim \mathcal{N}(0, t-s) \text{ for } 0 \leq s < t \tag{4.19}$$

$$W_t - W_s \text{ is independent of } W_u - W_r \text{ for } 0 \leq r \leq u \leq s < t \tag{4.20}$$

is called a **one-dimensional Brownian motion**.
(b) An **n-dimensional Brownian motion***is the* $\mathbb{R}^n$*-valued process*

$$W(t) = (W_1(t), ..., W_n(t)) \tag{4.21}$$

with components W_i *being independent one-dimensional Brownian motions.*

In the above definition, we could have relaxed the assumption that the Brownian motion has normally distributed increments by only requiring that it has stationary and independent increments. However, one can show that this together with the assumption of continuous paths implies that Brownian motion has the above normal distribution property. Thus, the above requirement can be made without loss of generality. Further, Brownian motion can be associated with its natural filtration,

$$F_t^W := \sigma\{W_s \mid 0 \leq s \leq t\}, \quad t \in [0, \infty). \tag{4.22}$$

For technical reasons we shall typically work with the $\mathbb{P}$-**augmentation of the natural filtration**,

$$F_t := \sigma\left\{F_t^W \cup N \,|N \in F, \mathbb{P}(N) = 0\right\}, \quad t \in [0, \infty) \tag{4.23}$$

and call it the **Brownian filtration**. This has some technical advantages which are not explained here (see Karatzas and Shreve [1991] or Korn and Korn [2001] for details). In the literature, the requirement of independent increments of a Brownian motion is often stated in a more general way as

$$W_t - W_s \text{ are independent of } F_s \text{ for } 0 \leq s < t \tag{4.24}$$

when the Brownian motion is associated with a given filtration $\{F_t\}_{t\geq 0}$. If $\{F_t\}_{t\geq 0}$ is either the natural filtration or the Brownian filtration then this is equivalent to the original requirement. When in the sequel we consider a Brownian motion $\{(W_t, F_t)\}_{t\geq 0}$ with an arbitrary filtration $\{F_t\}_{t\geq 0}$, we implicitly assume requirement (4.24) to be satisfied.

Correlated Brownian motion and the Cholesky decomposition

Above we have simply defined a multidimensional Brownian motion as a vector of independent one-dimensional Brownian motions. However, for some applications in finance sometimes a *correlated* multidimensional Brownian motion is assumed to be the underlying stochastic process. By this, one understands a Brownian motion where the components have a given correlation structure. We demonstrate that for the purpose of simulation we can always restrict ourselves to the independent case. To see this, consider first the two-dimensional situation. Assume that we are given a two-dimensional, independent Brownian motion $(W_1(t), W_2(t))$. From this, we can obtain a two-dimensional Brownian motion $(\tilde{W}_1(t), \tilde{W}_2(t))$ with

$$\mathbb{C}orr\left(\tilde{W}_1(t), \tilde{W}_2(t)\right) = \rho \tag{4.25}$$

by setting

$$\tilde{W}_1(t) = W_1(t), \quad \tilde{W}_2(t) = \rho W_1(t) + \sqrt{1-\rho^2}W_2(t). \tag{4.26}$$

Obviously, both components have the required correlation. Further, as the sum of two independent normal random variables is again normally distributed (where the means and the variances are simply added), it follows that the second component also has the required distribution. This method of construction can be generalized to the multidimensional setting. Indeed, if we want to generate an n-dimensional Brownian motion $\tilde{W}(t)$ with a given positive definite covariance matrix Σ, then by using its Cholesky decomposition

$$\Sigma = LL' \tag{4.27}$$

(with L a lower triangular matrix with diagonal elements that are equal to the square roots of the eigenvalues of Σ) and setting

$$\tilde{W}(t) = LW(t) \tag{4.28}$$

yields the desired process. L is obtained via Algorithm 2.19. It is usually built-in in standard mathematical software.

4.4.1 Properties of Brownian motion

A (one-dimensional) Brownian motion has remarkable properties. In particular, its paths are extremely irregular (see Karatzas and Shreve [1991] for proofs of most of what follows). More precisely, one can show that we have the following theorem.

THEOREM 4.10
(a) $\mathbb{P}$-almost all paths of the Brownian motion $\{W_t\}_{t\in[0,\infty)}$ are nowhere differentiable as functions of time t.
(b) With the definition of

$$Z_n(\omega) := \sum_{i=1}^{2^n} \left| W_{i/2^n}(\omega) - W_{(i-1)/2^n}(\omega)\right|, \quad n \in \mathbb{N}, \omega \in \Omega, \tag{4.29}$$

we obtain

$$Z_n(\omega) \xrightarrow{n\to\infty} \infty \quad \mathbb{P}\text{-a.s.}\ , \tag{4.30}$$

i.e. the paths $W_t(\omega)$ of the Brownian motion admit infinite variation on the interval $[0,1]$ $\mathbb{P}$-almost surely. Even more: the paths $W_t(\omega)$ have infinite variation on each nonempty interval $[s_1, s_2] \subset [0,\infty)$ $\mathbb{P}$-almost surely.

Before we simulate some paths of a Brownian motion we will show the important martingale property of the process and also introduce a simple generalization, the Brownian motion with drift and volatility.

THEOREM 4.11
(a) A one-dimensional Brownian motion W_t is a martingale.
(b) A **Brownian motion with drift** μ **and volatility** σ *with $\mu, \sigma \in \mathbb{R}$,*

$$X_t := \mu t + \sigma W_t \ , \ t > 0, \tag{4.31}$$

is a martingale if and only if $\mu = 0$, a super-martingale if and only if $\mu \le 0$, and a sub-martingale if and only if $\mu \ge 0$.

REMARK 4.12 1. The above theorem easily follows from the properties of Brownian motion. For part (a), simply use the representation

$$W_t = W_s + (W_t - W_s)\,, \tag{4.32}$$

and the fact that the independent increments have zero mean. The assertion for the Brownian motion with drift μ and volatility σ is a consequence of the linearity of (conditional) expectation. Note also that a Brownian motion X_t with drift μ and volatility σ satisfies $X_t \sim \mathcal{N}\left(\mu, \sigma^2\right)$.

2. Similarly one can show: each stochastic process with independent, centered increments (i.e. the increments have zero mean) is a martingale with respect to its natural filtration.

3. The independence of its increments ensures that a Brownian motion is a Markov process. The independent increments property and the decomposition (4.32) yield the joint distribution of $(W_{t_1}, ..., W_{t_n})$ for $0 < t_1 < .. < t_n$:

$$\begin{pmatrix} W_{t_1} \\ W_{t_2} \\ \vdots \\ W_{t_n} \end{pmatrix} \sim \mathcal{N}\left(\begin{pmatrix} 0 \\ 0 \\ \vdots \\ 0 \end{pmatrix}, \begin{pmatrix} t_1 & t_1 & \dots & t_1 \\ t_1 & t_2 & \dots & t_2 \\ \vdots & \vdots & \dots & \vdots \\ t_1 & t_2 & \dots & t_n \end{pmatrix} \right) \tag{4.33}$$

□

We will see more properties of a Brownian motion when we consider Itô integrals and stochastic differential equations, but will now turn to the simulation of its paths. Although the paths of a Brownian motion are quite irregular, the process itself is relatively easy to simulate as it has independent and stationary increments. As – in the spirit of our general discussion on the simulation of continuous-time stochastic processes – it is impossible to simulate a complete path of a Brownian motion, we can at least simulate it exactly at each point $t \in [0,T]$. So, choosing a sufficiently fine time grid on which we simulate the process exactly and then do a linear interpolation in between is an appropriate approximation. We present this as Algorithm 4.3.

Algorithm 4.3 Simulation of a Brownian motion

Let a partition $0 = t_0 < t_1 < ... < t_n = T$ of $[0,T]$ be given. Then, we obtain a path $W_.(\omega)$ of a one-dimensional Brownian motion via:

1. Set $W_0(\omega) = 0$.
2. For $k = 1$ to n
   (a) Simulate a standard normally distributed random number Z_k.
   (b) Set $W_{t_k}(\omega) = W_{t_{k-1}}(\omega) + \sqrt{t_k - t_{k-1}} Z_k$.
   (c) Between t_{k-1} and t_k obtain W_t via linear interpolation, i.e. for $t \in (t_{k-1}, t_k)$ set

$$W_t(\omega) = W_{t_{k-1}}(\omega) + \frac{t - t_{k-1}}{t_k - t_{k-1}} \left(W_{t_k}(\omega) - W_{t_{k-1}}(\omega) \right).$$

Figure 4.1 shows some simulated paths of a one-dimensional Brownian motion with $n = 500$ equally spaced points t_i. Actually, at this level it is hard to realize that the remaining points are obtained by linear interpolation. Also, the appearance of the paths gives a good impression of their (theoretical) nondifferentiability. According to the properties of Brownian motion, if we would zoom in on the time scale then the zoomed version of the paths would admit exactly the same typical behaviour of nondifferentiability.

REMARK 4.13 A d-dimensional Brownian motion is obtained by simulating d independent one-dimensional Brownian motions. In case a correlated Brownian motion should be simulated, a multiplication of the just generated independent Brownian motion with the triangular matrix L obtained from the Cholesky decomposition of the covariance matrix Σ yields the desired result. To use the above one-dimensional algorithm, only the simulation of the random number Z_k has to be modified. Simply replace it by

- Simulate a random number $Z_k \sim \mathcal{N}(0, I)$

in the standard, uncorrelated case (with I denoting the d-dimensional identity matrix) and by

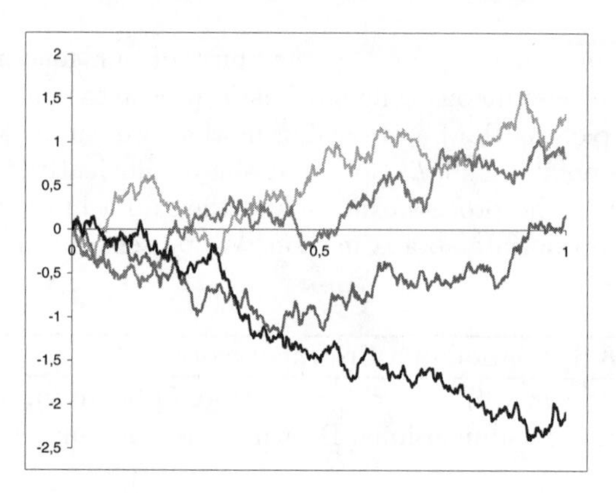

FIGURE 4.1: Simulated paths of a Brownian motion on $[0, 1]$, $n = 500$.

- Simulate a random number $Z_k \sim \mathcal{N}(0, \Sigma)$

in the correlated case (with Σ denoting the given covariance matrix). □

In light of the irregular path behaviour of the Brownian motion, linear interpolation between the time grid points seems to be a crude method. However, we can show convergence of the simulated and linearized Brownian motion to the **real one** if the diameter of the partition $I_n = \{t_0, t_1, ..., t_n\}$,

$$diam\,(I_n) = \sup_{i=1,..,n} \{|t_i - t_{i-1}|\}, \tag{4.34}$$

approaches zero for $n \to \infty$. For this, we first need a process version of weak convergence and a corresponding process central limit theorem, which will be the contents of the following excursion.

4.4.2 Weak convergence and Donsker's theorem

As we would like to have a convergence result that rests on easily checkable assumptions and is also valid for a wide class of approximation approaches for stochastic processes, we introduce the framework of weak convergence on metric spaces. The reader who is not interested in this framework can directly go to the considerations of the consequences of the announced result, Donsker's theorem (see Theorem 4.17).

The theory can be expressed most conveniently if we view a stochastic process $(X_t, F_t)_{t\in[0,T]}$ with continuous paths on $[0, T]$ as a **function-valued random variable**. This interpretation is quite a natural one: the outcome ω of the "experiment" is the corresponding path $X_.(\omega)$ of the stochastic process which assigns a continuous function on $[0, T]$ to each $\omega \in \Omega$. For ease of notation, we will from now on set $T = 1$ and introduce a probability space that supports this interpretation of a stochastic process as a function-valued random variable. More precisely, we look at the probability space $C[0,1]$ of continuous, real-valued functions on $[0,1]$ equipped with the corresponding Borel σ-field $\mathcal{B}(C[0,1])$ and a probability measure $\mathbb{P}$, i.e.

$$(\Omega, F, \mathbb{P}) = (C[0,1], \mathcal{B}(C[0,1]), \mathbb{P}). \tag{4.35}$$

The interpretation of a stochastic process X as a function-valued random variable on $(\Omega, F, \mathbb{P})$ is now realized by defining

$$X(\omega) := \omega \ \forall\ \omega \in C[0,1]. \tag{4.36}$$

To obtain the value $X_t(\omega)$ of the process at time $t \in [0,1]$ we use the projection on the t-th coordinate of ω,

$$X(t,\omega) = \omega(t). \tag{4.37}$$

Recalling equivalent characterizations of convergence in distribution (or **weak convergence**) for random variables and realizing that (Ω, F) is a metric space when endowed with the supremum metric (to ensure that the limit of a sequence with respect to this metric has continuous paths!)

$$\rho(x,y) = \sup_{0 \le t \le 1} |x(t) - y(t),| \tag{4.38}$$

it does not come as a surprise that we use the framework of weak convergence on metric spaces (see Billingsley [1968]).

DEFINITION 4.14
Let $(S, \mathcal{B}(S))$ be a metric space with metric ρ and the Borel-σ-field $\mathcal{B}(S)$ over S. Let further $\mathbb{P}, \mathbb{P}_n$, $n \in \mathbb{N}$ be probability measures on $(S, \mathcal{B}(S))$. Then we say that the sequence $\mathbb{P}_n$ **converges weakly** *towards $\mathbb{P}$ if for each continuous and bounded real valued function f on S we have*

$$\int_S f\, dP_n \xrightarrow{n\to\infty} \int_S f\, dP. \tag{4.39}$$

To translate this definition to the metric space we are interested in, the space of continous stochastic processes, we introduce $C(C[0,1], \mathbb{R})$, the space of uniformly continuous, bounded functionals on $C[0,1]$.

DEFINITION 4.15
Let $X_n = \{X_n(t)\}_{t\in[0,1]}$ be a sequence of continuous stochastic processes. We then say that X_n **converges weakly** *(or* **converges in distribution***) towards the continuous process X if we have*

$$\mathbb{E}(f(X_n)) \xrightarrow{n\to\infty} \mathbb{E}(f(X)) \tag{4.40}$$

for all $f \in C(C[0,1], \mathbb{R})$.

As we have

$$\mathbb{E}(f(X_n)) = \int f \, d\mathbb{P}_n, \quad \mathbb{E}(f(X)) = \int f \, d\mathbb{P}, \tag{4.41}$$

weak convergence of the stochastic processes means weak convergence of the underlying probability distributions $\mathbb{P}_n \to \mathbb{P}$. To show that the usual convergence in distribution of $\mathbb{R}^n$-valued random variables is implied by the convergence in distribution of stochastic processes, we cite the following theorem which is a special case of Theorem 5.1 of Billingsley (1968).

THEOREM 4.16

Let $\mathbb{P}, \mathbb{P}_n$, $n \in \mathbb{N}$ *be probability measures on the metric space* $(S, \mathcal{B}(S))$ *endowed with the metric* ρ. *Further, let* $h : S \to S'$ *be a measurable mapping into a metric space* S' *with metric* ρ' *and Borel-*σ*-field* $\mathcal{B}(S')$. *Let us further assume that the set* D_h *of points of discontinuity of* h *is a zero set, i.e.*

$$\mathbb{P}(D_h) = 0. \tag{4.42}$$

Then convergence in distribution is preserved under the mapping h*:*

$$\mathbb{P}_n \xrightarrow{n\to\infty} \mathbb{P} \text{ in distribution} \Rightarrow \mathbb{P}_n \cdot h^{-1} \xrightarrow{n\to\infty} \mathbb{P} \cdot h^{-1} \text{ in distribution.} \tag{4.43}$$

In particular, continuous mappings preserve convergence in distribution.

Note that $(\mathbb{R}^k, \mathcal{B}(\mathbb{R}^k))$ is also a metric space. Let then X_n, X be real-valued, continuous stochastic processes. For the set of k fixed time points $0 \leq t_1 < ... < t_k \leq 1$ we then obtain from Theorem 4.16:

$$\begin{aligned} &X_n \xrightarrow{n\to\infty} X \text{ in distribution} \\ &\quad \Rightarrow (X_n(t_1), ..., X_n(t_k)) \xrightarrow{n\to\infty} (X(t_1), .., X(t_k)) \text{ in distribution.} \end{aligned} \tag{4.44}$$

Thus, the usual convergence in distribution of $\mathbb{R}^k$-valued random variables is implied by the convergence in distribution of the corresponding stochastic processes. The converse is in general not true. So, it is not obvious that if we choose a sequence I_n of partitions $0 = t_0 < t_1 < ... < t_n = 1$ of $[0,1]$, that then our linear interpolation based-simulation approach of the Brownian motion X_n will indeed converge in distribution towards the Brownian motion $W = \{W_t\}_{t\in[0,1]}$. However, the following theorem implies that our approach indeed leads to the desired weak convergence (see Karatzas and Shreve [1991] or Donsker [1952]).

THEOREM 4.17 Donsker's theorem
Let $\{\xi_n\}_{n\in\mathbb{N}}$ be a sequence of independent and identically distributed random variables with $E(\xi_i) = 0$, $0 < Var(\xi_i) = \sigma^2 < \infty$. Let

$$S_0 = 0, \quad S_n = \sum_{i=1}^{n} \xi_n. \tag{4.45}$$

We construct a sequence X_n of stochastic processes by

$$X_n(t,\omega) = \frac{1}{\sigma\sqrt{n}} S_{[nt]}(\omega) + (nt - [nt]) \frac{1}{\sigma\sqrt{n}} \xi_{[nt]+1}(\omega) \tag{4.46}$$

for $t \in [0,1]$, $n \in \mathbb{N}$. Then this sequence converges weakly towards the one-dimensional Brownian motion $\{W(t)\}_{t\in[0,1]}$, i.e. we have

$$X_n \xrightarrow{n\to\infty} W \text{ in distribution.} \tag{4.47}$$

REMARK 4.18 1. The process X_n constructed in Equation (4.46) can be identified with our linear interpolation-based simulation approach to Brownian motion. Indeed, if we take all ξ_i to be independent and standard normally distributed, then for equally spaced partitions, i.e. $t_k = \frac{k}{n}$, we have

$$X_n\left(\frac{k}{n},\omega\right) = \frac{1}{\sigma\sqrt{n}} S_k(\omega) \sim \mathcal{N}\left(0, \frac{k}{n}\right) \tag{4.48}$$

while for $t \in \left(\frac{k}{n}, \frac{k+1}{n}\right)$ we obtain $X_n(t)$ by linear interpolation. This is exactly our approach specialized to an equally spaced partition. Thus, Donsker's theorem is the theoretical justification of our approach. Furthermore, it is clear that for a multidimensional Brownian motion with independent components Donsker's theorem also justifies the use of the linear interpolation-based approximation. Indeed, as weak convergence is preserved under continuous mappings, we then also have a justification for the linear interpolation-based approach for a general, correlated Brownian motion.

2. The convergence assertion and the limiting distribution in Donsker's theorem are independent of the particular choice of ξ_i. For this reason, the theorem is also called **Donsker's invariance principle**. It can thus be viewed as a **process version** of the central limit theorem. Extending it to arbitrary time intervals $[0,T]$ is only a question of notation.

3. Although we have used normally distributed random variables ξ_i as a basis for constructing the approximating sequence of processes X_n, one could imagine using simpler random variables. A possible suggestion would be

$$\xi_i = Y_i - q \tag{4.49}$$

with $Y_i \sim B(1,q)$, $0 \leq q \leq 1$. Its sum

$$S_n = \sum_{i=1}^{N} \xi_i \tag{4.50}$$

is called a **random walk**, in particular, a symmetric random walk for $q = 1/2$. One can thus think of the Brownian motion as a limit of suitably scaled random walks. Thus, Donsker's theorem also extends the de Moivre-Laplace theorem on the approximation of the normal distribution by binomial ones to the stochastic process setting.

4. One can use Donsker's theorem to obtain approximate distributions for stochastic processes S_n constructed as above when n is large. Donsker's theorem then shows that an appropriately scaled version X_n converges in distribution towards a Brownian motion, i.e. we have an asymptotic normal distribution in each $t \in [0,T]$ for $S_n(t)$. □

Brownian motion has a lot more remarkable properties. Their presentation is way beyond the scope of this book and we refer the interested reader to the monographs by Hida (1980) and Karatzas and Shreve (1991) for more on this topic. Here, we will state the following result, the law of iterated logarithm for the Brownian path by Hincin (1933) and a useful corollary.

THEOREM 4.19

Let $\{W_t\}_{t\geq 0}$ be a one-dimensional Brownian motion. Then, for $\mathbb{P}$-almost all $\omega \in \Omega$ we have:

$$\limsup_{t\to\infty} \frac{W_t(\omega)}{\sqrt{2t\, log\,(log\,(t))}} = 1, \tag{4.51}$$

$$\liminf_{t\to\infty} \frac{W_t(\omega)}{\sqrt{2t\, log\,(log\,(t))}} = -1 \tag{4.52}$$

COROLLARY 4.20

Let $X_t = \mu \cdot t + \sigma W_t$, $t \geq 0$ be a Brownian motion with drift μ and volatility σ. We then have

$$lim_{t\to\infty} \frac{X_t}{t} = \mu \;\; \mathbb{P}\text{-a.s.} \tag{4.53}$$

This corollary in particular tells us that in a Brownian motion with drift, the drift asymptotically dominates the fluctuations of the Brownian motion.

4.4.3 Brownian bridge

Here, we will get back to the question whether there are alternatives to linear interpolation between two time points t_{k-1} and t_k of a Brownian motion.

Consider the situation when we cannot observe the Brownian motion between these two points but can observe its values $W_{t_{k-1}}$ and W_{t_k}. Is it possible to plug in a suitable stochastic process that can be used as an interpolation from $W_{t_{k-1}}$ and W_{t_k} and that is similar to the unobserved Brownian motion? To answer this, we first give the following definition.

DEFINITION 4.21

Let $\{W_t\}_{t\in[0,T]}$ be a one-dimensional Brownian motion, let $a, b \in \mathbb{R}$ be two real numbers. Then, the process

$$B_t^{a,b} = a\frac{T-t}{T} + b\frac{t}{T} + \left(W_t - \frac{t}{T}W_T\right), \quad t \in [0,T] \tag{4.54}$$

is called a **Brownian bridge** *from a to b.*

Obviously, the process $B_t^{a,b}$ starts in a at time $t = 0$ and ends in b at time T. Some of its simulated paths are given in Figure 4.2. One can derive its

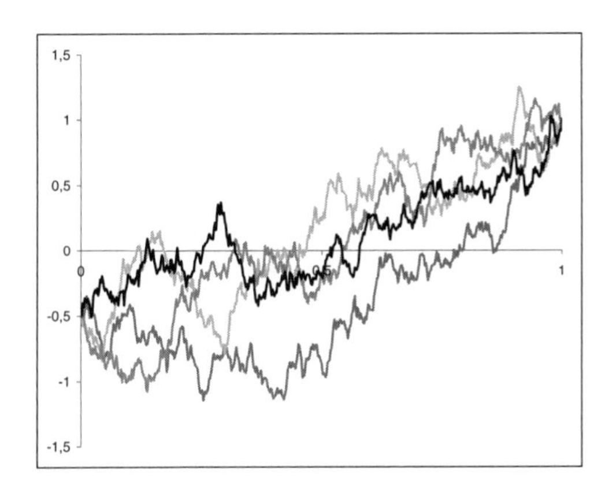

FIGURE 4.2: Simulated paths of a Brownian bridge from -0.5 to 1.

distribution by the independent increments property of Brownian motion.

PROPOSITION 4.22

A Brownian bridge from a to b satisfies

$$B_t^{a,b} \sim \mathcal{N}\left(a + \frac{t}{T}(b-a), t - \frac{t^2}{T}\right) \tag{4.55}$$

For a d-dimensional Brownian motion (with independent components) a definition of a **d-dimensional Brownian bridge** from a to b with $a, b \in \mathbb{R}^d$

is totally similar to the one in Equation (4.54). Also, a generalization of the preceding proposition is straightforward.

Before we describe how to simulate a path of a Brownian bridge, we want to relate it to a Brownian motion that is conditioned to pass through some given values. For this, we need the following formula about the conditional distribution in a multivariate normal setting.

PROPOSITION 4.23 Normal conditional distribution formula
Let $Z = \left(Z^{(1)}, ..., Z^{(d)}\right)$ be a d-dimensional random vector with $Z \sim \mathcal{N}(\mu, \Sigma)$. We partition Z into its first d_1 components X and the remaining $d - d_1$ components Y. Then, with the notation

$$\begin{pmatrix} X \\ Y \end{pmatrix} \sim \mathcal{N}\left(\begin{pmatrix} \mu_X \\ \mu_Y \end{pmatrix}, \begin{pmatrix} \Sigma_X & \Sigma_{XY} \\ \Sigma_{YX} & \Sigma_Y \end{pmatrix}\right) \tag{4.56}$$

and assuming that Σ_Y^{-1} exists, we have that the conditional distribution of $X\,|Y = y$ is d_1-dimensional normal with

$$X\,|Y = y \sim \mathcal{N}\left(\mu_X + \Sigma_{XY}\Sigma_Y^{-1}(y - \mu_Y), \Sigma_X - \Sigma_{XY}\Sigma_Y^{-1}\Sigma_{YX}\right). \tag{4.57}$$

By noting that we have

$$\begin{pmatrix} W_t \\ W_T \end{pmatrix} \sim \mathcal{N}\left(\begin{pmatrix} 0 \\ 0 \end{pmatrix}, \begin{pmatrix} t & t \\ t & T \end{pmatrix}\right), \tag{4.58}$$

a direct consequence of this proposition is that we have

$$W_t\,|W_T = b \sim \mathcal{N}\left(b\frac{t}{T}, t - \frac{t^2}{T}\right), \tag{4.59}$$

i.e. with the help of Proposition 4.22 we obtain that a Brownian bridge from 0 to b is nothing else than a **Brownian motion conditioned to arrive at b at time T**. By introducing a **Brownian motion starting at a** via $W_0^a = a$, we have $W_t^a \sim \mathcal{N}(a, t)$ and thus

$$W_t^a\,|W_T^a = b \sim \mathcal{N}\left(a + (b - a)\frac{t}{T}, t - \frac{t^2}{T}\right). \tag{4.60}$$

This means that a Brownian bridge from a to b is a Brownian motion starting at a conditioned to arrive at b at time T.

We have thus found a suitable process for our interpolation problem for an unobserved period between two known values of a Brownian motion: Simply use a Brownian bridge that starts at time t_{k-1} in $W_{t_{k-1}}$ and ends at time t_k in W_{t_k}. Of course, we then have to replace the total running time T of the

Algorithm 4.4 Forward simulation of a Brownian bridge

1. Simulate a path of a Brownian motion $W_t(\omega)$ on $[0,T]$.
2. Set $B_t^{a,b} = a\frac{T-t}{T} + b\frac{t}{T} + \left(W_t - \frac{t}{T}W_T\right) \quad \forall\, t \in [0,T]$.

Brownian bridge above by $t_k - t_{k-1}$. What remains is to present a simulation algorithm for a Brownian bridge. There are at least two techniques. Algorithm 4.4 simply simulates a path of a Brownian motion and then translates it into a Brownian bridge from a to b by suitably adding the remaining parts.

While this algorithm is easy to programme and straightforward to understand, it is the second construction method that can be used as an ingredient of variance reduction techniques in connection with Brownian motion. In contrast to the forward simulation method that starts at time 0, the second method starts by sampling W_T first and then constructs the remaining points between $W_0 = 0$ and W_T by sampling a finite set of points W_t followed by linear interpolation between them. For this, we need the conditional distribution of W_t given W_u and W_s. We obtain it as an application of Proposition 4.23 (simply use the joint distribution of (W_s, W_t, W_u)).

PROPOSITION 4.24

Let W be a one-dimensional Brownian motion, $a, b \in \mathbb{R}$, $0 < s < t < u$. Then, the conditional distribution of W_t given (W_u, W_s) is given by

$$W_t\,|(W_u = b, W_s = a) \sim \mathcal{N}\left(\tfrac{(u-t)a+(t-s)b}{u-s}, \tfrac{(u-t)(t-s)}{u-s}\right). \tag{4.61}$$

The backward simulation approach identifies a Brownian bridge as a Brownian motion starting in a and conditioned to end in b. Assume that we have constructed the k points $W_0, W_{t_1}, ..., W_{t_{k-2}}, W_T$ and that we want to fill in a further point at time s with $t_i < s < t_{i+1}$. If we assume that the set $t_1, ..., t_{k-2}$ is increasing, then due to the Markov property and the independent increments property of Brownian motion we have

$$\begin{aligned} W_s \,\big|\big(W_0 = a, ..., W_{t_i} = x, W_{t_{i+1}} = y, ..., W_T = b\big) \\ \sim W_s \,\big|\big(W_{t_i} = x, W_{t_{i+1}} = y\big). \end{aligned} \tag{4.62}$$

The conditional distribution on the right side is given by Proposition 4.24 as:

$$\begin{aligned} W_s \,\big|\big(W_{t_{i+1}} = y, W_{t_i} = x\big) \\ \sim \mathcal{N}\left(\frac{(t_{i+1}-s)\,x + (s-t_i)\,y}{t_{i+1}-t_i}, \frac{(t_{i+1}-s)\,(t_{i+1}-t_i)}{t_{i+1}-s}\right). \end{aligned} \tag{4.63}$$

Equipped with these results we can set up a simulation algorithm given we have fixed the rule to choose the time points where we do a conditional simulation. A particularly suitable strategy for practical purposes is the so-called **dyadic partition** of $[0,T]$. It is built on a successive halving of the grid size. More precisely, one first starts with T (which we call level 0), then continues with $T/2$, then $T/4, 3T/4$ followed by $T/8, 3T/8, 5T/8, 7T/8$ and so on. Hence, at level k we are in the situation that all the values of the Brownian bridge at times $jT/(2^k)$, $j = 0, 1, ..., 2^k$ are generated. As we do not generate points twice, only the values $(2j-1)/(2^k)$ are generated at level k of the iteration. Thus, in this special case, it follows from relation (4.63) that for all points which we have to generate at level k, we have:

$$W_s \,|\, \big(W_{t_{i+1}} = y, W_{t_i} = x\big) \sim \mathcal{N}\left(\frac{x+y}{2}, \frac{T}{2^{k+1}}\right) \tag{4.64}$$

with $s = (2j-1)T/(2^k)$ for a suitable index j and $t_i = 2(j-1)T/(2^k)$, $t_{i+1} = 2jT/(2^k)$. As expected, the mean is the mean of the already simulated neighbourhood values. The variance is just half of the length of the distance of the new time point to the neighbourhood points. In this special case, the corresponding Algorithm 4.5 has a particularly simple form.

Algorithm 4.5 Backward simulation of a Brownian bridge from a to b with $n = 2^K$ time points

---

1. Simulate a standard normally distributed random variable Z and set $W_T = \sqrt{T}Z$. Further set $W_0 = 0, h = T$.
2. For $k = 1$ to K do
   (a) Set $h = h/2$.
   (b) For $j = 1$ to 2^{k-1} do
      i. Simulate a standard normally distributed random variable Z and set
      ii. $W_{(2j-1)h} = \frac{1}{2}\left(W_{2(j-1)h} + W_{2jh}\right) + \sqrt{h}Z$.

REMARK 4.25 1. We can simulate a d-dimensional Brownian bridge in exactly the same way as in the above algorithm. The basis is the multidimensional version of Proposition 4.24:

$$W_t \,|\, (W_u = b, W_s = a) \sim \mathcal{N}\left(\frac{(u-t)\,a + (t-s)\,b}{u-s}, \frac{(u-t)\,(t-s)}{u-s} I_d\right) \tag{4.65}$$

with $W(.)$ a d-dimensional Brownian motion, $a, b \in \mathbb{R}^d$, $0 < s < t < u$, and I_d the d-dimensional identity matrix. In the case of a Brownian motion with covariance matrix Σ we simply have to replace I_d above by Σ. The only change in the above algorithm is to replace the standard normally distributed random variable Z in Step 2 (b) by a multivariate normally distributed random variable $Z \sim \mathcal{N}(0, I_d)$, respectively $Z \sim \mathcal{N}(0, \Sigma)$.

2. Of course, we can also simulate a Brownian bridge with a different sequence of points t_i than the above dyadic partition. Then, the relation defining the "new" value of the Brownian bridge in the algorithm has to be based on the more general conditional expectation representation (4.63) than on its special case (4.64). □

Variance reduction with the Brownian bridge construction

The Brownian bridge construction allows us to think about the methods of stratified sampling, conditional sampling, or importance sampling. To illustrate this, we consider the simplest case of variance reduction for the final value W_T of a Brownian motion by the two-step procedure of Algorithm 4.6.

Algorithm 4.6 Variance reduction by Brownian bridge

For $i = 1$ to N:

1. Apply a variance reduction method to W_T that results in the simulation of a realization $W_{T,i}(\omega) = z_i$.
2. Simulate a path $W_{.,i}(\omega)$ of the Brownian motion by simulating a Brownian bridge from 0 to z_i on $[0, T]$.

With this algorithm it is simple to simulate a Brownian path that ends in a given region. One can even stratify the terminal wealth of the Brownian motion via stratifying the uniform random numbers that are then transformed into the normal distribution of W_T.

REMARK 4.26 As the first elements of a sequence of quasirandom numbers are usually more evenly distributed than the later ones, it would be good – in terms of variance reduction – to use them for simulating the *important points* of a path of a stochastic process. Indeed, the Brownian bridge backward sampling method ensures this. It first samples the value W_T which has the highest variance of all the W_t. Then, it generates the middle point $W_{T/2}$ which has the highest (conditional) variance of the remaining points, and so on. Thus, using the backward simulation fits perfectly to using quasirandom number sequences. This is particularly true for the multidimensional case. □

4.5 Basics of Itô calculus

4.5.1 The Itô integral

For analyzing the performance of simulation methods for functionals of Brownian motion $g(W_t, t \in [0,T])$, we need to be able to work analytically with those functionals. For this purpose and for modelling purposes in financial and actuarial mathematics we introduce the so-called **Itô calculus** (see Karatzas and Shreve [1991] for a detailed introduction). It is based on the Itô integral and its properties. Before we develop it, we would like to point out that due to the properties of the Brownian paths (especially their nondifferentiability and infinite variation), a definition of an integral of the form

$$\int_0^t X_s(\omega)\, dW_s(\omega) \tag{4.66}$$

with a path of a Brownian motion as integrator does not exist in the usual sense of a Lebesgue-Stieltjes integral. To show that there is a reasonable way to define an integral $\int X dW$, we collect the main ideas of Itô's approach.

First, for $X_s \equiv 1$ the integral should just be the Brownian motion itself. The following steps are:

1. Introduce $\int X dW$ for suitable simple integrands in a direct way.
2. Show that more general integrands X can be approximated by a sequence of simple integrands X_n in a suitable norm.
3. Show that one can define the integral $\int X dW$ as a certain limit of the integrals $\int X_n dW$.

As understanding the main steps of the construction of the Itô integral is important for developing corresponding simulation approaches, we spend some time on presenting these steps in some detail.

We start with the class of **simple processes** and their Itô integral (we refer to Korn and Korn [2001] for the proofs of all results cited in this section). A simple process has paths of step functions where the jump times are all fixed, but the jump heights are random (see Figure 4.3).

DEFINITION 4.27
Let $\{(W_t, F_t) \mid t \in [0,T]\}$ be a one-dimensional Brownian motion on a probability space $(\Omega, F, \mathbb{P})$.
(a) A stochastic process $\{X_t\}_{t\in[0,T]}$ is called a **simple process** *if there exist real numbers $0 = t_0 < t_1 < ... < t_p = T$, $p \in \mathbb{N}$ and bounded random variables $\Phi_i : \Omega \to \mathbb{R}$, $i = 0, 1, ..., p$, with*

$$\Phi_0 \text{ is } F_0\text{-measurable}, \; \Phi_i \text{ is } F_{t_{i-1}}\text{-measurable}, \; i = 1, ..., p, \tag{4.67}$$

such that for each $\omega \in \Omega$ $X_t(\omega)$ has the following representation

$$X_t(\omega) = X(t,\omega) = \Phi_0(\omega) \cdot 1_{\{0\}}(t) + \sum_{i=1}^{p} \Phi_i(\omega) \cdot 1_{(t_{i-1},t_i]}(t). \tag{4.68}$$

(b) For a simple process $\{X_t\}_{t\in[0,T]}$ and $t \in (t_k, t_{k+1}]$ the **stochastic integral** *or* **Itô integral** *$I.(X)$ is defined by*

$$I_t(X) := \int_0^t X_s \, dW_s := \sum_{1\le i\le k} \Phi_i \left(W_{t_i} - W_{t_{i-1}}\right) + \Phi_{k+1}\left(W_t - W_{t_k}\right), \tag{4.69}$$

or more general for $t \in [0,T]$:

$$I_t(X) := \int_0^t X_s \, dW_s := \sum_{1\le i\le p} \Phi_i \left(W_{t_i\wedge t} - W_{t_{i-1}\wedge t}\right). \tag{4.70}$$

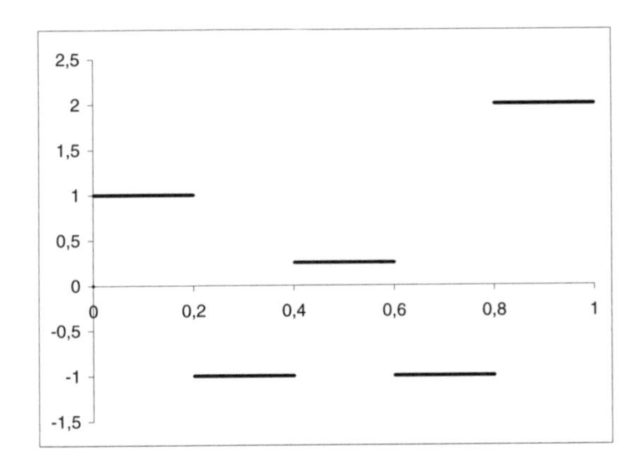

FIGURE 4.3: A path of a simple stochastic process.

REMARK 4.28 1. Note that X_t is $F_{t_{i-1}}$-measurable for all $t \in (t_{i-1}, t_i]$. Further, the paths $X(.,\omega)$ of the simple process X_t are left-continuous step functions with height $\Phi_i(\omega) \cdot 1_{(t_{i-1},t_i]}(t)$. This implies that the Itô integral retains the measurability properties of the Brownian motion.

2. On each interval where X is constant, the increments of the Brownian motion on that interval are multiplied with the corresponding value of X_t, namely Φ_i, to obtain the value of the corresponding Itô integral. Compare

this with the Lebesgue-Stieltjes integral for simple functions. Figure 4.4 illustrates that an Itô integral of a simple process has irregular paths although the integrand is simple. Here, we have plotted the integrand, the underlying Brownian motion, and the resulting stochastic integral in one diagramme. It is instructive to note the behaviour of the stochastic integral in dependence of both ingredients, the integrand and the Brownian path.

3. Assume that the simple process X is deterministic, i.e. its jump heights are all constant. Note then that by the construction of the Itô integral for simple processes and the properties of Brownian motion, we have:

$$I_t(X) \sim \mathcal{N}\left(0, \int_0^t X_s^2 \, ds\right). \tag{4.71}$$

We will sometimes make use of this property. □

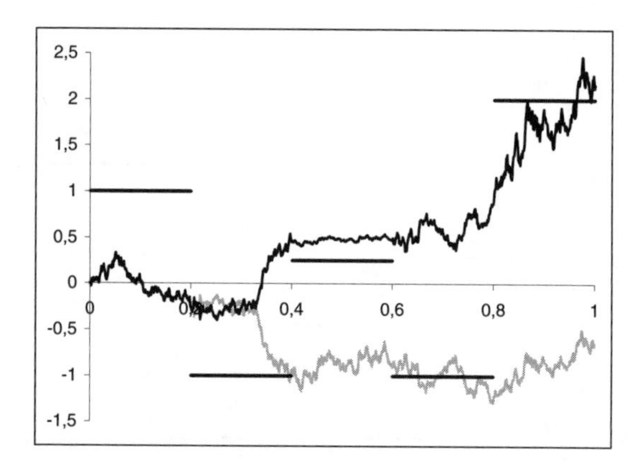

FIGURE 4.4: A simple process (step function), a path of a Brownian motion (grey), and their corresponding Itô integral.

The main properties of the above defined Itô integral are summarized in the following theorem.

THEOREM 4.29 Elementary properties of the stochastic integral
Let X be a simple process. Then we have:
(a) $\{(I_t(X), F_t)\}_{t\in[0,T]}$ is a continuous martingale. In particular, we have

$$\mathbb{E}(I_t(X)) = 0 \text{ for } t \in [0,T]. \tag{4.72}$$

(b) The variance of the Itô integral is finite with

$$\mathbb{E}\left(\int_0^t X_s \, dW_s\right)^2 = \mathbb{E}\left(\int_0^t X_s^2 \, ds\right) \text{ for } t \in [0,T]. \tag{4.73}$$

The theorem says that the Itô integral preserves the continuity and the martingale property of the underlying Brownian motion. The variance formula in the second part will be used to extend the integral to more general integrands with the help of results from L^2-theory. To introduce this class of stochastic processes, we need the concept of a progressively measurable stochastic process. Although this notion of measurability is very technical, one can always keep in mind that all processes with left-continuous (right-continuous) paths are progressively measurable.

DEFINITION 4.30

Let $\{(X_t, \mathcal{G}_t)\}_{t\in[0,\infty)}$ be a stochastic process. It will be called **progressively measurable** *if for all $t \geq 0$ the mapping*

$$[0,t] \times \Omega \to \mathbb{R}^n, \quad (s,\omega) \mapsto X_s(\omega) \tag{4.74}$$

is $\mathcal{B}([0,t]) \otimes \mathcal{G}_t$-$\mathcal{B}(\mathbf{R}^n)$-measurable.

Heuristically, an integrand that is progressively measurable with respect to the filtration $\{F_t\}_{t\geq 0}$, which corresponds to the Brownian motion, does not depend on the increments of the Brownian motion $W_{t+\epsilon} - W_t$ for all $\epsilon > 0$. Equipped with this notion of measurability, we introduce the following class of stochastic processes where $(\Omega, F, \mathbb{P})$ is a probability space on which a Brownian motion $(W_t, F_t)_{t\in[0,T]}$ is defined):

$$\begin{aligned} L^2[0,T] &:= L^2\left([0,T], \Omega, F, \{F_t\}_{t\in[0,T]}, \mathbb{P}\right) \\ &:= \Big\{\{(X_t, F_t)\}_{t\in[0,T]} \text{ real-valued stochastic process } | \\ &\qquad X \text{ progressively measurable, } \mathbb{E}\left(\int_0^T X_t^2 \, dt\right) < \infty\Big\} \end{aligned} \tag{4.75}$$

It can be shown that all those processes $X \in L^2[0,T]$ can be approximated by a sequence $X^{(n)}$ of simple processes such that we have

$$\lim_{n\to\infty} E \int_0^T \left(X_s - X_s^{(n)}\right)^2 ds = 0. \tag{4.76}$$

We can then define a sequence of Itô integrals $I.(X^{(n)})$ for those simple processes which converges with respect to the norm given by

$$\left\| I_{\cdot}\left(X^{(n)}\right)\right\|_{L^T}^2 := E\left(\int_0^T X_s \, dW_s\right)^2 . \tag{4.77}$$

The limit process Z of this sequence is called the **Itô integral**. We set

$$I_t(X) := \int_0^t X_s \, dW_s := Z_t. \tag{4.78}$$

It can be shown that this definition is independent of the choice of the approximating sequence $X^{(n)}$ of simple processes. If the integrand X is a simple process then the definition of the Itô just given coincides with the one for simple processes. The integral retains all the important properties of the stochastic integral, i.e. we have:

THEOREM 4.31
The Itô integral $(I_t(X), F_t)_{t\in[0,T]}$ as defined in Equation (4.78) is a continuous martingale with

$$I_t(aX + bY) = aI_t(X) + bI_t(Y) \text{ for } a, b \in \mathbb{R},\ X, Y \in L^2[0,T], \tag{4.79}$$

$$\mathbb{E}(I_t(X)) = 0\ ,\ \ \mathbb{E}\left(I_t(X)^2\right) = \mathbb{E}\left(\int_0^t X_s^2 \, ds\right). \tag{4.80}$$

A multidimensional version of the just defined stochastic integral can be obtained by a suitable reduction to a component-wise definition:

DEFINITION 4.32
Let $\{(W(t), F_t)\}_{t\in[0,T]}$ be an m-dimensional Brownian motion with components $W_i(t), i = 1,..,m$. Let $\{(X(t), F_t)\}_t$ be an $\mathbb{R}^{n,m}$-valued progressively measurable process with each component $X_{ij} \in L^2[0,T]$. Then the **multidimensional Itô integral** *of X with respect to W is defined by*

$$\int_0^t X(s)\, dW(s) := \begin{pmatrix} \sum_{j=1}^m \int_0^t X_{1j}(s)\, dW_j(s) \\ \vdots \\ \sum_{j=1}^m \int_0^t X_{nj}(s)\, dW_j(s) \end{pmatrix},\ t \in [0,T], \tag{4.81}$$

where all single integrals inside the sums of the right-hand side are one-dimensional Itô integrals as defined in Equation (4.78).

Note that by this definition all components of the multidimensional Itô integral are again martingales.

For applications in finance we need a further extension of the Itô integral to processes that are limits of $L^2[0,T]$-processes. We give it in the one-dimensional case (this extends to the multidimensional one as done above) and introduce:

$$\begin{aligned} H^2[0,T] &:= H^2\left([0,T], \Omega, F, \{F_t\}_{t\in[0,T]}, P\right) \\ &:= \Big\{ (X_t, F_t)_{t\in[0,T]} \text{ real-valued stochastic process } \Big| \\ &\qquad X \text{ progressively measurable, } \int_0^T X_t^2 \, dt < \infty \text{ a.s. } P \Big\} \end{aligned} \tag{4.82}$$

By introducing a sequence of stopping times τ_n, $n \in \mathbb{N}$, given by

$$\tau_n(\omega) := T \wedge \inf\left\{ 0 \le t \le T \,\middle|\, \int_0^t X_s^2(\omega)\, ds \ge n \right\}, \tag{4.83}$$

we define the sequence $X^{(n)}$ of stopped processes via

$$X_t^{(n)}(\omega) := X_t(\omega) \cdot 1_{\{\tau_n(\omega) \ge t\}}. \tag{4.84}$$

As the processes $X^{(n)}$, $n \in \mathbb{N}$ are members of $L^2[0,T]$, their stochastic integral is given by the definition in Equation (4.78). We use this to define the stochastic integral $I(X)$ for processes $X \in H^2[0,T]$ by

$$I_t(X) := I_t\left(X^{(n)}\right) \text{ for } 0 \le t \le \tau_n. \tag{4.85}$$

This is well-defined as we have

$$\tau_n \xrightarrow{n\to\infty} +\infty \ \mathbb{P}\text{-a.s.} \tag{4.86}$$

for $X \in H^2[0,T]$. It can be shown that the above definition is consistent for different members τ_n and τ_m of this sequence of stopping times. This so-defined stochastic integral is still linear and possesses continuous paths. However, as the expected value and the variance of the process X is only finite among the above introduced sequence of stopping times, we do in general not have the martingale property and the variance formula for the Itô integral of a process $X \in H^2[0,T]$. That these properties are satisfied along a sequence of stopping times ensures that the Itô integral is a (continuous) local martingale as defined below:

DEFINITION 4.33

A stochastic process $\{(X_t, F_t)\}_{t\ge 0}$ is called a **local martingale** *if there exists a sequence of stopping times $\tau_n, n \in \mathbb{N}$ with*

$$\tau_n(\omega) \overset{n\to\infty}{\to} \infty \tag{4.87}$$

for $\mathbb{P}$-almost all $\omega \in \Omega$ such that the stopped processes $\left\{\left(\hat{X}_t^{(n)}, F_t\right)\right\}_{t\geq 0}$ defined by

$$\hat{X}^{(n)}(\omega) = X_{t\wedge\tau_n(\omega)}(\omega) \tag{4.88}$$

are all martingales. Such a sequence of stopping times is called a **localizing sequence**.

With this generalization of the Itô integral we introduce the class of Itô processes which simply consist of a sum of a Lebesgue and an Itô integral and which will play the central role for the rest of the chapter.

DEFINITION 4.34
Let $\{(W(t), F_t)\}_{t\in[0,\infty)}$ be an m-dimensional Brownian motion, $m \in \mathbb{N}$.
(a) A stochastic process $\{(X(t), F_t)\}_{t\in[0,\infty)}$ of the form

$$X(t) = X(0) + \int_0^t K(s)\,ds + \sum_{j=1}^m \int_0^t H_j(s)\,dW_j(s) \tag{4.89}$$

with $X(0)$ F_0-measurable, $\{K(t)\}_{t\in[0,\infty)}$ and $\{H(t)\}_{t\in[0,\infty)}$ progressively measurable processes with

$$\int_0^t |K(s)|\;ds < \infty, \qquad \int_0^t H_i^2(s)\,ds < \infty \text{ } \mathbb{P}\text{-a.s.} \tag{4.90}$$

for all $t \geq 0$, $i = 1, ..., m$ is called a **real-valued Itô process**.
*(b) An n-***dimensional Itô process** *$X = (X^{(1)}, ..., X^{(n)})$ consists of a vector of components being real-valued Itô processes.*

Before we state the main result of Itô calculus, the Itô formula in its different versions, we introduce the notions of quadratic variation and covariation.

DEFINITION 4.35
Let X and Y be two real-valued Itô processes with representations

$$X(t) = X(0) + \int_0^t K(s)\;ds + \int_0^t H(s)\;dW(s), \tag{4.91}$$

$$Y(t) = Y(0) + \int_0^t L(s)\;ds + \int_0^t M(s)\;dW(s). \tag{4.92}$$

Then, the **quadratic covariation** *of X and Y is defined by*

$$\langle X, Y\rangle_t := \sum_{i=1}^m \int_0^t H_i(s)\cdot M_i(s)\;ds. \tag{4.93}$$

The special case of $\langle X\rangle_t := \langle X,X\rangle_t$ is called the **quadratic variation** *of X.*

The name **quadratic variation** can also be justified by the fact that indeed the pathwise quadratic variation from calculus coincides with the above defined quadratic variation in the case of Itô processes.

We further introduce the integral of a real-valued, progressively measurable process Y with respect to the one-dimensional Itô process X via

$$\int_0^t Y(s)\,dX(s) := \int_0^t Y(s)\,K(s)\,ds + \int_0^t Y(s)\,H(s)\,dW(s) \tag{4.94}$$

if all the integrals on the right-hand side are defined.

4.5.2 The Itô formula

Now we have put all ingredients together to state the Itô formula in different versions. We start with the one-dimensional case below.

THEOREM 4.36 One-dimensional Itô formula
Let W be a one-dimensional Brownian motion, X a real-valued Itô process with representation

$$X_t = X_0 + \int_0^t K_s\,ds + \int_0^t H_s\,dW_s. \tag{4.95}$$

Let $f:\mathbb{R}\to\mathbb{R}$ be a C^2-function. Then, for all $t\geq 0$ we have

$$\begin{aligned} f(X_t) &= f(X_0) + \int_0^t f'(X_s)\;dX_s + \tfrac{1}{2}\cdot\int_0^t f''(X_s)\,d\langle X\rangle_s \\ &= f(X_0) + \int_0^t \left(f'(X_s)\cdot K_s + \tfrac{1}{2}\cdot f''(X_s)\cdot H_s^2\right)\,ds \\ &\qquad + \int_0^t f'(X_s)\,H_s\,dW_s \quad \mathbb{P}\text{-a.s.}\,. \end{aligned} \tag{4.96}$$

In particular, $f(X_t)$ is again an Itô process and all integrals appearing above are defined.

REMARK 4.37 To state Itô's formula it is notationally convenient to use the symbolic differential notation

$$df(X_t) = f'(X_t)\;dX_t + \tfrac{1}{2}\cdot f''(X_t)\;d\langle X\rangle_t\,. \tag{4.97}$$

However, one should remember that this does not indicate that an Itô process has differentiable paths. □

REMARK 4.38 Although the concept of Itô calculus is highly technical, it is easy to work with the Itô formula. We highlight this by some simple examples and encourage the reader to do some more of this kind.

1. Using the process $X_t = t$ we can deduce from Itô's formula that we have

$$f(t) = f(0) + \int_0^t f'(s)\, ds \tag{4.98}$$

for a (twice) continuously differentiable function f. Hence, the fundamental theorem of calculus can be regarded as a special case of Itô's formula.

2. If we use the (still deterministic) process $X_t = h(t)$ with a C^1-function h, then application of Itô's formula leads to the well-known chain rule:

$$(f \circ h)(t) = (f \circ h)(0) + \int_0^t f'(h(s)) \cdot h'(s)\, ds \tag{4.99}$$

3. Finally, with $X_t = W_t$ and the choice of $f(x) = x^2$, we obtain

$$W_t^2 = \int_0^t 2 \cdot W_s\, dW_s + \tfrac{1}{2} \cdot \int_0^t 2\, ds = 2 \cdot \int_0^t W_s\, dW_s + t \tag{4.100}$$

by the Itô formula. The additional term "t" on the right-hand side of the equation has its origin in the nonvanishing quadratic variation of W_t. □

THEOREM 4.39 Multidimensional Itô formula

Let $X(t) = (X_1(t), ..., X_n(t))$ be an n-dimensional Itô process with

$$X_i(t) = X_i(0) + \int_0^t K_i(s) ds + \sum_{j=1}^{m} \int_0^t H_{ij}(s)\, dW_j(s),\ i = 1, ..., n, \tag{4.101}$$

and $W(t)$ an m-dimensional Brownian motion. Let further $f : [0, \infty) \times \mathbb{R}^n \to \mathbb{R}$ be a $C^{1,2}$-function, i.e. f is continuous, continuously differentiable with respect to the first variable (time) and twice continuously differentiable with *respect to the last n variables (space). We then have*

$$f(t, X_1(t), ..., X_n(t)) = f(0, X_1(0), ..., X_n(0)) + \int_0^t f_t(s, X_1(s), ..., X_n(s))\, ds + \sum_{i=1}^n \int_0^t f_{x_i}(s, X_1(s), ..., X_n(s)) dX_i(s) + \frac{1}{2}\sum_{i,j=1}^n \int_0^t f_{x_i x_j}(s, X_1(s), ..., X_n(s))\, d\langle X_i, X_j\rangle_s. \quad (4.102)$$

Finally, we state the very useful product rule as a corollary to the multidimensional Itô formula:

COROLLARY 4.40 Product rule or partial integration
Let X_t, Y_t be one-dimensional Itô processes with representations

$$X_t = X_0 + \int_0^t K_s ds + \int_0^t H_s dW_s,\ Y_t = Y_0 + \int_0^t \mu_s ds + \int_0^t \sigma_s dW_s. \quad (4.103)$$

Then we have

$$X_t \cdot Y_t = X_0 \cdot Y_0 + \int_0^t X_s dY_s + \int_0^t Y_s dX_s + \int_0^t d\langle X, Y\rangle_s$$
$$= X_0 \cdot Y_0 + \int_0^t (X_s\mu_s + Y_s K_s + H_s\sigma_s)\, ds + \int_0^t (X_s\sigma_s + Y_s H_s)\, dW_s. \quad (4.104)$$

4.5.3 Martingale representation and change of measure

In this section we present two more celebrated results on properties of the Brownian motion which are related to Itô integrals (see Korn and Korn [2001] for their proofs). The first one states that every so-called Brownian martingale can be represented as an Itô integral with respect to the corresponding Brownian motion.

DEFINITION 4.41
Let $(\Omega, F, \mathbb{P})$ be a probability space, $(W_t, F_t)_{t\geq 0}$ a one-dimensional Brownian motion defined on this space. Every real-valued martingale $(M_t, F_t)_{t\geq 0}$ with respect to the Brownian filtration F_t is called a **Brownian martingale**.

THEOREM 4.42 **Itô's martingale representation theorem**
Let M be a Brownian martingale which in addition satisfies

$$\mathbb{E}\left(M_t^2\right) < \infty \text{ for all } t \geq 0. \tag{4.105}$$

Then, there exists a stochastic process $\psi \in L^2[0,T]$ for all $T < \infty$ such that we have

$$M_t = \mathbb{E}(M_t) + \int_0^t \psi_s dW_s \text{ for } t \geq 0. \tag{4.106}$$

REMARK 4.43 1. The martingale representation theorem plays a decisive role for the concept of replication in a Black-Scholes market model as we will see in the next chapter. There exist multidimensional variants of it (i.e. both the Brownian motion and the integrand ψ are of the same dimension d). Further, one can relax the martingale requirement. However, if M is only a Brownian local martingale then the integrand ψ can only be guaranteed to be in $H^2[0,T]$ (for all finite T).

2. The integrand ψ in the theorem is unique in an L^2 sense. We refer the reader for details to Korn and Korn (2001).

3. As the martingale representation theorem says that each Brownian martingale equals a stochastic integral plus its expected value, this in particular implies that each Brownian martingale has continuous paths! □

The second result presented here shows in particular how a Brownian motion with drift can be transformed into a Brownian motion by changing to another probability measure. It is also a basic result for option pricing (see the next chapter).

We will consider a more general setting with a nonconstant drift. Let therefore $\{(X(t), \mathcal{F}_t)\}_{t\geq 0}$ be an m-dimensional progressively measurable process where $\{\mathcal{F}_t\}$ is the Brownian filtration with

$$\int_0^t X_i^2(s)\, ds < \infty \text{ a.s. } P \text{ for all } t \geq 0,\, i = 1, ..., m. \tag{4.107}$$

Further, let

$$Z(t, X) := \exp\left(-\sum_{i=1}^m \int_0^t X_i(s)\, dW_i(s) - \tfrac{1}{2}\int_0^t \|X(s)\|^2\, ds\right). \tag{4.108}$$

A sufficient condition for this process to be a martingale is the so-called **Novikov condition**:

$$E\left(\exp\left(\tfrac{1}{2}\int_0^t \|X(s)\|^2\, ds\right)\right) < \infty. \tag{4.109}$$

If $Z(t,X)$ is indeed a martingale, we have $E(Z(t,X)) = 1$ for all $t \geq 0$, and for all $T \geq 0$ we can define a probability measure $\mathbb{Q} = \mathbb{Q}_T$ on $\mathcal{F}_T$ via

$$\mathbb{Q}(A) := E\left(1_A \cdot Z(T,X)\right) \text{ for } A \in \mathcal{F}_T. \tag{4.110}$$

The following theorem shows how a $\mathbb{Q}$-Brownian motion $W^{\mathbb{Q}}(t)$ can be constructed from a $\mathbb{P}$-Brownian motion $W(t)$ with drift by a change of measure from $\mathbb{P}$ to $\mathbb{Q}$.

THEOREM 4.44 Girsanov's theorem
Let $Z(t,X)$ as above be a martingale. Define $\{(W^{\mathbb{Q}}(t), \mathcal{F}_t)\}_{t\geq 0}$ by

$$W_i^{\mathbb{Q}}(t) := W_i(t) + \int_0^t X_i(s)\, ds, \quad 1 \leq i \leq m, \quad t \geq 0. \tag{4.111}$$

Then, for each fixed $T \in [0,\infty)$ the process $\{(W^{\mathbb{Q}}(t), \mathcal{F}_t)\}_{t\in[0,T]}$ is an m-dimensional Brownian motion on $(\Omega, \mathcal{F}_T, Q)$ where the probability measure $\mathbb{Q}$ is defined in Equation (4.110).

4.6 Stochastic differential equations

4.6.1 Basic results on stochastic differential equations

Having introduced Itô processes and also their differential notation, it is now straightforward to introduce the notion of a stochastic differential equation (SDE) and its (strong) solution.

DEFINITION 4.45
A **(strong) solution** *$X(t)$ to the* **stochastic differential equation**

$$dX(t) = b(t, X(t))\, dt + \sigma(t, X(t))\, dW(t), \quad X(0) = x \tag{4.112}$$

for given functions $b : [0,\infty) \times \mathbb{R}^d \to \mathbb{R}^d$, $\sigma : [0,\infty) \times \mathbb{R}^d \to \mathbb{R}^{d,m}$ is a d-dimensional continuous process $\{(X(t), F_t)\}_{t\geq 0}$ on $(\Omega, F, \mathbb{P})$ that satisfies

$$X(0) = x, \tag{4.113}$$

$$X_i(t) = x_i + \int_0^t b_i(s, X(s))\, ds + \sum_{j=0}^{m} \int_0^t \sigma_{ij}(s, X(s))\, dW_j(s), \tag{4.114}$$

$$\int_0^t \left(|b_i(s, X(s)| + \sum_{j=1}^{m} \sigma_{ij}^2(s, X(s)) \right) ds < \infty \tag{4.115}$$

$\mathbb{P}$-a.s. for all $t \geq 0$, $i \in \{1, ..., d\}$.

REMARK 4.46 A strong solution thus is a stochastic process that has the integral representation (4.114) **and** is defined on the given probability space $(\Omega, F, \mathbb{P})$ where the Brownian motion W is also already given. For the notion of a weak solution, it is enough if there simply exists a process X satisfying Equation (4.114) on some probability space for some Brownian motion. We do not go into further details here but refer the interested reader to Karatzas and Shreve (1991). □

Two simple examples of SDEs with applications in finance include:

- the one-dimensional linear homogeneous equation

$$dX(t) = bX(t)\,dt + \sigma X(t)\,dW(t)\,, \quad X(0) = x\,, \tag{4.116}$$

  with $b, \sigma \in \mathbb{R}$ and $W(.)$ a one-dimensional Brownian motion,

- the one-dimensional linear equation with additive noise

$$dX(t) = (a + bX(t))\,dt + \sigma dW(t)\,, \quad X(0) = x\,, \tag{4.117}$$

  with $a, b, \sigma \in \mathbb{R}$ and $W(.)$ a one-dimensional Brownian motion.

Both these equations possess unique strong solutions that will be given below in Section 4.6.2 when we concentrate on linear SDEs.

As in the case of deterministic differential equations there is an existence and uniqueness result for SDEs similar to the well-known Picard and Lindelöf theorem (see Korn and Korn [2001] for proofs of the results of this section).

THEOREM 4.47 Existence and uniqueness of solutions for SDEs
Let the coefficients $b(t,x)$, $\sigma(t,x)$ of the SDE (4.112) be continuous functions satisfying both a Lipschitz and a growth condition

$$\|b(t,x) - b(t,y)\| + \|\sigma(t,x) - \sigma(t,y)\| \leq K\|x - y\| \tag{4.118}$$

$$\|b(t,x)\|^2 + \|\sigma(t,x)\|^2 \leq K^2\left(1 + \|x\|^2\right) \tag{4.119}$$

for all $t \geq 0$, $x, y \in \mathbb{R}^d$ and a constant $K > 0$ (where $\|.\|$ denotes the Euclidean norm of suitable dimension). Then there exists a continuous, strong solution $\{(X(t), F_t)_{t\geq 0}\}$ of (4.112) with

$$E\left(\|X(t)\|^2\right) \leq C \cdot \left(1 + \|x\|^2\right) \cdot e^{C \cdot T} \text{ for all } t \in [0,T] \tag{4.120}$$

for some constant $C = C(K,T)$ and $T > 0$. Further, $X(.)$ is unique up to indistinguishability, i.e. if $Y(.)$ would be another solution to (4.112) then we would have

$$\mathbb{P}(X(t) = Y(t)\,, \forall t \geq 0) = 1. \tag{4.121}$$

REMARK 4.48 It can be shown that the solution $\{(X(t), F_t)\}_t$ of (4.112) is a Markov process. This in particular means that for all Borel measurable, bounded functions f we have

$$\mathbb{E}\left(f\left(X\left(s\right)\right) \mid F_t\right) = \mathbb{E}\left(f\left(X\left(s\right)\right) \mid X\left(t\right)\right) = g\left(X\left(t\right)\right) \tag{4.122}$$

for fixed $t \leq s$ with $g(x) := \mathbb{E}(f(X^{t,x}(s)))$. Here, the use of the upper index t, x implies that the process $X^{t,x}$ solves the SDE (4.112) with initial condition $X(t) = x$. For simplicity of notation, we often omit the upper indices in the following, but mark corresponding expectations with this upper index instead:

$$\mathbb{E}\left(...X^{t,x}\left(s\right)...\right) = \mathbb{E}^{t,x}\left(...X\left(s\right)...\right). \tag{4.123}$$

□

4.6.2 Linear stochastic differential equations

As for ordinary differential equations, we know most about solutions in the linear setting. We will start with the one-dimensional case where we can present a completely explicit solution which is a generalization of the well-known variation of constants formula.

***THEOREM 4.49* Variation of constants**
Let $\{(W(t), F_t)\}_{t\in[0,\infty)}$ be an m-dimensional Brownian motion. Let $x \in \mathbb{R}$ and A, a, S_j, σ_j be progressively measurable, real-valued processes with

$$\int_0^t \left(|A(s)| + |a(s)|\right)\, ds < \infty, \quad \int_0^t \left(S_j^2(s) + \sigma_j^2(s)\right)\, ds < \infty \;\forall t \geq 0 \tag{4.124}$$

$\mathbb{P}$-a.s., $j = 1, ..., m$. Then the **general one-dimensional linear SDE**

$$dX\left(t\right) = \left(A\left(t\right)X\left(t\right) + a\left(t\right)\right)dt + \sum_{j=1}^{m}\left(S_j\left(t\right)X\left(t\right) + \sigma_j\left(t\right)\right)dW_j\left(t\right), \tag{4.125}$$

$$X(0) = x \tag{4.126}$$

possesses the unique solution $\{(X(t), F_t)\}_{t\in[0,\infty)}$

$$X(t) = Z(t)\cdot\left(x + \int_0^t \frac{1}{Z(u)}\left(a(u) - \sum_{j=1}^{m} S_j(u)\sigma_j(u)\right)\, du \right.$$
$$\left. + \sum_{j=1}^{m}\int_0^t \frac{\sigma_j(u)}{Z(u)}\, dW_j(u)\right) \tag{4.127}$$

where

$$Z(t) = \exp\left(\int_0^t \left(A(u) - \tfrac{1}{2} \cdot \|S(u)\|^2 \right) du + \int_0^t S(u)\; dW(u) \right) \tag{4.128}$$

is the unique solution of the homogeneous equation

$$dZ(t) = Z(t) \left(A(t)\, dt + S(t)'\, dW(t) \right), \quad Z(0) = 1. \tag{4.129}$$

Of particular interest for financial modelling will be the solution $Z(t)$ of the homogeneous equation in the variation of constants formula. Of course, if we change the initial condition to $Z(0) = z$, we simply have to multiply the solution by z. Another special case of interest is given by the equation

$$dX(t) = (a - AX(t))\, dt + \sigma dW_1(t), \quad X(0) = x \tag{4.130}$$

with nonzero A and real constants a, σ. Its unique solution

$$X(t) = xe^{-At} + \frac{a}{A}\left(1 - e^{At}\right) + \sigma \int_0^t e^{-A(t-u)}\, dW_j(u) \tag{4.131}$$

satisfies

$$X(t) \sim \mathcal{N}\left(xe^{-At} + \frac{a}{A}\left(1 - e^{-At}\right), \frac{\sigma^2}{2A}\left(1 - e^{-2At}\right) \right) \tag{4.132}$$

with an obvious limiting distribution for $t \to \infty$ in the case of $A > 0$.

In the general n-dimensional setting we have a similar result, but cannot always solve the homogeneous equation explicitly.

THEOREM 4.50 Multidimensional homogeneous linear SDE
Let $\{(W(t), F_t)\}_{t\in[0,\infty)}$ be an m-dimensional Brownian motion. Let $x \in \mathbb{R}^n$, let A, S^j be $n \times n-$matrices satisfying

$$AS^j = S^j A \text{ and } S^j S^k = S^k S^j \tag{4.133}$$

for $j, k = 1, ..., m$. Then the linear homogeneous SDE

$$dZ(t) = AZ(t)\, dt + \sum_{j=1}^{m} S^j Z(t)\, dW_j(t), \quad Z(0) = Z_0 \tag{4.134}$$

with constant coefficients possesses the unique solution

$$Z(t) = Z_0 \exp\left(\left(A - \frac{1}{2} \sum_{j=1}^{m} \left(S^j\right)^2 \right) t + \sum_{j=1}^{m} S^j W_j(t) \right) \tag{4.135}$$

$$= Z_0 \sum_{k=0}^{\infty} \frac{\left(\left(A - \frac{1}{2} \sum_{j=1}^{m} \left(S^j\right)^2 \right) t + \sum_{j=1}^{m} S^j W_j(t) \right)^k}{k!}. \tag{4.136}$$

THEOREM 4.51 Multidimensional linear SDE
Let $\{(W(t), F_t)\}_{t\in[0,\infty)}$ be an m-dimensional Brownian motion. Let $x \in \mathbb{R}^n$, let A, S^j be progressively measurable $n \times n-$matrix-valued processes, and a, σ^j $\mathbb{R}^n$-valued processes with

$$\int_0^t (|A_{ik}(s)| + |a_i(s)|)\, ds < \infty, \quad \int_0^t \left({S_{ik}^j}^2(s) + {\sigma_i^j}^2(s))\right) ds < \infty \qquad (4.137)$$

for all $t \geq 0$ $\mathbb{P}$-a.s., $i, k = 1, ..., n, j = 1, ..., m$. Then the **general n-dimensional linear SDE**

$$dX(t) = (A(t) X(t) + a(t))\, dt + \sum_{j=1}^{m} (S_j(t) X(t) + \sigma_j(t))\, dW_j(t) \qquad (4.138)$$

$$X(0) = x \qquad (4.139)$$

admits the unique solution $\{(X(t), F_t)\}_{t\in[0,\infty)}$

$$X(t) = Z(t) \cdot \left(x + \int_0^t Z(u)^{-1} \left(a(u) - \sum_{j=1}^{m} S_j(u)\sigma_j(u)\right) du + \sum_{j=1}^{m} \int_0^t Z(u)^{-1}\sigma_j(u) dW_j(u)\right) \qquad (4.140)$$

with $Z(t)$ the unique solution of the homogeneous equation

$$dZ(t) = A(t) Z(t)\, dt + \sum_{j=1}^{m} S^j(t) Z(t)\, dW(t), \quad Z(0) = I. \qquad (4.141)$$

4.6.3 The square-root stochastic differential equation

A particular SDE without an explicit solution but where the distribution of the solution is known is the so-called square-root equation

$$dX(t) = \kappa(\theta - X(t))\, dt + \sigma\sqrt{X(t)} dW(t), \quad X(0) = x \qquad (4.142)$$

with $x, \kappa, \theta, \sigma$ positive constants. It will play a prominent role in stock price modelling (especially in the local and the stochastic volatility models) and in interest rate modelling. Its first treatment, however, goes back to Feller (1951) and has no relation to finance whatsoever. The process has generated a lot of interest since then, see e.g. Cox et al. (1985) or Dufresne (2001). Note that the square-root SDE cannot be treated by our standard existence and uniqueness results as the square-root function is neither differentiable in the origin nor is it Lipschitz continuous on $[0, \infty)$. Thus, it typically needs a separate treatment when one wants to state assertions on its properties. We summarize the main results of the above mentioned research in the following theorem.

THEOREM 4.52
(a) For positive constants $x, \kappa, \theta, \sigma$ the square-root equation (4.142) has a unique nonnegative solution $X(t)$.
(b) With

$$d = \frac{4\kappa\theta}{\sigma^2}, \quad g(t) = \frac{4\kappa e^{-\kappa t}}{\sigma^2 \left(1 - e^{-\kappa t}\right)} \tag{4.143}$$

$e^{\kappa t} g(t) X(t)$ has a noncentral chi-square distribution with d degrees of freedom and noncentrality parameter $xg(t)$.
(c) We have

$$\mathbb{E}(X(t)) = \theta + (x - \theta) e^{-\kappa t}, \tag{4.144}$$

$$\mathbb{V}ar(X(t)) = \frac{x\sigma^2 e^{-\kappa t}}{\kappa}\left(1 - e^{-\kappa t}\right) + \frac{\theta\sigma^2}{2\kappa}\left(1 - e^{-\kappa t}\right)^2. \tag{4.145}$$

(d) For $2\kappa\theta \geq \sigma^2$ the solution $X(t)$ is strictly positive. For $2\kappa\theta \leq \sigma^2$ the origin can be attained by $X(t)$.

The simulation of this equation will be discussed further in Chapter 5.

4.6.4 The Feynman-Kac representation theorem

There is an in-depth relation between expectations of functionals of solutions of SDEs and solutions to a special type of partial differential equations (PDE). We first have to develop the necessary terms.

DEFINITION 4.53
Let $X(t)$ be the unique solution of the SDE (4.112) under the conditions (4.118) and (4.119). For $f : \mathbb{R}^d \to \mathbb{R}$, $f \in C^2(\mathbb{R}^d)$, the operator A_t, defined by

$$(A_t f)(x) := \tfrac{1}{2} \sum_{i=1}^{d} \sum_{k=1}^{d} a_{ik}(t, x) \frac{\partial^2 f}{\partial x_i \partial x_k}(x) + \sum_{i=1}^{d} b_i(t, x) \frac{\partial f}{\partial x_i}(x) \tag{4.146}$$

with

$$a_{ik}(t, x) = \sum_{j=1}^{m} \sigma_{ij}(t, x) \sigma_{kj}(t, x) \tag{4.147}$$

is called the **characteristic operator** *corresponding to $X(t)$.*

REMARK 4.54 It is now easy to assign characteristic operators to stochastic processes that are known to be explicit solutions of SDEs:

1. $X(t) = W(t)$ solves the equation $dX(t) = dW(t),\ X(0) = 0$. Hence,

$$\frac{1}{2}\Delta = \frac{1}{2}\sum_{i=1}^{d}\frac{\partial^2}{\partial x_i^2} \tag{4.148}$$

is the characteristic operator of the d-dimensional Brownian motion.

2. $X(t) = x \cdot e^{\left(b-\frac{1}{2}\sigma^2\right)t+\sigma W(t)}$ is the solution of the linear SDE

$$dX(t) = X(t)\left(b\,dt + \sigma\,dW(t)\right), \quad X(0) = x \tag{4.149}$$

and thus has the characteristic operator A_t given by

$$(A_t f)(x) = \tfrac{1}{2}\sigma^2 x^2 f''(x) + b\,x\,f'(x)\,. \tag{4.150}$$

□

To state our desired result that relates SDEs and PDEs we now set up the special type of problem, a so-called Cauchy problem.

DEFINITION 4.55

Let $T > 0$ be fixed. Then the **Cauchy problem** *corresponding to the operator A_t is to find a function $v(t,x) : [0,T]\times\mathbb{R}^d \to \mathbb{R}$ satisfying*

$$-v_t + kv = A_t v + g \ \text{ on } \ [0,T)\times\mathbb{R}^d \tag{4.151}$$

$$v(T,x) = f(x) \ \text{ for } \ x\in\mathbb{R}^d \tag{4.152}$$

for given functions $f:\mathbb{R}^d\to\mathbb{R}, g:[0,T]\times\mathbb{R}^d\to\mathbb{R}, k:[0,T]\times\mathbb{R}^d\to[0,\infty)$.

To ensure the uniqueness of a solution of the Cauchy problem, we additionally require that v obeys a polynomial growth condition:

$$\max_{0\le t\le T}|v(t,x)| \le M\left(1+\|x\|^{2\mu}\right) \ \text{with}\ M>0,\ \mu\ge 1. \tag{4.153}$$

Further, the functions f, g, k should be continuous. We assume that for suitable constants L, λ we have

$$|f(x)| \le L\left(1+\|x\|^{2\lambda}\right), \quad L>0,\ \lambda\ge 1 \quad \underline{\text{or}} \quad f(x)\ge 0\,, \tag{4.154}$$

$$|g(t,x)| \le L\left(1+\|x\|^{2\lambda}\right), \quad L>0,\ \lambda\ge 1 \quad \underline{\text{or}} \quad g(t,x)\ge 0. \tag{4.155}$$

THEOREM 4.56 The Feynman-Kac representation

Under assumptions (4.154) *and* (4.155), *let $v(t,x) : [0,T]\times\mathbb{R}^d\to\mathbb{R}$ be a continuous solution of the Cauchy problem* (4.151) *with $v\in C^{1,2}([0,T)\times\mathbf{R}^d)$.*

Denote by A_t in Equation (4.151) the characteristic operator of the unique solution $X(t)$ of the SDE (4.112) with continuous coefficients b, σ satisfying condition (4.118), $b_i(t,x), \sigma_{ij}(t,x) : [0,\infty) \times \mathbb{R}^d \to \mathbb{R}$ for $i = 1, ..., d$, $j = 1, ..., m$. If then $v(t,x)$ satisfies the polynomial growth condition (4.153), we have the representation

$$v(t,x) = E^{t,x}\left(f(X(T)) \cdot \exp\left(-\int_t^T k(\theta, X(\theta))\, d\theta \right) + \int_t^T g(s, X(s)) \cdot \exp\left(-\int_t^s k(\theta, X(\theta))\, d\theta \right) ds \right). \quad (4.156)$$

In particular, $v(t,x)$ is the unique solution of the PDE (4.151) which satisfies the polynomial growth condition (4.153).

REMARK 4.57 1. Note the exact assertion of the theorem: If we can show that the PDE (4.151) possesses a classical (i.e. sufficiently smooth) solution satisfying the polynomial growth condition (4.153) then it is given by the above expectation as a function of the initial parameters of the solutions to the SDE (4.112). However, in general we do not have the opposite direction. More precisely, the above expectation need not necessarily be the solution of PDE (4.151) as it simply may not be smooth enough. If on the other hand we can calculate this expectation, then we can check if it solves the Cauchy problem. If this is indeed the case, then it is the unique solution of PDE (4.151) satisfying the polynomial growth condition (4.153).

2. It is further important to see that we now even have a possibility to solve the Cauchy problem by Monte Carlo simulation. To do so, we have to:

1. Show that the Cauchy problem admits a classical solution that satisfies the growth condition (4.153).

2. Approximate the expectation in (4.156) by the Monte Carlo method:

   (a) Simulate N paths of the underlying SDE (4.112).

   (b) Calculate the corresponding values of the functionals in Equation (4.156).

   (c) Estimate $v(t,x)$ by the arithmetic mean over those values.

□

4.7 Simulating solutions of stochastic differential equations

4.7.1 Introduction and basic aspects

As with ordinary differential equations (ODEs), most stochastic differential equations do not admit explicitly given solutions. This makes it necessary to consider numerical methods to solve them. However, there are fundamental differences between deterministic ODEs and SDEs which are mainly due to the following facts:

- A solution to an SDE is a (function-valued) random variable, and thus we obtain different solutions for different $\omega \in \Omega$.
- A (strong) solution to an SDE is not smooth, as the underlying Brownian motion is not smooth at all.

The second fact yields that numerical schemes for ODEs that rest on smoothness properties of the solution are not automatically good when adapted to the situation of an SDE. The first fact has an important consequence: it depends on the purpose for what we need the solution of an SDE. Are we interested in

- obtaining a path $Y(t,\omega), t \in [0,T]$ that is **as close as possible** to the (unknown) solution path $X(t,\omega), t \in [0,T]$,
- or in computing an expectation of a functional $\mathbb{E}(g(X))$ of the SDE?

In the first case, the path Y obtained by a numerical scheme should perfectly mimic the behaviour of X. In the second case, we have already seen in Section 3.3 that it might be useful to simulate a random variable that might be totally different from X, but lead to a very efficient and accurate computation of $\mathbb{E}(g(X))$. These two aspects lead to two different notions of convergence of numerical schemes for SDEs, the so-called strong and weak convergence. Before we consider this, we make a short remark on the simulation of a solution to an SDE when it has an explicitly known solution.

Simulating with explicit solutions

In the rare case of an SDE with a unique explicit solution it is typically enough to simulate an underlying Brownian motion and plug it into the explicit formula. A standard example is the one-dimensional linear homogeneous equation with constant coefficients

$$dX(t) = X(t)(a\,dt + b\,dW(t)), \quad X(0) = x \tag{4.157}$$

with $x > 0$. The unique solution

$$X(t) = x\exp\left(\left(a - \frac{1}{2}b^2\right)t + bW(t)\right) = f(t, W(t)) \tag{4.158}$$

is exactly of this form with a suitable function f. For each single time instant t the exact distribution of $X(t)$ can be simulated by drawing $N(0,t)$-distributed random numbers and plugging them into the explicit formula. More generally, if the distribution of $X(t)$ is explicitly known, then simulating the stochastic process at a particular point of time is not different from simulating random numbers as in Chapter 2.

The task becomes more involved when a functional of the whole path $\{X(t,\omega) : t \in [0,T]\}$ is considered. Even in the example of the geometric Brownian motion above, we cannot simulate the whole path of a process. Only a discretized version of it can be generated. But this problem has already been dealt with when we introduced Brownian motion. As the geometric Brownian motion is a continuous function of the Brownian motion, we still have weak convergence of a discretized version combined with linear interpolation. We simply choose an equally spaced partition $0 = t_0 < T/n < 2T/n... < nT/n = T$ of the interval $[0,T]$, simulate the values of the Brownian motion $W(t_i)$, obtain the values $X(t_i)$ from it, and interpolate linearly in between those time points. Then, Donsker's theorem ensures weak convergence if we choose a suitable sequence of partitions of $[0,T]$. This means that (at least) for bounded functionals g we have the following.

PROPOSITION 4.58
Assume that the real-valued SDE

$$dX(t) = a(t, X(t))\,dt + \sigma(t, X(t))\,dW(t) \tag{4.159}$$

has an explicit solution of the form $X(t) = f(t, W(t))$ with f a continuous, real-valued function. Let Y_n be the approximation to X that is constructed by

$$Y_n(t) = f(t, W(t)) \quad \text{if } t = iT/n \text{ for some } i = 0, 1, ..., n \tag{4.160}$$

and extended to all $t \in [0,T]$ by linear interpolation. Then, for each bounded, measurable functional $g : C[0,t] \to \mathbb{R}$ we have

$$\mathbb{E}(g(Y_n)) \xrightarrow{n\to\infty} \mathbb{E}(g(X)). \tag{4.161}$$

4.7.2 Numerical schemes for ordinary differential equations

If there is no explicit solution to the stochastic differential equation under study, then – as in the deterministic case – we have to rely on numerical discretization schemes. For a survey on such numerical schemes, the monograph by Kloeden and Platen (1999) is still the authoritative source. We refer the reader to most of the proofs in the upcoming sections (if not otherwise stated) to this excellent reference.

To obtain an idea about how to construct numerical discretization schemes, we will survey some commonly used discretization methods for deterministic

ODEs. More precisely, we will have a look at the initial value problem

$$x'(t) = a(t, x), \quad x(t_0) = x_0 \tag{4.162}$$

where we assume the function a to satisfy smoothness and growth conditions to ensure both existence and uniqueness for the solution of the initial value problem. To solve it numerically on $[0, T]$, we consider a time discretization $t_0 < t_1 < t_2 < ... < t_k \leq T$. Although it is often convenient to assume an equidistant time spacing, we will not explicitly assume it and set

$$h_n := t_{n+1} - t_n. \tag{4.163}$$

As $x(t_0)$ is explicitly known, we also know $x'(t_0) = a(t_0, x_0)$. This could be used to find an approximation for $x(t_1)$. With this approximation we could then – again using the differential equation – try to compute an approximation of $x(t_2)$ and so on. To obtain an approximation for $x(t_1)$ we could

- approximate $x'(t_0)$ in a suitable way including $x(t_1)$ and solve for this,
- or use the Taylor expansion of $x(t)$ in $t = t_0$ and try to infer $x(t_1)$ from it in an approximate way.

In doing so, making errors in our approximation cannot be avoided. To make this more precise, we introduce the notation of

- $x(t; t_0, x_0)$ = the true solution of the initial value problem starting in t_0 with x_0,
- y_n = an approximation for $x(t_n; t_0, x_0)$ generated by a numerical discretization algorithm that started with $y_0 = x_0$,
- $x(t; t_n, y_n)$ = the true solution of the initial value problem if it would start at time t_n with initial value y_n.

DEFINITION 4.59

With the above notation and an approximation scheme for the initial value problem that produces the value y_n at step n of the iteration, we call

$$l_{n+1} := x(t; t_{n+1}, y_n) - y_{n+1} \tag{4.164}$$

the **local discretization error** *at step n and*

$$e_{n+1} := x(t; t_{n+1}, x_0) - y_{n+1} \tag{4.165}$$

the **global discretization error** *at step n.*

Approximation procedures – I: Explicit one-step methods

The first obvious idea to construct an approximation scheme has already been indicated above as approximating the derivative $x'(t_0)$ via

$$\frac{x(t_1) - x(t_0)}{t_1 - t_0} \approx x'(t_0) = a(t_0, x_0). \tag{4.166}$$

Taking this as an equality and solving for $x(t_1)$ leads to the **Euler method** given by

$$y_1 = y_0 + a(t_0, y_0) \cdot h_0. \tag{4.167}$$

Of course, for general n we just replace the subscripts 1 and 0 by $n+1$ and n in the above equation. An obvious generalization of this is stated below.

DEFINITION 4.60
A numerical discretization procedure for solving the initial value problem that is given by the iteration procedure

$$y_{n+1} = y_n + g(t, y_n, h_n) \cdot h_n \tag{4.168}$$

is called an explicit **one-step procedure** *with incremental function g.*

Examples of such one-step methods are

- the Euler method given by $g(t, x, h) = a(t, x)$,
- the second-order Runge-Kutta method given by

$$g(t, x, h) = \alpha a(t, x) + \beta a(t + \gamma h, x + \gamma a(t, x) h) \tag{4.169}$$

  with $\alpha + \beta = 1$, $\gamma\beta = \frac{1}{2}$. This in particular means that we have one free parameter. The choice $\alpha = \beta = \frac{1}{2}$, $\gamma = 1$ is called the Heun method. The more popular fourth-order Runge-Kutta method is given by

$$g(t, x, h) = \frac{1}{6}\left(k_1^{(n)} + 2k_2^{(n)} + 2k_3^{(n)} + k_4^{(n)}\right) \tag{4.170}$$

  with

$$k_1^{(n)} = a(t_n, y_n), \quad k_2^{(n)} = a\left(t_n + \tfrac{1}{2}h_n, y_n + \tfrac{1}{2}k_1^{(n)} h_n\right)$$
$$k_3^{(n)} = a\left(t_n + \tfrac{1}{2}h_n, y_n + \tfrac{1}{2}k_2^{(n)} h_n\right), \quad k_4^{(n)} = a\left(t_{n+1}, y_n + k_3^{(n)} h_n\right).$$

To explain the intuition behind the construction of the Runge-Kutta methods and to introduce the fundamental tool for determining the order of the local discretization error, we look at the **Taylor expansion** up to order p

$$x(t_{n+1}) = x(t_n) + x'(t_n) \cdot h_n + \ldots + \frac{1}{p!} x^{(p)}(t_n) \cdot h_n^p + O\left(h^{(p+1)}\right) \tag{4.171}$$

which is valid if x is sufficiently differentiable in t_n. As we know the function $a(t,x)$ explicitly and using the differential equation $x'(t) = a(t,x)$, we can express all the derivatives of x in t in terms of derivatives of a. In particular, for the first three derivatives we have

$$x'(t) = a(t,x), \quad x''(t) = a_t(t,x) + a_x(t,x)\,a(t,x)\ ,$$
$$x'''(t) = a_{tt}(t,x) + a_x(t,x)\,a_t(t,x) + a(t,x)\left(2a_{tx}(t,x) + a_x(t,x)^2\right).$$

Replacing these terms in the Taylor formula yields discretization methods with a local discretization error determined by the order of h of the remainder. Truncating the Taylor expansion after the first derivative leads to the Euler method (which thus has a local discretization error of 2). The Taylor method truncated after the second derivative yields a one-step method of the form

$$y_{n+1} = y_n + a(t_n, y_n)\cdot h_n + \frac{1}{2}\left(a_t(t_n, y_n) + a_x(t_n, y_n)\,a(t_n, y_n)\right)\cdot h_n^2. \quad (4.172)$$

Thus, in principle, we could obtain discretization schemes of any desired order of h for the local discretization error (given sufficient smoothness of a). However, the formula gets very lengthy as a depends on t and x. And, even more important, evaluating all the derivatives of a can become very costly from a computational point of view. One therefore tries to approximate the derivatives of a in a suitable way. The intuition behind the second-order Runge-Kutta method highlights this. Note therefore that a Taylor expansion in (t,x) of the second term in this method yields

$$\begin{aligned} g(t,x,h) &= \alpha a(t,x) + \beta a(t+\gamma h, x + \gamma(t,x)\,h) \\ &= (\alpha+\beta)\,a(t,x) + \gamma\beta h\left(a_t(t,x) + a_x(t,x)\,a(t,x)\right) + O\left(h^2\right) \end{aligned}$$

By comparing this with the Taylor formula truncated after the second derivative, we see that the requirements on α, β, γ in the second-order Runge-Kutta method ensure an order of h^3 of the local discretization error.

Approximation procedures – II: Implicit methods

If the right-hand side of an approximation scheme also depends on y_{n+1} we speak of an **implicit method**, as we have to solve this equation for y_{n+1} (typically by numerical methods) to obtain the next iterate. A simple example would be the implicit Euler scheme

$$y_{n+1} = y_n + a(t_{n+1}, y_{n+1})\,h_n \quad (4.173)$$

or the so-called **trapezoidal method**

$$y_{n+1} = y_n + \frac{1}{2}\left(a(t_n, y_n) + a(t_{n+1}, y_{n+1})\right)h_n. \quad (4.174)$$

In this last method, the average slope of $x(t)$ on $[t_n, t_{n+1}]$ is approximated by the mean of the slope in the two endpoints of the interval. Although implicit methods require a solution procedure for a (typically) nonlinear equation, they are applied to real problems due to their good behaviour with respect to numerical stability (see Stoer and Bulirsch [1993]).

Approximation procedures – III: Multistep methods

While one-step methods try to achieve a high order of the local discretization error by including (at least approximately) higher order derivatives to describe the evolution of $x(t)$, multistep methods try to achieve a good description by instead including information of the past evolution of $x(t)$. For simplicity, we assume an equidistant spacing, i.e. $h_n = h$ for all n:

DEFINITION 4.61
A numerical discretization method of the form

$$y_{n+1} = \sum_{j=1}^{k} \alpha_j y_{n+1-j} + \sum_{j=0}^{k} \beta_j a\left(t_{n+1-j}, y_{n+1-j}\right) h \tag{4.175}$$

with $\alpha_i, \beta_i \in \mathbf{R}$ is called a **multistep method** *to solve the initial value problem. It is called explicit if we have $\beta_0 = 0$ and implicit otherwise.*

Note that these kinds of methods need k starting points $y_1, ..., y_k$ which have to be computed by another method. Examples are

- $y_{n+1} = y_{n-1} + 2a\left(t_n, y_n\right) h$, the so-called **midpoint method**, which can be derived via approximating $x'\left(t_n\right)$ by a central difference quotient
$$\frac{x\left(t_{n+1}\right) - x\left(t_{n-1}\right)}{2h}.$$
From this, it is easy to see via Taylor expansion that this method has a local discretization error of order $O\left(h^3\right)$.
- $y_{n+1} = y_n + \frac{1}{12}\left(5a\left(t_{n+1}, y_{n+1}\right) + 8a\left(t_n, y_n\right) - a\left(t_{n+1}, y_{n+1}\right)\right)$, the (implicit) Adams-Moulton method.

REMARK 4.62 A possibility to avoid implicit methods is to first calculate an approximation $\bar{y}_{n+1}$ to y_{n+1} by some explicit method (a **prediction** of y_{n+1}) and then use this at the right-hand side of the discretization scheme to calculate the final value of y_{n+1} (the **correction**). We will not go into detail for such **predictor-corrector methods** as we will not use them in the stochastic differential equation part. □

Convergence considerations

Of course, the heuristical statements above concerning the order of the local error of the discretization schemes considered so far do not ensure convergence of the methods. We will therefore introduce some theoretical background:

DEFINITION 4.63

(a) Let y_n be a sequence generated by a discretization scheme starting with $y_0 = x_0$. We say that the scheme is **convergent** *if we have*

$$\lim_{max_n h_n \to 0} |x(t_{n+1}; t_0, x_0) - y_{n+1}| = 0 \tag{4.176}$$

on any finite interval $[t_0, T]$.
(b) We call a one-step method **consistent** *for the initial value problem if its incremental function g satisfies*

$$g(t; x, 0) = a(t, x). \tag{4.177}$$

Consistency implies in particular that if we plug the exact solution $x(t)$ of the initial value problem into the discretization scheme, then for $h \to 0$ convergence of the generated sequence towards $x(t)$ is ensured. We only cite one result characterizing the convergence behaviour of one-step methods for initial value problems for ODEs (see Kloeden and Platen [1999]).

THEOREM 4.64

Assume that a one-step method given by

$$y_{n+1} = y_n + g(t_n, y_n, h_n) h_n \tag{4.178}$$

satisfies the global Lipschitz condition in (t, x, h) :

$$|g(t', x', h') - g(t, x, h)| \leq K(|t' - t| + |x' - x| + |h' - h|). \tag{4.179}$$

(a) If in addition the one-step method obeys a global bound of the form

$$|g(t, x, 0)| \leq L \tag{4.180}$$

then it is convergent if and only if it is consistent.
(b) If the one-step method has a local discretization error of order $O(h^{p+1})$ then it admits a global discretization error of order $O(h^p)$.

REMARK 4.65 As we have $g(t, x, h) = a(t, x)$ in the Euler scheme, it is obviously consistent. Of course, the Lipschitz and boundedness conditions in the theorem depend on the function a in the case of the Euler scheme. □

4.7.3 Numerical schemes for stochastic differential equations

In principle, the discretization methods for deterministic differential equations can also be used for solving stochastic differential equations of the type

$$dX(t) = a(t, X(t)) dt + \sigma(t, X(t)) dW(t) \tag{4.181}$$

numerically. Here, for simplicity, we consider a one-dimensional SDE; modifications for the multidimensional case are given below. The main difference is that besides replacing dt by a time difference Δt, one also has to replace the infinitesimal increment of the Brownian motion dW_t by the finite difference

$$\Delta W(t) := W(t + \Delta t) - W(t). \tag{4.182}$$

By this simple modification we already obtain the most basic numerical scheme for solving SDEs, namely the Euler-Maruyama method.

The Euler-Maruyama method

We state it in Algorithm 4.7 and then comment on its usefulness.

Algorithm 4.7 The Euler-Maruyama scheme

Let $\Delta t := T/N$ for a given N. Then approximate the SDE (4.181) via:

1. Set $Y_N(0) = X(0) = x_0$.
2. For $j = 0$ to $N - 1$ do
   (a) Simulate a standard normally distributed random number Z_j.
   (b) Set $\Delta W(j\Delta t) = \sqrt{\Delta t} Z_j$ and

$$\begin{aligned} Y_N((j+1)\Delta t) = Y_N(j\Delta t) &+ a(j\Delta t, Y_N(j\Delta t))\Delta t \\ &+ \sigma(j\Delta t, Y_N(j\Delta t))\Delta W(j\Delta t). \end{aligned}$$

The Euler-Maruyama method is very popular in the applications in finance due to its simplicity. We will see later that there are also good theoretical reasons for its popularity when considering convergence aspects. A more formal motivation for the method and also for a whole class of methods is based on the **Itô-Taylor expansion**. This expansion formally resembles the look of a classical Taylor expansion but is based on the Itô formula as the underlying processes are Itô processes. We give its application for the one-dimensional case when even the coefficient functions are of the autonomous forms $a(X(t))$, $\sigma(X(t))$. Applying the Itô formula separately to the two coefficient functions of the SDE (4.181) and using the notations

$$L^0 = a(x)\frac{\partial}{\partial x} + \frac{1}{2}\sigma^2(x)\frac{\partial^2}{\partial x^2}, \quad L^1 = \sigma(x)\frac{\partial}{\partial x} \tag{4.183}$$

we obtain:

$$
\begin{aligned}
X(t) &= X(0) + \int_0^t a(X(s))\,ds + \int_0^t \sigma(X(s))\,dW(s) \\
&= X(0) + \int_0^t \left\{ a(X(0)) + \int_0^s L^0 a(X(u))du + \int_0^s L^1 a(X(u))dW(u) \right\} ds \\
&\quad + \int_0^t \left\{ \sigma(X(0)) + \int_0^s L^0 \sigma(X(u))du + \int_0^s L^1 \sigma(X(u))dW(u) \right\} dW(s) \\
&= X(0) + a(X(0)) \cdot t + \sigma(X(0)) \cdot \int_0^t dW(s) + R.
\end{aligned} \tag{4.184}
$$

Here, R denotes the rest containing all the terms of the second but last line which are missing in the last one. By hoping that a small value of t yields a negligible R, the last line above is then taken as an approximation for $X(t)$. We obtain the general step of the Euler-Maruyama method by replacing the time interval $[0,t]$ by $[t, t+\Delta t]$.

REMARK 4.66 1. **Multidimensional Euler-Maruyama scheme:** The extension to a d-dimensional setting of the Euler-Maruyama method is straightforward. One simply treats each component as a one-dimensional approximation procedure of the corresponding component of the underlying SDE. If the underlying Brownian motion is m-dimensional, the following two modifications are needed:

- In each step generate an m-dimensional standard normal variable $Z_i \sim \mathcal{N}(0, I_d)$ instead of just a univariate one.
- Replace the iteration procedure by the following d component iterations (the upper index denotes the component of the corresponding vector)

$$
\begin{aligned}
Y_N^{(i)}((j+1)\Delta t) &= Y_N^{(i)}(j\Delta t) + a^{(i)}(j\Delta t, Y_N(j\Delta t))\,\Delta t \\
&\quad + \sum_{k=1}^{m} \sigma^{(i,k)}(j\Delta t, Y_N(j\Delta t))\,\Delta W^{(k)}(j\Delta t).
\end{aligned} \tag{4.185}
$$

We will look at the performance of the Euler-Maruyama method and its convergence properties after we have introduced more general schemes below.

2. The above Itô-Taylor expansion can be generalized to the case of:

- Time-dependent coefficients: Then all coefficients occurring in the L-operators are time-dependent and the operator L^0 has to include a time derivative leading to

$$
L^0 = \frac{\partial}{\partial t} + a(t,x)\frac{\partial}{\partial x} + \frac{1}{2}\sigma^2(t,x)\frac{\partial^2}{\partial x^2}. \tag{4.186}
$$

- A multivariate setting: Then the L-operators have the forms that correspond to the application of the n-dimensional Itô formula to $C^{1,2}$-functions.

3. For the validity of the Itô-Taylor expansion we need suitable smoothness of the coefficient functions a and σ. However, the Euler-Maruyama method in its final form can be applied without any smoothness requirement at all. □

The Milstein method

The simplest idea to extend the Euler-Maruyama scheme is to include the term of the biggest order in the remainder R (of course when t is small) of Equation (4.184) for the approximation yielding the next iteration point. To identify it, note that $\Delta W(t)$ is of order $\sqrt{\Delta t}$ as can be seen from

$$\mathbb{V}ar\left(\Delta W(t)\right) = \Delta t. \tag{4.187}$$

Thus, the double stochastic integral

$$\begin{aligned}
&\int_0^t \int_0^s \sigma\left(X(u)\right)\sigma'\left(X(u)\right) dW(u) dW(s) \\
&\qquad = \sigma\left(X(0)\right)\sigma'\left(X(0)\right) \int_0^t \int_0^s dW(u) dW(s) + \tilde{R} \\
&\qquad = \frac{1}{2}\sigma\left(X(0)\right)\sigma'\left(X(0)\right)\left(W(t)^2 - t\right) + \tilde{R}
\end{aligned} \tag{4.188}$$

is the dominating term in the remainder of the Itô-Taylor approximation for the Euler-Maruyama approximation. If we add the last line to the iteration procedure for obtaining the new value of Y_n in each step (ignoring the new remainder $\tilde{R}$) then we obtain the so-called Milstein scheme (see Milstein [1978]). As the application of the Itô-Taylor formula for time-dependent coefficients does not influence the fact that the double stochastic integral is the dominating term of the remainder, we can write the corresponding Algorithm 4.8 in the more general form for time-dependent coefficients.

As with the Euler-Maruyama method, the Milstein method is very easy to implement which is a reason that it is also quite popular among practitioners in finance. However, there is one particular problem. While one has the intuitive feeling that by using a more sophisticated approximation function (by including higher order terms) one could expect a better convergence behaviour, the implicit requirement of a differentiable function $\sigma(.,.)$ makes its direct application to the square root process questionable. This fact and the performance of both methods – Milstein and Euler-Maruyama – are illustrated by applying them to the SDE

$$dX(t) = -0.5a^2 X(t)\, dt + a\sqrt{1 - X(t)^2} dW(t), \quad X(0) = x \tag{4.189}$$

Algorithm 4.8 The Milstein scheme

Let $\Delta t := T/N$ for a given N. Then solve the SDE (4.181) via:

1. Set $Y_N(0) = X(0) = x_0$.
2. For $j = 0$ to $N-1$ do
   (a) Simulate a standard normally distributed random number Z_j.
   (b) Set $\Delta W(j\Delta t) = \sqrt{\Delta t} Z_j$ and

$$\begin{aligned} Y_N((j+1)\Delta t) &= Y_N(j\Delta t) + a(j\Delta t, Y_N(j\Delta t))\Delta t \\ &\quad + \sigma(j\Delta t, Y_N(j\Delta t))\Delta W(j\Delta t) \\ &\quad + \frac{1}{2}\sigma(j\Delta t, Y_N(j\Delta t))\sigma'(j\Delta t, Y_N(j\Delta t))\left(\Delta W(j\Delta t)^2 - \Delta t\right). \end{aligned}$$

(with $a \in \mathbb{R}$, $-1 < x < 1$) which has the explicit solution

$$X(t) = \sin(aW(t) + \arcsin(x)). \tag{4.190}$$

In Figure 4.5 the Milstein method seems to have problems whenever the real process gets close to 1. Of course, this is the point where $\sigma(.)$ is nondifferentiable and the derivative of $\sigma(.)$ for values of $X(t)$ close to 1 gets very large. This actually results in the fact that the approximating sequence of the Y_n of the Milstein scheme eventually leaves the region $[-1, 1]$ where the process $X(t)$ is defined and where we terminate its application. Clearly, we could have capped Y_n by 1, but for demonstrational purposes did not do that. In Chapter 5 we will deal in more detail with a comparable situation when the simulation of the Heston model will be considered. The Euler-Maruyama method performs much more satisfying, although we should remark that it can produce values of Y_n outside $[-1, 1]$, too. However, as it is a derivative-free method, it does not suffer as much here as the Milstein method when $X(t)$ gets large.

REMARK 4.67 It is tempting to say that a d-dimensional version of the Milstein scheme can be obtained by extending it to a component-wise iteration procedure as in the case of the Euler-Maruyama method. There is however one point that prevents this. In a multidimensional version we would also have to evaluate double stochastic integrals of the form

$$I^{(i,k)} = \int_{j\Delta t}^{(j+1)\Delta t} \int_{j\Delta t}^{s} dW^{(i)}(u) dW^{(k)}(s). \tag{4.191}$$

As they can be easily evaluated in a simple closed form only for the case of $i = k$, the only direct generalization of the Milstein scheme to the mul-

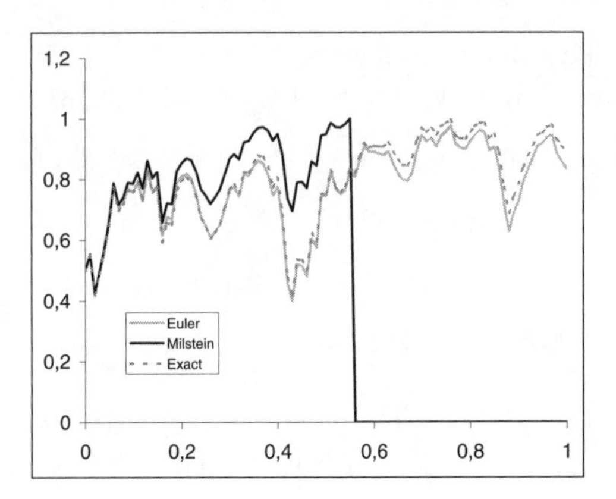

FIGURE 4.5: Comparison between the Euler and Milstein methods.

tidimensional setting would occur if component i of the corresponding SDE would solely depend on component i of the underlying Brownian motion and on no other one (of course, here we are working with a standard Brownian motion where the components are independent). In this special case, the multidimensional Milstein scheme can be identified as d one-dimensional Milstein schemes working in parallel. □

4.7.4 Convergence of numerical schemes for SDEs

We will now turn to the question of convergence and speed of convergence of our proposed numerical schemes and also consider general theory of convergence for numerical schemes for SDEs. Depending on the purpose of the simulation of the solution of an SDE, we consider two fundamentally different notions of convergence.

Strong convergence and pathwise approximation

Let δ be the maximum stepsize in a numerical scheme, i.e.

$$\delta = \max_{i=1,\ldots,n} \left(t_i - t_{i-1}\right), \quad t_0 = 0 < t_1 < \ldots < t_n = T. \tag{4.192}$$

We then denote by $Y^\delta = Y_n^\delta$ the corresponding numerical scheme given by

$$Y^\delta \left(t_0\right) = f_0 \left(X \left(t_0\right)\right), \tag{4.193}$$

$$Y^\delta \left(t_{j+1}\right) = f_{j+1} \left(Y^\delta \left(t_0\right), \ldots, Y^\delta \left(t_j\right)\right) + g_j \left(Z_{j+1}\right). \tag{4.194}$$

Here, f_i, g_i are measurable functions and Z_i is assumed to be F_{t_i}-measurable. Note in particular that we must have $n \geq T/\delta$. So, although we are in the following omitting the index n, we always have an implicit n-dependence.

DEFINITION 4.68
We say that a numerical scheme for solving the SDE (4.181) *or more generally* (4.112) **converges strongly** *on* $[0,T]$ *to the solution* X *of the SDE if for the final time* T *we have*

$$\lim_{\delta \to 0} \mathbb{E}\left(\left|X(T) - Y^{\delta}(T)\right|\right) = 0. \tag{4.195}$$

A strongly convergent scheme is said to have **convergence rate** γ *if for some constants* C *and* $\delta_0 > 0$ *we have*

$$\mathbb{E}\left(\left|X(T) - Y^{\delta}(T)\right|\right) \leq C \cdot \delta^{\gamma} \ \forall \delta \in [0, \delta_0]. \tag{4.196}$$

Thus, strong convergence means in particular that for almost every path of the SDE, the numerical scheme approximates the (final value of the) SDE as close as possible if the maximum stepsize approximates zero. We cite a well-known result that completely determines the strong convergence behaviour of the Euler-Maruyama method (see Kloeden and Platen [1999]).

THEOREM 4.69 Strong convergence: Euler-Maruyama scheme
Under the assumptions of Theorem 4.47 (i.e. Lipschitz coefficients and linear growth conditions on the coefficients of the SDE) and additionally

$$|a(t,x) - a(s,x)| + |\sigma(t,x) - \sigma(s,x)| \leq K(1 + |x|)\,|t - s|^{1/2} \tag{4.197}$$

for some suitable constant K*, the Euler-Maruyama scheme converges strongly with a convergence rate of* $\gamma = 1/2$*. Even more, if we define* $Y^{\delta}(t)$ *as the process generated by the Euler-Maruyama method with a time stepsize of* δ *and defined by linear interpolation at nontime-grid points, then we have*

$$\mathbb{E}\left(\sup_{0 \leq t \leq T} \left|X(t) - Y^{\delta}(t)\right|\right) < C\delta^{1/2}. \tag{4.198}$$

The uniform convergence result (4.198) is important for applications in finance such as the computation of the prices of path-dependent options.

As we do not explicitly consider the multidimensional Milstein scheme, we only state a theorem which is a special case of a result given in Kloeden and Platen (1999) on the strong convergence behaviour of the Milstein scheme.

THEOREM 4.70 Strong convergence: Milstein scheme
In addition to the assumptions of Theorems 4.47 and 4.69, let $\sigma(t,x)\frac{\partial \sigma(t,x)}{\partial x}$ *satisfy the conditions on the coefficients of Theorem 4.69. If further we have*

$$a \in C^{1,1}, \quad \sigma \in C^{1,2}, \tag{4.199}$$

then the Milstein scheme converges strongly with a convergence rate of $\gamma = 1$.

Thus, under stronger assumptions on the coefficient functions the Milstein scheme shows indeed a significantly better performance with respect to the speed of convergence than the Euler-Maruyama method. To state it in compact form: For the Euler-Maruyama method we have

$$\mathbb{E}\left(\left|X(T) - Y^{\delta}(T)\right|\right) = O\left(\delta^{1/2}\right) \quad \text{for } \delta \to 0 \tag{4.200}$$

while for the Milstein method we have

$$\mathbb{E}\left(\left|X(T) - Y^{\delta}(T)\right|\right) = O(\delta) \quad \text{for } \delta \to 0. \tag{4.201}$$

However, note that if the requirements of differentiability of the coefficients are not satisfied then the Milstein method might show irregular performance. Under suitable assumptions, it outperforms the Euler-Maruyama method and is recommended when pathwise approximations are needed.

Weak convergence and approximation of expectations

For most of our applications in estimating expectations of functionals of an SDE we do not need its exact paths. Then, we can drop the requirement that the random numbers in a numerical scheme are connected to the paths of the underlying SDE, i.e. we no longer require the F_{t_i}-measurability of the random numbers Z_i. In such a situation a convergence in the following weak sense is sufficient for our quantity to estimate on one hand, and is the most we can in general expect on the other hand.

DEFINITION 4.71
We say that a numerical scheme for the SDE (4.181) *or more generally* (4.112) **converges weakly** *on* $[0,T]$ *to the solution* X *of the SDE with respect to the class of functions* H *if at the final time* T *we have*

$$\lim_{\delta \to 0} \left|\mathbb{E}(g(X(T))) - \mathbb{E}\left(g\left(Y^{\delta}(T)\right)\right)\right| = 0 \tag{4.202}$$

for all $g \in H$. *A weakly convergent scheme with respect to the class* H *is said to have* **convergence rate** γ *if for some constants* C *and* $\delta_0 > 0$ *we have*

$$\left|\mathbb{E}(g(X(T))) - \mathbb{E}\left(g\left(Y^{\delta}(T)\right)\right)\right| \leq C \cdot \delta^{\gamma} \;\forall \delta \in [0,\delta_0] \;\forall g \in H. \tag{4.203}$$

The above notion of weak convergence allows us to consider a much wider class of numerical schemes than the one corresponding to strong convergence. As an example, we can introduce a **weak Euler scheme** by simply replacing the standard normal random variable Z_j in the Euler-Maruyama scheme by

$$Z_j = \begin{cases} 1 \\ -1 \end{cases} \text{each with a probability of } 0.5. \tag{4.204}$$

For most of our problems, it is reasonable that the class of test functions H in the definition of weak convergence contains the polynomials. Of course, the rate of weak convergence depends on the class of test functionals. Here, we only state a result for the Euler-Maruyama method, which is a corollary of Theorem 14.5.1 in Kloeden and Platen (1999).

THEOREM 4.72 Weak convergence: Euler-Maruyama scheme
If we have Lipschitz continuous and polynomially bounded autonomous coefficient functions $a(x), \sigma(x)$ which are in C_P^4 (i.e. they are four times differentiable and together with their derivatives at most polynomially growing) then the Euler-Maruyama scheme is weakly convergent with a convergence rate of $\gamma = 1$ with respect to the class H of all polynomials, i.e. we have

$$\left|\mathbb{E}\left[g\left(X\left(T\right)\right)\right] - \mathbb{E}\left[g\left(Y^{\delta}\left(T\right)\right)\right]\right| = O\left(\Delta t\right) \quad \text{for } \delta \to 0 \tag{4.205}$$

for any polynomial g.

REMARK 4.73 1. Surprisingly, we gain an additional order of convergence of 1/2 for the Euler-Maruyama scheme when considering weak convergence. This is **not true** for the Milstein scheme which also has a weak convergence order of 1 (see Kloeden and Platen [1999]). Thus, with respect to weak convergence it will be more efficient to use the Euler-Maruyama scheme as it is simpler than the Milstein method and obtains the same rate of convergence.

2. The assumptions needed for the validity of the weak convergence theorem above are quite strong. They can be somewhat relaxed. As this requires additional technicalities, we refer the reader to Kloeden and Platen (1999).

3. We would also like to point out that the SDEs considered in finance and insurance mathematics are often of a quite simple form. For simulating them, more complicated discretization methods do not necessarily yield big improvements in performance.

4. **Consistency and convergence:**
We have similar relations between consistency and convergence for numerical schemes for SDEs as in the case of ODEs. They are again of the form

$$\textit{consistency} + \textit{regularity assumptions} \implies \textit{convergence}$$

with respect to both weak and strong convergence. As the concept of consistency is more technical than in the ODE setting, we do not introduce it here (see Kloeden and Platen [1999] for the definition). □

4.7.5 More numerical schemes for SDEs

So far, we have just considered two explicit examples of numerical schemes for solving SDEs. There is however a vast amount of literature on this topic that deals with a lot of different approaches and methods. We cannot give a full survey here, but mention just some aspects.

Strong Taylor approximations

If one performs an Itô-Taylor expansion such as (4.184) to a higher order, then numerical schemes can be designed that converge strongly with higher rates of convergence. In principle, every rate can be achieved by using an Itô-Taylor expansion of appropriate order and using it to construct a corresponding numerical scheme. However, the schemes get very lengthy already for orders 1.5 and 2. As they all include multiple stochastic integrals, they are also not easy to evaluate at each time discretization. Usually, the SDEs considered in finance and insurance mathematics do not require their use.

A particular one-dimensional example that (under some regularity assumptions) is of strong convergence order 2 is another scheme due to Milstein (1978) and described in Algorithm 4.9. It is a direct consequence of an Itô-Taylor expansion up to order 2 in the case of autonomous coefficients. Its performance is analyzed in Duffie and Glynn (1995).

Algorithm 4.9 The Milstein order 2 scheme

Let $\Delta t := T/N$ for a given N. Then solve the SDE (4.181) via:

1. Set $Y_N(0) = X(0) = x_0$.
2. For $j = 0$ to $N-1$ do
   (a) Simulate a standard normally distributed random number Z_j.
   (b) Set $\Delta W(j\Delta t) = \sqrt{\Delta t} Z_j$ and

$$\begin{aligned} Y_N((j+1)\Delta t) = Y_N(j\Delta t) &+ a(Y_N(j\Delta t))\Delta t \\ &+ \sigma(Y_N(j\Delta t))\Delta W(j\Delta t) \\ + \sigma(Y_N(j\Delta t))\sigma'(Y_N(j\Delta t)) &\left(\frac{1}{2}\Delta W(j\Delta t)^2 - \Delta t\right) \\ &+ \nu(Y_N(j\Delta t))\Delta t\Delta W(j\Delta t) + \eta(Y_N(j\Delta t))(\Delta t)^2 \end{aligned}$$

   with

$$\begin{aligned} \nu(x) &= \frac{1}{2}(a(x)\sigma'(x) + a'(x)\sigma(x)) + \frac{1}{4}\sigma^2(x)\sigma''(x), \\ \eta(x) &= \frac{1}{2}a(x)a'(x) + \frac{1}{4}a''(x)\sigma^2(x). \end{aligned}$$

Note that this scheme requires many function evaluations per iteration step. Also, we need the derivatives of a and σ. Using it will of course mainly depend on the exact form of the SDE under consideration.

Implicit schemes for SDEs

As we already mentioned, for deterministic ODE often implicit schemes are used due to their numerical stability. As there, an implicit scheme is characterized by the fact that the right side of the iteration step also depends on the (yet unknown) new iteration point $Y((j+1)\Delta t)$. In the case of the **fully implicit Euler-Maruyama method** we would have

$$\begin{aligned} Y_N((j+1)\Delta t) = Y_N(j\Delta t) &+ a((j+1)\Delta t, Y_N((j+1)\Delta t))\Delta t \\ &+ \sigma((j+1)\Delta t, Y_N((j+1)\Delta t))\Delta W(j\Delta t). \end{aligned} \quad (4.206)$$

One major problem here is that the solution for $Y_N((j+1)\Delta t)$ might require a division by an expression containing the Brownian increment $\Delta W(j\Delta t)$. As, however, this expression might be 0 or very close to 0, one often uses the term **implicit scheme** only for such schemes where the Brownian part is of an explicit form. In the Euler-Maruyama case, this would then be

$$\begin{aligned} Y_N((j+1)\Delta t) = Y_N(j\Delta t) &+ a((j+1)\Delta t, Y_N((j+1)\Delta t))\Delta t \\ &+ \sigma(j\Delta t, Y_N(j\Delta t))\Delta W(j\Delta t). \end{aligned} \quad (4.207)$$

Application of such an implicit Euler-Maruyama scheme to the example SDE

$$dX(t) = -0.5a^2 X(t)\,dt + a\sqrt{1 - X(t)^2}\,dW(t), \quad X(0) = x \quad (4.208)$$

(with $a \in \mathbb{R}$, $-1 < x < 1$) results in the scheme

$$Y_N((j+1)\Delta t) = \frac{2}{2 + a^2\Delta t}\left(Y_N(j\Delta t) + a\sqrt{1 - Y_N(j\Delta t)^2}\,\Delta W(j\Delta t)\right). \quad (4.209)$$

An application of an implicit Milstein scheme is given by Kahl and Jäckel (2006) for the simulation of the so-called Heston model.

Runge-Kutta type and predictor-corrector methods

Also, these types of methods can be generalized to applying them to SDEs. This is of particular interest if one wants to avoid working with derivatives of the coefficient functions. However, as they do not play a big role in the applications in finance and insurance, we will not present them here but refer the interested reader to the special literature (see Kloeden and Platen [1999]).

4.7.6 Efficiency of numerical schemes for SDEs

Although we looked at the convergence and the rate of convergence of the different numerical schemes for SDEs, this is only one component of the error we are interested in. If we want to estimate $\mathbb{E}(g(X))$ by the Monte Carlo method, then indeed we have to consider the difference between the Monte Carlo estimator and $\mathbb{E}(g(X))$. A popular measure for this difference is the **mean-squared error**, MSE, defined by

$$MSE\left(\bar{g}_N\left(Y^\delta\right)\right) = \mathbb{E}\left(\bar{g}_N\left(Y^\delta\right) - \mathbb{E}\left(g\left(X\right)\right)\right)^2 \tag{4.210}$$

with Y^δ being the approximating stochastic process generated by a particular discretization scheme with a (maximum) stepsize of δ and

$$\bar{g}_N\left(Y^\delta\right) = \frac{1}{N}\sum_{i=1}^{N} g\left(Y_i^\delta\right) \tag{4.211}$$

denoting the usual crude Monte Carlo estimator based on N independent copies $g\left(Y_i^\delta\right)$ of $g\left(Y^\delta\right)$. Note that due to the fact that $\bar{g}_N\left(Y^\delta\right)$ is not an unbiased estimator for $g\left(X\right)$ (due to the discretization error), we cannot take its variance as an indicator for its performance but have to use the MSE instead. However, we can decompose the MSE via

$$\begin{aligned}
MSE\left(\bar{g}_N\left(Y^\delta\right)\right) &= \mathbb{E}\left(\bar{g}_N\left(Y^\delta\right) - \mathbb{E}\left(g\left(X\right)\right)\right)^2 \\
&= \mathbb{E}\left(\bar{g}_N\left(Y^\delta\right) - \mathbb{E}\left(\bar{g}_N\left(Y^\delta\right)\right) + \mathbb{E}\left(\bar{g}_N\left(Y^\delta\right)\right) - \mathbb{E}\left(g\left(X\right)\right)\right)^2 \\
&= \mathbb{E}\left(\left(\bar{g}_N\left(Y^\delta\right) - \mathbb{E}\left(\bar{g}_N\left(Y^\delta\right)\right)\right)^2\right) + \left(\mathbb{E}\left(\bar{g}_N\left(Y^\delta\right)\right) - \mathbb{E}\left(g\left(X\right)\right)\right)^2 \\
&= \mathbb{V}ar\left(\bar{g}_N\left(Y^\delta\right)\right) + bias\left(\bar{g}_N\left(Y^\delta\right)\right)
\end{aligned} \tag{4.212}$$

into the sum of the variance of the crude Monte Carlo estimator plus the bias caused by the use of a numerical scheme to approximate the stochastic process X by Y^δ. We also see by this decomposition of the MSE that we have two possibilities to reduce the MSE:

- Reduce the Monte Carlo variance by using a higher number N of sampled paths of the stochastic process.
- Reduce the bias by choosing a smaller stepsize δ.

Note that we are facing a trade-off between these two possibilities. If the amount of computational time is given, then reducing the bias results in a smaller stepsize, which on the other hand makes each sample path more costly. Thus, the smaller bias can only be realized if N is reduced, which then increases the Monte Carlo variance. Consequently, for a given amount of computational time, the orders of the two errors should coincide to balance out this situation, i.e. if we would like to **achieve a given accuracy** expressed by an MSE of $O(\epsilon^2)$ then we require:

$$O\left(\epsilon^2\right) = O\left(1/N\right) + O\left(\delta^2\right). \tag{4.213}$$

We thus need a stepsize of $\delta = \epsilon$ and $N = \lceil \epsilon^{-2} \rceil$ sample paths. As the computational work for each sample path is of order $O(1/\delta)$, we obtain a total amount of computational work of $O(\epsilon^{-3})$, a fact that is often paraphrased as follows.

COROLLARY 4.74
The computational complexity for estimating $E(g(X))$ by the crude Monte Carlo method combined with the Euler-Maruyama scheme equals $O(\epsilon^{-3})$ for a required MSE of $O(\epsilon^2)$.

In the following sections we will present some methods to obtain a lower computational complexity by using the same ingredients, the Euler-Maruyama scheme and the Monte Carlo estimator, in a more sophisticated way.

4.7.7 Weak extrapolation methods

Talay-Tubaro extrapolation

The principle of extrapolation is based on the idea that one is not only looking at a particular result from a discretization scheme but at the sequence of those results for different, decreasing stepsizes. Information from previous schemes with larger stepsizes should also be considered to give an impression of the improvement of the error as a function of the stepsize. This principle has been introduced to Monte Carlo methods for estimating expectations of functions of an SDE by Talay and Tubaro (1990). It is based on the following expansion result that we state in our terminology.

PROPOSITION 4.75
Let g be a C^∞ function, let the coefficient functions of the SDE also be in C^∞. Then, for the approximating process $Y^\delta(t)$ calculated by the Euler-Maruyama scheme with stepsize δ, there exists a constant C such that we have

$$\mathbb{E}\left(g\left(Y\left(T\right)\right)\right) - \mathbb{E}\left(g\left(Y^\delta\left(T\right)\right)\right) = C\delta + O\left(\delta^2\right). \tag{4.214}$$

The regularity requirements on both the functional g and the coefficient functions of the SDE can be relaxed. As an example, if in addition to the assumptions in the proposition above, we assume that the coefficient functions a, σ of the SDE are bounded and σ is uniformly elliptic, i.e. it satisfies

$$x'\sigma(x)\sigma(x)'x \geq cx'x \ \ \forall x \in \mathbb{R}^d \tag{4.215}$$

for a suitable positive c, then the assertion of Proposition 4.75 stays valid if g is only a bounded Borel measurable function (see also Bally and Talay [1996]).

A direct consequence of Proposition 4.75 is that the functional

$$Q_g^\delta = \mathbb{E}\left(2g\left(Y^\delta(T)\right) - g\left(Y^{2\delta}(T)\right)\right) \tag{4.216}$$

obtained from an extrapolation of the two estimation functionals with stepsizes δ and 2δ, respectively, has a weak convergence order of 2 as we have

$$Q_g^\delta = \mathbb{E}(g(Y(T)) - \mathbb{E}(g(Y^{2\delta}(T))) - 2(\mathbb{E}(g(Y(T)) - \mathbb{E}(g(Y^\delta(T)))). \tag{4.217}$$

Using the extrapolation functional Q_g^δ thus results in a significantly smaller bias of $O(\delta^2)$ for a given stepsize of δ compared to the Euler-Maruyama scheme. As, further, the computational amount per sample path is comparable to that of the Euler-Maruyama scheme (at most not higher than twice the amount for generating a sample path of stepsize δ), we obtain a big gain in efficiency. Actually, one could use independent Brownian motions for calculating the two different approximations. However, it will be more efficient to use one discretely sampled Brownian motion with stepsize δ to compute a realization of $2g\left(Y_i^\delta(T)\right) - g\left(Y_i^{2\delta}(T)\right)$ in one attempt. This will in general also reduce the variance of the corresponding Monte Carlo estimator. For simplicity we choose a stepsize with $2\delta = T/K$ in Algorithm 4.10.

We comment on the numerical performance of the Talay-Tubaro method in Section 4.7.8.

The statistical Romberg method

The so-called **statistical Romberg method** is conceptually related to the Talay-Tubaro extrapolation, but it has been introduced by Kebaier (2005) with the aim to reduce the computational complexity of the combination of Monte Carlo methods and numerical discretization schemes to compute $\mathbb{E}(g(X(T)))$. Here, we consider a d-dimensional SDE and a Lipschitz-continuous function $g : \mathbb{R}^d \to \mathbb{R}$ that satisfies the usual growth condition. Kebaier introduced a control variate approach in the following way:

$$Q = g\left(Y^{1/n}(T)\right) - g\left(Y^{1/m}(T)\right) + \mathbb{E}\left(g\left(Y^{1/m}(T)\right)\right) \tag{4.218}$$

with $m << n$. The two functions above are expected to be positively correlated which is the basis for a good control variate approach. However, the expectation on the right-hand side of (4.218) still has to be calculated. The key observation of Kebaier (2005) is that the variance of Q is decreasing linearly in m, i.e. we have

$$\sigma_Q^2 = \mathbb{V}ar(Q) = O(1/m). \tag{4.219}$$

If we further note that we have $\mathbb{E}(Q) = \mathbb{E}(g(Y^{1/n}(T)))$ then we directly obtain

$$\mathbb{V}ar\left(\bar{Q}_N\right) = O\left(\frac{1}{mN}\right) \tag{4.220}$$

Algorithm 4.10 Talay-Tubaro extrapolation

Let $\delta = T/(2K)$, N and x be given.

For $i = 1$ to N

- Set $Y_i^\delta(0) = Y_i^{2\delta}(0) = x$.
- For $j = 1$ to $2K$ simulate independent $N(0,1)$-random numbers $Z_{i,j}$.
- Simulate a path with stepsize δ, i.e. for $j = 1$ to $2K$:

$$Y_i^\delta(j\delta) = Y_i^\delta((j-1)\delta) + a\left((j-1)\delta, Y_i^\delta((j-1)\delta)\right)\delta + \sigma\left((j-1)\delta, Y_i^\delta((j-1)\delta)\right)\sqrt{\delta}Z_{i,j}.$$

- Simulate a path with stepsize 2δ, i.e. for $j = 1$ to K:

$$Y_i^{2\delta}(2j\delta) = Y_i^{2\delta}(2(j-1)\delta) + 2a\left(2(j-1)\delta, Y_i^{2\delta}(2(j-1)\delta)\right)\delta + \sigma\left(2(j-1)\delta, Y_i^{2\delta}(2(j-1)\delta)\right)\sqrt{\delta}\left(Z_{i,2j} + Z_{i,2j-1}\right).$$

- Set $\tilde{g}\left(\tilde{Y}_i(T)\right) = 2g\left(Y_i^\delta(T)\right) - g\left(Y_i^{2\delta}(T)\right)$.

Calculate the Talay-Tubaro Monte Carlo estimator

$$\bar{g}_N\left(Y^\delta\right) = \frac{1}{N}\sum_{i=1}^N \tilde{g}\left(\tilde{Y}_i(T)\right).$$

---

which indeed is smaller by a factor $1/m$ compared to the crude Monte Carlo estimator of $\mathbb{E}(g(Y^{1/n}(t)))$. However, there remain two problems:

- How to compute $\mathbb{E}\left(g\left(Y^{1/m}(T)\right)\right)$?
- How to choose n and m?

The answer to the first question is that we are going to estimate the expectation by the crude Monte Carlo approach, i.e. by

$$\bar{g}_{N_m}\left(\left(Y^{1/m}(T)\right)\right) = \frac{1}{N_m}\sum_{i=1}^{N_m} g\left(Y_i^{1/m}(T)\right). \tag{4.221}$$

If we have then also chosen N_n, the number of samples related to Q, then we can introduce the **statistical Romberg estimator**

$$\bar{Q}_{N_n,N_m} = \frac{1}{N_n} \sum_{i=1}^{N_n} \left(g\left(Y_i^{1/n}(T)\right) - g\left(Y_i^{1/m}(T)\right)\right) + \bar{g}_{N_m}\left(\left(\hat{Y}^{1/m}(T)\right)\right). \quad (4.222)$$

Here, we should emphasize that using $\hat{Y}$ in the last part of the right-hand side of the equation means that to estimate $\mathbb{E}(g(Y^{1/m}(T)))$, we should use new paths of the SDE that are independent of those used for the first average. The optimal choices for this approach are given in the following theorem, which is a special case of Kebaier (2005).

THEOREM 4.76 Complexity of the statistical Romberg method
Let $\epsilon = 1/n$ be the required MSE of the statistical Romberg method (with $n \in \mathbb{N}$). Assume that the function f and the coefficients of the SDE (4.112) satisfy conditions such that the Euler-Maruyama scheme has a weak convergence order of 1. Then the minimal computational effort to obtain this MSE by the statistical Romberg method based on the Euler-Maruyama scheme is achieved for the choices of

$$m = \sqrt{n}, \quad N_m = n^2, \quad N_n = n^{1.5}. \quad (4.223)$$

In particular, we obtain a total computational complexity C_{SR} for this optimally designed statistical Romberg method of

$$C_{SR} = O\left(\epsilon^{-2.5}\right). \quad (4.224)$$

REMARK 4.77 1. At first sight, it seems to be counterintuitive that the number of sample paths $N_m = n^2$ needed to estimate $\mathbb{E}(g(Y^{1/m}(T)))$ is higher than $N_n = n^{1.5}$, the number of sample paths that is generated to estimate the expectation on the finer scale n. However, this small N_n is a consequence of the control variate approach. In total, the effort for simulating $\mathbb{E}(g(Y^{1/m}(T)))$ equals $O(m \cdot N_m) = O(n^{2.5})$, which is the same order of effort $O(n \cdot N_n)$ to estimate $\mathbb{E}((g(Y^{1/m}(T)) - g(Y^{1/m}(T)))$. We are thus working with the same discretization $\epsilon = 1/n$ as in the corresponding crude Monte Carlo method, but require much fewer of the expensive paths with stepsize ϵ.

2. The results of Kebaier (2005) are presented in a more general setting than the one above. We refer the interested reader to this article. □

The numerical performance of Algorithm 4.11 is described in Section 4.7.8.

Algorithm 4.11 The statistical Romberg method

Assume that we are given an accuracy requirement of $\epsilon = 1/n$ (with $n = k^2$ for some $k \in \mathbb{N}$).

1. Set $m = n^{0.5}$, $N_m = n^2$, $N_n = n^{1.5}$.

2. Simulate N_m paths of the Y-process by the Euler-Maruyama scheme with stepsize $\delta = 1/m$, and compute $\bar{g}_{N_m}\left(\left(\hat{Y}^{1/m}(T)\right)\right)$ from it.

3. Simulate N_n further paths of the Y-process by the Euler-Maruyama scheme with stepsize $\delta = 1/n$, each resulting in $g\left(Y_i^{1/n}(T)\right)$. From those paths of the Y-process use the underlying Brownian motion at times $jT/m, j = 0, 1, ..., m$ to obtain $g\left(Y_i^{1/m}(T)\right), i = 1, .., N_n$.

4. Calculate the statistical Romberg estimator

$$\bar{Q}_{N_n,N_m} = \frac{1}{N_n}\sum_{i=1}^{N_n}\left(g\left(Y_i^{1/n}(T)\right) - g\left(Y_i^{1/m}(T)\right)\right) + \bar{g}_{N_m}\left(\left(\hat{Y}^{1/m}(T)\right)\right).$$

4.7.8 The multilevel Monte Carlo method

The multilevel Monte Carlo method is a recent development introduced to finance by Giles (2008) (see Heinrich [2001] for its introduction to parametrical integration) and takes up ideas from multigrid methods for PDE. Its aim is to reduce the computational complexity of the combined Monte Carlo and discretization scheme approach when the expectation of a function(al) of an SDE should be computed. It uses information from a sequence of computations with decreasing stepsizes. By this, as many as possible expensive samples simulated with the finest grid size should be avoided given a required order of accuracy. We consider a d-dimensional SDE of the usual form

$$dX(t) = a(t, X(t))\,dt + \sigma(t, X(t))\,dW(t) \tag{4.225}$$

and assume that the coefficients satisfy conditions ensuring the existence of a unique strong solution. Our aim is to compute $\mathbb{E}(g(X(T)))$ by the Monte Carlo method where $g : \mathbb{R}^d \to \mathbb{R}$ is Lipschitz-continuous. For a decreasing sequence of stepsizes δ_i, let $Y^{\delta_i}(T)$ be approximations of $X(T)$ generated by the Euler-Maruyama method. Then, a standard control variate approach similar to the statistical Romberg method would use the relation

$$\mathbb{E}\left(g\left(Y^{\delta_L}(T)\right)\right) = \mathbb{E}\left(g\left(Y^{\delta_L}(T)\right) - g\left(Y^{\delta_{L-1}}(T)\right)\right) + \mathbb{E}\left(g\left(Y^{\delta_{L-1}}(T)\right)\right). \tag{4.226}$$

As, however, we do not have an analytical expression for the expected value corresponding to the stepsize δ_{L-1}, we can again use a control variate technique to estimate it. By iterating this we obtain:

$$\mathbb{E}\left(g\left(Y^{\delta_L}(T)\right)\right) = \mathbb{E}\left(g\left(Y^{\delta_0}(T)\right)\right) + \sum_{i=1}^{L} \mathbb{E}\left(g\left(Y^{\delta_i}(T)\right) - g\left(Y^{\delta_{i-1}}(T)\right)\right). \quad (4.227)$$

This relation can be seen as the basis for a **multiple control variate approach** to obtain the desired expected value by Monte Carlo simulation. In line with our analysis of the MSE and the statistical Romberg method note:

- To achieve a required accuracy of $O(\epsilon^2)$ we need $\delta_L = O(\epsilon)$.
- The expectations of the above differences should have a small variance.
- The information obtained from larger stepsizes δ_i should help to save samples on the δ_L-level.

The expectations on the right-hand side of Equation (4.227) are all estimated by crude Monte Carlo estimators

$$I_0(g) = \frac{1}{N_0} \sum_{i=1}^{N_0} g\left(Y_i^{\delta_0}(T)\right), \quad (4.228)$$

$$I_j(g) = \frac{1}{N_j} \sum_{i=1}^{N_j} \left(g\left(Y_i^{\delta_j}(T)\right) - g\left(Y_i^{\delta_{j-1}}(T)\right)\right) \quad (4.229)$$

for $j = 1, ..., L$. Here, it is important to point out that all these $L+1$ Monte Carlo estimators have to be based on different, independent samples. We then introduce the **multilevel Monte Carlo estimator** as

$$I(g) = \sum_{j=0}^{L} I_j(g). \quad (4.230)$$

Due to the above independence assumption for the paths to obtain the I_j, the variance of the multilevel estimator is given by

$$\sigma_I^2 = \mathbb{V}ar\left(I(g)\right) = \sum_{j=0}^{L} N_j^{-1} \sigma_j^2 \quad (4.231)$$

with σ_j^2 denoting the variance of $g(Y_j^{\delta_j}(T)) - g(Y_i^{\delta_{j-1}}(T))$. As the multilevel Monte Carlo estimator requires a computational cost of roughly $\sum_{j=0}^{L} N_j \delta_j^{-1}$, simple Lagrangian optimization shows that given a fixed computational cost, the variance of the estimator will be minimal when we choose N_j to be proportional to $\sqrt{\sigma_j^2 \delta_j}$. Giles (2008) suggests using a stepsize of

$$\delta_j = M^{-j}T \text{ for some integer } M \geq 2 \tag{4.232}$$

(he actually chooses $M = 4$). Further, to obtain a desired MSE of $O\left(\epsilon^2\right)$ Giles (2008) proves a so-called **complexity theorem**. It states, in particular, that for payoff functionals where the Euler-Maruyama method obtains its usual weak and strong convergence rates, the multilevel Monte Carlo method has a computational complexity of $O\left(\epsilon^{-2}\left(\ln\left(\epsilon\right)\right)^2\right)$ for the choices of

$$L = \left\lceil \frac{\ln\left(\epsilon^{-1}\right)}{\ln\left(M\right)} \right\rceil, \quad N_l = \left\lceil 2\epsilon^{-2}\sqrt{\sigma_l^2 \delta_l}\left(\sum_{i=0}^{L} \sqrt{\sigma_i^2/\delta_i}\right) \right\rceil. \tag{4.233}$$

There is, however, one drawback of the above explicit formula for N_l, the number of Monte Carlo paths at level l: the variances σ_i^2 of the differences $g(Y_j^{\delta_j}(T)) - g(Y_i^{\delta_{j-1}}(T))$ are not known *a priori*. Thus, one has to include an **estimation loop** on each level l to obtain a sufficiently reliable estimator for σ_l^2 on which the subsequent calculations are based. Further, to calculate the Monte Carlo estimator $I_j(g)$, $j \geq 1$, one uses the same normal random numbers to generate the fine and the coarse paths (compare to the Talay-Tubaro Algorithm 4.10; however, note that we here have to sum up four normally distributed random numbers to obtain the coarse path).

Algorithm 4.12 Multilevel Monte Carlo simulation

Assume that ϵ and M are given; L is chosen according to Equation (4.233).

1. Variance estimation loop:
   For each $l = 0, 1, ..., L$ simulate $N_l = 10^4$ sample paths with stepsize δ_l to obtain the usual estimate of $\hat{\sigma}_l^2$ for σ_l^2.

2. Define the optimal sample sizes $N_l, l = 0, ..., L$ from Equation (4.233).

3. At each level $l = 0, ..., L$ simulate N_l new, independent sample paths to obtain the Monte Carlo estimators $I_j(g)$ of Equations (4.228) and (4.229).

4. Obtain the multilevel Monte Carlo estimator $I(g) = \sum_{l=0}^{L} I_j(g)$.

REMARK 4.78 1. **Algorithmic refinements**: Instead of using an already fixed L in Algorithm 4.12, Giles (2008) suggests building up the multilevel Monte Carlo estimator starting from the coarser levels. Then, by estimating the remaining bias of the estimator, one can decide after each level if it is already fine enough to be used as the finest stepsize δ_L. When using such a method, the required sample sizes N_l at each level are not *a priori* determined, as L is not yet fixed. However, the numbers N_l from Equation (4.233) are increasing in L. We can therefore compute the necessary extra samples when L is increased in each iteration step below. As a check for convergence, Giles suggests stopping the iteration with $L = l \geq 2$, if we have

$$\max\left\{M^{-1}\left|I_{l-1}(g)\right|, \left|I_l(g)\right|\right\} < \frac{1}{\sqrt{2}}(M-1)\,\epsilon. \tag{4.234}$$

With this modification, we restate a new, adaptive version of the multilevel Monte Carlo method in Algorithm 4.13 below.

2. In Giles (2007) a variant of the multilevel Monte Carlo algorithm based on the Milstein method for one-dimensional SDEs is given. □

Algorithm 4.13 Adaptive multilevel Monte Carlo simulation

---

1. Start with $L = 0$.
2. Simulate $N_L = 10^4$ samples to estimate σ_L^2.
3. Define the optimal sample sizes $N_l, l = 0, ..., L$ from Equation (4.233).
4. Evaluate the extra samples at each level $l = 0, ..., L$ as needed for N_l and update the Monte Carlo estimators from Equations (4.228) and (4.229), accordingly.
5. If $L \geq 2$ then check the convergence condition (4.234). If it is satisfied then stop.
6. If $L < 2$ or if the convergence check has failed, set $L = L + 1$ and go to Step 2.

Numerical performance of Talay-Tubaro extrapolation, statistical Romberg, and multilevel Monte Carlo method

From a complexity point of view, the multilevel Monte Carlo method seems to be the most promising among the recently introduced ones. We will highlight this here in comparing the crude method, Talay-Tubaro extrapolation, statistical Romberg extrapolation, and the multilevel Monte Carlo approach.

As a first test example, we take a two-dimensional SDE from Kebaier (2005):

$$dX(t) = -\frac{1}{2}X(t)\,dt - Y(t)\,dW(t), \tag{4.235}$$

$$dY(t) = -\frac{1}{2}Y(t)\,dt + X(t)\,dW(t) \tag{4.236}$$

with $Z(0) = (X(0), Y(0)) = (\cos(\theta), \sin(\theta))$ for some $\theta \in [0, 2\pi]$. It can directly be checked that the unique solution to this equation is given by

$$Z(t) = (X(t), Y(t)) = (\cos(\theta + W(t)), \sin(\theta + W(t))). \tag{4.237}$$

Our aim will be to estimate the expectation

$$\mathbb{E}\left(\left(\|Z(T)\|^2 - 1\right)\right) = 0 \tag{4.238}$$

with the above four methods and compare their performance. Note in particular that although $Z(t)$ lives on the unit sphere, the discretized versions obtained by one of the numerical schemes need not result in simulated paths on the unit sphere at all.

For the choice of $\theta = 0$, we obtained the following results for different accuracy requirements given by ϵ (which is related to the choice of $M = 4$ in the multilevel Monte Carlo method and which determines $N = 1/\epsilon^2$):

Method (time)	$\epsilon = 4^{-2}$	CPU	$\epsilon = 4^{-5}$	CPU
Crude MC	0.04	(1)	$2 \cdot 10^{-4}$	(1)
Talay-Tubaro	-0.05	(2.76)	$1 \cdot 10^{-6}$	(1.82)
Stat. Romberg	0.004	(0.53)	$2 \cdot 10^{-3}$	(0.27)
Multilevel MC	0.01	(80.29)	$9 \cdot 10^{-4}$	(4.23)

Table 4.1: Comparison of Extrapolation MC Methods for SDEs (4.235) and (4.236), Relative CPU Times (CPU), $T = 1$ (Exact Value = 0)

The results show that the Talay-Tubaro method seems to converge faster, which indeed it should as it has a higher order of convergence. Also, the statistical Romberg method is by far the fastest one. This is expected when compared to the crude Monte Carlo method and to the Talay-Tubaro method. However, the disappointing performance of the multilevel Monte Carlo method calls for an explanation. First, it is clear that it is slow for a comparably large ϵ, as then the extra runs for estimating the variances on each level l need more computing time than the actual estimation steps for obtaining the desired expectation. For $\epsilon = 4^{-5}$, the performance is comparably better, but still the method is far away from being comparable. A close analysis shows that the variance of the estimator on the coarsest level is much smaller than those on all other levels. This has the consequence that only very few observations are used at this level which is in contrast to the idea that one should use more paths here and save more expensive paths. In this sense, this explains why there is no saving of CPU time. Such an effect cannot harm the statistical Romberg method as it has already determined the distribution of the load between fine and coarse grid paths before the problem is tackled.

For a more detailed analysis of the behaviour of the multilevel Monte Carlo method in such examples we refer to Imkeller (2009).

To show the potential of the multilevel Monte Carlo method and to underline the above comments, we look at the computation of

$$\mathbb{E}\left(\left(\int_0^1 \exp\left(0.2W\left(t\right)+0.03t\right)dt-1\right)^+\right) \tag{4.239}$$

for a one-dimensional Brownian motion $W(.)$. The corresponding SDE of the integrand which we denote by $S(t)$ reads as

$$dS\left(t\right)=S\left(t\right)\left(0.05dt+0.2dW\left(t\right)\right),\ S\left(0\right)=1.$$

We now use the adaptive variant of the multilevel Monte Carlo method for $M=4$ (the results for $M=2$ are comparable but slightly inferior) and compare its performance to the crude Monte Carlo method. In addition to the CPU time relative to the crude Monte Carlo method, we also give the level L on which the multilevel Monte Carlo method terminates in Table 4.2.

Method (time)	$\epsilon=0.01$	CPU	L	$\epsilon=0.001$	CPU	L
Crude MC	0.0608	(1)		0.0605	(1)	
Multilevel MC	0.0603	(2.89)	2	0.0604	(0.015)	2

Table 4.2: Comparison of Crude MC and Adaptive Multilevel MC, Relative CPU Times (CPU) (Exact Value = 0.0606)

Again, for a low accuracy requirement, the crude Monte Carlo method is slightly faster. The reason is the dominance of the preestimation procedure for the variance on the different levels in the multilevel method. However, this time, there is an impressive improvement in efficiency by a factor of more than 60 on the higher accuracy level by the multilevel method. And on top of that, it needed the same time on the level of $\epsilon=0.0001$ as the crude method for $\epsilon=0.001$. While in the two examples above, a level of $L=2$ was sufficient, for this very small ϵ a level of $L=3$ was necessary. As a conclusion, the multilevel Monte Carlo method has an enormous potential to reduce the computational complexity, at least when the variance of the estimators on the coarse level are not too small compared to the ones on the finer levels.

4.8 Which simulation methods for SDE should be chosen?

As in Section 3.4.2, it is now a natural question to ask which of the different simulation and discretization methods should be used. Although there is no universally valid recommendation, a rule of thumb is that whenever an exact simulation is possible (i.e. the [conditonal] distribution of the increments of the solution to the SDE under study are known) then one should use the exact simulation. Of course, if the required distribution cannot be generated in a convenient way or is not known in closed form, then one has to rely on numerical schemes for discretizing the SDE.

If in this case the coefficient functions of the SDE are smooth enough to apply a high order method such as the Milstein second order one, then it is recommended to use it. However, in our applications in finance we will often encounter situations where the coefficient functions of the SDE do not satisfy the assumptions that ensure the successful application of even the Milstein first order scheme. In such a situation the simple Euler-Maruyama method is a robust alternative. This is in particular true when one is not interested in the approximation of the solution paths of the SDE. Then, only weak convergence is what we are looking for. Here, the Euler-Maruyama and the Milstein scheme (of first order) have the same order of convergence and there is no need to use the more complicated Milstein scheme.

If many paths of the underlying SDE are needed to obtain our MC estimator (such as e.g. when high accuracy is required) then it is advisable to use variance reduction methods such as the statistical Romberg method or the multilevel Monte Carlo method. While the first one is conceptually much simpler, the second one yields better performance with respect to variance reduction. However, the statistical Romberg method is also more robust than the multilevel method. The application of the multilevel method requires a careful analysis of the problem to be dealt with if a bad performance, as in the first numerical example of Section 4.7.8, should be avoided. Further, if one has to perform a second simulation on a finer grid if one is not satisfied with the performance of the actually done Monte Carlo estimation, then using the Talay-Tubaro estimator is a very convenient way to reuse the already obtained results and at the same time increase the order of convergence.

More aspects of various algorithms, computational examples, or convergence considerations can be found in Kloeden and Platen (1999).

Chapter 5

Simulating Financial Models: Continuous Paths

5.1 Introduction

Modern financial mathematics is definitely among the most popular subjects of applied mathematics today from both the academic and the industry point of view. The attractivity of the research problems and methods paired with enormous interest by students resulted in the creation of many master programmes in financial mathematics including new chairs in this area. The increasing complexity of derivatives traded at stock exchanges, the big demand for sophisticated products, and the exact valuation methods required by the banks, insurance companies, and investors at the financial markets created a huge demand for mathematical methods and models that had not been encountered before in this area.

Financial mathematics is mainly concerned with

- Modelling of the evolution of financial processes such as stock prices, interest rates, inflation, exchange rates, or commodity prices.
- Pricing derivatives on basic underlyings such as stock prices, interest rates, or commodities.
- Portfolio optimization, i.e. the search for optimal investment strategies.
- Risk measurement and management.

Many research papers have been written on the various subjects of financial mathematics during recent decades. Its acceptance in both theory and practise is underlined by the Nobel prizes awarded to H. Markowitz in 1990, to R. Merton and M. Scholes in 1997, and to R. Engle and C. Granger in 2003. Various current monographs on the topics of financial mathematics exist. Just to mention a few, we cite Björk (2004), Bingham and Kiesel (1998), Duffie (2001), Karatzas and Shreve (1998), and Korn and Korn (2001).

In this chapter, we will present the main ideas of stock price modelling, option pricing, and interest rate modelling together with applications of Monte Carlo methods. We restrict ourselves to the case where the underlying driving uncertainty is modelled by a (multidimensional) Brownian motion. In many situations, this makes the analysis of the corresponding problems tractable. However, there are also arguments for considering models for stock prices or interest rates that allow for noncontinuous changes, so-called jumps. We will devote Chapter 7 to these models.

5.2 Basics of stock price modelling

In this section, we will explain the basics of stock price modelling. We concentrate on continuous-time stock price models admitting continuous paths, i.e. the stock prices as a function of time have no jumps.

When we look at the evolution of a stock price or of a stock price index over time we recognize certain remarkable features. Among them are:

- Stock prices do not change in a smooth way over time.
- Locally, (seemingly) random fluctuations dominate a clear tendency.
- The evolution looks similar over various parts of the time interval but has no cycle or seasonality.

This also has implications for the modelling of stock prices. Indeed, if we insist on a stock price model with continuous paths by economical (!) reasoning, the stock price path should be nowhere differentiable as a function of time. If – in contrast – the path would be differentiable at some point, then it would be possible to predict with certainty that the stock price increases or decreases (depending on the sign of the derivative) in the next time instant. Thus, one could predict for sure whether an investment would lead to a gain or to a loss over a short time horizon. Consequently, no one would buy the sure losers and no one would sell the sure winners. Hence, we need a model where the stock price path has no point of differentiability. An obvious ingredient for such a model is the Brownian motion $W(t)$ as introduced in the last chapter.

In fact, an (incomplete) history of stock price modelling starts with the PhD thesis of Bachelier (1900), **Théorie de la Spéculation**. Ahead of his time, he modelled stock prices as a Brownian motion with drift (although at this time the technical term Brownian motion had not yet been coined) via deducing a Fokker-Planck equation for their transition density. The fact that a Brownian motion with drift can attain negative values was a clear argument against this model. Further, the evolution of a stock price should be multiplicative and not additive as its returns should be proportional to the current value. Therefore, in the 1960s, Samuelson (1965) came up with the introduction of the **geometric Brownian motion** as an appropriate stock price model. Here, the problems of the Bachelier model were overcome by simply taking its exponential. The main breakthrough of this model came in the 1970s when the famous Black-Scholes formulae for the price of European call options and put options were developed by Black and Scholes (1973) and by Merton (1973).

During the 1980s the imperfections of the geometric Brownian motion model became clear. This was not only due to extreme events such as the 1987 crash but also due to the characteristics of the prices observed at the option markets. To explain these prices in a better way, various new classes of stock price models entered the scene in the 1990s. The two main streams were the class of **local volatility models** (see e.g. Dupire [1997]) and the class of **stochastic volatility models** (see e.g. Heston [1993]). Both approaches were developed with the aim to explain the nonuniformity in the intensity of the price fluctuations (also known as **volatility clustering**). During the last decade, models based on Lévy processes have been a popular subject in academia (we will refer to this in detail in Chapter 7). The increasing complexity of traded options (such as highly path-dependent options or options on realized stock price volatility) led practitioners to develop generalizations of the local and stochastic volatility models (see e.g. Bergomi [2005]).

As for many problems in finance (such as option pricing or risk management) no explicit analytical solution formulae exist, the use of suitable Monte Carlo methods is often the easy and obvious choice. Therefore, we will explain the different stock price models and the way to simulate them below.

5.3 A Black-Scholes type stock price framework

The model framework we will present in this section is a slight generalization of the famous Black-Scholes model. We will call it the **linear model**. We assume that the dynamics of the prices of n different stocks are given by the following n-dimensional stochastic differential equation (SDE):

$$dS_i(t) = \mu_i(t) S_i(t) dt + \sum_{j=1}^{n} \sigma_{i,j}(t) S_i(t) dW_j(t), \quad S_i(0) = s_i \tag{5.1}$$

for $i = 1, ..., n$ with $\{(W(t), F_t), t \in [0, T]\}$ an n-dimensional Brownian motion. Here, the market coefficients μ (the **drift**) and σ (the **volatility matrix**) are assumed to be F_t-progressively measurable, bounded processes. We also assume that the matrix σ is uniformly positive definite, i.e. we have

$$x'\sigma(t,\omega)\sigma(t,\omega)' x \geq c \cdot x'x \;\forall (t,\omega) \in [0,T] \times \Omega \tag{5.2}$$

for some positive constant c. As a linear SDE, the stock price equation has the unique solution $S_i(t)$ given by

$$S_i(t) = s_i \exp\left(\int_0^t \left(\mu_i(s) - \frac{1}{2}\sum_{j=1}^{n}\sigma_{i,j}^2(s)\right) ds + \sum_{j=1}^{n}\int_0^t \sigma_{i,j}(s)\, dW_j(s)\right). \quad (5.3)$$

In addition to the risky investment in stocks, there is the possibility of a riskless investment in a bond (or better: a bank account). Its evolution over time is governed by the equation

$$dB(t) = r(t)\, B(t)\, dt, \quad B(0) = 1 \quad (5.4)$$

which has the unique solution

$$B(t) = \exp\left(\int_0^t r(s)\, ds\right). \quad (5.5)$$

Here, the interest rate process $r(t)$ is assumed to be bounded and progressively measurable with respect to the filtration F_t.

With this first stock price model, we will introduce the investors (or traders) into our market by specifying their actions and behaviour. The possible actions of an investor are:

1. Rebalancing of his holdings, i.e. he can sell securities and invest the money in other securities. This action will be modelled by the **portfolio process** or by the **trading strategy**.

2. Consuming parts of his wealth which will be incorporated in our setting via the **consumption process**.

Further, an investor should not have insider information. In particular, he is not allowed to have knowledge of future prices. We consider only price takers (so-called **small investors**) which are characterized by the fact that their actions do not influence the stock price behaviour. Our investors are endowed with an initial wealth and then have to act in a self-financing way. Hence, their wealth only changes due to gains/losses from trading and due to consumption. We ignore that stocks are not perfectly divisible and that there are transaction costs. Also, we assume that our investors can hold negative positions in bond and stocks. The possibility for a negative bond position implies that we assume the same interest rate for borrowing and lending. We formalize these standard assumptions in the following definition.

DEFINITION 5.1

Let $\{W(t), F_t\}_{t\in[0,T]}$ be an n-dimensional Brownian motion. Assume that we are in a market where stocks and a bond are traded with price dynamics given

by Equations (5.1) and (5.4).
(a) A **trading strategy** *φ is an $\mathbb{R}^{n+1}$-valued progressively measurable process*

$$\varphi(t) := (\varphi_0(t), \varphi_1(t), ..., \varphi_n(t))' \tag{5.6}$$

such that the integrals

$$\int_0^T \varphi_0(t)\, dB(t), \quad \int_0^T \varphi_i(t)\, dS_i(t), \; i = 1, ..., n \tag{5.7}$$

are all defined and finite. The value $x := \varphi_0(0) + \sum_{i=1}^n \varphi_i(0) s_i$ is called **initial value** *of φ or* initial wealth *of the investor.*
(b) Let φ be a trading strategy with initial value $x > 0$. The process

$$X(t) := \varphi_0(t) B(t) + \sum_{i=1}^n \varphi_i(t) S_i(t) \tag{5.8}$$

is called the **wealth process** *corresponding to φ with* **initial wealth** *x.*
(c) A nonnegative progressively measurable process $c(t)$ with

$$\int_0^T c(t)\, dt < \infty \text{ } \mathbb{P}\text{-a.s.} \tag{5.9}$$

is called a **consumption rate process** *(for short:* **consumption process***).*
(d) A pair (φ, c) consisting of a trading strategy φ and a consumption rate process c is called **self-financing** *if the corresponding wealth process $X(t)$, $t \in [0, T]$ satisfies $\mathbb{P}$-a.s.*

$$X(t) = x + \int_0^t \varphi_0(s)\, dB(s) + \sum_{i=1}^n \int_0^t \varphi_i(s)\, dS_i(s) - \int_0^t c(s)\, ds. \tag{5.10}$$

(e) Let (φ, c) be a self-financing pair consisting of a trading strategy and a consumption process with corresponding wealth process $X(t) > 0$ $\mathbb{P}$-a.s. for all $t \in [0, T]$. Then the $\mathbb{R}^n$-valued process

$$\pi(t) := (\pi_1(t), ..., \pi_n(t))', \; t \in [0, T] \text{ with } \pi_i(t) = \frac{\varphi_i(t) \cdot S_i(t)}{X(t)} \tag{5.11}$$

is called a **self-financing portfolio process corresponding to** *(φ, c).*

REMARK 5.2 1. The portfolio process denotes the fractions of total wealth invested in the different stocks. Therefore, the fraction of wealth invested in the bond is given by

$$\left(1 - \pi(t)' \underline{1}\right) = \frac{\varphi_0(t) \cdot B(t)}{X(t)} \quad \text{where } \underline{1} := (1, ..., 1)' \in \mathbb{R}^n. \tag{5.12}$$

2. Knowing the wealth $X(t)$ and prices $S_i(t)$, the description of the trading/consumption activities of an investor by a self-financing pair (π, c) is equivalent to the use of the pair (φ, c). We will always use the alternative (portfolio process or trading strategy) which is more convenient.

3. The requirement that the integrals in Equation (5.7) are defined and finite simply is a technical condition that ensures that the changes in wealth from gains/losses of investment are defined. The requirements are in particular satisfied under our assumptions on the market coefficients if we have $\mathbb{P}$-a.s.

$$\int_0^T |\varphi_0(t)|\, dt < \infty, \tag{5.13}$$

$$\sum_{j=1}^{n} \int_0^T (\varphi_i(t) \cdot S_i(t))^2\, dt < \infty \text{ for } i = 1, ..., n. \tag{5.14}$$

□

DEFINITION 5.3
A self-financing pair (φ, c) or (π, c) consisting of a trading strategy φ or a portfolio process π and a consumption process c will be called **admissible for the initial wealth** $x > 0$, *if the corresponding wealth process satisfies*

$$X(t) \geq 0 \;\mathbb{P} \text{ a.s. } \forall t \in [0, T]. \tag{5.15}$$

The set of admissible pairs (π, c) with initial wealth x will be denoted by $\mathcal{A}(x)$.

5.3.1 An important special case: The Black-Scholes model

In the Black-Scholes model the market coefficients μ_i, σ_{ij} are assumed to be constant, which leads to bond and stock prices of the form:

$$B(t) = \exp(rt), \tag{5.16}$$

$$S_i(t) = s_i \exp\left(\left(\mu_i - \frac{1}{2}\sum_{j=1}^{n} \sigma_{i,j}^2\right) t + \sum_{j=1}^{n} \sigma_{i,j} W_j(t)\right). \tag{5.17}$$

Here, we can further determine

$$\mathbb{E}(S_i(t)) = s_i \exp(\mu_i t), \tag{5.18}$$

$$\mathbb{V}ar(S_i(t)) = {s_i}^2 \exp(2\mu_i t)) \left(\exp\left(\textstyle\sum_{j=1}^{n} \sigma_{ij}^2 t\right) - 1\right), \tag{5.19}$$

$$\mathbb{C}ov(ln(S_i(t)), ln(S_j(t))) = \textstyle\sum_{k=1}^{n} \sigma_{ik}\sigma_{jk} t. \tag{5.20}$$

Due to the explicit form of the stock price as a function of time and the Brownian motion, $f(t, W(t))$, the simulation of the stock price $S(t)$ or of a component $S_i(t)$ in the Black-Scholes model causes no problems at all:

- If we only need the value of $S(t)$ at a particular time t then it is enough to simulate $N(0, t \cdot I)$-distributed random variables, obtain $W(t)$ from it, and evaluate $f(t, W(t))$.
- If the whole path of $S(t)$ is needed then simulate a path of the Brownian motion $W(t)$ and obtain $S(t)$ from it.

Formally, we have the simple Algorithm 5.1.

Algorithm 5.1 Simulating a stock price path in the Black-Scholes model

---

Let $0 = t_0 < t_1 < ... < t_k = T$ be a partition of $[0, T]$, let $S(0)$ be given.

1. Set $a_j = \mu_j - 0,5 \sum_{i=1}^n \sigma_{ji}^2$.
2. For $j = 1$ to k do
   (a) $\delta = t_j - t_{j-1}$.
   (b) Simulate $Z \sim N(0, \delta \cdot I)$.
   (c) For $i = 1$ to n do $S_i(t_j) = S_i(t_{j-1}) \cdot \exp(a_j \delta + \sum_{m=1}^n \sigma_{im} Z_m)$.
3. Interpolate linearly in between the points of the partition.

We return to the linear model and derive a simple SDE for the wealth process. It is based on switching from the trading strategy to the portfolio process. For this, let (φ, c) be a self-financing pair consisting of a trading strategy and a consumption process. We then obtain

$$
\begin{aligned}
dX(t) &= \varphi_0(t)\, dB(t) + \sum_{i=1}^{n} \varphi_i(t)\, dS_i(t) - c(t)\, dt \\
&= \varphi_0(t)\, B(t)\, r(t)\, dt + \\
&\sum_{i=1}^{n} \varphi_i(t)\, S_i(t) \left(\mu_i(t)\, dt + \sum_{j=1}^{m} \sigma_{ij}(t)\, dW_j(t) \right) - c(t)\, dt \\
&= \left(1 - \pi(t)' \underline{1}\right) X(t)\, r(t)\, dt \\
&+ \sum_{i=1}^{n} X(t)\, \pi_i(t) \left(\mu_i(t)\, dt + \sum_{j=1}^{m} \sigma_{ij}(t)\, dW_j(t) \right) - c(t)\, dt \\
&= \left(1 - \pi(t)' \underline{1}\right) X(t)\, r(t)\, dt \\
&+ X(t) \left(\pi(t)' \mu(t)\, dt + \pi(t)' \sigma(t)\, dW(t)\right) - c(t)\, dt. \qquad (5.21)
\end{aligned}
$$

As this is again a linear SDE, the variation of constants formula ensures that it admits a unique solution given some suitable integrability requirements for $\pi(t)$. As μ, r, σ are uniformly bounded and as the consumption process c is assumed to be integrable, we only have to require

$$\int_0^T \pi_i^2(t)\ dt < \infty\ \mathbb{P}\text{-a.s. for } i = 1, ..., n \tag{5.22}$$

to ensure uniqueness and existence of the wealth Equation (5.21). This allows a definition of the portfolio process without referring to a trading strategy.

DEFINITION 5.4

The progressively measurable $\mathbb{R}^n$-valued process $\pi(t)$ is called a **self-financing portfolio process** *for the consumption process $c(t)$ if the wealth Equation (5.21) has a unique solution $X(t) = X^{\pi,c}(t)$ with*

$$\int_0^T (X(t) \cdot \pi_i(t))^2\, dt < \infty\ \mathbb{P}\text{-a.s. for } i = 1, ..., d. \tag{5.23}$$

Note that the integrability condition (5.23) on the portfolio process and the wealth process is indeed exactly the integrability condition (5.7) on the trading strategy. Further, note that if the portfolio process satisfies the condition (5.22) then continuity of the corresponding wealth process yields that condition (5.23) is satisfied, too. Thus, condition (5.23) is weaker than condition (5.22). In particular, if there is no consumption then condition (5.22) implies strict positivity of the wealth process (simply look at the explicit form of the solution of Equation (5.21) under assumption (5.22)). However, the weaker condition (5.23) allows for portfolio processes that can lead to the ruin of the investor (i.e. for $X(t) = 0$ for some $t \in [0, T]$ or even negative values of $X(t)$). This will be a typical situation in the replication approach to option pricing.

Constant portfolio process and proportional consumption

A simple but practically relevant example of a self-financing pair $(\pi, c) \in \mathcal{A}(x)$ is that of a constant portfolio process and a consumption rate that is proportional to the current wealth, i.e.

$$\pi(t) \equiv \pi \in \mathbb{R}^n \text{ constant},\ c(t) = \gamma \cdot X(t), \tag{5.24}$$

for some $\gamma > 0$ and $X(t)$ the wealth process corresponding to (π, c). Hence, the investor rebalances his holdings in such a way that the fractions of wealth invested in the different stocks and in the bond remain constant over time. Further, the velocity of the increase in consumption ("the consumption rate") is always proportional to the current wealth of the investor. The wealth equation corresponding to this example has the form

$$dX(t) = X(t)\left((r(t) - \gamma)\, dt + \pi'\left((\mu(t) - r(t)\underline{1})\, dt + \sigma(t)\, dW(t)\right)\right) \tag{5.25}$$

resulting in

$$X(t) = x \cdot \exp\left(\int_0^t \left[r(s) - \gamma + \pi' \left(\mu(s) - r(s) \cdot \underline{1}\right) - \frac{1}{2} \left\| \pi' \sigma(s) \right\|^2 \right] ds + \int_0^t \pi' \sigma(s)\, dW(s) \right). \quad (5.26)$$

In particular, $X(t)$ is always strictly positive, and we have $(\pi, c) \in \mathcal{A}(x)$.

5.3.2 Completeness of the market model

We close this section on the linear market by stating the **Theorem on Complete Markets**. For this, we need some abbreviations:

$$\gamma(t) := \exp\left(-\int_0^t r(s)\, ds \right), \theta(t) := \sigma^{-1}(t) \left(b(t) - r(t)\, \underline{1} \right), \quad (5.27)$$

$$Z(t) := \exp\left(-\int_0^t \theta(s)'\, dW(s) - \frac{1}{2} \int_0^t \|\theta(s)\|^2 ds \right), \quad (5.28)$$

$$H(t) := \gamma(t) \cdot Z(t). \quad (5.29)$$

$\theta(t)$ can be interpreted as a kind of relative risk premium for stock investment. The process $H(t)$ will play a crucial role in connection with option pricing. Note that $H(t)$ is positive, continuous, and progressively measurable. Further, it is the unique solution of the SDE

$$dH(t) = -H(t) \left(r(t)\, dt + \theta(t)'\, dW(t) \right), \quad H(0) = 1. \quad (5.30)$$

THEOREM 5.5 Completeness of the market

Assume that we are in the linear market model of this section.
(a) Let $(\pi, c) \in \mathcal{A}(x)$. Then the corresponding wealth process $X(t)$ satisfies

$$\mathbb{E}\left(H(t) X(t) + \int_0^t H(s) c(s)\, ds \right) \leq x \ \forall t \in [0, T]. \quad (5.31)$$

(b) Let $B \geq 0$ be an F_T-measurable random variable and $c(t)$, $t \in [0, T]$ a consumption process satisfying

$$x := \mathbb{E}\left(H(T) B + \int_0^T H(s) c(s)\, ds \right) < \infty. \quad (5.32)$$

Then, there exists a portfolio process $\pi(t)$, $t \in [0, T]$, with $(\pi, c) \in \mathcal{A}(x)$ and the corresponding wealth process $X(t)$ satisfies

$$X(T) = B \ \mathbb{P}\text{-a.s.}. \quad (5.33)$$

REMARK 5.6 1. In view of Part (a) of the theorem the process $H(t)$ can be regarded as the appropriate discounting process that determines the initial wealth at time $t = 0$,

$$\mathbb{E}\left(\int_0^T H(s)\,c(s)\,ds + H(T)\,B\right),$$

which is necessary to be able to attain future aims (such as living according to a given consumption process or obtaining a wealth of B at time $t = T$). Thus, Part (a) puts bounds on the desires of an investor given his initial capital $x \geq 0$. Part (b) proves that future aims which are feasible in the sense of Part (a) can indeed be realized. It thus says that each desired final wealth B in $t = T$ can be **exactly attained** via trading according to an appropriate self-financing pair (π, c) if one possesses sufficient initial capital. This, however, is exactly what we will call a **complete market**.

2. The main tool to derive the above theorem is the so-called Martingale Representation theorem of Chapter 4 (see Korn and Korn [2001] for a proof).

3. By the Complete Markets Theorem we will obtain an explicit solution of the option pricing problem in the linear market setting (see Section 5.5.2). □

5.4 Basic facts of options

The star area of modern continuous-time finance is indeed that of option pricing. It contains the most famous result of financial mathematics, the Black-Scholes formula for pricing European put and call options. The importance of this formula for theory and practical applications is underlined by the Nobel Prize for Economics awarded to Robert Merton and Myron Scholes in 1997 to honour their contributions to option pricing. Fischer Black had passed away two years before and therefore did not receive Nobel prize fame.

What are options?

Options are **derivative securities**, i.e. securities which are derived from underlying assets. They have been traded for centuries, but finally gained economic importance in the last century. This was mainly due to the start of organized option trading with the opening of the Chicago Board Options Exchange in 1973. Simple examples of options are call and put options.

A **call option** (call) is a contract that gives its holder the right (but not the obligation!) to buy a certain fixed amount of an asset during a specified future time period for an already agreed price, the **strike price** or **exercise price**, from the seller or **writer** of the option. Its counterpart is a **put option** (put). Here, the holder has the right to sell a fixed amount of an asset to the writer of the put for the strike price. One distinguishes between so-called **American options** and **European options**. The holder of an American option is free to sell or to buy the asset during the whole time span of the contract, in contrast to a European option where the holder can only exercise his option at the end of the lifetime of the option, the so-called **expiry** or **maturity**.

Today, options are a trademark of modern financial markets, and they are a kind of all-purpose securities. There exist options on equities, bonds, goods such as oil, energy, weather, metals, corn, pork bellies, currencies, or even on options (just to mention a few). They are traded in enormous volumes on stock exchanges all over the world. Also, their explicit contractual form can be very different from the above **plain vanilla** puts and calls. We will present many examples of these **exotic options** below. We first give a formalization of European calls and puts:

The European call and the European put as basic examples

A European call on one share of a stock gives its holder the right to buy this share at time $t = T$ for the strike price $K \geq 0$ which was fixed at time $t = 0$. Hence, if the final share price $S_1(T)$ exceeds K, the holder of the option buys the share for a price of K and can then sell it immediately at the market for the price of $S_1(T)$. This leads to a gain of $S_1(T) - K$ (ignoring transaction costs). In the case of $S_1(T) < K$, the holder does not make use of his right to buy the share for a price K. Thus, in this case, there is no gain from holding the option. Combining the two cases leads to a final payment of

$$(S_1(T) - K)^+ \quad \text{in } t = T. \tag{5.34}$$

The holder of a European put has the right to sell one share at time $t = T$ for the price $K > 0$. Hence, similar as in the case of a European call, one can show that the possession of the European put leads to a payment of

$$(K - S_1(T))^+ \quad \text{in } t = T. \tag{5.35}$$

Practitioners often think of options in terms of their **payoff diagrams**. The payoff diagram of an option is the graph of the final gain through this option as a function of the underlying stock price $S_1(T)$. This of course requires that the final payment is a function of $S_1(T)$. Figure 5.1 shows the payoff diagram for the European call and the European put.

An easy way to generate new types of options is to combine these two types of payoff diagrams to obtain different types of payoff profiles. This is very popular among traders. Mathematically, it corresponds to holding a certain combination of puts and calls.

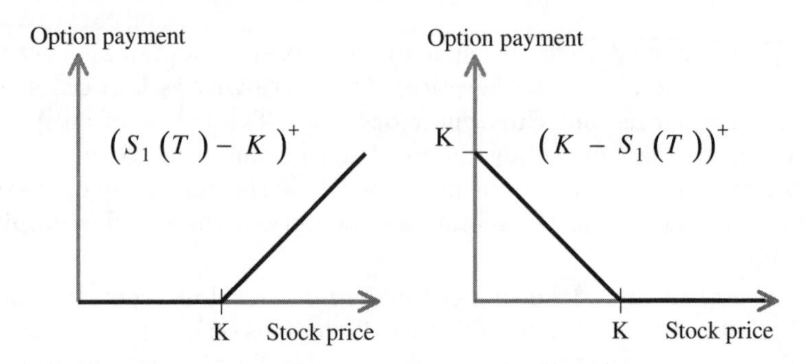

FIGURE 5.1: Payoffs of a call option (left) and of a put option (right).

A short history of option trading

At the beginning of the 17th century, the first use of option type contracts took place in the Netherlands. During the so-called tulip mania, some kind of put options were bought by tulip growers to insure themselves against high price fluctuations. However, when in 1637 the Dutch tulip market crashed, the option sellers were not able to keep their part of the agreements. This resulted in a serious economic crisis in the Netherlands. To guarantee the safety of both the writer and the seller in option contracts, it needs an institution that acts as an intermediate that steps in if one of the two sides defaults. Organized trading of options started in London in the 18th century. Still, due to the absence of strict laws for option trading, irregularities frequently occurred. This ended when in 1930, option trading received a legal framework. In the beginning of the 1970s of the 20th century, trading in options started gaining the economic importance it has today. With the opening of the Chicago Board Options Exchange in 1973 as a starting point, organized option trading spread all over the world. Nowadays, options are ubiquitous in financial markets.

Reasons for trading options

The trading of options is mainly justified by two reasons, **protection** and **speculation**. An easy application is the protection of a stock position. An investor that holds one share of a certain stock and wants to ensure that the value of this position at time T will not fall below a specified value K simply buys a European put on this stock with maturity T and strike K. The value of the portfolio made up of the share and the put at time T then equals

$$S(T) + (K - S(T))^+ \geq K. \tag{5.36}$$

Of course, the investor has to pay a certain price for the put option to obtain this protection. Further examples to insure oneself against unfavourable price evolutions can be easily constructed.

Options are also traded by speculators who hope for an overproportional increase of the option value compared to the price of the underlying asset. For example, it is obvious that the price of an option increases by less than one Euro if the price of the underlying stock increases by one Euro. However, the relative price increase of the option will typically be higher than that of the stock in this case. This is the so-called **leverage effect**. Further, options are much cheaper and more liquid than the underlying asset itself which is again very attractive for speculators. On the other hand, speculation with options can lead to a total loss as options only have a finite lifetime, and a zero-payoff is a natural event in options such as calls or puts.

5.5 An introduction to option pricing

5.5.1 A short history of option pricing

The modern theory of option pricing started with the dissertation **Théorie de la Spéculation** of L. F. Bachelier (see Bachelier [1900]). There, with stock prices modelled as a Brownian motion with drift, he wanted to derive theoretical prices for options on these stocks to compare them with the actual market prices. He suggested using the expected value of the discounted corresponding option payments as the option price. However, it is one of the most spectacular results of modern option pricing theory that this suggestion does in general not (!) yield a reasonable option price!

The decisive breakthrough of option pricing in its modern form was obtained by Fischer Black and Myron Scholes in Black and Scholes (1973). They derived a partial differential equation for the option price and solved it. At the same time, Robert Merton obtained a generalization of the results of Black and Scholes in Merton (1973). The perhaps most elegant approach to option pricing is the so-called **replication approach** which is based on pure arbitrage considerations in a complete market setting. It simply states that two financial assets should have the same price if their future payments coincide in all states of the world. This approach, which is a direct application of martingale theory, started with the work of Harrison and Pliska (1981). This will be described in the next section.

5.5.2 Option pricing via the replication principle

Motivation: Option pricing in the one-period binomial model

We introduce the main idea of the replication approach in a one-period **binomial model**. There, the market consists of a bond and a stock with trading dates 0 and T. The bond price process is given by $B(0) = 1, B(T) = \exp(rT)$. The stock price starts with an initial value of $S(0)$ and can attain two possible values, $dS(0)$ or $uS(0)$ with $d < u$. The probability for $S(T) = uS(0)$ is assumed to equal $p \in (0,1)$. Note further that we must have

$$d < \exp(rT) < u \tag{5.37}$$

to avoid possibilities for riskless gains without the need for investing initial capital, so-called **arbitrage opportunities**. To see this, assume first that we have $d \geq \exp(rT)$. Then by borrowing money and investing it in the stock, we always receive at least the amount of money at the terminal time T to pay back the credit. However, in the case of $S(T) = uS(0)$, we even have money left after paying back our initial credit. Hence, everyone would enter such an investment. Therefore, the market would increase the initial stock price until this possibility disappears. With a similar argument, we can show that the relation $\exp(rT) \geq u$ cannot be satisfied.

We now consider a call option with strike $K = 100$ and maturity $T = 1$ in the binomial model. Further, we choose $u = 1.2$, $d = 0.95$, and $r = 0$. This results in the payment streams given in Figure 5.2.

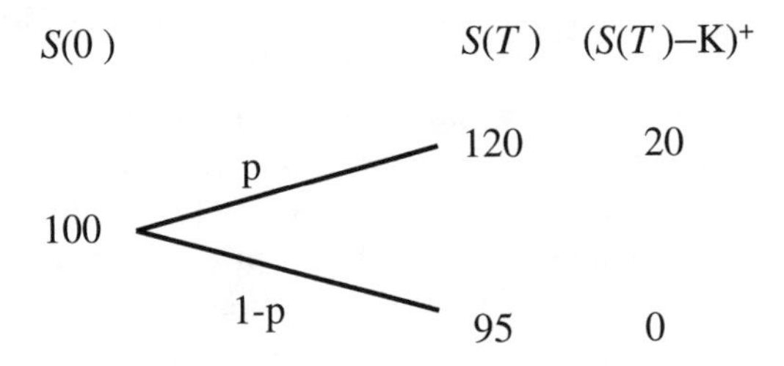

FIGURE 5.2: Payment streams for stock and call in the binomial model.

Bachelier's suggestion to use the **net present value**

$$\mathbb{E}\left(\exp(-rT)\left(S_1(T) - K\right)^+\right) \tag{5.38}$$

as the option price would lead to a price of the above call of

$$\mathbb{E}\left(\left(S_1(T) - K\right)^+\right) = (20) \cdot p + 0 \cdot (1-p) = 20 \cdot p. \tag{5.39}$$

As expected, this suggested price depends heavily on the probability of success, p. However, two different traders would in general not agree on the actual value of p. It is thus a nice feature of the replication approach to option pricing that p is not needed for calculating the option price at all. The main reason for this is that the final payment of the option can be obtained by following a suitable self-financing trading strategy in stock and bond. This principle of synthetically constructing the option is called the **replication principle**. For this we have to determine $(\varphi_0(0), \varphi_1(0))$ such that we obtain

$$X(T) = \varphi_0(0)\,B(T) + \varphi_1(0)\,S(T) = (S(T) - K)^+. \tag{5.40}$$

Then, we define the option price $\hat{C}$ of the call as the initial capital needed in $t = 0$ to buy the replication strategy $(\varphi_0(0), \varphi_1(0))$,

$$\hat{C} = \varphi_0\,(0)\,B\,(0) + \varphi_1\,(0)\,S\,(0)\,. \tag{5.41}$$

This is the only reasonable price for the option. To see this, assume first that the option price $\tilde{C}$ would be smaller than $\hat{C}$. Then we could buy the option for $\tilde{C}$ and sell $(\varphi_0(0), \varphi_1(0))$ for $\hat{C}$ (i.e. we hold the position $(-\varphi_0(0), -\varphi_1(0))$. At time $t = T$ the payments obtained from the option and from holding the position $(-\varphi_0(0), -\varphi_1(0))$ neutralize each other. Hence, we would have realized a gain of $\hat{C} - \tilde{C}$ in $t = 0$ without the use of our own capital. In case of $\tilde{C} > \hat{C}$ we would sell the call and hold the position $(\varphi_0(0), \varphi_1(0))$ which only costs $\hat{C}$. Again, we would realize a riskless gain without the need for investing initial capital. Thus, in both cases there exist arbitrage opportunities.

As an arbitrage opportunity would be realized by all market participants, it would lead to immediate price adjustments wiping out the arbitrage opportunity. Hence, it is reasonable to assume the absence of arbitrage.

In our example requirement (5.40) yields the system of equations

$$\varphi_0\,(0) \cdot 1 + \varphi_1\,(0) \cdot 120 = 20 \tag{5.42}$$

$$\varphi_0\,(0) \cdot 1 + \varphi_1\,(0) \cdot 95 = 0 \tag{5.43}$$

with the unique solution

$$(\varphi_0\,(0)\,, \varphi_1\,(0)) = \left(-76, \tfrac{4}{5}\right), \tag{5.44}$$

leading to an option price of

$$\hat{C} = -76 \cdot 1 + \tfrac{4}{5} \cdot 100 = 4. \tag{5.45}$$

This price is independent of the unknown probability p. Note further that the above calculated call price coincides with the expected value of the discounted terminal payment of the call if and only if we have $p = 1/5$. In this case, $S_1(t)$ is a martingale. This is no coincidence as we will see in Section 5.5.2.

Option pricing in the linear diffusion market model

We take up the idea of finding the option price in a complete market by replication in the linear market model. Before we can do so, we introduce the mathematical definition of an option, or more generally, a contingent claim, and the notion of an arbitrage opportunity.

DEFINITION 5.7

A self-financing and admissible pair (φ, c), consisting of a trading strategy φ and a consumption process c, is called an **arbitrage opportunity** *if the* *corresponding wealth process satisfies:*

$$X(0) = 0\,, \ \ X(T) \geq 0 \ \mathbb{P}\text{-a.s.}, \tag{5.46}$$

$$\mathbb{P}(X(T) > 0) > 0 \ \text{ or } \ \mathbb{P}\left(\int_0^T c(t)\, dt > 0\right) > 0. \tag{5.47}$$

Thus an arbitrage possibility possibly generates money out of nothing, but never leads to a negative payment. One can show that the linear market model is free of arbitrage opportunities (see Korn and Korn [2001]). We introduce a contingent claim as a generalization of an option.

DEFINITION 5.8
A (European) **contingent claim** (g, B) *consists of an* $\{F_t\}_t$*-progressively measurable payout rate process* $g(t)$*,* $t \in [0, T]$ $g(t) \geq 0$*, and an* F_t*-measurable terminal payment* $B \geq 0$ *at time* $t = T$ *with*

$$\mathbb{E}\left(\left(\int_0^T g(t)\ dt + B\right)^{\mu}\right) < \infty \ \text{ for some } \mu > 1. \tag{5.48}$$

REMARK 5.9 1. Note that in addition to a general final payment B, we have also introduced a payment process which can be used as a model for a continuous dividend flow, a so-called dividend yield. As a slight misuse, we shall often use the name option as a synonym for contingent claim.

2. A European contingent claim only generalizes European options as the payment time is fixed. American type options will be treated separately. □

To introduce the replication approach, we define a replication strategy and the notion of the fair price.

DEFINITION 5.10
(a) $(\pi, c) \in A(x)$ *is called a* **replication strategy** *for the contingent claim* (g, B) *if we have*

$$g(t) = c(t) \ \ \mathbb{P}\text{-a.s.} \forall t \in [0, T]\,, \ \ X(T) = B \ \mathbb{P}\text{-a.s.} \tag{5.49}$$

where $X(t)$ *is the wealth process corresponding to* (π, c)*.*
(b) The set of replication strategies of price x *is defined by*

$$\mathcal{D}(x) := \mathcal{D}(x; (g, B)) := \{(\pi, c) \in \mathcal{A}(x) \mid (\pi, c) \text{ replication strategy for } (g, B)\} \tag{5.50}$$

(c) The **fair price** $\hat{p}_{g,B}$ *of the contingent claim* (g, B) *is defined as*

$$\hat{p}_{g,B} := \inf\{p \mid \mathcal{D}(p) \neq \emptyset\}\,. \tag{5.51}$$

The existence of a replication strategy is ensured by the Theorem on Complete Markets. Its second part also suggests a candidate for the fair price,

$$\tilde{x} = \mathbb{E}\left(H(t)\, B + \int_0^T H(t)\, g(t)\, dt \right). \tag{5.52}$$

Indeed, one can show (see Korn and Korn [2001]) the following.

THEOREM 5.11 Fair price of a contingent claim
The fair price of the contingent claim (g, B) *is given by*

$$\hat{p}_{g,B} = \mathbb{E}\left(H(T)\, B + \int_0^T H(t)\, g(t)\, dt \right) < \infty, \tag{5.53}$$

and there exists a unique replication strategy $(\hat{\pi}, \hat{c}) \in \mathcal{D}(\hat{p}_{g,B})$. *Its wealth process* $\hat{X}(t)$ *(also called* **valuation process** *of* (g, B)*) is given by*

$$\hat{X}(t) = \frac{1}{H(t)} \mathbb{E}\left(H(T)\, B + \int_t^T H(s)\, g(s)\; ds \,|F_t \right). \tag{5.54}$$

This theorem leaves many aspects to work on:

- The explicit computation of the fair price for special choices of (g, B).
- The use of Monte Carlo methods for calculating the fair price if there is no explicit formula.
- The interpretation of the pricing formula.

We will now deal with these topics.

The Black-Scholes formula

For the special cases of European call and put options, the expectations in the fair price theorem can be calculated in a completely explicit form, the so-called Black-Scholes formula (see Black and Scholes [1973]).

COROLLARY 5.12 Black-Scholes formula
Consider the Black-Scholes market model with $n = m = 1$, $r(t) \equiv r$, $b(t) \equiv b$, $\sigma(t) \equiv \sigma > 0$ *for all* $t \in [0, T]$, $T > 0$, $r, b, \sigma \in \mathbb{R}$. *Then, we have:*
(a) The price $X_C(t)$ *at time* t *of a European call option with strike* $K > 0$ *and maturity* T *is given by*

$$X_C(t) = S_1(t)\, \Phi(d_1(t)) - K e^{-r(T-t)} \Phi(d_2(t)), \tag{5.55}$$

$$d_1(t) = \frac{\ln\left(\frac{S_1(t)}{K}\right) + \left(r + \frac{1}{2}\sigma^2\right)(T-t)}{\sigma\sqrt{T-t}}, \quad d_2(t) = d_1(t) - \sigma\sqrt{T-t} \tag{5.56}$$

where Φ is the distribution function of the standard normal distribution.
(b) The price $X_P(t)$ of a European put option with strike price $K > 0$ and maturity T is given by (with $d_i(t)$ as in Equation (5.56))

$$X_P(t) = Ke^{-r(T-t)}\Phi(-d_2(t)) - S_1(t)\Phi(-d_1(t)). \tag{5.57}$$

Understanding the Black-Scholes formula and its scope

The Black-Scholes formula is a cornerstone of modern financial mathematics. It is a sound mathematical basis for option trading and paved the way for the rapid growth of option markets and of the number of different, much more complicated, so-called **Exotic options**. The main reason for its success is also its main mystery: While the (in principle) observable riskless interest rate r enters the pricing formula, the mean rate of stock return b does not appear. As b is not observable at all (and also hard to estimate from time series data!), it has been greatly appreciated by traders that no view on b is needed to price calls and puts. Further, as b can be interpreted as a preference parameter characterizing the attractiveness of stock investment, the above valuation formulae for the European call and put are also called a **preference free valuation**.

Of course, this fact needs an explanation. The answer is similar to the explanation in the one-period binomial setting. There exists a probability measure $\mathbb{Q}$ underlying the pricing mechanism that is different from the personal **subjective** probability measure $\mathbb{P}$. We will give two ways to explain this, the idea of **market consistent pricing** and the concept of **change of measure** to the equivalent martingale measure.

To understand the idea behind the consistent pricing explanation, we look at the net present value of the stock held at the future time T,

$$\mathbb{E}\left(e^{-rT}S_1(T)\right) = s_1 e^{(b-r)T}. \tag{5.58}$$

This net present value equals today's price s_1 of the stock if and only if we choose $b = r$. Hence, if the market as a total has computed the actual price of the stock via the principle of net present value, the market must have assumed equality between b and r. It is now easy to verify by explicit integration that for $b = r$ the option pricing formulae obtained by Black and Scholes equal

$$\mathbb{E}\left(e^{-rT}(S_1(T) - K)^+\right), \ \mathbb{E}\left(e^{-rT}(K - S_1(T))^+\right).$$

Thus, the Black-Scholes formula can be explained by the fact that the market simply uses the same assumption $b = r$ for computing option prices that has already been used for calculating the underlying stock price.

The concept of change of measure can be explained with the help of Girsanov's theorem (see Chapter 4). We introduce a new Brownian motion

$$W^{\mathbb{Q}}(t) := W(t) + \theta \cdot t \tag{5.59}$$

with $\theta = (b-r)/\sigma$ under the probability measure $\mathbb{Q}$ obtained via

$$\mathbb{Q}(A) = \mathbb{E}(1_A Z(T)) \ \forall A \in F_T \tag{5.60}$$

with Z defined as in Equation (5.28). We then have

$$S_1(t) = s_1 \exp\left(\left(r - \frac{1}{2}\sigma^2\right)t + \sigma W^{\mathbb{Q}}(t)\right). \tag{5.61}$$

Thus, we obtain (remember the definition of $H(T)$)

$$\begin{aligned} X_C(0) &= \mathbb{E}\left(\left(s_1 \exp\left[\left(b - \tfrac{1}{2}\sigma^2\right)T + \sigma W(T)\right] - K\right)^+ Z(T)\right) \\ &= \mathbb{E}_{\mathbb{Q}}\left(\exp(-rT)\left(s_1 \exp\left[\left(r - \tfrac{1}{2}\sigma^2\right)T + \sigma W^{\mathbb{Q}}(T)\right] - K\right)^+\right) \end{aligned} \tag{5.62}$$

where $\mathbb{E}_{\mathbb{Q}}(.)$ denotes the expected value with respect to the measure $\mathbb{Q}$. This last representation explains the independence of the Black-Scholes price from b. Further, from Equation (5.61) we obtain that

1. The discounted stock price $S_1(t)/B(t) = s_1 \cdot \exp\left(\sigma W^{\mathbb{Q}}(t) - \frac{1}{2}\sigma^2 t\right)$ is a $\mathbb{Q}$-martingale. Therefore, $\mathbb{Q}$ is called an **equivalent martingale measure**, and we have

$$dS_1(t) = S_1(t)\left(r\,dt + \sigma\,dW^{\mathbb{Q}}(t)\right). \tag{5.63}$$

2. The option price equals its net present value in the so-called **risk-neutral market** (given by the use of $\mathbb{Q}$) where all normalized security prices $S_i(t)/S_i(0)$ have the same expectation.

Option pricing and the equivalent martingale measure

As the use of equivalent martingale measures is a central concept in option pricing, we give a general definition of an equivalent martingale measure.

DEFINITION 5.13

Consider a probability space $(\Omega, F, \mathbb{P})$ equipped with a filtration $\{F_t : t \in [0,T]\}$. Assume that on this probability space a financial market model consisting of $n+1$ securities with price processes $S_0(t), ..., S_n(t)$ is defined where the price process $S_0(t)$ is strictly positive. Then, a probability measure $\mathbb{Q}$ which is equivalent to $\mathbb{P}$ (i.e. both probability measures have the same zero sets) is called an **equivalent martingale measure** *(EMM) for this market if all discounted price processes*

$$\tilde{S}_i(t) = \frac{S_i(t)}{S_0(t)} \tag{5.64}$$

are martingales with respect to $\mathbb{Q}$ (and $\{F_t : t \in [0,T]\}$).

REMARK 5.14 One can show that the existence of an EMM implies that the corresponding financial market is free of arbitrage opportunities. Contrarily, the absence of arbitrage opportunities only implies the existence of an EMM under some (mild) additional technical conditions. In a complete market (i.e. a market where each contingent claim can be replicated) there exists only one EMM (see e.g. Björk [2004]). □

As our linear diffusion market model is complete, there is exactly one such EMM $\mathbb{Q}$. It can be calculated as in the Black-Scholes model above by

$$\mathbb{Q}(A) = \mathbb{E}(1_A Z(T)) \ \forall A \in F_T \tag{5.65}$$

With its help we obtain a representation of the option price which is more suitable for applying the Monte Carlo method:

THEOREM 5.15 Option pricing with the EMM
Assume that we are in the linear market model with EMM $\mathbb{Q}$. Let (g, B) be a contingent claim. Then for $0 \leq t \leq T$ its price process $\hat{X}(t)_{g,B}$ is given by

$$\hat{X}_{g,B}(t) = E_{\mathbb{Q}}\left(e^{-\int_t^T r(s)ds} B + \int_t^T e^{-\int_t^s r(u)du} g(s)ds \,|F_t \right). \tag{5.66}$$

A very important consequence of this theorem is the following: **for the purpose of option pricing we can always assume**

$$b_i(t) = r(t) \ \text{ for } i = 1, ..., n. \tag{5.67}$$

For completeness, we also present the **Black-Scholes partial differential equation** (Black-Scholes PDE) as used in Black and Scholes (1973) to derive the Black-Scholes formula in their original paper. It is based on the relation between PDEs and SDEs given by the Feynman-Kac representation theorem.

THEOREM 5.16 Option pricing and the Black-Scholes PDE
Assume that we are in the one-dimensional Black-Scholes framework.
(a) The Black-Scholes PDE

$$\frac{1}{2}\sigma^2 s^2 C_{ss} + rsC_s + C_t - rC = 0, \quad \text{for } (t,s) \in [0,T) \times (0,\infty) \tag{5.68}$$

$$C(T,s) = (s-K)^+, \quad \text{for } s \geq 0 \tag{5.69}$$

has a unique solution $C \in C([0,T] \times (0,\infty)) \cap C^{1,2}([0,T) \times (0,\infty))$ given by

$$C(t, S_1(t)) = S_1(t)\,\Phi(d_1(t)) - Ke^{-r(T-t)}\Phi(d_2(t)) \tag{5.70}$$

with $d_i(t)$ as in Equation (5.56). *In particular, the unique solution coincides with the Black-Scholes formula.*
(b) The unique replication strategy (φ_0, φ_1) for the European call is given by

$$\varphi_0(t) = (C(t, S_1(t)) - C_s(t, S_1(t)) S_1(t)) e^{-rt}, \tag{5.71}$$

$$\varphi_1(t) = C_s(t, S_1(t)). \tag{5.72}$$

REMARK 5.17 1. To price other European options with a final payment of the form $f(S_1(T))$ with the help of the Black-Scholes PDE, only the boundary condition (5.69) has to be changed to

$$C(T, s) = f(s), \quad \text{for } s \geq 0. \tag{5.73}$$

2. The above theorem yields the representation of the stock part of the replication strategy as the partial derivative of the option price with respect to the underlying, the so-called **delta** of the option. □

A very useful result for computing option prices is the so-called **Log-normal valuation formula.**

PROPOSITION 5.18
Let $X \sim N(0,1)$, $m \in \mathbb{R}$, $v, K \geq 0$. Then we have:

$$\mathbb{E}\left(\left(ye^{m+vX} - K\right)^+\right) = ye^{\tilde{m}}\Phi(d_1) - K\Phi(d_1 - v), \tag{5.74}$$

$$\tilde{m} = m + \tfrac{1}{2}v^2, \quad d_1 = \frac{\ln(y/K) + (m + v^2)}{v}. \tag{5.75}$$

5.5.3 Dividends in the Black-Scholes setting

In real markets, dividends are one attractive feature for a stock investment. They are usually paid in lump sum at (approximately) fixed dates. For the discussion of realistic modelling of dividend payments we refer to Korn and Rogers (2005). A popular approximation used by practitioners is that of a continuous dividend stream, i.e. there is a continuous payment stream given by $\delta S_1(t)dt$. This then leads to a stock price equation of

$$dS_1(t) = S_1(t)((r - \delta)dt + \sigma dW(t)) \tag{5.76}$$

in the risk-neutral market. As the dividend stream is paid to the holder of the stock and not to the holder of the option, one can show that the modified **Black-Scholes formula with continuous dividends** is obtained by replacing $S_1(t)$ by $\exp(-\delta t)S_1(t)$:

$$X_C(t) = e^{-\delta t} S_1(t) \Phi(d_1(t)) - Ke^{-r(T-t)}\Phi(d_2(t)) \tag{5.77}$$

$$d_1(t) = \frac{\ln\left(\frac{S_1(t)}{K}\right) + \left((r-\delta) + \frac{1}{2}\sigma^2\right)(T-t)}{\sigma\sqrt{T-t}}, \quad d_2(t) = d_1(t) - \sigma\sqrt{T-t}. \tag{5.78}$$

As the replacement of $S_1(t)$ by $\exp(-\delta t)S_1(t)$ is also valid for other models, we will not consider a dividend rate separately in the following. For simulation reasons, one can show that the correction term of $-S_1(t)\delta dt$ appearing in the SDE (5.76) is also valid in other models, a reason for the popularity of the use of continuous dividend streams.

5.6 Option pricing and the Monte Carlo method in the Black-Scholes setting

Although the Black-Scholes model cannot explain all characteristics of real market prices, it is still used as the benchmark in practical applications. Further, it often serves as an orientation for the price of a very complicated, high-dimensional option. Therefore, we introduce the Monte Carlo method in a Black-Scholes setting. Note that due to the use of the unique EMM $\mathbb{Q}$ for pricing options we can always assume

$$b_i = r \tag{5.79}$$

which is equivalent to assume $\mathbb{P} = Q$, i.e. we are directly modelling under the equivalent martingale measure. Indeed, for the purpose of applying Monte Carlo methods to option pricing it is only necessary to know that **an option price is a (discounted) expected value of a random variable**. We will also mainly neglect the payout stream g for simplicity. For the payment B of a (European) option the Monte Carlo task is to approximate $\mathbb{E}_{\mathbb{Q}}\left(e^{-rT}B\right)$.

Algorithm 5.2 Option pricing via Monte Carlo simulation

1. Simulate n independent realizations B_i of the final payoff B.

2. Choose the discounted mean as an approximation for the option price,

$$\left(\frac{1}{n}\sum_{i=1}^{n} B_i\right) \cdot e^{-rT} \approx \mathbb{E}_{\mathbb{Q}}\left(e^{-rT}B\right).$$

While the second step of this algorithm causes no problems, simulating the final payment B of the option depends on many aspects such as

- The form of the payoff: Does it depend only on one single point in time or on the (whole) path of the underlying?

- The underlying: Do we have one or many underlying stocks?

We will therefore consider multi- and single-asset options separately and will also distinguish between the Monte Carlo valuation of path-dependent and path-independent options.

The use of the Monte Carlo method for option pricing has been pioneered by Boyle (1977). Boyle et al. (1997) can be viewed as the beginning of the more recent research in applying sophisticated Monte Carlo techniques to the pricing of complicated options.

5.6.1 Path-independent European options

Here, we always assume to have an option payoff of the form:

$$B = f(S(T)) = g(W(T)) \tag{5.80}$$

with $S(T) = (S_1(T), ..., S_n(T))$. To obtain g note that the explicit form of the stock prices in the Black-Scholes model implies $S(T) = h(W(T))$ with

$$h_i(x) = s_i \exp\left(\left(r - \frac{1}{2}\sum_{j=1}^{n} \sigma_{ij}^2\right) T + \sum_{j=1}^{n} \sigma_{ij} x_j\right). \tag{5.81}$$

Of course, it is only worth using a Monte Carlo method for calculating the option price if it has no closed-form analytical representation. In such a situation, a simple algorithm for obtaining a Monte Carlo price $\hat{p}_{B,N}$ is:

Algorithm 5.3 MC pricing of path-independent options

Let $f(S(T)) = g(W(T))$ be the final payoff of an option.

1. Set $\hat{p}_{B,N} = 0$.
2. For $i = 1$ to N do
   (a) Simulate $Z^{(i)} \sim N(0, I)$.
   (b) Calculate $B^{(i)} = g\left(\sqrt{T} Z^{(i)}\right)$.
   (c) Set $\hat{p}_{B,N} = \hat{p}_{B,N} + B^{(i)}$.
3. Set $\hat{p}_{B,N} = \frac{1}{N} e^{-rT} \hat{p}_{B,N}$.

This is only a crude Monte Carlo framework. If, however, we consider a particular class of options then we can use all kinds of variance reduction methods from Chapter 3. We look at basket options as an example:

Pricing basket options with moment matching and control variate methods

The payment of a basket option depends on an average of a **basket** of stock prices, i.e. a (European) basket call has a final payment of

$$B = \left(\sum_{i=1}^{n} w_i S_i (T) - K \right)^+ \tag{5.82}$$

where the weights w_i usually sum up to 1. As the distribution of this sum of stock prices is not known, there is no closed formula for the basket call price. There exist various approximation methods for pricing a basket option, but they all have weaknesses (see Krekel et al. [2004]). The most popular one is **Lévy's moment matching method** (see Lévy [1992]). It replaces the basket payment by a log-normally distributed random variable Z with the same mean and variance, i.e. we define

$$Z = \exp (m + vX) \tag{5.83}$$

with $X \sim N(0,1)$ and m, v given by

$$m = 2\ln(M) - \tfrac{1}{2}\ln\left(V^2\right),\ v = \ln\left(V^2\right) - 2\ln(M), \tag{5.84}$$

$$M = \mathbb{E}\left(\textstyle\sum_{i=1}^n w_i S_i(T)\right) = e^{rT} \textstyle\sum_{i=1}^n w_i s_i, \tag{5.85}$$

$$V^2 = \mathbb{E}\left(\textstyle\sum_{i=1}^n w_i S_i(T)\right)^2 = e^{2rT} \textstyle\sum_{i,j=1}^n s_i s_j e^{\left(\sum_{k=1}^n \sigma_{ik}\sigma_{jk}\right)T} \tag{5.86}$$

With this notation the approximation for the basket price is given by a Black-Scholes type formula

$$\hat{p}_{basket} \approx e^{-rT}\left(M\Phi(d_1) - K\Phi(d_2)\right), \tag{5.87}$$

$$d_1 = \tfrac{m+v^2-\ln(K)}{v},\ d_2 = d_1 - v. \tag{5.88}$$

While the Lévy approximation yields good results for a variety of parameters r, σ, it performs weakly if the volatilities of the stocks are quite different. In such a case, a Monte Carlo approach with a well-chosen control variate is a good choice. There are two obvious candidates for a control variate:

- the geometric mean call with $B_{geo} = \left(\left(\prod_{i=1}^n n w_i S_i(T)\right)^{1/n} - K\right)^+$,
- the weighted sum $B_w = \sum_{i=1}^n w_i \left(S_i(T) - K\right)^+$ of the single stock calls.

For both choices we can compute the explicit option prices. For the weighted mean of the calls, the Black-Scholes formula yields

$$p_{B_w} = \textstyle\sum_{i=1}^n w_i \left(s_i \Phi\left(d_1^{(i)}\right) - K e^{-rT}\Phi\left(d_2^{(i)}\right)\right) \tag{5.89}$$

$$d_1^{(i)} = \frac{\ln(s_i/K) + \left(r + \frac{1}{2}\nu_i^2\right)T}{\nu_i\sqrt{T}},\ \ d_2^{(i)} = d_1^{(i)} - \nu_i\sqrt{T},\ \ \nu_i^2 = \textstyle\sum_{j=1}^n \sigma_{ij}^2. \tag{5.90}$$

The log-normal valuation formula (see Proposition 5.18) yields the price for the geometric average call as follows.

THEOREM 5.19 Option price of a geometric average basket call
The price of a geometric average basket call with $\omega_i = 1/n$ in the Black-Scholes model is given by

$$p_{B_{geo}} = e^{-rT}\left(\tilde{s}e^{\tilde{m}}\Phi\left(\tilde{d}_1\right) - K\Phi\left(\tilde{d}_2\right)\right), \tag{5.91}$$

$$\nu = \tfrac{1}{n}\sqrt{\textstyle\sum_{j=1}^{n}\left(\sum_{i=1}^{n}\sigma_{ij}^2\right)^2},\ m = rT - \tfrac{1}{2n}\textstyle\sum_{i,j=1}^{n}\sigma_{ij}^2 T, \tag{5.92}$$

$$\tilde{m} = m + \tfrac{1}{2}\nu^2, \tilde{s} = \left(\textstyle\prod_{i=1}^{n} s_i\right)^{1/n}, \tilde{d}_1 = \tfrac{\ln(\tilde{s}/K)+m+\nu^2}{\nu}, \tilde{d}_2 = \tilde{d}_1 - \nu. \tag{5.93}$$

As the geometric mean often differs a lot from the arithmetic mean, we also apply a certain way of moment matching: We use a modified strike $\tilde{K}$ in the geometric average basket option such that the moments of the linear versions of the payoffs coincide, i.e. we require

$$\mathbb{E}\left(\left(\frac{1}{n}\sum_{i=1}^{n} S_i(T) - K\right)\right) = \mathbb{E}\left(\left(\prod_{i=1}^{n} S_i(T)\right)^{1/n} - \tilde{K}\right) \tag{5.94}$$

which – with the notations of Theorem 5.19 – results in

$$\tilde{K} = K - e^{rT}\frac{1}{n}\sum_{i=1}^{n} s_i + \tilde{s}e^{\tilde{m}}. \tag{5.95}$$

Example 5.20
We look at a basket call on four stocks with $T = 5$, $K = 100$, $S_i(0) = 100$, and equal weights of $w_i = 0.25$. We assume $r = 0$, equal correlations between the log-returns of 0.5, equal volatilities for each stock of $0.4 = (\sigma_{i1}^2 + ... + \sigma_{i4}^2)^{1/2}$, and $i = 1, .., 4$. Table 5.1 clearly demonstrates the variance reduction obtained by the above presented control methods. Although the use of the single calls as control variate performs well here, it can lead to skewed results if the stock prices have very different volatilities.

5.6.2 Path-dependent European options

For the pricing of path-dependent options we need simulations of (parts of the) paths of the underlying price processes. We give examples of those options and demonstrate the application of variance reduction techniques.

Asian options and moment matched control variates

Asian options have the feature that their payoff contains an averaging procedure over the path of the price of a stock. They are typically single-stock

Method N	10,000		1,000,000	
	Price	95%-CI	Price	95%-CI
Crude MC	27.07	[25.84, 28.29]	27.99	[27.86, 28.12]
MC with geom. mean	27.86	[27.46, 28.26]	27.99	[27.95, 28.03]
MC with corr. geom. mean	28.00	[27.68, 28.33]	27.97	[27.93, 28.00]
MC with single calls	28.13	[27.83, 28.43]	28.00	[27.97, 28.03]

Table 5.1: Monte Carlo Prices for a Basket Call (Exact Value 28.00)

options but are also traded on baskets. Typical examples of Asian options (or: Average options) based on continuous averages are:

$$B = \left(S_1(T) - \frac{1}{T} \int_0^T S_1(s)\,ds \right)^+ \quad \text{continuous } \textbf{Asian option}, \tag{5.96}$$

$$B = \left(\frac{1}{T} \int_0^T S_1(s)\,ds - K \right)^+ \quad \text{continuous } \textbf{fixed-strike average}. \tag{5.97}$$

In actually traded options the continuous-time averages are replaced by discretized versions of the form

$$B = \left(S_1(T) - \frac{1}{n} \sum_{i=1}^{n} S_1(t_i) \right)^+ \quad \text{discrete } \textbf{Asian option}, \tag{5.98}$$

$$B = \left(\frac{1}{n} \sum_{i=1}^{n} S_1(t_i) - K \right)^+ \quad \text{discrete } \textbf{fixed-strike average}. \tag{5.99}$$

The main problem in pricing these options is the same as in basket option pricing: the sum of log-normally distributed random variables is no longer log-normal. This also implies that the integral over a set of log-normals is in general not log-normal. Consequently, there is no closed analytical pricing formula, for both the continuous and the discrete versions.

A first naive way is to simulate various paths of the stock prices, compute the resulting payoffs, and estimate the option price by a discounted arithmetic mean. As in the basket option case, we can use geometric mean-based versions of the options as control variates. As the product of log-normally distributed random variables is again log-normally distributed, the log-normal valuation result, Theorem 5.18, yields the following.

THEOREM 5.21 Option price for geometric averages

In the one-dimensional Black-Scholes model the price of the discrete fixed-

strike average option on the geometric mean with payoff

$$B = \left(\left(\prod_{i=1}^{n} S_1(t_i) \right)^{1/n} - K \right)^+ \quad \text{geometric } \textbf{fixed-strike average} \tag{5.100}$$

and $0 = t_0 < t_1 < ... < t_n$ *is given by*

$$p_{GFA} = e^{-rT} S_1(0) e^{m + \frac{1}{2}\nu^2} \Phi\left(\frac{\left(\ln(S_1(0)/K) + m + \nu^2\right)}{\nu} \right) - e^{-rT} K \Phi\left(\frac{(\ln(S_1(0)/K) + m)}{\nu} \right), \tag{5.101}$$

$$m = \left(r - \tfrac{1}{2}\sigma^2\right) \tfrac{1}{n} \sum_{i=1}^{n} t_i, \quad \nu = \tfrac{\sigma}{n} \sqrt{\sum_{i=1}^{n} (n+1-i)^2 (t_i - t_{i-1})}. \tag{5.102}$$

Kemna and Vorst (1990) used this formula and the fact that the arithmetic mean is at least as big as the geometric mean to obtain lower bounds for the prices of the corresponding Asian options. Again, one can also use a corrected strike $\tilde{K}$ in the geometric version to have a matching mean, i.e.

$$\mathbb{E}\left(\frac{1}{n} \sum_{i=1}^{n} S_1(t_i) - K \right) = \mathbb{E}\left(\left(\prod_{i=1}^{n} S_1(t_i) \right)^{1/n} - \tilde{K} \right). \tag{5.103}$$

This value is given by

$$\tilde{K} = K + S_1(0) \left(\exp\left(\left(r - \tfrac{1}{2}\sigma^2\right) \bar{t}_n + \tfrac{1}{2} \tfrac{\sigma^2}{n^2} \tilde{t}_n \right) - \tfrac{1}{n} \textstyle\sum_{i=1}^{n} e^{rt_i} \right), \tag{5.104}$$

$$\bar{t}_n = \tfrac{1}{n} \textstyle\sum_{i=1}^{n} t_i, \quad \tilde{t}_n = \textstyle\sum_{i=1}^{n} (n+1-i)(t_i - t_{i-1}), \ t_0 = 0. \tag{5.105}$$

For equidistant time spacing we have $\bar{t}_n = (n+1)\Delta t/2$. This idea is summarized in Algorithm 5.4 with $p_{GFA}(\tilde{K})$ denoting the price with strike $\tilde{K}$. In contrast to the moment matching of Chapter 3, the control variate and **not the sample** is corrected for the mean. Hence, we instead talk of **moment matched control variates**.

Algorithm 5.4 Corrected geometric mean control for fixed-strike average options

---

1. Simulate N payoffs: B_i="Asian option − Geometric mean option".
2. Approximate the Asian price by $\hat{p}_{FSA}^N := p_{GFA}(\tilde{K}) + e^{-rT} \sum_{i=1}^{N} B_i$.

The numerical performance of both control variate methods is comparable to the performance of the similar control variates in the basket option case.

REMARK 5.22 1. Note that we do not need a discretization method for simulating the relevant stock price paths above, as their values are only needed at a finite number of time points t_i where we can sample from the exact distribution of $S(t_i)$ (respectively from the increments $S(t_i) - S(t_{i-1})$). The situation would be different for the continuous-time average options.

2. There is much literature on Asian option pricing such as Rogers and Shi (1995) who used a PDE approach which has been taken up by other authors later. Further popular methods are a moment matching approach by Turnbull and Wakeman (1991) and numerical inversion of the Laplace transform by Geman and Yor (1993). A comparison between these analytical approaches and Monte Carlo methods is given in Fu et al. (1999). □

Barrier option pricing, importance sampling, and Brownian bridge techniques

Barrier options are popular derivatives. They only provide a final payment, usually a call or a put on the underlying, if the path of the underlying crosses certain barriers (or not). Simple examples of barrier options are the (one-sided) **knock-out barrier option** given by the final payments

$$B_{DOC} = (S_1(T) - K)^+ 1_{\{S_1(t) > H \forall t \in [0,T]\}} \quad \textit{down-and-out call}, \tag{5.106}$$
$$B_{UOC} = (S_1(T) - K)^+ 1_{\{S_1(t) < H \forall t \in [0,T]\}} \quad \textit{up-and-out call}, \tag{5.107}$$
$$B_{DOP} = (K - S_1(T))^+ 1_{\{S_1(t) > H \forall t \in [0,T]\}} \quad \textit{down-and-out put}, \tag{5.108}$$
$$B_{UOP} = (K - S_1(T))^+ 1_{\{S_1(t) < H \forall t \in [0,T]\}} \quad \textit{up-and-out put} \tag{5.109}$$

where K is the strike and $H \geq 0$ the barrier of the option. By noting that, e.g., the final payoff of a call (a put) is given as the sum of a down-and-out call and a down-and-in call with the same strike K, we can obtain the price of an in-barrier option as the difference of the corresponding option without barrier and the out-barrier option ("in-out parity"). We can therefore in the following concentrate on out-options.

Variants of the above barrier options with upper and lower barrier H_1, H_2 are also traded at the financial market and have payments similar to the one-sided barrier options. As an example we consider the **double-barrier knock-out call** with payoff

$$B_{DBKOC} = (S_1(T) - K)^+ 1_{\{H_2 > S_1(t) > H_1 \forall t \in [0,T]\}}. \tag{5.110}$$

While the above presented options contain a permanent barrier condition in their payoff, we speak of **continuous** barrier options. In contrast, we will also consider **discrete** barrier options where the barrier condition only has to hold at a finite set of time instants $0 \leq t_1 < ... < t_m \leq T$. That is, the discrete down-and-out call has a payoff of

$$B_{DOC}^{N} = (S_1(T) - K)^+ 1_{\{S_1(t_i) > H \ \forall t_i, \ i=1,\dots,m\}} \tag{5.111}$$

As the joint distribution of the running maximum/minimum of a Brownian motion with drift and its terminal value $W(T)$ at time T is explicitly known (see Korn and Korn [2001]), there exist simple explicit pricing formulae for one-sided (or single-barrier) options in a Black-Scholes type market.

PROPOSITION 5.23
In a Black-Scholes market the price of a down-and-out call on a single stock a) with barrier $H < S(0)$ and strike $K < H$ is given by

$$\begin{aligned} X_{do}^{Call}(0) &= S(0)\Phi(d_1) - Ke^{-rT}\Phi\left(d_1 - \sigma\sqrt{T}\right) \\ &- S(0)\left(\frac{H}{S(0)}\right)^{2\frac{r}{\sigma^2}+1}\Phi(d_2) + e^{-rT}K\left(\frac{H}{S(0)}\right)^{2\frac{r}{\sigma^2}-1}\Phi\left(d_2 - \sigma\sqrt{T}\right), \end{aligned} \tag{5.112}$$

$$d_1 = \frac{\ln\left(\frac{S(0)}{H}\right) + \left(r + \frac{1}{2}\sigma^2\right)T}{\sigma\sqrt{T}}, \quad d_2 = \frac{\ln\left(\frac{H}{S(0)}\right) + \left(r + \frac{1}{2}\sigma^2\right)T}{\sigma\sqrt{T}}. \tag{5.113}$$

b) with barrier $H < S(0)$ and strike $K \geq H$ is given by

$$\begin{aligned} X_{do}^{Call}(0) &= S(0)\Phi(d_3) - Ke^{-rT}\Phi\left(d_3 - \sigma\sqrt{T}\right) \\ &- S(0)\left(\frac{H}{S(0)}\right)^{2\frac{r}{\sigma^2}+1}\Phi(d_4) + e^{-rT}K\left(\frac{H}{S(0)}\right)^{2\frac{r}{\sigma^2}-1}\Phi\left(d_4 - \sigma\sqrt{T}\right), \end{aligned} \tag{5.114}$$

where we get d_3 and d_4 out of d_1 and d_2 by substituting H by K.

There are similar explicit formulae for all other types of single-barrier options (see Reiner and Rubinstein [1991]). They can be used to obtain approximations for discrete barrier options later on. For double-barrier options such explicit pricing formulae do in general not exist.

Monte Carlo (MC) pricing of discrete barrier options: The standard method

As for discrete barrier options we only have to check the barrier conditions at time points $t_1 < ... < t_m$, we only need to simulate the values of the price path $S(t)$ at those times t_i and of course at the final time T. Even more, in the case of an out-option, we only have to simulate until time T if the barrier condition is never violated. Otherwise, we stop the simulation path at the first time when the barrier condition is violated and set the corresponding payoff to zero. In the case of an in-option, we only have to simulate the final payoff if the barrier condition has been fulfilled at some time instant t_i before T. Note that this standard approach can easily be adapted to every kind of barrier conditions, whether they are double barriers, time-dependent barriers, or a combination of in- and out-criteria. Further, it makes no problem to apply it to a multiasset setting in a d-dimensional model of the Black-Scholes type. To give a general formulation, we leave out the possibility to stop the simulation if for an out-option the barrier condition is violated for the first time.

Algorithm 5.5 Monte Carlo for discrete barrier options

For $i = 1$ to N

1. Simulate the stock prices at the barrier times $S^{(i)}(t_j), j = 1, ..., m$.
2. If the barrier condition is satisfied at all times $t_1, ..., t_m$ then compute the final payoff $B^{(i)} = f\left(S^{(i)}(T)\right)$, else set $B^{(i)} = 0$.

Obtain the Monte Carlo estimate for the discrete barrier option price as

$$\hat{p}_{B,N} = \frac{1}{N} e^{-rT} \sum_{i=1}^{N} B^{(i)}.$$

MC pricing of discrete barrier options: Conditional survival

The standard method often delivers a quite satisfying performance in the one-dimensional setting. However, when the initial stock price is close to the barrier of an out-option, we are likely to simulate a lot of price paths that actually do not lead to the computation of a final payoff. They typically violate the barrier condition early. To avoid this problem, we can use an importance sampling approach that consists of generating only price paths that are conditioned to survive until T (see also Glasserman and Staum [2001]). We achieve this by stepwise conditional sampling.

We consider a discrete out-barrier option with a final payoff of $B = f(S(T))$ and assume that at time t_i the boundary condition is not violated if we have

$$S(t_i) \in (H_1(t_i), H_2(t_i)), \ i = 1, ..., m. \tag{5.115}$$

Note that we allow for time-dependent barriers. If we can explicitly compute the conditional survival probabilities $p_i(S(t_i))$ defined by

$$p_i(s) = \mathbb{P}\left(S(t_{i+1}) \in (H_1(t_{i+1}), H_2(t_{i+1}))\right) | S(t_i) = s) \tag{5.116}$$

then with the notation of the likelihood

$$L_j = \prod_{i=0}^{j} p_i(S(t_i)) \tag{5.117}$$

suitable conditioning yields the representation

$$\mathbb{E}\left(e^{-rT} B \prod_{i=1}^{m} 1_{\{S(t_i) \in (H_1(t_i), H_2(t_i))\}}\right) = \mathbb{E}\left(e^{-rT} L_{m-1} B\right). \tag{5.118}$$

We will use the right side of this representation and simulate only paths that are conditioned on survival and at the same time compute their conditional survival probability along the path. Then, we average over the resulting payoffs. We do this explicitly for double-barrier knock-out options with time-dependent barriers. Due to Equation (5.118) this estimator is unbiased.

Conditional on $S(t_i) = s$ with $\Delta_i = t_{i+1} - t_i$ and using

$$S(t_{i+1}) = S(t_i)\, e^{\left(r - 1/2\sigma^2\right)\Delta_i + \sigma\sqrt{\Delta_i}\Phi^{-1}(U)} \tag{5.119}$$

with U uniformly distributed on $[0, 1]$, we obtain

$$S(t_{i+1}) \in (H_1(t_{i+1}), H_2(t_{i+1})) \Longleftrightarrow U \in \left(1 - p_i^-, 1 - p_i^+\right) \tag{5.120}$$

with

$$\begin{aligned} p_i^- &= \mathbb{P}\left(S(t_{i+1}) > H_1(t_{i+1}) | S(t_i) = s\right) \\ &= \Phi\left(\frac{ln\left(S(t_i)/H_1(t_{i+1})\right) + \left(r - 1/2\sigma^2\right)\Delta_i}{\sigma\sqrt{\Delta_i}}\right) \end{aligned} \tag{5.121}$$

and p_i^+ obtained from p_i^- by replacing the lower barrier $H_1(t_{i+1})$ by the upper one $H_2(t_{i+1})$. Thus, to ensure that the survival condition (5.120) is always satisfied, we replace U by the conditional random number

$$\tilde{U} = \left(1 - p_i^-\right) + V\left(p_i^- - p_{i+1}^+\right) \tag{5.122}$$

and obtain the conditional survival probability as

$$p_i(s) = p_i^- - p_i^+. \tag{5.123}$$

From this, we obtain Algorithm 5.6.

REMARK 5.24 1. Glasserman and Staum (2001) prove that using the conditional MC estimate results in a variance reduction compared to the standard method. This reduction can be high if the initial stock price is close to

Algorithm 5.6 Conditional MC for double-barrier knock-out options

Set $S(t_0) = S(0) = s$, $L_{-1} = 1$.
For $i = 1$ to N

1. Simulate a stock price path conditioned on survival, i.e.
   For $j = 0$ to $m - 1$ do
   - Calculate p_j^-, p_j^+ according to Equation (5.121).
   - Simulate a random number $\tilde{U}$ according to Equation (5.122).
   - Set $S^{(i)}(t_{j+1}) = S^{(i)}(t_j)\, e^{\left(r - 1/2\sigma^2\right)\Delta_j + \sigma\sqrt{\Delta_j}\Phi^{-1}(\tilde{U})}$.
   - Set $L_j^{(i)} = L_{j-1}^{(i)} \cdot \left(p_j^- - p_j^+\right)$.
2. Simulate the final payoff at time T, i.e.
   - Set $S^{(i)}(T) = S^{(i)}(t_m)\, e^{\left(r - 1/2\sigma^2\right)(T - t_m) + \sigma\sqrt{T - t_m}\Phi^{-1}(U)}$ with U uniformly distributed on [0,1].
   - Set $B^{(i)} = f\left(S^{(i)}(T)\right)$.

Obtain the conditional MC estimate for the double-barrier knock-out option

$$\hat{p}_{B,N}^{cond,DBKNO} = \frac{1}{N} e^{-rT} \sum_{i=1}^{N} L_{m-1}^{(i)} B^{(i)}.$$

---

the barrier (judged with respect to the time to maturity and the volatility of the stock). However, conditioning only reduces the variance caused by the barrier condition. It does not affect the variance of the final payoff.

2. The conditional method can also be generalized to the multidimensional setting. We only have to modify the definition and computation of the probabilities $p_j^{\pm}$ and the generation of the – now multidimensional – random number $\tilde{U}$ in the algorithm in a suitable way. This, however, can be quite tedious when looking at the details. □

MC pricing of continuous barrier options: Brownian bridge techniques

If we want to calculate the price of a continuous double-barrier knock-out call by the standard Monte Carlo approach (i.e. simulate a discrete stock price path, check the boundary conditions, compute the final payoff if the boundary condition is satisfied, and average) then our estimator is no longer unbiased. Indeed, we will systematically overestimate the option price as due to the discretization of the price paths, the option knocks out too seldom in the standard approach. To make this even more precise, imagine that we generate two prices $S(t_i)$ and $S(t_{i+1})$ which are both very close to the barrier. Then, in the standard approach we make no corrections at all as we have not seen a crossing of the barrier. However, the price process could have crossed the barrier between time t_i and t_{i+1} or not. We also have a similar problem for in-options. However, then the price is systematically underestimated by the standard method. In general, we face a so-called **monitoring bias** which is only disappearing slowly, i.e. it is of order $O(m^{-1/2})$ if m is the number of discretization points in the standard method (see Gobet [2009]).

An alternative to the (implicit) linear interpolation between the two simulated values of the standard approach is to use a Brownian bridge for filling the gap (for a survey on the full generality of this method see Gobet [2009]). In simple examples this gap can be the whole interval $[0,T]$. Let $f(S(T))$ be the final payoff of the (continuous) barrier option. We then have

$$\begin{aligned}
&\mathbb{E}\left(1_{\{S(t)\in(H_1,H_2)\ \forall t\in[0,T]\}} f\left(S\left(T\right)\right)\right) = \\
&\qquad = \mathbb{E}\left(\mathbb{E}\left(1_{\{S(t)\in(H_1,H_2)\ \forall t\in[0,T]\}}\left|S\left(T\right), S\left(0\right)\right) f\left(S\left(T\right)\right)\right) \\
&\qquad = \mathbb{E}\left(\left(1-p\left(S\left(0\right), S\left(T\right), H_1, H_2, T, \sigma\right)\right) f\left(S\left(T\right)\right)\right) \qquad (5.124)
\end{aligned}$$

with

$$\begin{aligned}
&p\left(s_1, s_2, H_1, H_2, T, \sigma\right) = \\
&\qquad = \mathbb{P}\left(\exists t \in [0,T] : S\left(t\right) \notin \left(H_1, H_2\right) \left|S\left(0\right) = s_1, S\left(T\right) = s_2\right.\right). \qquad (5.125)
\end{aligned}$$

Due to Equation (5.124), Algorithm 5.7 produces an unbiased estimate for the price of a down-and-out option.

Algorithm 5.7 MC pricing of out-barrier options with the bridge technique

---

Consider a double-barrier knock-out option with a final payoff of $B = f(S(T))$ and barriers H_1, H_2.
For $i = 1$ to N

- Simulate values $S^{(i)}(T)$ for the stock price at time T.
- Compute $B^{(i)} = \left(1 - p\left(S\left(0\right), S^{(i)}\left(T\right), H_1, H_2, T,\right)\right) f\left(S^{(i)}\left(T\right)\right)$

Obtain the bridge MC estimate for the double-barrier knock-out option

$$I_B^{bridge,DBKO} = \frac{1}{N} e^{-rT} \sum_{i=1}^{N} B^{(i)}.$$

Of course, this algorithm only works if we can compute the relevant probability explicitly. This can be done in some examples (see Gobet [2009] or

Karatzas and Shreve [1998] for the calculation of the probabilities).

Example 1: Single-barrier option
Here, we either have $H_1 = 0$ or $H_2 = +\infty$. For simplicity we consider the case of $H_2 = \infty$. Then, we need to compute the probability that a Brownian motion with drift that starts from $\ln(s_1)$ and reaches $\ln(s_2)$ at time T falls below $\ln(H_1)$ between 0 and T. On the one hand, this conditional process is a Brownian bridge (we therefore talk of the **Brownian bridge technique**) and on the other hand the probability is then given by

$$p\left(s_1, s_2, H_1, +\infty, T, \sigma\right) = \begin{cases} 1 & , \text{ if } s_1 < H_1 \text{ or } s_2 < H_1 \\ \exp\left(-2\frac{ln(s_1/H_1)ln(s_2/H_1)}{\sigma^2 T}\right) & , \text{ else} \end{cases}. \tag{5.126}$$

Again, this probability is obtained from the joint distribution of the final value of a Brownian motion with drift and its running maximum. In the case of an upper bound (i.e. $H_1 = 0$) we have

$$p\left(s_1, s_2, 0, H_2, T, \sigma\right) = \begin{cases} 1 & , \text{ if } s_1 > H_2 \text{ or } s_2 > H_2 \\ \exp\left(-2\frac{ln(s_1/H_2)ln(s_2/H_2)}{\sigma^2 T}\right) & , \text{ else} \end{cases}. \tag{5.127}$$

We can also use the above probabilities to obtain the price of the corresponding in-option. We only have to modify the left side of Equation (5.124) in the obvious way and on the right side replace $1 - p(.)$ by $p(.)$.

Example 2: Piecewise constant single-barrier case
In this situation, we can still apply the methods of Example 1. As we have a single barrier that is piecewise constant, we obtain the relevant knock-out probability as a product of the knock-out probabilities computed along the intervals of constancy of the barrier: Let the time-dependent barrier $H_1(t)$ take on the values $H_{1,1}, ..., H_{1,m}$ on the time intervals given by $0 = t_1 < .. < t_m < t_{m+1} = T$. Then, we have to simulate the values $S(t_i)$ of the stock price process and obtain the knock-out probability (conditional on this discrete stock price path) as the product of the probabilities $(1 - p(S(t_i), S(t_{i+1}), H_{1,i}, +\infty, t_{i+1} - t_i, \sigma))$, $i = 1, ..., m$. Compared to the standard method the Brownian bridge technique is unbiased. Further, the stock price only has to be simulated at $m + 1$ time instants.

Example 3: Double-barrier option
While in Example 1 we also have a simple explicit pricing formula, it is definitely not available in the situation of a double-barrier option. However, one has at least an explicit formula for the probability required in the Brownian

bridge algorithm above (see Gobet [2009]):

$$p(s_1, s_2, H_1, H_2, T, \sigma) = \\ = \sum_{k=-\infty}^{+\infty} \left(\exp\left(-2\frac{k \ln(H_2/H_1)\left(k \ln(H_2/H_1) + \ln(s_2/s_1)\right)}{\sigma^2 T} \right) \right. \\ \left. - \exp\left(-2\frac{\left(k \ln(H_2/H_1) + \ln(s_1/H_2)\right)\left(k \ln(H_2/H_1) + \ln(s_2/H_1)\right)}{\sigma^2 T} \right) \right) \tag{5.128}$$

if neither s_1 nor s_2 are outside (H_1, H_2). If that would be the case the above probability would equal 1. Of course, the numerical evaluation of the above series is not trivial.

Moon (2008) considers a slight variant of the above approach and also applies it to single-asset double-barrier options and to multiasset barrier options where the barrier condition is only relevant for one asset. An example of such an option would be given by the payout of

$$B_{out} = (S_1(T) - K)^+ \cdot 1_{S_2(t) > H \ \forall t \in [0,T]}. \tag{5.129}$$

Moon simulates each stock price path exactly on a discrete grid until the maturity or until the first knock-out time. Moon's idea is to treat each barrier separately for computing the knock-out (or knock-in) probabilities. These probabilities are explicitly given by Equation (5.126). Of course, the sum over the probabilities of crossing a specific barrier is slightly higher than the probability of crossing at least one barrier. However, given that the time discretization is sufficiently fine, this difference can be ignored as the probability of the price process crossing the whole barrier interval in a short time step is extremely small. After having calculated the knock-out probabilities, Moon suggests drawing a $\mathcal{U}[0,1]$-distributed random number for each probability. If all probabilities are below their corresponding $\mathcal{U}[0,1]$ numbers then the simulation of the stock price path continues. If this is not the case the path is considered to be knocked out and the next path simulation will start.

We give Moon's algorithm for pricing a double-barrier knock-out call option with piecewise constant barriers $(L_1, U_1), \ldots, (L_m, U_m)$ as Algorithm 5.8.

REMARK 5.25 Modifications of the above algorithm for in-options, puts, or other path-independent final payoffs B are straightforward and can be done by modifying the relevant steps (in-out criteria, bridge probability computation, or computation of the final payoff). A variant of a multiasset final payoff (such as a basket option) combined with a single-asset barrier criterion can also easily be obtained (see Moon [2008]). Later, we will comment on the use of the algorithm in more general stock price settings. □

Algorithm 5.8 Double-barrier knock-out call pricing using Moon's method for piecewise constant barriers

Let the strike K, barriers $L_i < U_i$ valid on $[t_i, t_{i+1}]$ with $0 = t_1 < \ldots < t_{m+1} = T$, and maturity T be given. Set $\delta_i = t_{i+1} - t_i$.

For $i = 1$ to N do

1. Set $S_0^i = S(0)$.
2. For $j = 1$ to m do
   (a) Set $V_i = 0$ and generate $Y_{ij} \sim N(0,1)$.
   (b) Set $S_j^i = S_{j-1}^i e^{\left((r-1/2\sigma^2)\delta+\sigma\sqrt{\delta}Y_{ij}\right)}$.
   (c) If $S_j^i \notin (L_j, U_j)$ then go to Step 1.
   (d) Else $p_{low} = e^{-2\frac{\ln\left(S_j^i/L_1\right)\ln\left(S_{j+1}^i/L_1\right)}{|\sigma^2|\delta_i}}$, $p_{up} = e^{-2\frac{\ln\left(S_j^i/U_1\right)\ln\left(S_{j+1}^i/U_1\right)}{|\sigma^2|\delta_i}}$.
   (e) Simulate two independent random variables X_1, X_2 with $X_i \sim \mathcal{U}[0,1]$.
   (f) If $p_{low} \geq X_1$ or $1 - p_{up} \geq X_2$ then go to Step 1.
   (g) If $j = m$ then $V_i = (S_m^i - K)^+$.

Obtain the Monte Carlo estimate for the option price as

$$I_B^{Moon,DBKO} = \frac{1}{N} e^{-rT} \sum_{i=1}^{N} V_i.$$

REMARK 5.26 Broadie et al. (1997, 1999) have shown that by suitably shifting the barrier of a continuous single-barrier option, one obtains a good approximation for the price of a corresponding discrete single-barrier option. One can also use it in a reverse way, i.e. to price a continuous barrier option approximately by a suitable discrete barrier option. This technique is further developed in Gobet (2009) and in Gobet and Menozzi (2007) for the multiasset case. As it is of a technical nature and beyond the scope of this book we refer the reader to these references. □

5.6.3 More exotic options

Due to the enormous number of different versions of exotic options, in this book, we can only deal with some popular examples in great detail. However, it should be clear that there is at least always a crude way to deal with any kind of option:

- Simulate the stock price paths many times and calculate the corresponding payments along the paths.
- Average the simulated payments and discount them in a suitable way to obtain a crude Monte Carlo estimate for the option price.

While this method always works, there are always possibilities for cleverly adapted algorithms that make use of special properties of the particular option type or that use suitable variance reduction methods. The literature on these methods is far too big to give an exhaustive survey. We mention just some types of options that are widespread:

- Lookback options which have payments based on the minimum/maximum of a stock price on a certain time span can be of the form

$$B_{lookback} = \left(\max_{t \in [0,T]} S(t) - K \right)^+ .$$

- Cliquet options which typically consist of a guarantee against downside movements, but still have some upside potential. They appear in various forms such as e.g.

$$B_{Nap} = (a + \min_{i=1,\ldots,n} r_i)^+, \text{ a so-called } \textbf{Napoleon},$$

$$B_{rc} = (b + \sum_{i=1}^{n} r_i^-)^+, \text{ a } \textbf{reverse cliquet}, \textit{or}$$

$$B_{accu} = (c + \sum_{i=1}^{n} \max(\min(r_i, cap), floor))^+, \text{ an } \textbf{accumulator}$$

where r_i is the return of a stock or an index over period i, a, b, c and $cap, floor$ are constants. Cliquet options are known to be particularly sensitive with respect to changes in the volatility of the stock.

5.6.4 Data preprocessing by moment matching methods

We already mentioned a certain kind of moment matching when pricing basket options and Asian options above. There, we concentrated on the equality of the expected value of the arithmetic average and of the approximating geometric average of stock prices. Here, we will take up the approach (already discussed in connection with variance reduction methods in Chapter 3) of modifying the generated sample to match its theoretical moments. The usual way of doing this leaves some possibilities of choice such as:

- **Matching of the moments of the underlying Brownian motion**: If the stock price processes are driven by a d-dimensional Brownian motion $W(t)$, then matching of the first two moments of its components with the empirical counterparts of the sampled values $W^{(i)}(t)$ at the fixed time t are ensured by using the modified sample values

$$\tilde{W}_j^{(i)}(t) = \frac{W_j^{(i)}(t) - \bar{W}_j(t)}{\bar{s}(t)/\sqrt{t}} \tag{5.130}$$

for each component j of the Brownian motion and where $\bar{W}(t)$ and $\bar{s}(t)$ are the sample mean and standard deviation for fixed t.

- **Matching of the moments of the asset prices**: Here, one matches the empirical moments of the stock prices that enter the relevant option payment to their theoretical expectations. This is done by using for each stock price the modified sample values

$$\tilde{S}_j^{(i)}(t) = \frac{S_j^{(i)}(t)\,\mathbb{E}(S(t))}{\bar{S}_j(t)} \tag{5.131}$$

  where $\bar{S}_j(t)$ is the sample mean of the generated stock prices at time t.

- **Matching of the mean by antithetic sampling**: The simplest way to obtain a sample mean that matches the desired expectation is the method of antithetic variates (see Section 3.3.1).

In contrast to the above presented methods, our aim here is to present a new approach of Wang (2008) that focuses on matching the correlation of the underlying Brownian motion in an exact way by the empirical means. We call this sample data preprocessing. This is particularly interesting as for some types of options the correlation structure has an enormous influence on their price. A typical example is a maximum-call on n stocks given by

$$B_{max} = \left(\max_{k=1,\ldots,n}\{S_k(T)\} - K\right)^+. \tag{5.132}$$

It is immediately clear that the price of this option would be much higher for independent stocks than for totally dependent stocks.

The sample preprocessing by Wang considers the sample of the uncorrelated, d-dimensional Brownian motion before it is used to generate the corresponding stock prices. Its main aim is to create a sample of size N for a d-dimensional standard normal distribution with sample mean 0 and unit sample covariance matrix I. The method is described in Algorithm 5.9.

One can then use the preprocessed data to construct a Brownian motion sample from it. This can then be used for Monte Carlo option pricing. Wang (2008) reports a significant improvement of the performance of Monte Carlo option pricing for various option types when the above preprocessing method is used. It performs particularly well when the exact correlation between the log-returns of the stocks is small.

Algorithm 5.9 Preprocessing a d-dimensional normal sample

1. Generate d independent samples of size N of independent standard normally distributed random numbers $(z_1, \ldots, z_d)$.

2. Introduce the mean-corrected samples $\tilde{z}_i = z_i - \bar{z}_i \cdot \underline{1}$ with $\bar{z}_i$ the sample mean over the elements z_{ij}, $j = 1, ..., N$ and $\underline{1} = (1, \ldots, 1) \in \mathbb{R}^N$.

3. Calculate the empirical covariance matrix $\tilde{C}$ of the $\tilde{z}$-vectors, i.e.

$$\tilde{C}_{ij} = \frac{1}{N-1} \sum_{k=1}^{N} \tilde{z}_{ik} \tilde{z}_{jk}, \quad i, j = 1, \ldots, d.$$

4. Compute the Cholesky decomposition $\tilde{C} = \tilde{A}' \tilde{A}$.

5. Obtain the sample with the desired properties $Z' = (z_1', \ldots, z_d')$ via

$$Z' = (\tilde{z}_1, \ldots, \tilde{z}_d') \tilde{A}^{-1}.$$

5.7 Weaknesses of the Black-Scholes model

We have so far demonstrated the main methods and principles in option pricing in a (multidimensional) Black-Scholes model. This model is still an industry benchmark. However, there is common agreement among practitioners and academics that the Black-Scholes model is an oversimplification of the real movements of stock and option prices. One can verify this by performing statistical tests for the normality of the log-returns of stocks. There, the null hypothesis of normally distributed log-returns is usually rejected with a small p-value. Also other properties such as existing variances of the log-returns or their independence are often questioned by looking at statistical properties of suitable financial time series. We will not go into detail about those statistical issues. However, we will present another way of demonstrating that the assumptions of the Black-Scholes model are not satisfied and which is an important tool for practitioners. It is to use so-called **volatility surfaces**.

Implied volatility surfaces and the Black-Scholes formula

In the Black-Scholes formula for a call option on a single stock,

$$X_C(t) = S_1(t) \cdot \Phi(d_1(t)) - K \cdot e^{-r(T-t)} \Phi(d_2(t)) \tag{5.133}$$

with $d_i(t)$, $i = 1, 2$ as given in Equation (5.56) the only parameter that is not directly observable is the volatility σ. One could of course try to estimate it from log-returns of real data via

$$\mathbb{V}ar\left(\ln\left(\frac{S(t+\Delta t)}{S(t)}\right)\right)=\sigma^2\Delta t, \tag{5.134}$$

but one can also get it from a market price of a call option. As the Black-Scholes formula is strictly increasing in σ (for positive values of σ), there is a unique value σ^* such that the Black-Scholes formula with this value delivers a theoretical price that equals the market price of this particular call (of course under the assumption that all other parameters such as K, r, T are fixed). We then call σ^* the **implied volatility** of this call price. Consequently, if the Black-Scholes model would indeed describe the reality in an appropriate way, we could take a market price of any other call option, invert the Black-Scholes formula, and should obtain (at least approximately) the same implied volatility σ^*. To judge how well the option pricing world is explained by the Black-Scholes model, one can have a look at so-called **implied volatility curves** or at **implied volatility surfaces**. For an implied volatility curve, one typically considers calls (or puts) with either a fixed maturity T and varies the strike K or fixes the strike K and varies the maturities. Then, one takes all observable market prices and uses the Black-Scholes formula to calculate the implied volatilities for – say – calls with a fixed maturity T given by

$$p_{call}^{market}(K_i;T)=X_c(0;\sigma^*(K_i),K_i,T). \tag{5.135}$$

On the left-hand side of this equation we have the market price of a call with a strike of K_i. On the right-hand side we calculate the price by the Black-Scholes formula for a strike of K_i, a maturity of T, and determine $\sigma^*(K_i)$ implicitly by requiring equality between the left and the right sides. Then the function

$$f(K)=\sigma^*(K) \tag{5.136}$$

is called a **volatility curve** for a fixed maturity T. Of course, we only have points $\sigma^*(K_i)$ and usually have to interpolate between those points, but we do not address this issue here. To illustrate the different behaviour of the implied volatility as a function of the normalized strike K/s_1, we show some typical curves from the different markets of foreign exchange, commodities, and stocks together with a constant Black-Scholes model curve in Figure 5.3. Neither of the real curves is constant!

Depending on the form of the curve, one is speaking of a **volatility skew** or a **volatility smile**. If instead of just a volatility curve, we also look at the behaviour of the implied volatility as a function of the second variable (in our case above, the maturity) then one obtains a volatility surface if one also lets this variable vary. Figure 5.4 shows such a volatility surface.

In all the cases we have seen, the volatility curves or surfaces do not look as if they were generated from a Black-Scholes model. To cope with this fact, more complicated models have been introduced. The first two major streams

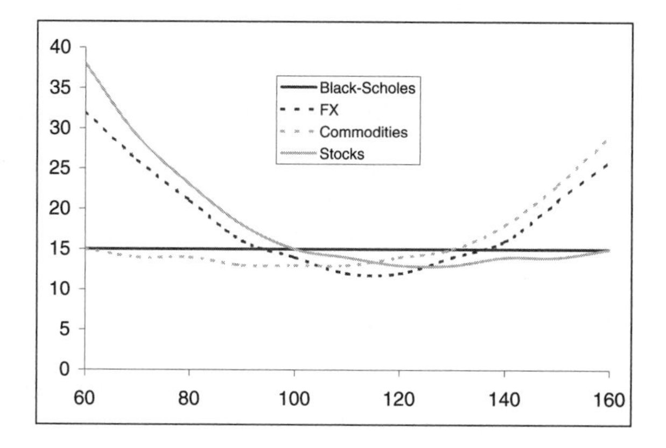

FIGURE 5.3: (Schematic) implied volatility curves from different markets.

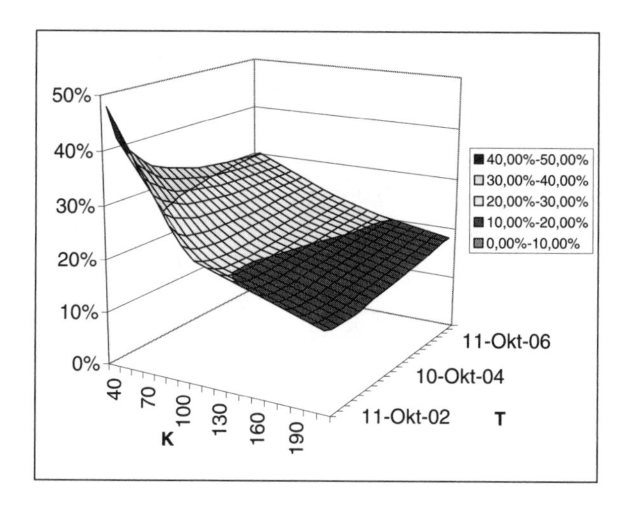

FIGURE 5.4: (Schematic) implied volatility surface from an equity market.

of such models were the local volatility and the stochastic volatility models. The aim of those models is to generate option prices that produce implied volatility curves or surfaces that mimic the really existing ones. As a by-product the volatility surfaces are another (however, artificial) justification for the importance of the Black-Scholes formula as it is needed to invert both the market prices and the theoretical option prices from the new models.

5.8 Local volatility models and the CEV model

In local volatility models one replaces the constant volatility parameter σ of the Black-Scholes model by a **volatility function** that depends on both the running time and the stock price. We will only consider the case of a single stock with dynamics given as the solution of

$$dS(t) = rS(t)\,dt + \sigma(t, S(t))\,dW(t), \quad S(0) = s \tag{5.137}$$

with $W(t)$ a one-dimensional Brownian motion. The function $\sigma(t,s)$ is assumed to have a form that ensures the existence of a unique solution to the SDE (5.137). One can show that for a sufficiently regular volatility function the market consisting of the usual bond and one stock is still complete and that there is a unique EMM $\mathbb{Q}$. As always, we assume that we are directly modelling under this EMM, a fact that is expressed by using a drift of r. Of course, the general form of the above SDE can also include a drift of μ.

Before we present a specific example of a volatility function, we state a famous result by Dupire (1997). It says that given any set of observable market prices of call options, there exists a volatility function such that these prices coincide with the corresponding theoretical call prices.

THEOREM 5.27 (Dupire [1997])
If today's market call prices $p_c^{market}(0, S; K, T)$ are known for all possible strikes $K \geq 0$ and all maturities $T \geq 0$ then with the choice of

$$\sigma(K,T) = \sqrt{\frac{2\frac{\partial p_c^{market}}{\partial T} + rK\frac{\partial p_c^{market}}{\partial K}}{K^2\frac{\partial^2 p_c^{market}}{\partial K^2}}} \tag{5.138}$$

the market prices coincide with the theoretical call prices in the corresponding local volatility model, i.e. we have

$$p_c^{market}(0, S; K, T) = \mathbb{E}\left(e^{-rT}(S(T) - K)^+\right) \; \forall (T, K) \in [0, \infty)^2. \tag{5.139}$$

In particular, we assume that all required partial derivatives of today's market price curves exist.

Although this result is very impressive, there are some problems with its practical application:

- To obtain the volatility function, we need a continuous set of market prices, which obviously is not available.
- The volatility function thus has to be obtained by inter- and extrapolation, which can cause a lot of problems.

- There are no closed-form solutions even for prices of simple options.

We refer the interested reader to Dupire (1997) for further discussion of this general setting.

Instead, we would like to present the so-called constant elasticity of variance CEV model as the most prominent parametric local volatility model. There, the stock price is given as the unique solution to the SDE

$$dS(t) = rS(t)\,dt + \sigma S(t)^{\alpha}\, dW(t)\,,\ S(0) = s \tag{5.140}$$

for $\alpha \in [0,1]$ (for values of $\alpha \notin [0,1]$ we refer to Davydov and Linetsky [2001]), r, σ given real constants. For special choices of α we obtain:

- $\alpha = 1$: this is the Black-Scholes setting, i.e. the stock price is log-normally distributed.
- $\alpha = 0$: we can solve the stock price equation explicitly and obtain

  $$S(t) = s\exp(rt) + \sigma \int_0^t \exp(r(t-u))\, dW(u) \tag{5.141}$$

  which implies that the stock price is normally (!) distributed with

  $$\mathbb{E}(S(t)) = s\exp(rt)\,,\quad \mathbb{V}ar(S(t)) = \frac{\sigma^2}{2r}(\exp(2rt) - 1)\,. \tag{5.142}$$

  We will meet a generalizaton of this model (the Vasicek model) in the area of interest rate modelling.
- $\alpha = 0.5$: here, the SDE cannot be solved explicitly, but the stock price stays nonnegative (see the section on the square-root equation in Chapter 4). We will look in more detail at a generalization of this model (the Cox-Ingersoll-Ross model) when modelling interest rates.

The CEV model thus is a generalization of the Black-Scholes model. The extra parameter α can be used for approximating the shape of observed implied volatility curves better than the Black-Scholes model. The name CEV model stems from the fact that for the variance function $\sigma(t,S)^2 = \sigma^2 S^{2\alpha}$ the **elasticity of variance** given by

$$\frac{d\sigma^2/dS}{\sigma^2/S} = 2\alpha \tag{5.143}$$

is constant. Besides the special choices for α mentioned above, the stock price is not given by an explicit analytical formula. Even more, for $\alpha \in [0,1)$ there is a positive probability that the stock price will attain the value of 0. In the case of $\alpha = 0$ we have already discussed this above. In the case of $\alpha \in (0,1)$ this simply means that the process stays at 0 from that time onwards due to the form of SDE (5.140). However, given this situation, it is nearly a surprise that there exists an explicit formula for the price of European calls in the CEV model (see Schroder [1989] or Davydov and Linetsky [2001]).

THEOREM 5.28

For $\alpha \in (0,1)$ the price of a European call with strike K and maturity T in the CEV model is given by

$$C(0,s;\alpha,\sigma,T,K) = sQ(y;z,\zeta) - e^{-rT}KQ(\zeta;z-2,y), \tag{5.144}$$

$$z = 2 + \tfrac{1}{1-\alpha}, \tag{5.145}$$

$$\zeta = \frac{2rs^{2(1-\alpha)}}{\sigma^2(1-\alpha)\left(1-e^{-2r(1-\alpha)T}\right)},\ y = \frac{2rK^{2(1-\alpha)}}{\sigma^2(1-\alpha)\left(e^{2r(1-\alpha)T}-1\right)}, \tag{5.146}$$

and $Q(x;u,v)$ the complimentary noncentral chi-square distribution with u degrees of freedom and noncentrality parameter v evaluated at point x.

Of course, a closed form solution for liquidly traded options such as calls allows the **calibration** of the input parameters α, σ of the CEV model: Use those parameters α^*, σ^* that minimize the sum of squared differences

$$\sum_{i=1}^{m} \left(p_c^{market}(0,s;K_i) - C(0,s;\alpha,\sigma,r,T,K_i)\right)^2 \tag{5.147}$$

over all admissible values for σ and α. Figure 5.5 illustrates the skewness behaviour of implied volatility curves for different values of α. Here, we have always chosen $S(0) = 100, r = 0$ and a variable volatility $\sigma(\alpha) = \sigma S(0)^{1-\alpha}$ such that the initial volatility for all different values of α coincide.

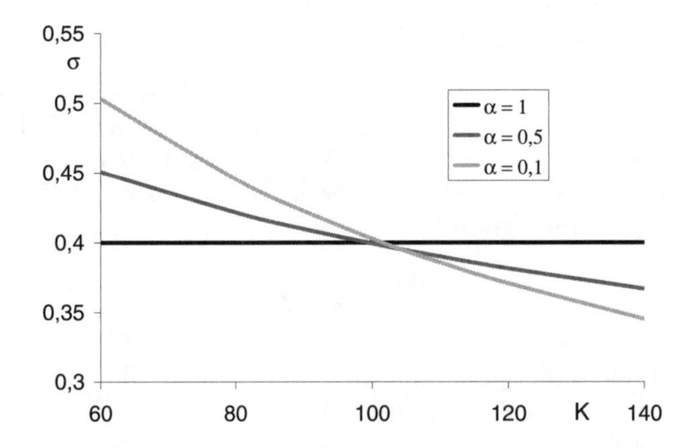

FIGURE 5.5: Implied volatility curves of the CEV model.

5.8.1 CEV option pricing with Monte Carlo methods

Although we are now in a more complicated model than the Black-Scholes one, calculating an option price still means calculating an expectation. Therefore, the principle Monte Carlo methods used for this stay the same. This is also true for the discussion of the variance reduction methods applied to the pricing of certain exotic options in the Black-Scholes setting.

The main new aspect introduced by using the CEV model is that to obtain values of $S\left(t\right)$ we now have to rely on discretization schemes such as the Euler-Maruyama or the Milstein one. Although the volatility function in the CEV model is typically not Lipschitz-continuous in $S=0$, one can still apply both these methods as long as the stock price is far away from zero. Only when the stock price is very close to 0, we might encounter problems in using these discretization methods. Then, in the discretized versions it cannot be avoided that the discretized price will go negative. For the case of $\alpha \in (0,1)$ the stock price path is then simply set to 0 from that time step onward.

As all the single steps (generating a stock price path, calculating a Monte Carlo estimator with or without variance reduction) are already described in detail, here we only give a rough description of an algorithm.

Algorithm 5.10 Monte Carlo pricing in the CEV model

Let the payoff B of the option be given. Then:
For $i=1$ to N

1. Simulate CEV stock price paths $S^{(i)}(t), t \in [0,T]$.
2. Calculate the option payoffs $B^{(i)} = B(S^{(i)}(t), t \in [0,T])$.

Calculate a Monte Carlo estimate $\frac{1}{N}e^{-rT}\sum_{i=1}^{N} B^{(i)}$ for the option price.

REMARK 5.29 1. To make the framework more precise, one first has to decide on the discretization scheme for simulating the stock price paths. Then, one chooses the fineness of the discretization equal to ϵ, the required accuracy (in the sense of the square root of the mean square error). Of course, one also has to take care of the type of option payoff B. For example, in the case of an average strike option the discretization has to include at least all those points over which we are averaging. Finally, as we are in general not able to perform an exact solution, extrapolation methods such as the multilevel Monte Carlo method are well suited for this setting. It then only remains to calculate $N=N(\epsilon)$, the number of stock price paths needed.

2. Note that neither the variance reduction methods we used in the Black-Scholes setting nor our valuation methods for American options explicitly

required the characteristics of this model. We can therefore use all these methods developed in the Black-Scholes setting in the CEV case, too, only the simulation of the stock prices that enter the options gets more involved.

3. **A particular variance reduction method:**
As long as we are considering an option with a closed pricing formula in the Black-Scholes setting, it is convenient to use the Black-Scholes case as a control variate, i.e. choose the volatility $\sigma_{BS} = \sigma s^{\alpha-1}$ such that at $t = 0$ the volatilities in both models agree, compute the option price in the Black-Scholes setting with this volatility, p_{BS}, and use the control variate estimator

$$\frac{1}{N} e^{-rT} \sum_{i=1}^{N} B^{(i)} - B_{BS}^{(i)} + p_{BS} \tag{5.148}$$

where the payoffs $B_{BS}^{(i)}$ are calculated from the geometric Brownian motion paths with volatility σ_{BS}. Note in particular that we have to use the same Brownian motion for both paths, $B^{(i)}$ and $B_{BS}^{(i)}$. □

A numerical example: Pricing a digital call

To demonstrate the variance reduction method of Remark 5.29 above, we compare it to the crude Monte Carlo method. We look at the cases of a simple digital call option that pays one unit of money if and only if the stock price at maturity $S(T)$ is above the strike K. The parameters used are:

$$r = 0, \ \alpha = 0.1, \ \sigma = 25, \ \sigma_{BS} = 0,3962, \ S(0) = K = 100, \ T = 1.$$

For both methods we choose the Euler-Maruyama scheme to simulate the stock price paths and require an accuracy of order $\epsilon = 0.01$ which also requires a number of simulation runs of N =10,000. The exact price of the digital call equals 0.4915(±0.0005). Our simulation results are given in Table 5.2.

Method	Price	95%-CI
Crude MC	0.4896	[0.4798, 0.4994]
MC with BS control	0.4920	[0.4869, 0.4970]

Table 5.2: MC Pricing of a Digital Call in the CEV Model ($\alpha = 0.1$)

Note that the simple control variate method already leads to a variance reduction of a factor of 2. This becomes more pronounced the closer the CEV model is to a Black-Scholes model, i.e. for higher values of α. For an otherwise identical digital call but with $\alpha = 0.5$, $\sigma = 3$, $\sigma_{BS} = 0.3$, and an exact value of 0.4698(±0.0004), we obtain a variance reduction of a factor of 3 (see Table 5.3).

Method	Price	95%-CI
Crude MC	0.4735	[0.4637, 0.4833]
MC with BS control	0.4685	[0.4652, 0.4717]

Table 5.3: MC Pricing of a Digital Call in the CEV Model ($\alpha = 0.5$)

5.9 An excursion: Calibrating a model

One might ask why we have given the complicated call option pricing formula in the CEV model. Obviously, it is straightforward to obtain the call price by a crude Monte Carlo simulation with a sufficient accuracy which is comparable to the numerical error that one is making when using the exact formula. The main reason for this is a question that is also relevant for general models.

How do we get the input parameters of a model?

The answer is that we have to **calibrate** the model. By this we mean that we take market prices of securities and derivatives as input variables. We then try to determine the parameters of our preferred model (e.g. for stock prices) in such a way that the difference between the observed market prices and the theoretical model prices are as close as possible. For this, an explicit pricing formula for the securities we are looking at is extremely helpful. As an example, we look at the Black-Scholes model and describe how to calibrate the volatility parameter σ from call option prices in Algorithm 5.11.

Algorithm 5.11 Calibrating σ in the Black-Scholes model

Denote by $C^M(0, S(0), T_1, K_1)$, ..., $C^M(0, S(0), T_n, K_n)$ the market prices of (European) call options on the stock with varying strikes and maturities.

Denote by $C(0, S(0); T, K, \sigma)$ the prices of European calls with maturity T and strike K obtained by the Black-Scholes formula under the assumption that the volatility parameter is σ (r is assumed to be known).

Let $\omega_1, ..., \omega_n$ be positive weights with $\sum_{i=1}^{n} \omega_i = 1$.

Solve the least-squares problem

$$\min_{\sigma>0} \sum_{i=1}^{n} \omega_i \left(C^M(0, S(0), T_i, K_i) - C(0, S(0); T_i, K_i, \sigma)\right)^2$$

to obtain the parameter σ that explains the observed market prices as good as possible when using the Black-Scholes model.

REMARK 5.30 1. The least-squares problem above is a highly nonlinear problem due to the nonlinear dependence of the call prices on the volatility σ. Its solution requires a nonlinear solver. One should in particular be aware of the fact that one can end up in a local minimum. It is therefore strongly advised that one should check the calibrated parameters for plausibility.

For other, more complicated models such as the CEV model or some interest rate models, we need more model parameters than just σ. However, the principal form of the algorithm remains the same. The minimization in the least-squares problem then has to be done with respect to all those parameters.

2. Of course, one can also include other market prices apart from the usual calls to calibrate a model. However, if the corresponding derivative does not admit an explicit pricing formula as a function of the underlying parameters, we need numerical methods (such as the Monte Carlo method) to compute the model prices. This can then be very time-consuming. Therefore, closed pricing formulae are the basis of a convenient model calibration.

3. In practical applications, some market derivative prices have different importance (due to the liquidity or the size of the bid-ask spread of the contract) for the calibration of the parameters. One takes care of this by assigning the squared differences different weights (which add up to 1) according to their importance in the calibration procedure. □

5.10 Aspects of option pricing in incomplete markets

By definition, in an **incomplete market** there exists a contingent claim that cannot be replicated by trading in the basic assets. A simple example of an incomplete market is the **one-period trinomial model** with a stock price that can change from s at time 0 to $S(T) \in \{us, s, ds\}$ at time T with

$$u > e^{rT} > d \text{ and } u > 1 > d. \tag{5.149}$$

It is now an easy exercise to show that for a call with strike K satisfying $s < K < us$ there is no replication strategy (φ_0, φ_1). Indeed, the three equations that are equivalent to the replication property have no solution:

$$\varphi_0 e^{rT} + \varphi_1 us = us - K, \quad \varphi_0 e^{rT} + \varphi_1 s = 0, \quad \varphi_0 e^{rT} + \varphi_1 s = 0. \tag{5.150}$$

Obviously, the unique solution to the last two equations is $(\varphi_0, \varphi_1) = (0, 0)$, which does not solve the first equation.

In such a situation the replication approach is no longer a great help to determine an option price. Here and in the general case, however, one can determine so-called **arbitrage bounds** for the option price. Let us therefore consider an option with a final payment B at time T. Then, we have the following arbitrage bounds:

- the **sub-hedging price** p_B^- defined as the maximal amount of money such that a trading strategy $\varphi(.)$ exists that leads to a final wealth of $X^\varphi(T)$ that never exceeds the final payments of the option.
- the **super-hedging price** p_B^+ defined as the minimal amount of money such that a trading strategy $\varphi(.)$ exists that leads to a final wealth of $X^\varphi(T)$ that never is below the final payments of the option.

Obviously, a possible price p_B of the option B that does not allow for an arbitrage opportunity has to satisfy

$$p_B^- \leq p_B \leq p_B^+. \tag{5.151}$$

So, the first consequence of these considerations is that the arbitrage approach in general only yields bounds for option prices in an incomplete market. However, if the option is attainable (i.e. there exists a replication strategy for it) then we have equality in relation (5.151) and the price of B is unique.

Our second tool for valuing contingent claims in a complete market was the equivalent martingale measure. Again, in an incomplete market the situation is more subtle. While in the one-period binomial model, only the probability p of a stock price increase determined the probability measure, in the one-period trinomial model we have to specify two probabilities to uniquely determine the probability measure. However, the martingale condition in this example only requires that the discounted future expected stock price equals its starting value. This determines just one probability; the other one is at our disposal. So, we have infinitely many EMM.

These two facts that replication is in general impossible (and should be replaced by sub- and super-hedging to get price bounds) and that there exist infinitely many equivalent martingale measures are the generic situation for incomplete markets. For a survey of the rigorous treatment of the different aspects of option pricing in incomplete markets we refer to Delbaen and Schachermayer (2006). There, in particular the Fundamental Theorems of Asset Pricing are stated in a precise form. As this requires a lot of technical details, we here only state their essential contents that we need:

- The existence of an EMM ensures the absence of arbitrage opportunities.
- If $\mathbb{Q}$ is an EMM then if we choose $\mathbb{E}_{\mathbb{Q}}(e^{-rT}B)$ as an option price we do not introduce an arbitrage opportunity by that.

Indeed, one can show that in all the models we are considering, there exists at least one EMM. Thus, arbitrage-free option prices of the form $\mathbb{E}_{\mathbb{Q}}(e^{-rT}B)$ can be used. Of course, if this EMM is no longer unique, one has to decide which EMM to use for option pricing. There are various suggestions in the literature. It is beyond the scope of this book to discuss them all and in particular to give advice on which one to use. However, for the purpose of Monte Carlo option pricing, it is enough that we assume that the choice of the pricing EMM $\mathbb{Q}$ has already been made.

5.11 Stochastic volatility and option pricing in the Heston model

Let us again consider the case of just one stock. While in the local volatility models one tries to obtain a more realistic behaviour of the stock price by introducing a nonlinear volatility function, in the stochastic volatility setting the volatility is assumed to follow a separate stochastic process. This can be motivated by the fact that on stock markets price changes are caused by trading activities. Considering the volatility as a measure for the price variations, one can think of it as being caused by the intensity or volume of trading or by the demand for the stock.

We can put this in our diffusion framework by looking at a two-dimensional stock price equation. However, to distinguish between stock price and volatility evolution, we will denote the stock price by $S(t)$ and the volatility process by $\nu(t)$. A stochastic volatility model is usually given by

$$dS(t) = \mu S(t)\,dt + \sqrt{\nu(t)}S(t)\,dW_1(t), \ S(0) = s, \tag{5.152}$$

$$d\nu(t) = \alpha(t)\,dt + \beta(t)\left(\rho dW_1(t) + \sqrt{1-\rho^2}dW_2(t)\right), \ \nu(0) = \nu. \tag{5.153}$$

Here, $\alpha(t)$, $\beta(t)$ are stochastic processes that are progressively measurable with respect to the filtration generated by the two-dimensional Brownian motion $(W_1(t), W_2(t))$. Further, we assume that they have a form such that a unique solution to the two-dimensional SDE for price and volatility exists.

There are various specifications for the coefficient functions in the volatility equation above (such as the model in Hull and White [1987] or the model in Stein and Stein [1991]). However, the most popular one – especially in real world applications – is the **Heston model**. To ensure that the volatility process remains nonnegative, Heston (1993) used a square-root process (see Section 4.6.3 for details on the square-root-diffusion process) for the volatility. Thus, we consider the model given by

$$dS(t) = rS(t)\,dt + \sqrt{\nu(t)}S(t)\,dW_1(t), \tag{5.154}$$

$$d\nu(t) = \kappa(\theta - \nu(t))\,dt + \sigma\sqrt{\nu(t)}dW_2(t) \tag{5.155}$$

where the two Brownian motions have a correlation of

$$\mathbb{C}orr(W_1(t), W_2(t)) = \rho. \tag{5.156}$$

In reality this correlation between the Brownian motions driving the stock price and the volatility is typically negative (sometimes close to -1!), an effect which is also referred to as a **leverage effect**. θ is the long-term limit of the volatility, κ determines the speed of the drift toward that long-term value, and σ is the **volatility of the volatility**.

A main reason for the success of the Heston model with practitioners is a (semi-)explicit pricing formula for European calls and puts (see Andersen [2007] for this version).

THEOREM 5.31 Heston call price formula
(a) In the Heston model specified by Equations (5.154) *to* (5.156) *the price of a European call with strike* K *and maturity* T *is given by*

$$p_C = S(0) - Ke^{-rT} \cdot \frac{1}{2\pi} \int_{-\infty}^{\infty} \frac{\exp\left(\left(\frac{1}{2} - iu\right) \ln\left(S(0)\, e^{rT}/K\right) + h_1 - \left(u^2 + \frac{1}{4}\right) h_2\right) \nu(0)}{u^2 + \frac{1}{4}} du \tag{5.157}$$

where i *is the complex unit and where we have*

$$h_1 = -\tfrac{\kappa\theta}{\sigma^2}\left(\delta_+ T + 2\ln\left(\tfrac{\delta_- + \delta_+ e^{-\xi T}}{2\xi}\right)\right), \quad h_2 = \tfrac{1 - e^{-\xi T}}{\delta_- + \delta_+ e^{-\xi T}}, \tag{5.158}$$

$$\hat{\kappa} = \kappa - \tfrac{\rho\sigma}{2}, \quad \delta_\pm = \xi \mp (iu\rho\sigma + \hat{\kappa}), \tag{5.159}$$

$$\xi = \sqrt{u^2\sigma^2\left(1 - \rho^2\right) + 2iu\sigma\rho\hat{\kappa} + \hat{\kappa}^2 + \sigma^2/4}. \tag{5.160}$$

(b) Let

$$\varphi(u, w) = \mathbb{E}\left(e^{iu\nu(T) + iwx(T)}\right) \tag{5.161}$$

be the joint characteristic function of $x(T) = ln(S(T)/S(0))$ *and of* $\nu(T)$*. Then we have:*

$$\varphi(u, w) = e^{iwrT + C(T;u,w) + D(T;u,w)\nu(0)} \tag{5.162}$$

with

$$d(w) = \sqrt{(iw\rho\sigma - \kappa)^2 + w^2\sigma^2 + \sigma^2 iw}, \tag{5.163}$$

$$Q(u, w) = \tfrac{\alpha_+(w) - iu}{\alpha_-(w) - iu}, \quad \alpha_\pm(w) = \tfrac{\kappa - iw\rho\sigma \pm d(w)}{\sigma^2}, \tag{5.164}$$

$$D(T; u, w) = \alpha_+(w) \frac{1 - Q(u,w)e^{d(w)T}\frac{\alpha_-(w)}{\alpha_+(w)}}{1 - Q(u,w)e^{d(w)T}}, \tag{5.165}$$

$$C(T; u, w) = \kappa\theta\left[\alpha_+(w)\, T + \tfrac{\alpha_-(w) - \alpha_+(w)}{d(w)} \ln\left(\tfrac{Q(u,w)e^{d(w)T} - 1}{Q(u,w) - 1}\right)\right] \tag{5.166}$$

Although there is no closed-form solution for the volatility equation, we know that $\nu(t)$ has a noncentral chi-square distribution with suitable degrees of freedom and noncentrality parameter (see Theorem 4.52). It would therefore be tempting to use a direct exact simulation approach for the volatility. Such an exact algorithm for simulating $\nu(t)$ is described in Broadie and Kaya (2006). Then, given the simulated volatility process, one could use a suitable discretization scheme (such as e.g. the Euler or the Milstein schemes) for the stock price process. This will work fine as long as the two Brownian motions driving the stock price and the volatility process are independent. However, in the general case there is one problem: How to obtain the required correlation between the Brownian motion in the stock price and the volatility equation? Broadie and Kaya (2006) describe a procedure that includes an inversion of a Fourier transform and which is unbiased but also slow when compared to simpler schemes (see Lord et al. [2008]).

A straightforward approach is to use a discretization algorithm to simulate paths of both the stock price and the volatility. An obvious choice for an algorithm based on the Euler-Maruyama scheme is the following algorithm.

Algorithm 5.12 Simulating price paths in the Heston model (naive way)

---

1. Initialize the volatility and the stock price process: $\nu(0) = \nu_0,\ S(0) = s$.
2. Choose $\Delta = \frac{T}{n}$ with n the number of steps, T the maturity.
3. For $j = 1$ to n do
   (a) Simulate independent random numbers $Z \sim N(0,1)$,$Y \sim N(0,1)$.
   (b) Set $W = \rho Z + \sqrt{1-\rho^2}Y$.
   (c) Update the volatility:
   $$\nu(j\Delta) = \nu((j-1)\Delta) + \kappa(\theta - \nu((j-1)\Delta))\Delta + \\ + \sigma\sqrt{\nu((j-1)\Delta)}\sqrt{\Delta}W.$$
   (d) Update the log-stock price $X(t) = \ln(S(t))$:
   $$X(j\Delta) = X((j-1)\Delta) + \left(r - \frac{1}{2}\nu((j-1)\Delta)\right)\Delta + \\ \sqrt{\nu((j-1)\Delta)}\sqrt{\Delta}Z.$$
4. Interpolate $X(t)$ linearly between the times $j\Delta,\ j = 0, 1, ..., n$.

Though at first sight the above algorithm seems to be a textbook application of the Euler-Maruyama scheme, it contains an obvious flaw: While in the continuous-time setting the volatility process is always nonnegative as the solution of the above volatility equation, the discretized version in Step 3 (c) of the algorithm above might attain negative values. Then, one cannot use its square root in Step 3 (c) and (d) of the next iteration. To get around this problem, there are various suggestions in the literature (see Lord et al. [2008] for a systematic treatment of them):

1. **Absorption**: Use the positive part of the previous iterate, i.e.

$$\nu(j\Delta) = \nu((j-1)\Delta)^{+} + \kappa\left(\theta - \nu((j-1)\Delta)^{+}\right)\Delta + \\ + \sigma\sqrt{\nu((j-1)\Delta)^{+}}\sqrt{\Delta}W,$$

   and use $\nu((j-1)\Delta)^{+}$ in the simulation step for $X(j\Delta)$.

2. **Reflection**: Use the absolute value of the previous iterate, i.e.

$$\nu(j\Delta) = |\nu((j-1)\Delta)| + \kappa(\theta - |\nu((j-1)\Delta)|)\Delta + \\ + \sigma\sqrt{|\nu((j-1)\Delta)|}\sqrt{\Delta}W,$$

   and use $|\nu((j-1)\Delta)|$ in the simulation step for $X(j\Delta)$.

3. Use the absolute value $|\nu((j-1)\Delta)|$ only in the square-root part in both the computation of the next iterate for the volatility and the stock price. Otherwise, use the standard Euler scheme for both processes (method of Higham and Mao [2005]).

4. **Partial truncation**: Use the positive part $\nu((j-1)\Delta)^{+}$ for the square-root part in both the computation of the next iterate for the volatility and the stock price. Otherwise, use the standard Euler scheme for both processes (see Deelstra and Delbaen [1998]).

5. **Full truncation**: Use the positive part of the previous iterate only in the drift and the diffusion term, i.e.

$$\nu(j\Delta) = \nu((j-1)\Delta) + \kappa\left(\theta - \nu((j-1)\Delta)^{+}\right)\Delta + \\ + \sigma\sqrt{\nu((j-1)\Delta)^{+}}\sqrt{\Delta}W,$$

   and use $\nu((j-1)\Delta)^{+}$ in the simulation step for $X(j\Delta)$ (see Lord et al. [2008]).

Lord et al. (2008) report that the full truncation method performs best. Thus, if one wants to use the above algorithm in a way ensuring nonnegativity of the volatility process one should simply use the full truncation modification in Step 3 (c) and (d) of the algorithm. We will demonstrate its performance at the end of the next section.

5.11.1 The Andersen algorithm for the Heston model

The intensive application of the Heston model in the financial world led to a search for more accurate methods than the above Euler-Maruyama scheme. While an algorithm in Kahl and Jäckel (2006) based on an implicit variant of the Milstein scheme promises positivity of the volatility process for a certain parameter constellation, it performed poorly in a test series by Andersen (2007). On the other hand the exact simulation algorithm by Broadie and Kaya (2006) is quite slow and also has further problems when applied to the Heston model. However, there is a recent algorithm by Andersen (2007) that combines ideas of exact simulation with the speed of an Euler-Maruyama discretization scheme. The main idea is to

- use a suitable (Gaussian) approximation of the noncentral chi-square distribution for simulating the volatility process,
- use a suitable discretization scheme for the subsequent simulation of the stock price.

Here, great care is taken to achieve a correct correlation between both simulated processes, volatility and stock price paths. The presentation of the algorithm is quite involved. A detailed presentation of the derivation of all results and modifications is beyond the scope of a textbook. We present the main ideas step by step as they also give interesting insights.

Step 1: Approximation of the volatility process

As the volatility process is always nonnegative, the approximation should also be. Further, as the volatility process has a (noncentral) chi-square distribution, one is tempted to use a displaced squared Gaussian as an approximation. Indeed, Andersen suggests the use of a combined method. For large values $\hat{\nu}(t)$ he uses a square of a Gaussian distribution as an approximation while for small values $\hat{\nu}(t)$ an expression obtained from an asymptotic expansion of the density of $\hat{\nu}(t+\Delta)$ is the choice. More precisely, let

$$\hat{\nu}(t+\Delta) = a\left(b + Z_\nu\right)^2 , \tag{5.167}$$

with $Z_\nu \sim N(0,1)$ and a, b determined via matching the relevant first two moments. The following lemma shows that a moment matching with this choice can be obtained if the first two moments of the conditional distribution for $\hat{\nu}(t+\Delta)$ given $\hat{\nu}(t)$ are suitably related. Indeed, it can be shown that this requires that $\hat{\nu}(t)$ should not be too small. We need the explicit form of the first two moments of the volatility process (see Theorem 4.52):

$$m := m\left(\hat{\nu}(0)\right) := \mathbb{E}\left(\nu(\Delta)\,|\nu(0) = \hat{\nu}(0)\right) = \theta + \left(\hat{\nu}(0) - \theta\right) e^{-\kappa\Delta} \tag{5.168}$$

$$\begin{aligned} s^2 &:= s^2\left(\hat{\nu}(0)\right) := \mathbb{V}ar\left(\nu(\Delta)\,|\nu(0) = \hat{\nu}(0)\right) \\ &= \frac{\hat{\nu}(0)\,\sigma^2 \exp(-\kappa\Delta)}{\kappa}\left(1 - \exp(-\kappa\Delta)\right) + \frac{\theta\sigma^2}{2\kappa}\left(1 - \exp(-\kappa\Delta)\right)^2 \end{aligned} \tag{5.169}$$

LEMMA 5.32
With the above notation let $\Psi := s^2/m^2$. *In the case of* $\Psi \leq 2$ *we set*

$$b^2 = 2\Psi^{-1} - 1 + \sqrt{2\Psi^{-1}}\sqrt{2\Psi^{-1} - 1} > 0, \tag{5.170}$$

$$a = \frac{m}{1+b^2}. \tag{5.171}$$

Then, for $\hat{\nu}(t+\Delta)$ *generated as in Equation* (5.167), *we have matching of the first two moments of the approximating and the exact distribution, i.e.*

$$m = E(\hat{\nu}(\Delta)) \ , \ s^2 = Var(\hat{\nu}(\Delta)). \tag{5.172}$$

As the above lemma is only valid for a restricted choice of Ψ, we also have to look for an approximation of the exact (conditional) distribution of $\nu(t+\Delta)$ given $\hat{\nu}(t)$ is small. We therefore look at the density of the chi-square distribution with η degrees of freedom, which has the form

$$f_{\chi^2}(x;\eta) = const \cdot e^{-x/2} x^{\eta/2-1}. \tag{5.173}$$

By noting that as the relevant value of the degrees of freedom for the distribution of the volatility process

$$\eta = 4\kappa\theta/\sigma^2 \tag{5.174}$$

typically satisfy

$$\eta < 2 \tag{5.175}$$

in practical applications, in this case the above density attains large values around 0 (which corresponds to the behaviour of the noncentral chi-square density for small values of $\hat{\nu}(t)$). Indeed, the cumulative distribution function can in this case be well approximated by a distribution function having a point mass in 0. Andersen (2007) chooses an approximation of the form

$$F(x) = p + (1-p)\left(1 - e^{-\beta x}\right), \quad x \geq 0 \tag{5.176}$$

with p and β again determined by moment matching. With the inverse of F,

$$F^{-1}(u) = \begin{cases} 0 & , \quad 0 \leq u \leq p \\ \frac{1}{\beta}\ln\left(\frac{1-p}{1-u}\right), & p < u < 1 \end{cases}, \tag{5.177}$$

we can use the inverse transformation method to simulate the next step value of the volatility process as

$$\hat{\nu}(t+\Delta) = F^{-1}(U_V), \quad U_V \sim \mathcal{U}(0,1). \tag{5.178}$$

The moment matching to obtain p and β can be done for sufficiently small values of $\hat{\nu}(t)$ as ensured by the following lemma.

LEMMA 5.33
We use the notation of Lemma 5.32. In the case of $\Psi \geq 1$ *we set*

$$p = \frac{\Psi - 1}{\Psi + 1} \in [0,1) \ , \ \beta = \frac{1-p}{m} = \frac{2}{m(\Psi+1)} > 0. \tag{5.179}$$

Then, for $\hat{\nu}(t+\Delta)$ *generated as in Equation* (5.178) *we have matching of the first two moments of the approximating and the exact distribution, i.e.*

$$m = \mathbb{E}(\hat{\nu}(\Delta)) \ , \ s^2 = \mathbb{V}ar(\hat{\nu}(\Delta)). \tag{5.180}$$

By the two lemmas above, for all possible values of Ψ (i.e. for all possible values of $\hat{\nu}(t)$) we have a method to generate the next step value $\hat{\nu}(t+\Delta)$ of the volatility process. To decide which one to use in the overlapping area $[1,2]$, Andersen (2007) suggests using $\Psi_{sw} = 1.5$ as a switching point, i.e. above Ψ_{sw} we use the first and below Ψ_{sw} we use the second method.

Algorithm 5.13 Quadratic exponential (QE) method for simulating the volatility in the Heston model

---

1. Given $\hat{\nu}(t)$, compute $m = m(\hat{\nu}(t))$ and $s^2 = s^2(\hat{\nu}(t))$ as given in Equations (5.168) and (5.169).
2. Compute $\Psi = s^2/m^2$.
3. Draw a uniform random number U_v.
4. If $\Psi \leq \Psi_{sw}$ then
   - Compute a, b as in Lemma 5.32
   - Compute $Z_\nu = \Phi^{-1}(U_v)$
   - Set $\hat{\nu}(t+\Delta) = a(b+Z_v)^2$
5. Otherwise
   - Compute b, p as in Lemma 5.33
   - Set $\hat{\nu}(t+\Delta) = F^{-1}(U_V)$

As the variance s^2 decreases with decreasing stepsize Δ, the parameter Ψ approaches 0 with decreasing stepsize. Thus, for small stepsizes usually the first method in Algorithm QE above is always used.

REMARK 5.34 Andersen (2007) also gives a volatility approximation by a truncated Gaussian distribution which performs worse than QE. □

Step 2: Simulation of the stock price process

We now derive the recursion for the stock price process. It is clear that due to the nonlinear appearance of the normal random number Z_ν in each step of the volatility simulation, a standard log-stock price simulation of the form

$$\ln\left(\hat{S}(t+\Delta)\right) = \ln\left(\hat{S}(t)\right) - \frac{1}{2}\hat{\nu}(t)\Delta + \sqrt{\hat{\nu}(t)}\sqrt{\Delta}Z_s \tag{5.181}$$

will lead to a correlation problem. Indeed, choosing $Corr(Z_s, Z_\nu) = \rho$ will not produce the required correlation between the two processes. We thus need another approach, which relies on the exact correlation between the log-stock price and the volatility process. From the joint characteristic function in Theorem 5.31, Andersen (2007) inferred a limit for small values of Δ:

$$Corr(\ln(S(\Delta)), \nu(\Delta)) = \rho + o(\Delta). \tag{5.182}$$

Thus, the conditional correlation between $\ln(\hat{S}(t+\Delta))$ and $\hat{\nu}(t+\Delta)$ given $\ln(\hat{S}(t))$ and $\hat{\nu}(t)$ should also equal ρ. Using Itô's formula we obtain

$$\begin{aligned} d\ln(S(t)) &= \left(r - \frac{1}{2}\nu(t)\right)dt + \sqrt{\nu(t)}dW_1(t) \\ &= \left(r - \frac{1}{2}\nu(t)\right)dt + \rho\sqrt{\nu(t)}dW_2(t) + \sqrt{1-\rho^2}\sqrt{\nu(t)}d\tilde{W}(t) \end{aligned}$$

with $\tilde{W}(t)$ independent of $W_2(t)$. Plugging in the representation for $\nu(t)$ solved for $\int_t^{t+\Delta} \sqrt{\nu(u)}dW_2(u)$ we arrive at

$$\begin{aligned} \ln(S(t+\Delta)) = \ln(S(t)) + r\Delta + \frac{\rho}{\sigma}(\nu(t+\Delta) - \nu(t) - \kappa\theta\Delta) + \\ + \left(\frac{\kappa\rho}{\sigma} - \frac{1}{2}\right)\int_t^{t+\Delta} \nu(u)\,du + \sqrt{1-\rho^2}\int_t^{t+\Delta}\sqrt{\nu(u)}d\tilde{W}(u). \end{aligned} \tag{5.183}$$

In this formula, the two integrals involving the volatility have to be approximated. For the first one, the simplest form would be

$$\int_t^{t+\Delta} \nu(u)\,du \approx \Delta\left(\gamma_1\nu(t) + \gamma_2\nu(t+\Delta)\right). \tag{5.184}$$

Here, $\gamma_1 = 1, \gamma_2 = 0$ (left approximation) or $\gamma_1 = \gamma_2 = \frac{1}{2}$ (central approximation) are simple choices. The normal variable Z_ν used to simulate the volatility step has to be independent of $\tilde{W}$. This is essential for our method below to generate the correct correlation between the two processes $\hat{S}$ and $\hat{\nu}$.

Further, note that given $\hat{\nu}(t)$ and given $\int_t^{t+\Delta} \nu(u)\,du$, we obtain

$$\left(\int_t^{t+\Delta} \sqrt{\nu(u)} d\tilde{W}_u \,\middle|\, \nu(t), \int_t^{t+\Delta} \nu(u)\,du \right) \sim N\left(0, \int_t^{t+\Delta} \nu(u)\,du\right) \quad (5.185)$$

which allows an easy simulation of this integral. Thus, we can suggest a discretization step for the stock price as

$$\begin{aligned} \ln\left(\hat{S}(t+\Delta)\right) = \ln\left(\hat{S}(t)\right) + r\Delta + \frac{\rho}{\sigma}\left(\hat{\nu}(t+\Delta) - \hat{\nu}(t) - \kappa\theta\Delta\right) + \\ + \Delta\left(\frac{\kappa\rho}{\sigma} - \frac{1}{2}\right)\left(\gamma_1\hat{\nu}(t) + \gamma_2\hat{\nu}(t+\Delta)\right) + \\ + \sqrt{\Delta}\sqrt{1-\rho^2}\sqrt{\gamma_1\hat{\nu}(t) + \gamma_2\hat{\nu}(t+\Delta)} \cdot Z \end{aligned} \quad (5.186)$$

with $Z \sim N(0,1)$ independent of the simulated volatility process. In Proposition 10 of Andersen (2007) a weak consistency property of the algorithm is shown. This in particular implies that for small values of Δ the conditional correlation between the increments of the approximating processes for stock price and the volatility are (approximately) of the right order. Thus, we can give Algorithm 5.14.

Algorithm 5.14 Stock price paths in the Heston model

1. Given $\hat{\nu}(t)$ simulate $\hat{\nu}(t+\Delta)$ by the algorithm QE given above.
2. Generate a standard normally distributed random number Z.
3. Given $\hat{\nu}(t)$, $\hat{\nu}(t+\Delta)$ and $\ln\left(\hat{S}(t)\right)$ simulate $\ln\left(\hat{S}(t+\Delta)\right)$ as in Equation (5.186) with γ_1, γ_2 both nonnegative and $\gamma_1 + \gamma_2 = 1$.

Further refinements (such as martingale corrections, a variant of the algorithm for dealing with time-dependent coefficients in the Heston model, or refined approximations of the volatility process) are possible, but will not be presented here. We refer the reader to Andersen (2007) for these details.

Some numerical illustrations will be given in Section 5.11.2 after the introduction of the Heath-Platen estimator.

5.11.2 The Heath-Platen estimator in the Heston model

An algorithm that is particularly suited for barrier options (or other options that contain payments related to exit or entry events) is the approach by Heath and Platen (see Heath and Platen [2002]). Although it can deal with a more general setting, we will here present it in the framework of the Heston model. The approach tries to approximate the local dynamics of the Heston model as closely as possible while preserving analytical tractability of the approximating process. To introduce it, we need some notation.

Let the stock price and the volatility process (under an already chosen EMM) be given by

$$dS(t) = S(t)\left(rdt + \sqrt{\nu(t)}dW_1(t)\right), \tag{5.187}$$

$$d\nu(t) = \kappa(\theta - \nu(t))\,dt + \sigma\sqrt{\nu(t)}\left(\rho dW_1(t) + \sqrt{1-\rho^2}dW_2(t)\right) \tag{5.188}$$

and introduce the operator

$$\begin{aligned} L^0 f(t,s,\nu) &= f_t(t,s,\nu) + rsf_s(t,s,\nu) + \frac{1}{2}\nu s^2 f_{ss}(t,s,\nu) \\ &+ \kappa(\theta-\nu) f_\nu(t,s,\nu) + \frac{1}{2}\nu\sigma^2 f_{\nu\nu}(t,s,\nu) + \nu s\sigma\rho f_{\nu s}(t,s,\nu). \end{aligned} \tag{5.189}$$

For a suitably integrable and smooth function $g(t,x,y)$ Itô's formula implies

$$\mathbb{E}\left(g(\tau, S(\tau), \nu(\tau))\right) = g(0,s,\nu) + \mathbb{E}\left(\int_0^\tau L^0 g(t, S(t), \nu(t))\,dt\right) \tag{5.190}$$

where τ is some exit time defined by

$$\tau = \inf\{t \geq 0 \,|(t, S(t)) \notin [0,T) \times \Gamma\} \tag{5.191}$$

for Γ an interval (which is possibly unbounded). Examples of such stopping times related to options are:

- A simple European option corresponding to $\tau = T$.
- Barrier options corresponding to the general case of $\tau = \inf\{t \geq 0 \,|(t, S(t)) \notin [0,T) \times (H_1, H_2)\}$.

For an option with a final payoff given by $B = h(\tau, S(\tau))$, we consider its nondiscounted value (again, we assume that we are working under an EMM)

$$u(t,s,\nu) = \mathbb{E}^{(t,s,\nu)}\left(h(\tau, S(\tau))\right) = \mathbb{E}\left(h(\tau, S(\tau))\,|F_t\right) \tag{5.192}$$

which is a martingale (with respect to the filtration F_t generated by $(S(.), \nu(.))$) given that we have $t \leq \tau(\omega)$ and that the option payoff $h(.,.)$ is sufficiently integrable (which we will always assume). This in particular implies

$$L^0 u(t,s,\nu) = 0, \quad (t,s,\nu) \in (0,T) \times \Gamma \times [0,\infty). \tag{5.193}$$

The main idea of Heath and Platen (2002) now is to find a function $\tilde{u}(t,s,\nu)$ such that we have

$$h(\tau, S(\tau)) = u(\tau, S(\tau), \nu(\tau)) = \tilde{u}(\tau, S(\tau), \nu(\tau)) \tag{5.194}$$

and that allows for an easy calculation of $\bar{u}(0, s, \nu)$. Given sufficient integrability of $\tilde{u}(.,.,.)$, we obtain by Itô's formula and Equations (5.192), and (5.194):

$$\begin{aligned} u(0, s, \nu) &= \mathbb{E}\left(h\left(\tau, S\left(\tau\right)\right)\right) \\ &= \tilde{u}\left(0, s, \nu\right) + \mathbb{E}\left(\int_0^\tau L^0 \tilde{u}\left(t, S\left(t\right), \nu\left(t\right)\right) dt\right). \end{aligned} \tag{5.195}$$

Now the idea for a good choice of $\tilde{u}(.,.,.)$ in the Heston setting is coupled to find a good approximating price process that mimics some properties of the Heston model, but is analytically tractable. Therefore, one introduces a Black-Scholes type Heston approximation with deterministic volatility

$$d\tilde{S}\left(t\right) = \tilde{S}\left(t\right)\left(rdt + \sqrt{\nu\left(t\right)}dW_1\left(t\right)\right), \tag{5.196}$$

$$d\tilde{\nu}\left(t\right) = \kappa\left(\theta - \tilde{\nu}\left(t\right)\right) dt \tag{5.197}$$

together with the operator

$$\begin{aligned} &\tilde{L}^0 f\left(t, s, \nu\right) = \\ &f_t\left(t, s, \nu\right) + rsf_s\left(t, s, \nu\right) + \frac{1}{2}\nu s^2 f_{ss}\left(t, s, \nu\right) + \kappa\left(\theta - \nu\right) f_\nu\left(t, s, \nu\right). \end{aligned} \tag{5.198}$$

For a given initial value of $\tilde{\nu}(0) = \nu$, we now have

$$\tilde{\nu}\left(t\right) = \theta + \left(\nu - \theta\right) e^{-\kappa t}. \tag{5.199}$$

For the special case of a European call option we directly obtain the following.

PROPOSITION 5.35
With the choice of $\tau = T$, $h(T, s) = (s - K)^+$ for some $K \geq 0$ we obtain

$$\tilde{u}\left(0, s, \nu\right) := \mathbb{E}\left(\left(\tilde{S}\left(T\right) - K\right)^+\right) \tag{5.200}$$

$$= e^{rT} BS\left(s, K, r, \bar{\sigma}_0, T\right) \tag{5.201}$$

where $BS(s, K, r, \sigma, T)$ denotes the Black-Scholes price of a European call with strike K and maturity T when the underlying starts with initial price s and has a drift of r and a volatility of σ. We further have

$$\bar{\sigma}_t := \sqrt{\tfrac{1}{T-t}\int_t^T \tilde{\nu}(y)dy} = \sqrt{\theta - \left(\nu\left(t\right) - \theta\right)\tfrac{\exp(-\kappa(T-t))-1}{\kappa(T-t)}}, \tag{5.202}$$

$$\tilde{L}^0 \tilde{u}\left(t, s, \nu\right) = 0, \quad \left(t, s, \nu\right) \in \left(0, T\right) \times \left[0, \infty\right)^2. \tag{5.203}$$

With the choice of $\tilde{u}$ in the proposition we could now use relation (5.195) to construct an unbiased estimator for $u(0, s, \nu)$ (and thus for the option price). However, relation (5.203) allows us to subtract $\tilde{L}^0\tilde{u}(t,s,\nu)$ under the integrand and preserve the unbiasedness of the estimator. Even more, by subtracting it, some terms cancel out compared to Equation (5.195). We therefore define the **Heath-Platen estimator** in this setting via the right-hand side of

$$\begin{aligned} u(0,s,\nu) &= \mathbb{E}\left(\left(\tilde{S}(T)-K\right)^+\right) \\ &= \tilde{u}(0,s,\nu) + \mathbb{E}\left(\int_0^T \left(L^0-\tilde{L}^0\right)\tilde{u}(t,S(t),\nu(t))\,dt\right). \end{aligned} \tag{5.204}$$

To use this estimator we still have to compute the integrand:

$$\begin{aligned} \left(L^0-\tilde{L}^0\right)\tilde{u}(t,s,\nu) &= \sigma\nu\left(s\rho\tilde{u}_{s\nu}(t,s,\nu) + \tfrac{1}{2}\sigma\tilde{u}_{\nu\nu}(t,s,\nu)\right) \\ &= \sigma\nu e^{r(T-t)}\left(s\rho BS_{s\bar{\sigma}_t}(s,K,r,\bar{\sigma}_t,T-t)\tfrac{d\bar{\sigma}_t}{d\nu}\right. \\ &\quad +\tfrac{1}{2}\sigma\left[BS_{\bar{\sigma}_t\bar{\sigma}_t}(s,K,r,\bar{\sigma}_t,T-t)\left(\tfrac{d\bar{\sigma}_t}{d\nu}\right)^2\right. \\ &\quad \left.\left.+\ BS_{\bar{\sigma}_t}(s,K,r,\bar{\sigma}_t,T-t)\tfrac{d^2\bar{\sigma}_t}{(d\nu)^2}\right]\right) \end{aligned} \tag{5.205}$$

where all the subscripts denote the partial derivatives with respect to the indicated variables. These partial derivatives are well known by practitioners and are often calculated anyway. Due to their popularity they even have special names and are known as **the Greeks** (see Section 5.15 for more on them). In particular, they can be obtained in closed form for a European call in the Black-Scholes setting. We will not state them explicitly here. However, one should remark that this is the extra work needed for the Heath-Platen estimator. To use it for a particular option requires that the relevant Greeks for this option in the Black-Scholes model have to be calculated. In particular, for barrier options this amounts to some additional work.

To put all this together in the form of an algorithm, one has to simulate N paths of the stock price, the volatility process, and of

$$Z(t) = \tilde{u}(0,s,\nu) + \mathbb{E}\left(\int_0^T \left(L^0-\tilde{L}^0\right)\tilde{u}(t,S(t),\nu(t))\,dt\right) \tag{5.206}$$

where the explicit form of (5.205) has to be used. For the simulation of this trivariate process (S,ν,Z), Heath and Platen (2002) use a weak predictor-corrector scheme. We will use an Euler-Maruyama scheme with full truncation below to keep the volatility process nonnegative, but of course any other suitable scheme can be used.

Some numerical comparisons of crude MC, Heath-Platen, and Andersen estimators

We will illustrate the different performance of the standard method (in particular, the Euler-Maruyama simulation with full truncation), the Andersen

Algorithm 5.15 Call pricing with the Heath-Platen estimator

Let N and $\Delta = T/n$ be given.

For $i = 1$ to N do

1. $S^{(i)}(0) = s,\ \nu^{(i)}(0) = \nu,\ Z^{(i)}(0) = e^{rT} BS(s, K, r, \bar{\sigma}_T, T)$.
2. For $j = 1$ to n do
   (a) Generate two independent $N(0,1)$-random numbers $Y_1^{(ij)}, Y_2^{(ij)}$.
   (b) Update (S, ν, Z):

$$S^{(i)}(j\Delta) = S^{(i)}((j-1)\Delta)\left(1 + r\Delta + \sqrt{\nu^{(i)}((j-1)\Delta)}\sqrt{\Delta}Y_1^{(ij)}\right)$$

$$\nu^{(i)}(j\Delta) = \nu^{(i)}((j-1)\Delta) + \kappa\left(\theta - \nu^{(i)}((j-1)\Delta)^+\right)\Delta + \\ + \sigma\sqrt{\nu^{(i)}((j-1)\Delta)^+}\sqrt{\Delta}\left(\rho Y_1^{(ij)} + \sqrt{1-\rho^2}Y_2^{(ij)}\right)$$

$$Z^{(i)}(j\Delta) = Z^{(i)}((j-1)\Delta) + \\ + \left(L^0 - \tilde{L}^0\right)\tilde{u}\left((j-1)\Delta, S((j-1)\Delta), \nu((j-1)\Delta)\right)\Delta$$

   Use relation (5.205) for the computation of the Z-update!

Estimate the call price by the Heath-Platen estimator

$$I_{HP,N} = \frac{1}{N} e^{-rT} \sum_{i=1}^{N} Z^{(i)}(T)$$

---

algorithm, and the Heath-Platen estimator by two simple examples. For an intensive numerical study in particular of the Andersen estimator we refer to Andersen (2007). There, the excellent performance of the Andersen algorithm is demonstrated especially for extreme settings of the market coefficients.

First, we will look at a short running European call (maturity $T = 0.5$) with moderate market coefficients. We vary both the correlation between stock price and volatility and the volatility of the volatility. For purposes of seeing differences, we have chosen a moderate discretization of $\delta = 0.005$ and N=10,000 simulation runs. As one can see from Table 5.4, the Heath-Platen estimator typically performs best and leads to a big variance reduction that decreases with increasing volatility of volatility. However, for the last case, the Andersen algorithm is closer to the true value which is even outside the confidence bound for the Heath-Platen estimator. This can be explained by the fact that due to the discretization scheme used for simulating the price path, all the estimators are biased. Note in particular the short confidence intervals of the Heath-Platen estimator.

Method	$\sigma = 0.2$ $\rho = -0.15$	$\sigma = 0.2$ $\rho = -0.8$	$\sigma = 0.7$ $\rho = -0.8$
Heath-Platen estimator	6.54602	6.59101	5.78902
Lower 95% bound	6.54542	6.58864	5.76593
Upper 95% bound	6.54661	6.59338	5.81211
Crude MC estimator	6.53499	6.58481	5.76985
Lower 95% bound	6.47678	6.53440	5.66188
Upper 95% bound	6.59319	6.63522	5.87781
Andersen estimator	6.49917	6.54820	5.83626
Lower 95% bound	6.43985	6.49720	5.72666
Upper 95% bound	6.55849	6.59920	5.94586
Exact value	6.54730	6.59440	5.82040

Table 5.4: Heston Call Prices with $T = 0.5, S(0) = K = 100, r = 0.04$, $\theta = \nu(0) = 0.04$, and $\kappa = 0.6$

As a second example, we have chosen a longer-running European call (maturity $T = 5$), $\delta = 0.01$, N=50,000, a high volatility of volatility, and a negative correlation between stock and volatility. Again, the Heath-Platen estimator performs best, keeping the discretization bias in mind (see Table 5.5).

Method	Estimator	Lower 95%	Upper 95%
Heath-Platen	34.8026	34.7736	34.8316
Crude MC	34.9887	34.5887	35.3886
Andersen	34.5734	34.1791	34.9677

Table 5.5: Heston Call Prices with Exact Value 34.8348, $T = 5$, $S(0) = K = 100, r = 0.05, \theta = \nu(0) = 0.09, \kappa = 2, \sigma = 1$, and $\rho = -0.7$

To judge the three algorithms, these two examples are not enough. However, the advantages of the Heath-Platen estimator are obvious (high variance reduction, best accuracy), but they come on the expense of having to calculate option price sensitivities, which can be quite tedious and need to be adapted to each special class of options that should be priced. Therefore, the other two methods are good alternatives whereas for moderate coefficient choices we could not see a clear advantage for the Andersen algorithm.

5.12 Variance reduction principles in non-Black-Scholes models

We would like to point out some general variance reduction principles in stock price models that differ from the Black-Scholes one. They can usually not compete with methods tailored to the particular stock price model (such as the Heath-Platen estimator for the Heston model). However, they are often useful for obtaining a first step in the direction of variance reduction.

Principle 1: Use a simpler, but not too simple, approximating price process in the construction of control variates

When computing an option price by Monte Carlo methods, an obvious candidate for a control variate is to use the option payments based on an approximating Black-Scholes model. For this, the first two moments of the approximating Black-Scholes price processes should coincide with the first two moments of the considered stock price process. As the Heath-Platen estimator has demonstrated, it will be advantageous to imitate as much as possible of the dynamics of the more complicated stock price process while still keeping analytical tractability.

Principle 2: Use localized versions of methods developed for simpler price processes

As generalized stock price processes behave locally like a Black-Scholes model, localized versions of specialized Monte Carlo methods for pricing particular types of options should also deliver good results. Even more, as option pricing in more sophisticated stock price models typically require path simulations by a numerical discretization scheme, they are dealt with as being local Black-Scholes models. We give an application of this idea below.

An application: Barrier option pricing with approximating exit probabilities

The standard method for barrier option pricing is to simulate discretized versions of stock price path with checks of the barrier condition at the simulated time instants. As in the Black-Scholes case, this method overestimates the prices of out-options as it is based on linear interpolation between the simulated stock prices. Instead, we could argue that between two discretization points the (log-)stock price process can be well approximated by a Brownian bridge. One could therefore use one of the Brownian bridge methods presented for barrier options in the Black-Scholes model in Section 5.6.2. For example, the suitable variant of Moon's algorithm (see Algorithm 5.8) would simply contain a barrier condition check in Step 2 (d)–(f) between every discretization point of the stock price path. However, it is clear that the absolute volatility $|\sigma|$ has to be replaced by the modulus of the volatility that is valid for the discretized stock price process on the corresponding interval.

5.13 Stochastic local volatility models

For purposes of pricing highly path-dependent options, models have been developed that combine the local and the stochastic volatility approaches. One such model is the **Bergomi model** (see Bergomi [2005]). The Bergomi model was developed to price options where the payment depends on different parts of the stock price path, especially when the price of the option depends highly on the stock volatility. It admits the following features:

- The volatility of the stock process can be determined by implied variances of forward variance swaps corresponding to the chosen timediscretization ("market-conform volatility behaviour").
- It is possible to decouple the evolution of these implied variance processes from the spot price evolution.

The stock price model is based on an underlying time structure that is adapted to the characteristics of the special type of option that has to be valued. Given a time discretization $0 = T_0 < T_1 < ... < T_n$ with $T_i = i\Delta$, the stock price is assumed to follow a CEV model of the type

$$dS(t) = S(t)\,(r_t - q_t)\,dt + S(t)\sigma_0^{(i)} \left(\frac{S(t)}{S_{T_i}}\right)^{1-\beta^{(i)}} dW_t \tag{5.207}$$

for each $t \in [T_i, T_i + \Delta]$. The volatility parameters $\sigma_0^{(i)}$, $\beta^{(i)}$ are piecewise constant on $[T_i, T_i + \Delta]$. They are readjusted at the beginning of the time interval. To obtain a price of a corresponding option, we have to model the distribution of these volatility parameters. The suggestion of Bergomi is to introduce **volatility dynamics** via the functions

$$\sigma_0^{(i)} = \sigma_0\left(\xi^{(i)}\left(T_i\right)\right),\; \beta^{(i)} = \beta\left(\xi^{(i)}\left(T_i\right)\right) \tag{5.208}$$

where the underlying processes $\xi^{(i)}(t)$ are assumed to follow weighted exponential Ornstein-Uhlenbeck (OU) processes, i.e.

$$\begin{aligned}\xi^{(i)}(t) = \xi^{(i)}(0)\exp&\left(\omega\left\{e^{-k_1(T_i-t)}X_t + \theta e^{-k_2(T_i-t)}Y_t\right\}\right)\cdot\\ \exp&\left(-\frac{\omega^2}{2}\left\{e^{-2k_1(T_i-t)}E\left(X_t^2\right) + \theta^2 e^{-2k_2(T_i-t)}E\left(Y_t^2\right)\right\}\right)\cdot\\ \exp&\left(-\frac{\omega^2}{2}\left\{2\theta e^{-(k_1+k_2)(T_i-t)}E\left(X_tY_t\right)\right\}\right),\end{aligned} \tag{5.209}$$

$$X_t = \int_0^t \exp\left(-k_1\,(t-u)\right) dU_u, \quad Y_t = \int_0^t \exp\left(-k_2\,(t-u)\right) dZ_u, \tag{5.210}$$

$$\mathbb{C}orr\,(U_t, Z_t) = \rho. \tag{5.211}$$

One could of course use as much Brownian components as one has time intervals, but the use of a two-dimensional instead of an n-dimensional one is a compromise with respect to tractability. However, we need at least a two-dimensional Brownian motion to be able to model short-term and long-term influences on the volatility dynamics. In the above model the choice of

$$k_1 > k_2 \tag{5.212}$$

implies that X_t can be seen as the **short factor** and Y_t as the **long factor**. The characteristic property of the Bergomi model is that the combination of the CEV stock price model and the exponential OU volatility models allows a decoupling of producing a good explanation of today's skew/smile and the dynamic evolution of the volatility structure over time. We will not go into further details here but refer the interested reader to Bergomi (2005).

5.14 Monte Carlo option pricing: American and Bermudan options

Until now, the options we have looked at were those of the European type. However, at real option markets many traded options can be exercised at any time before the maturity date (i.e. **American** type) or can be exercised at a finite set of time instants (so-called **Bermudan** options). As the buyer of an American or a Bermudan option is allowed to choose the exercise time of the option, the exact time of the payment is not a priori clear to the seller of the option. However, the payment B_τ for each fixed exercise strategy τ (where τ is a stopping time with values in $[0, T]$) is uniquely determined, as e.g.

$$B_\tau = \left(K - S\,(\tau)\right)^+ \tag{5.213}$$

for an American put with exercise strategy τ. So, the option price for this fixed exercising strategy τ is given by

$$p_{B_\tau} = \mathbb{E}\left(\exp^{-r\tau} B_\tau\right) \tag{5.214}$$

where the expected value is computed under the unique EMM in the complete market. As the seller of the option should be prepared against the worst such strategy (from his point of view) and the buyer should choose that exercise strategy that maximizes the value of his option, the price of an American option is given by

$$\sup_{\tau \in \mathcal{S}[0,T]} \mathbb{E}\left(\exp^{-r\tau} B_\tau\right) \tag{5.215}$$

where $\mathcal{S}[0,T]$ is the set of all stopping times (adapted to the filtration corresponding to our market model) with values in $[0,T]$ almost surely.

REMARK 5.36 It is important to understand that as the buyer of the option can choose any kind of – even strange or suboptimal – exercise strategy, it is not possible for the seller to perfectly replicate the payment arising from this strategy. However, it can be shown that in a complete market (such as the Black-Scholes market), there is an admissible trading strategy with initial wealth equalling the above defined price of the American option and admitting a wealth process $X^*(t)$ satisfying

$$X^*(t) \geq B_t \;\; \forall t \in [0,T] \text{ a.s.} \tag{5.216}$$

(see Karatzas and Shreve [1998]). Here, B_t denotes the payment resulting from the American option exercised at time t. Note that we typically cannot have equality at each time $t \in [0,T)$. This can be seen by an American put option which at time 0 is out of the money: There $K < S(0)$ yields

$$(K - S(T))^+ = 0. \tag{5.217}$$

If we would insist on replication then the wealth process – and therefore the price of the option – would have to equal zero. However, as it is still possible that the put option ends up in the money, but never leads to a negative wealth, a price of zero would therefore create an arbitrage opportunity! □

Before continuing, let us define an American contingent claim.

DEFINITION 5.37

An **American contingent claim** *consists of a progressively measurable stochastic process $B = \{(B_t, F_t)\}_{t\in[0,T]}$ with $B_t \geq 0$ and a final payment B_τ at the exercise time $\tau \in [0,T]$ chosen by the holder of the contingent claim. We assume in addition that τ is a stopping time, that $\{(B_t, F_t)\}_{t\in[0,T]}$ possesses continuous paths, and that*

$$\mathbb{E}\left(\sup_{0\leq s\leq T} (B_s)^\mu\right) < \infty \text{ for some } \mu > 1. \tag{5.218}$$

The suitable analogue to a replication strategy is the term of a **hedging strategy** for American contingent claims.

DEFINITION 5.38
(a) A portfolio process $\pi \in \mathcal{A}(x)$ *with corresponding wealth process* $X^{\pi}(t) \geq B_t$ *for all* $t \in [0,T]$ *is called a* **hedging strategy** *with price* $x > 0$ *for the American contingent claim* B. *Let* $\mathcal{H}(x) = \mathcal{H}(x; B)$ *be the set of hedging strategies for* B *with price* $x > 0$.
(b) $\hat{p} = \inf\{x > 0 \mid \mathcal{H}(x) \neq \emptyset\}$ *is called the* **fair price** *of the American contingent claim.*

Equipped with this technical framework, we can now state a result on the fair price of an American contingent claim.

THEOREM 5.39
The fair price $\hat{p}$ *of an American contingent claim* B *is given by*

$$\hat{p} = \sup_{\tau \in \mathcal{S}[0,T]} \mathbb{E}\left(e^{-r\tau} B_{\tau}\right), \tag{5.219}$$

and there exists a stopping time τ^* *such that the supremum will be attained for the hedging strategy* π^* *corresponding to* τ^*.

REMARK 5.40 1. Showing the existence of the optimal stopping strategy τ^* and the form of the valuation process $X^*(t)$ (see Karatzas and Shreve [1998]) is technically involved. However, simple arbitrage arguments yield that every price for the American contingent claim below or above $\hat{p}$ generates an arbitrage opportunity (see Korn and Korn [2001]).

2. By Theorem 5.39 the optimal strategy is to exercise the contingent claim at the first time τ^* when the intrinsic value B_{τ^*} of the option coincides with the option price $X^*(\tau^*)$. Although this seems to be an explicit solution, in general neither $X^*(t)$ nor τ^* have explicit representations. Even in the simplest case of an American put, numerical methods are needed for their computation. For further results we refer to Myneni (1992). □

Bermudan options are discrete versions of American contingent claims. Their owner has the right to exercise the option at a finite set of times $t_1 < ... < t_m$.

DEFINITION 5.41
Consider time instants $0 \leq t_1 < ... < t_m = T$. *A* **Bermudan option** *consists of a set of* F_{t_i}*-measurable random variables* $B_{t_i} \geq 0$ *and a final payment* B_{τ} *at the exercise time* $\tau \in \{t_1, ..., t_m\}$ *chosen by the holder of the option. Here,* τ *is assumed to be a stopping time and that*

$$\mathbb{E}\left(\sup_{s \in \{t_1,...,t_m\}} (B_s)^{\mu}\right) < \infty \text{ for some } \mu > 1. \tag{5.220}$$

As for American contingent claims, one can state the corresponding theorem on the fair price (defined similarly to the fair price of American contingent claims) and the existence of an optimal exercising strategy.

THEOREM 5.42

The fair price $\hat{p}$ of a Bermudan option B is given by

$$\hat{p} = \sup_{\tau \in \mathcal{S}\{t_1,\ldots,t_m\}} \mathbb{E}\left(\exp^{-r\tau} B_\tau\right), \tag{5.221}$$

where $\mathcal{S}\{t_1, ..., t_m\}$ is the set of stopping times with values in $\{t_1, ..., t_m\}$, and there exists a stopping time τ^* such that the supremum will be attained for the hedging strategy $(\pi^*, 0)$ corresponding to τ^*.

So, if we want to calculate the fair price of an American or a Bermudan option, we **cannot** simply generate a large number of price paths by a suitable Monte Carlo procedure. We also have to know at each time instant if exercising of the option would be profitable. But this can only be decided if the optimal exercise strategy would be known in advance. This leads to Algorithm 5.16.

Algorithm 5.16 Algorithmic framework for pricing American (Bermudan) options by Monte Carlo methods

---

1. Determine the optimal exercise strategy τ^* for the contingent claim B.
2. Determine the option price $\mathbb{E}\left(\exp^{-r\tau^*} B_{\tau^*}\right)$ by Monte Carlo simulation.

While the second step is straightforward (and similar to the pricing of European options via Monte Carlo), the necessary action in the first step is new. We will see in the following parts how to do this.

5.14.1 The Longstaff-Schwartz algorithm and regression-based variants for pricing Bermudan options

The algorithm by Longstaff and Schwartz (2001) is the most popular one used in real-life applications if we consider Bermudan options on more than one underlying (for calculating American option prices on one underlying binomial tree methods are typically the much more efficient and easier choice). The algorithm makes use of the **dynamic programming principle** of stochastic control (also called **backwards induction**). To understand this principle corresponding to the valuation problem of a Bermudan option B with possible exercise times $\{t_1, ..., t_m = T\}$, we introduce:

- $\mathcal{S}(i)$ as the set of stopping times τ with values in $\{i, ..., m\}$,
- $V(i) = e^{-rt_i} B_{t_i}$.

As we do not know the optimal exercise strategy *a priori*, we start at time T. Given that the option has not been exercised until then, its value at that time simply equals its intrinsic value, B_T. The net present value of it equals $\mathbb{E}(V(m))$. Note also that at time T the set $\mathcal{S}(m)$ consists of the constant time m which then yields the optimal stopping time $t_m = T$.

If we go one step backwards in time to the second but last possible exercise time t_{m-1}, we have to decide at each possible value of the underlying $S(t_{m-1})$ if we exercise the option or not. This decision is easy, as we only have to comparing the value of keeping the option until time T with the value to exercise it immediately and receive its intrinsic value $B_{t_{m-1}}$. The value of keeping the option is simply given by the net present value at time t_{m-1} of the payment B_T given the current stock price equals $S(t_{m-1})$, i.e. it equals

$$\mathbb{E}\left(e^{-r(T-t_{m-1})} B_T \left| S\left(t_{m-1}\right)\right.\right). \tag{5.222}$$

Comparing this value of keeping the option with its intrinsic value at time t_{m-1} is equivalent to comparing $V(m-1)$ with $\mathbb{E}\left(V(m)\left|S(t_{m-1})\right.\right)$. Depending on this comparison, the optimal exercise time $t_{\tau^*(m-1)}$ (conditioned on not having exercised before t_{m-1} and on the actual price $S_{t_{m-1}}$) equals m or $m-1$. As we now know how we can proceed in an optimal way from each possible price vector $S(t_{m-1})$, we can again go one step backwards in time and do the same considerations to obtain the optimal strategy at time t_{m-2} and so on. Thus, we arrive at Algorithm 5.17.

Algorithm 5.17 Dynamic programming for calculating the price of Bermudan options

---

1. Set $i = m$, $\tau(i) = m$.
2. At each time t_i with $i = m-1, ..., 0$ determine the optimal (conditional) exercise strategy $\tau(i) \in \mathcal{S}(i)$ via:

$$\tau(i) = \begin{cases} i & , \text{if } V(i) \geq \mathbb{E}\left(V\left(\tau\left(i+1\right)\right)\left|S\left(t_i\right)\right.\right) \\ \tau(i+1), & \text{else} \end{cases}$$

3. At time $t = 0$ the stopping time $t_{\tau(0)}$ is the optimal exercising strategy and the fair price of the Bermudan option is given by $\mathbb{E}(V(\tau(0)))$.

If we could calculate all the required conditional expectations in this algorithm, we would have indeed solved the pricing problem for the Bermudan option B. However, this is typically not the case. On the other hand, as the stock price $S(t)$ is a Markov process, by assuming that B_t is of the form

$$B_t = f\left(S(t)\right) \tag{5.223}$$

for some suitable function f, we know that we have the relations

$$V\left(i\right) = g\left(i, S\left(t_i\right)\right) = e^{-rt_i} f\left(S(t_i)\right), \tag{5.224}$$

$$\mathbb{E}\left(V\left(j\right) \middle| S\left(t_i\right)\right) = u\left(S\left(t_i\right)\right) \quad \text{for } i < j \tag{5.225}$$

with a suitable measurable function u. Note that this function depends on i and j, but we omit this for ease of notation. If we choose a parametric family U of functions u, we can set up a regression model to approximate the conditional expectation in the least-squares sense by solving the problem

$$\min_{u \in U} \mathbb{E}\left[\mathbb{E}\left(g\left(i+1, S\left(t_{i+1}\right)\right) \middle| S\left(t_i\right)\right) - u\left(S\left(t_i\right)\right)\right]^2. \tag{5.226}$$

We thus need a specification of the function space U and simulated data to solve the above regression problem. Popular choices for the family U are:

- $U := \left\{u : \mathbb{R}^d \to \mathbb{R} \middle| u\left(x\right) = \sum_{i=1}^{\infty} a_i x^i, a_i \in \mathbb{R}\right\}$,
- $U := \left\{u : \mathbb{R}^d \to \mathbb{R} \middle| u\left(x\right) = \sum_{i=1}^{k} a_i H_i\left(x\right), a_i \in \mathbb{R}, \right\}$ where $H_i : \mathbb{R}^d \to \mathbb{R}$ are **basis functions** and $k \in \mathbb{N}$.

The Laguerre polynomials used by Longstaff and Schwartz (2001) are examples of such basis functions. Note that both parameterizations of U are linear in the coefficients a_i. Therefore, the above least-squares problem is indeed a linear regression problem. Given we have N independent copies $S_{t_j}^{(n)}$, $n = 1, ..., N$, $j = 1, ..., m$ of simulated stock price paths, we can explicitly solve the regression problem

$$\min_{a \in \mathbb{R}^k} \frac{1}{N} \sum_{j=1}^{N} \left(g\left(i+1, S^{(j)}\left(t_{i+1}\right)\right) - \sum_{l=1}^{k} a_l H_l\left(S^{(j)}\left(t_i\right)\right)\right)^2. \tag{5.227}$$

Its solution is the optimal coefficient vector $(a_1^*, \ldots, a_N^*)$ of the linear regression problem. We thus compute the pseudoinverse of the design matrix via

$$H\left(i,j\right) = \left(H_1\left(S^{(j)}\left(t_i\right)\right), ..., H_k\left(S^{(j)}\left(t_i\right)\right)\right), \tag{5.228}$$

$$H\left(i\right) = \left(H\left(i,1\right)', ..., H\left(i,N\right)'\right)', \tag{5.229}$$

$$H^+\left(i\right) = \left(H\left(i\right)' H\left(i\right)\right)^{-1} H\left(i\right)', \tag{5.230}$$

and obtain

$$a^*\left(i\right) = H^+\left(i\right) g\left(i+1\right) \tag{5.231}$$

where $g(i+1)$ is the data vector, i.e. the vector containing all entries $g(i+1, S^{(j)}(t_{i+1}))$. The solution of this problem also yields an estimate $\hat{C}(S;i)$ for a functional description of the continuation value of the Bermudan option at time t_i (i.e. the value of not exercising the option and holding it further),

$$\hat{C}(S;i) = \sum_{l=1}^{k} a_l^*(i) H_l(S) \tag{5.232}$$

with $H_l(S)$ being component l in notation (5.228) with argument S.

In their original paper, Longstaff and Schwartz (2001) use only points $S^j(t_i)$ in the regression problem (5.227) with a positive intrinsic value $f(S^j(t_i))$. As stated in Wendel (2009), numerical experiments show that it is advisable to follow this strategy.

We have now put all the ingredients together to formulate the Longstaff-Schwartz algorithm for pricing Bermudan options, Algorithm 5.18.

Variants, convergence, and additional aspects of the LS-algorithm

1. Showing convergence of the Longstaff-Schwartz algorithm is a subtle business. There are two different sources for differences between the approximating option price $\hat{V}(0)$ and the actual one:

- A discretization error due to the estimation of the conditional expectation by a projection on a finite set of basis functions $H_1, ..., H_k$.
- A Monte Carlo error as the expected value (= the option price) is estimated by an arithmetic mean.

This convergence issue is dealt with in a rigorous and detailed way in Clément et al. (2002). There, the authors introduce

$$V^k(0) = \sup_{\tau \in \mathcal{S}(H_1,...,H_k)} \mathbb{E}\left(e^{-r\tau} B_\tau\right) \tag{5.233}$$

where the set $\mathcal{S}(H_1, ..., H_k)$ only contains exercise strategies based on the solution of the regression problems with the basis functions $H_1, ..., H_k$. As a second value, they introduce $V^{k,N}(0)$ which equals the Longstaff-Schwartz estimate $\hat{V}(0)$ as computed above with the k basis functions and N simulated stock price paths. Under some technical conditions Clément et al. (2002) prove:

- With growing number k of basis functions the approximating option price $V^k(0)$ converges towards the real option price, i.e. we have

$$V^k(0) \xrightarrow{k\to\infty} V(0) \tag{5.234}$$

if the sequence of basis functions is total in a suitable L^2-function space.

Algorithm 5.18 Longstaff-Schwartz algorithm for pricing Bermudan options

1. Choose a set of basis functions $H_1, ..., H_k$.

2. Generate N independent paths $S^j(t_1), ..., S^j(t_m)$, $j = 1, ..., N$ of the stock price at the possible exercise times of the Bermudan option.

3. Fix the terminal values of the Bermudan option for each path, i.e. set

$$\hat{V}(m, j) := e^{(-rT)} f\left(S^j(T)\right), \ j = 1, ..., N.$$

4. Continue backward in time: For $i = m - 1, ..., 1$

   - Solve the regression problem (5.227) at time t_i, i.e. calculate the vector of optimal weights $\hat{a}^*(i)$ at time t_i by

$$\hat{a}^*(i) = H^+(i)\,\hat{V}(i+1)$$

   - Calculate the estimates of the continuation values as

$$\hat{C}\left(S^j(t_i); i\right) = \sum_{l=1}^{k} a^*(i)_l\, H_l\left(S^j(t_i)\right), \ j = 1, ..., N.$$

   - For $j = 1$ to N set

$$\hat{V}(i, j) := \begin{cases} e^{-rt_i} f\left(S^j(t_i)\right), & \text{if } e^{-rt_i} f\left(S^j(t_i)\right) \geq C\left(S^j(t_i); i\right) \\ \hat{V}(i+1, j), & \text{else} \end{cases}$$

5. Set $\hat{V}(0) := \frac{1}{N} \sum_{j=1}^{N} \hat{V}(1, j)$.

- With growing number N of simulated stock price paths the Longstaff-Schwartz value $V^{k,N}(0)$ converges almost surely towards the approximating option price $V^k(0)$, i.e. we have

$$V^{k,N}(0) \xrightarrow{N \to \infty} V^k(0) \ \mathbb{P} \text{ a.s.} \tag{5.235}$$

To state it again: For a fixed number k of basis functions the LS-algorithm converges to the solution of the optimal stopping problem (5.233) and **not** to the option price if we let the number N of simulated paths tend to infinity. There remains the bias generated by the finite number k. Although the two convergence results are very nice from a mathematical point of view, they make it hard to give a direct implication how many basis functions to choose. At least, the results in Clément et al. (2002) ensure that we have a convergence rate of $O(1/\sqrt{N})$ for the Longstaff-Schwartz value towards $V^k(0)$.

However, we do not have a similar result for the number of basis functions. We therefore try to give some advice when we consider numerical examples below.

2. Longstaff and Schwartz (2001) claim that the algorithm yields a lower bound for the price of the Bermudan option as the computed exercise strategy (given by the estimates of the continuation functions) is not necessarily optimal. Given the convergence results of Clément et al. (2002), this is certainly asymptotically correct when the number N of simulated stock price paths is large. However, as Glasserman (2004) points out, the LS-algorithm above contains a mixing of a high and a low bias: As the optimization of the exercise strategy is only done on a finite set of paths, the obtained option value based on exactly these paths might be higher than the real value. Therefore, to ensure a low biased estimate of the option price, the LS-algorithm should contain a modified last step that is based on new, independently simulated stock price paths:

- Simulate M new stock price paths and calculate

$$\hat{V}(0) := \frac{1}{M} \sum_{j=N+1}^{M+N} \hat{V}(1,j). \tag{5.236}$$

In practical applications the LS-algorithm often already shows a low bias.

3. In Tsitsiklis and van Roy (1999, 2001) the authors introduced a variant of the LS-algorithm where in Step 4 the following value iteration procedure is used for $V(i,j)$:

$$\hat{V}(i,j) := \begin{cases} e^{-rt_i} f\left(S^j(t_i)\right), & \text{if } e^{-rt_i} f\left(S^j(t_i)\right) \geq \hat{C}_j\left(S^j(t_i);i\right) \\ \hat{C}_j\left(S^j(t_i);i\right), & \text{else} \end{cases} \tag{5.237}$$

Using new stock price paths for estimating $\hat{V}(0)$ yields a low biased estimator.

4. **Further improvements of the LS-algorithm**

a. The regression step in the LS-algorithm could also consist of nonlinear or nonparametric regression methods that might better capture the functional form of the conditional expectation (see Egloff [2005]). Indeed, this is a current research topic.

b. If the value of a corresponding European option is available at each point where a decision about exercising or not then this European value is a lower bound for the real continuation value and can be higher than the one obtained by the calculations based on the simulated paths. One should therefore only exercise the option if the intrinsic value is above both, the computed continuation and the European value.

c. Rasmussen (2005) uses European options as control variates.

5. **Computational effort and choices of basis functions**: As we will see in the numerical example below, the LS-algorithm often works quite well with

a simple choice of basis functions such as all monomials of the underlyings up to the third order. Including the payoff function together with its square also is often helpful. One could further recommend including the pricing formula for the European version of the option under study. However, as this formula might be quite hard to evaluate itself, this could lead to extremely long computing times even for low dimensions.

6. **Pricing with a parametric exercise boundary**: An alternative simple method that can be used to value American or Bermudan options is a parameterization of the exercise boundary. More precisely, we suggest a parametric family for the form of the optimal exercise boundary and then try to determine the parameter that delivers the highest estimate of the option price. We will demonstrate this method in more detail when dealing with Bermudan interest rate options in Section 5.19.4.

7. **Use of the LS-algorithms for pricing American options**: In Bally et al. (2005) a convergence result for the approximation of American option prices by Bermudan option prices is given. It states that L^p-error bounds are of the order $1/\sqrt{m}$.

Numerical illustration of the LS-algorithm: A Bermudan max-call

We highlight the behaviour of the LS-algorithm by looking at the following example of a Bermudan max-call on three stocks given by the payoff function

$$h(s_1, s_2, s_3) = (\max\{s_1, s_2, s_3\} - K)^+ \tag{5.238}$$

where we have used a time horizon of $T = 3$, an interest rate of $r = 0.1$, and a dividend rate of $div = 0.1$. Note that this results in an additional discount factor of $\exp(-0.1t)$ that enters the stock price (compare to Section 5.5.3). All stocks have a volatility of 0.2. Further, the log-returns between the stocks have correlations of $\rho_{12} = -0.25 = -\rho_{13}$, $\rho_{23} = 0.3$. This is achieved by using

$$\sigma = \begin{pmatrix} \frac{1}{5} & 0 & 0 \\ -\frac{1}{20} & \frac{\sqrt{15}}{20} & 0 \\ \frac{1}{20} & \frac{29}{100\sqrt{15}} & \frac{\sqrt{3659}}{100\sqrt{15}} \end{pmatrix} \approx \begin{pmatrix} 0.2 & 0 & 0 \\ -0.05 & 0.1936 & 0 \\ 0.05 & 0.0749 & 0.1561 \end{pmatrix}$$

Our aim is to calculate price estimates for varying stepsizes m of the Bermudan options. We also want to demonstrate the effect of different choices of the basis functions. Although we have tested more basis sets (see Wendel [2009] for a detailed study), we here only give the results for those listed in Table 5.6.

The first test case is for $S_1(0) = S_2(0) = S_3(0) = K = 100$ and $m = 4$, $m = 50$, i.e. the option is at the money. The results for $N = 50{,}000$ paths (together with 95%-confidence intervals) are given in Table 5.7. The numbers show some remarkable facts:

- The differences between the prices for different basis sets are much more pronounced for large m.

Set	Basis functions
0	1 1 1
I	$s_1,\ s_2,\ s_3$ + set 0
II	all monomials up to power 3
III	all monomials up to power 3 + $h(s_1, s_2, s_3)$
IV	all monomials in one variable up to power 3 + $h(s_1, s_2, s_3)$
	+ $s_1 s_2,\ s_1 s_3,\ s_2 s_3,\ s_1 s_2 s_3$
V	set IV + $h(s_1, s_2, s_3)^2,\ h(s_1, s_2, s_3)^3$
VI	all monomials in one variable up to power 7

Table 5.6: Sets of Basis Functions

- The two very simple basis sets 0 and I are clearly not suitable.
- The basis sets without mixed monomials yield smaller prices than those with mixed monomials.
- The basis sets with mixed monomials and the payoff function (i.e. III, IV, V) yield similar prices.

This behaviour returned through all our test sets and was even more pronounced for higher values of m such as $m = 250$ or of $N = 500{,}000$ (see also Wendel [2009]). For $m = 4, N = 500{,}000$, we also computed the price for a basis that included all monomials up to power 3 and the European option price. It could be verified for larger values of N (where the good basis sets lead to nearly exactly the same value) that the resulting price was indeed the best estimate for 50,000 paths. As it is already computationally intensive to calculate this European price, the computing times for the Bermudan price were often more than an hour (compared to some seconds for other basis sets).

The main consequence out of the numerical experiment is that the choice of the basis set should include functions that resemble the form of the payoff function and of the European option pricing function. However, the inclusion of the European option pricing function hinges critically on the possibility of an efficient computation of this function.

5.14.2 Upper price bounds by dual methods

As the suitable variant of the Longstaff-Schwartz algorithm yields a lower bound for the Bermudan option price, there remains the question how much it actually deviates from the exact price. Therefore, an upper bound to the price is needed to judge the quality of the estimate. If then the difference between those two estimates would be small, the price of the Bermudan option can be predicted with high accuracy. Here, we will present the idea of upper bounds via considering a dual optimization problem to the optimal stopping problem which yields the price of the Bermudan option. This idea goes back to Rogers

Set	Price	$m = 4$ 95%CI	Price	$m = 50$ 95%CI
0	16.838	[16.692, 16.983]	15.795	[15.699, 15.8903]
I	17.002	[16.649, 17.155]	16.732	[16.615, 16.850]
II	17.318	[17.163, 17.472]	17.842	[17.717, 17.966]
III	17.261	[17.104, 17.417]	18.172	[18.022, 18.323]
IV	17.446	[17.287, 17.605]	17.979	[17.829, 18.129]
V	17.479	[17.322, 17.637]	18.088	[17.939, 18.236]
VI	17.247	[17.095, 17.399]	17.620	[17.494, 17.740]

Table 5.7: Bermudan LS-Option Prices for Different m and Different Basis Functions (at the Money Option)

(2002) and to Haugh and Kogan (2004) who looked at the dual optimization problem of the optimal stopping problem which characterizes the price of a Bermudan option.

We will continue to use the discounted terms of the previous section. The starting point is the observation that due to the fact that not exercising the Bermudan option at time t_i may not be the optimal strategy, we obtain

$$\hat{V}(i) \geq \mathbb{E}\left(\hat{V}(i+1)\,|S(t_i)\right) \tag{5.239}$$

which means that the (discounted) value function process is a super-martingale. Due to e.g. Myneni (1992) we even know that the value function is the minimal super-martingale dominating the (discounted) payoff process $g(i, S(t_i))$, the so-called **Snell envelope** of the payoff process. However, as the Snell envelope is usually hard to compute, we have to look for another dominating super-martingale. With it, we would have an upper bound for the option price. Consider therefore a discrete-time martingale $M(i)$, $i = 0, 1, .., m$ that is defined at exactly the possible exercise times of the Bermudan option and that starts with $M(0) = 0$. Using the optional sampling theorem results in

$$\begin{aligned}\mathbb{E}\left(g\left(\tau, S\left(t_\tau\right)\right)\right) &= \mathbb{E}\left(g\left(\tau, S\left(t_\tau\right)\right) - M\left(\tau\right)\right) \\ &\leq \mathbb{E}\left(\max_{k=1,..,m}\left\{g\left(k, S\left(t_k\right)\right) - M\left(k\right)\right\}\right).\end{aligned} \tag{5.240}$$

As the inequality remains correct when we take the infimum over all possible martingales on the right-hand side, we can then take the supremum over all possible stopping times τ on the left part of the inequality and arrive at:

$$V(0) = \sup_{\tau}\mathbb{E}\left(V(\tau)\right) \leq \inf_{M}\mathbb{E}\left(\max_{k=1,..,m}\left\{V(k) - M(k)\right\}\right). \tag{5.241}$$

Hoping that our lower bound for the option price is a good one, we introduce the following martingale (which is also inspired by the Doob-Meyer decomposition of the super-martingale $\hat{V}$):

$$\hat{M}(k) := \sum_{i=1}^{k} \Delta(i) := \sum_{i=1}^{k} \left(\hat{V}(i) - \mathbb{E}\left(\hat{V}(i) \,|S(t_{i-1})\right)\right). \tag{5.242}$$

Note that if we had used the optimal exercise strategy for the computation of $\hat{V}(i)$, then it would be a martingale and $\hat{M}$ would be zero. Thus, the $\Delta(i)$ measures the quality of the lower bound for the option price at each time. The main problem of this approach is that we have to compute the conditional expectations in relation (5.242). This will lead to a nested simulation (see the Andersen-Broadie algorithm below). What remains to be done is to

- use the Longstaff-Schwartz stopping times $\hat{\tau}(i)$,
- estimate the expectations in relation (5.242) to obtain estimates for the martingale $\hat{M}(.)$,
- use the estimations for a (crude) Monte Carlo estimation of the expectation on the right-hand side of Equation (5.241) to estimate an upper bound for the option price.

While it is a straightforward Monte Carlo task to estimate $\mathbb{E}(\hat{V}(i)|S(t_{i-1}))$, we can obtain $\hat{V}(i)$ in a nearly identical way, by noting that we have

$$\begin{aligned}
V(i) &= \mathbb{E}\left(g\left(\tau_i, S\left(t_{\tau(i)}\right)\right)|S(t_i)\right) \\
&= \begin{cases} g(i, S(t_i)), & \text{if } g(i,S(t_i)) \geq \hat{C}(S(t_i);i), \\ \mathbb{E}\left(g\left(\tau(i+1), S\left(t_{\tau(i+1)}\right)\right)|S(t_i)\right) & \text{if } g(i,S(t_i)) < \hat{C}(S(t_i);i). \end{cases} \\
&= \begin{cases} g(i, S(t_i)), & \text{if } g(i,S(t_i)) \geq \hat{C}(S(t_i);i), \\ \mathbb{E}\left(V(i+1)|S(t_i)\right) & \text{if } g(i,S(t_i)) < \hat{C}(S(t_i);i). \end{cases}
\end{aligned} \tag{5.243}$$

Having made these considerations we can set up the **primal-dual algorithm** of Andersen and Broadie (2004), Algorithm 5.19. We use the same notation and data as in the Longstaff-Schwartz algorithm. In particular: Let $\hat{\tau}(i)$ be the stopping times obtained from the Longstaff-Schwartz algorithm characterized by the continuation functions $\hat{C}(S;i) = \sum_{l=1}^{k} a_l^*(i) H_l(S)$.

Further aspects of the Andersen-Broadie algorithm

1. Note that as we estimated the conditional expectations in the relation defining the martingale M, we cannot use the martingale argument to deduce that $\bar{Y}_{N_1}^{up}$ is indeed an estimator for an upper bound of the option price. However, Andersen and Broadie (2004) show that we are actually estimating $M(i)$ by $M(i) + \epsilon(i)$. The origin of $\epsilon(i)$ is the Monte Carlo error that we can

Algorithm 5.19 Andersen-Broadie algorithm to obtain upper bounds for the price of a Bermudan option

Repeat the following iteration independently for $j = 1$ to N_1:

1. Simulate a path of the stock price $S^j(t_1), ..., S^j(t_m)$ at the possible exercise times of the Bermudan option. Set $S^j(t_0) = S(0)$ and $M^j(0) = 0$.

2. For $i = 0$ to $m - 1$:

   - Compute $\hat{V}(i, j)$ according to (5.243) if $i > 0$.
   - Simulate N_2 stock price subpaths $\hat{S}^k(t_i),...,\hat{S}^k(t_{\tau(i+1)})$ that start with $\hat{S}^k(t_i) = S^j(t_i)$.
   - Estimate $\mathbb{E}(\hat{V}(i+1)\,|S^j(t_i)))$ by

$$\mathbb{E}\left(\hat{V}(i+1,j)\,|S^j(t_i))\right) = \frac{1}{N_2}\sum_{k=1}^{N_2} g\left(\tau(i+1), \hat{S}^k\left(t_{\tau(i+1)}\right)\right).$$

   - For $i > 0$ set

$$\Delta^j(i) = \hat{V}(i,j) - \mathbb{E}\left(\hat{V}(i,j)\,|S^j(t_{i-1})\right),\ M^j(i) = M^j(i-1) + \Delta^j(i).$$

3. Set $\hat{V}(m, j) = g(m, S^j(T))$ and

$$\Delta^j(m) = \hat{V}(m,j) - \mathbb{E}\left(\hat{V}(m,j)\,|S^j(t_{m-1}))\right),$$
$$M^j(m) = M^j(m-1) + \Delta^j(m).$$

4. Calculate $Y^{up}(j) = \max_{i=1,..,m}\left\{g\left(i, S^j\left(t_i\right)\right) - M^j\left(i\right)\right\}.$

Finally, estimate the upper bound for the option price by

$$\bar{Y}^{up}_{N_1} = \frac{1}{N_1}\sum_{j=1}^{N_1} Y^{up}(j).$$

---

assume to be normally distributed with a zero mean. By noting that we have

$$\mathbb{E}\left(\max_{i=1,\dots,m}\left(g\left(i, S\left(t_i\right)\right) - M\left(i\right) - \epsilon\left(i\right)\right)\right) \geq \mathbb{E}\left(g\left(i^*, S\left(t_{i^*}\right)\right) - M\left(i^*\right) - \epsilon\left(i^*\right)\right)$$
$$= \mathbb{E}\left(g\left(i^*, S\left(t_{i^*}\right)\right) - M\left(i^*\right)\right) = \mathbb{E}\left(\max_{i=1,\dots,m}\left(g\left(i, S\left(t_i\right)\right) - M\left(i\right)\right)\right) \qquad (5.244)$$

(where i^* is the random index where the maximum is attained) by the optional sampling theorem, we see that the estimator is indeed biased high.

2. **Confidence intervals for the Bermudan option price**
As we now have upper and lower Monte Carlo estimators for the option price, we could give a confidence interval that is based on the usual asymptotic normality assumption for large numbers of simulation runs. Let therefore $\bar{L}_N$ be a lower Monte Carlo estimator based on N simulated stock price paths (such as the suitably modified Longstaff-Schwartz value) with a corresponding sample variance of $\bar{\sigma}_L^2$. Further, let $\bar{\sigma}_{Y^{up}}^2$ denote the sample variance of the Andersen-Broadie upper bound based on N_1 simulated stock price paths. We then obtain an approximate 95%-confidence interval for the price of the considered Bermudan option as

$$\left[\bar{L}_N - 1.96\frac{\bar{\sigma}_L}{\sqrt{N}},\ \bar{Y}_{N_1}^{up} + 1.96\frac{\bar{\sigma}_{Y^{up}}}{\sqrt{N_1}}\right]. \tag{5.245}$$

Note that as both estimators are biased in the relevant direction, the confidence interval is clearly conservative. With a lower and an upper estimator at hand, one can also define the "real" estimator as the mean $(\bar{L}_N + \bar{Y}_{N_1}^{up})/2$.

3. **Computational effort for computing the upper bounds**
It is clear that the nested simulation to obtain the Anderson-Broadie upper bound is the bottleneck in the computation of the two bounds for the Bermudan option price. Andersen and Broadie (2004) report that this computation is responsible for 60% to 95% of the total CPU time. One can also obtain an upper bound of simulation steps as

$$N_{max} = N_1 \cdot N_2 \cdot m \tag{5.246}$$

where of course the subpaths typically have a much smaller length than m.

4. **Numerical illustration: A Bermudan put option on a single stock**
We consider the simplest possible Bermudan option, a put on a stock with just $m = 2$ exercise times. This is of course a toy example, but one which gives a lot of insight. As parameters we took

$$T = 1 = t_2,\ t_1 = 0.5,\ r = 0.1,\ \sigma = 0.2,\ S(0) = K = 100.$$

First, we should think about a simple but efficient basis for the regression approach. As the payoff is bounded, polynomials might not be a good choice. Including the payoff itself seems to be a good idea. However, the payoff is 0 for $S > K$ while the option value at time t_i is strictly positive, even if we have $S(t_1) > K$. To take care of this, we introduced a multiple of a Gaussian density with a mean of K into the basis. This is getting very small for large $S(t_1)$, but remains positive and adds the additional value needed at the strike of K. Not surprisingly, the very simple basis of

$$H_1(s) = 1,\ H_2(s) = (K - s)^+,\ H_3(s) = \exp\left(\frac{1}{2}(x - K)^2\right)$$

already showed an excellent performance. We look at the different steps:

Step 1: Determining the regression coefficients
Here, we distinguished between the Longstaff-Schwartz approach of basing the calculations only on those paths that are in the money in some time instants and the approach that takes into account all paths. As for the first set of paths the average value of payoffs is higher, the constant regression coefficient is higher as in the second method. This is compensated in the second method by higher coefficient for the other two basis functions, in particular for the third one. Two typical representatives obtained by 100,000 paths are:

$$a^{LS} = (1.8094, 0.6923, 2.1532) \text{ LS-approach}$$
$$a^{AP} = (1.0882, 0.7458, 2.8550) \text{ all paths used}$$

Step 2: Obtaining the lower bound
With the above regression coefficients the lower bounds are very close to the exact value of 4.313. We again present some typical values based on 100,000 paths, but the two methods performed similar:

$$price^{LS}_{low} = 4.3006, \ 95\%\text{-confidence interval } [4.2594, 4.3418]$$
$$price^{AP}_{low} = 4.3092, \ 95\%\text{-confidence interval } [4.2681, 4.3483]$$

Step 3: Obtaining the upper bound
For the upper bound, we used 1,000 paths that included the simulation of 1,000 "subpaths" at each time instant t_1, t_2 for obtaining the inner conditional expectations. Again, both methods performed similarly well:

$$price^{LS}_{up} = 4.3294, \ 95\%\text{-confidence interval } [4.3155, 4.3434]$$
$$price^{AP}_{up} = 4.3188, \ 95\%\text{-confidence interval } [4.3060, 4.3317]$$

It is very important to point out that for more complicated options results of such a quality cannot be obtained so easily. However, even in our simple setting, a naive approach might lead to inferior results. If we consider the basis consisting of all monomials up to the power of 3 then we obtain the following, very bad lower bounds (based on the same number of paths, computed with all paths), but acceptable upper ones:

$$price^{AP}_{low} = 3.4146, \ 95\%\text{-confidence interval } [3.3701, 3.4512]$$
$$price^{AP}_{up} = 4.3064, \ 95\%\text{-confidence interval } [4.2445, 4.3684]$$

An explanation for the bad performance (even in this simple example) is the use of a basis of unbounded functions for explaining a nonlinear bounded payoff.

6. **Upper bounds with nonnested simulation by the variant of Belomestny et al. (2009)**
As already pointed out by Andersen and Broadie (2004), the nested simulation for estimating the conditional expectations $\mathbb{E}(\hat{V}(i+1)\,|S^j(t_i)))$ is the main consumer of CPU time in the algorithm. The idea of Belomestny et al.

(2009) is an alternative approach to compute the martingale M as defined in relation (5.242). They assume that the payoff-function B_t of the Bermudan option is Lipschitz continuous in the price and 1/2-Hölder continuous in the time variable. With this, they refer to the martingale representation theorem that results in the representation

$$M(i) = \int_0^{t_i} U(s)\, dW(s) \text{ for } i = 1, ..., m \tag{5.247}$$

with a suitable progressively measurable integrand U. Instead of approximating the conditional expectations in Equation (5.242) of M, the idea is to approximate the integrand U in Equation (5.247) by a simple process $U^{\mathcal{P}}$ with $\mathcal{P}$ a partition $0 = T_0 < T_1 < T_{\mathcal{J}} = T$ that includes at least all the possible exercise times t_i of the Bermudan option. The suggestion for the process $U^{\mathcal{P}}$ now is

$$U_d^{\mathcal{P}}(T_j) = \mathbb{E}\left(\frac{W_d(T_{j+1}) - W_d(T_j)}{T_{j+1} - T_j} g\left(\tau(i), S\left(t_{\tau(i)}\right)\right) \mid S(T_j)\right) \tag{5.248}$$

for $t_{i-1} \leq T_j < t_i$. Here, d denotes component d of the relevant stochastic processes. The stopping times $\tau(i)$ are the Longstaff-Schwartz stopping times that we assume to be already given. The conditional expections in Equation (5.248) are now estimated in the following way:

- Generate N_3 paths of the n-dimensional Brownian motion $W(T_j)$, $j = 1, ..., \mathcal{J}$ and infer the stock prices at the exercise times from them.
- For all $j = 1, ..., \mathcal{J}$ estimate the integrand process $U^{\mathcal{P}}(T_j)$ by a least-squares method from the simulated paths of the Brownian motion corresponding to the stock price.
- From the so-obtained estimator $\hat{U}_d^{\mathcal{P}}(T_j, S)$ get an estimate for M by

$$\hat{M}^{\mathcal{P}}(i, S) = \sum_{T_j \in \mathcal{P}, T_j < t_i} \sum_{d=1}^{n} \hat{U}_d^{\mathcal{P}}(T_j, S)\left(W_d(T_{j+1}) - W_d(T_j)\right). \tag{5.249}$$

Then it only remains to perform a final estimation step:

- Simulate N_4 paths $S^j(t_1), ..., S^j(t_m)$ of the stock price at the exercise times and estimate the upper bound for the Bermudan option price by

$$\bar{Y}_{N_4}^{up,BBS} = \frac{1}{N_4} \sum_{j=1}^{N_4} \max_{i=1,...,m} \left(g\left(i, S^j(t_i)\right) - \hat{M}^{\mathcal{P}}\left(i, S^j\right)\right). \tag{5.250}$$

The resulting algorithm is reported to be fast and accurate. For further details and numerical examples the reader is refered to Belomestny et al. (2009) where also convergence considerations and the possibility to use their algorithm as a control variate in the Andersen-Broadie algorithm are discussed.

5.15 Monte Carlo calculation of option price sensitivities

5.15.1 The role of the price sensitivities

Calculating the price of an option is the central task when it comes to purchasing or selling an option. However, after this initial transaction has been performed, the main question is to judge and to hedge the risks that one has incurred by entering this option position. If one ignores the risk of having chosen a completely wrong model, then there still remains the risk of having chosen wrong input parameters and the risk of the random evolution of the security price itself. To obtain a feeling for the effect of small changes in these input parameters on the value of an option, traders are calculating the so-called **Greeks**. This name is reminiscent of the fact that most (but not all!) option price sensitivities (i.e. the partial derivatives of the option price with respect to the input parameters) are abbreviated by Greek letters. The standard option price sensitivities are:

$$
\begin{array}{llll}
\frac{\partial}{\partial S_1(t)} X(t) \quad \Delta \quad \text{“delta”}, & & \frac{\partial^2}{\partial S_1(t)^2} X(t) \quad \Gamma \quad \text{“gamma”} \\
\frac{\partial}{\partial t} X(t) \quad \Theta \quad \text{“theta”}, & & \\
\frac{\partial}{\partial r} X(t) \quad \rho \quad \text{“rho”}, & & \frac{\partial}{\partial \sigma} X(t) \quad \text{“vega”}.
\end{array}
$$

Here, $X(t)$ can also represent the price of a portfolio of options. Note that Δ and Γ measure the impact of price changes, Θ that of the decreasing time to maturity. vega and ρ, on the other hand, are measures for the consequence of possible errors in the input parameters volatility and interest rate.

To eliminate the influence of these possible errors or changes at least locally (i.e. for small changes/errors), traders try to make a portfolio of options neutral against changes in these parameters. They obtain this by buying/selling appropriate further securities or derivatives such that the (portfolio) sensitivities with respect to the different parameters are zero. This can always be obtained by setting up suitable linear portfolios.

While for all options with explicit price formulae one can do this analytically, a lot of exotic options (even in the Black-Scholes model) exist where this cannot be done. Even more, in local or stochastic volatility models or even more advanced models, there is only the chance for obtaining the sensitivities numerically. We will therefore discuss some methods below.

REMARK 5.43 1. The most popular Greeks are of course those that are derived from the Black-Scholes formula. We will only state the Δ of a

European call option explicitly:

$$\Delta_{call} = \Phi\left(d_1\left(t\right)\right) = \Phi\left(\frac{ln\left(S_1\left(t\right)/K\right) + \left(r + \frac{1}{2}\sigma^2\right)\left(T-t\right)}{\sigma\sqrt{T-t}}\right). \tag{5.251}$$

2. In the Black-Scholes model the Δ of an option always determines the stock part in the replication strategy for the option.

3. The concept of the Greeks can be generalized to options on multiple underlyings. Then we have a delta for each partial derivative with respect to one of the underlyings. The same modification applies to the now even mixed second partial derivatives with respect to the underlyings. □

We will now present some methods to obtain the option price sensitivities, but will compare their numerical performance in a separate subsection.

5.15.2 Finite difference simulation

The straightforward method to calculate a derivative $f'(x)$ numerically is to replace the corresponding differential quotient by a difference quotient

$$\frac{f\left(x+h\right) - f\left(x\right)}{f\left(x\right)}$$

for a nonzero value of h. As h is small, but does not go to zero, one talks of a **finite difference**. To state the method formally, let us assume that the price of an option $X(t)$ can be written as

$$X\left(t\right) = X\left(t; S\left(t\right), r, \sigma\right). \tag{5.252}$$

Note that not all exotic options allow such a representation, as there are options such as the Asian options whose value depends on the whole past of the stock price movement before time t. They need a more involved notation, but can in principle be treated similarly. However, having the above assumed dependence, we can now approximate the sensitivities by a suitable difference quotient. For example, the Δ of an option can be approximated by either

- the **forward difference**

$$\Delta = \frac{\partial X\left(t\right)}{\partial s} \approx \tilde{\Delta}_{for} := \frac{X\left(t; S\left(t\right)+h, r, \sigma\right) - X\left(t; S\left(t\right), r, \sigma\right)}{h} \tag{5.253}$$

- or the **central difference**

$$\Delta \approx \tilde{\Delta}_{cen} := \frac{X\left(t; S\left(t\right)+h, r, \sigma\right) - X\left(t; S\left(t\right)-h, r, \sigma\right)}{2h}. \tag{5.254}$$

From a heuristical point of view, one would believe that the central difference is a better approximation to the partial derivative as it takes into account information of the option price from both sides of $S(t)$ while the forward difference only looks to the right. Indeed, the following estimates for the bias for both methods support such an argument. For ease of notation, we consider a function $f(x)$ and assume it to be differentiable up to the third order with bounded derivatives. Then, a Taylor series expansion in x yields

$$f(x+h) = f(x) + f'(x)h + \frac{1}{2}f''(x)h^2 + \frac{1}{6}f'''(x)h^3 + o(h^3), \quad (5.255)$$

$$f(x-h) = f(x) - f'(x)h + \frac{1}{2}f''(x)h^2 - \frac{1}{6}f'''(x)h^3 + o(h^3). \quad (5.256)$$

Using these two relations we obtain the estimates for the bias of the forward difference as

$$B_{for}(f) = \left| \frac{f(x+h) - f(x)}{h} - f'(x) \right| + O(h) \quad (5.257)$$

and those of the central difference as

$$B_{cen}(f) = \left| \frac{f(x+h) - f(x-h)}{2h} - f'(x) \right| + O(h^2). \quad (5.258)$$

Of course, this advantage of the central difference having a higher order of approximation for the derivative than the forward difference has to be compared to the effort of getting the additional term $f(x-h)$. If this term (which in our setting is an option price calculated with a different input parameter) is easily available then central differences are certainly preferable.

How to compute the ingredients?

If we have decided which variant of the finite difference estimator we should use, then the next question concerns the computation of the two ingredients. Let us again focus on the computation of Δ. All relevant values $X(t; S(t), r, \sigma)$, $X(t; S(t)+h, r, \sigma)$, and $X(t; S(t)-h, r, \sigma)$ are expected values. If we can calculate them analytically then no Monte Carlo methods are needed at all. If, however, we need the Monte Carlo method to calculate them, then there are multiple ways to do that. Note also that the finite difference estimator already has a bias. This should not be increased by the Monte Carlo error. On the other hand, a Monte Carlo estimator that is more precise than the order of the bias cannot correct for it. Of course, one can still try to reduce the variance of the Monte Carlo estimator. If we look at the relevant quantities to estimate,

$$\tilde{\Delta}_{for} = \frac{X(t; S(t)+h, r, \sigma) - X(t; S(t), r, \sigma)}{h}, \quad (5.259)$$

$$\tilde{\Delta}_{cen} = \frac{X(t; S(t)+h, r, \sigma) - X(t; S(t)-h, r, \sigma)}{2h} \quad (5.260)$$

then we can hope for a small variance if the two expressions in the denominator are highly correlated. If $X(.)$ is monotonic in S then by an argument that we already used in the antithetic variate section, we could hope to have a small variance if we use the same random numbers (i.e. the same simulated Brownian paths) for the computation of both expectations. This principle is called the **principle of common random numbers** (or **path recycling**) and is also sometimes considered explicitly as a method for variance reduction.

Indeed, convergence of the finite difference estimators are well-studied (see e.g. Glasserman [2004] and specialized literature on numerical methods for partial differential equations). We will only highlight the fact that with the choice of the central finite difference and a stepsize h_n with

$$h_n = O\left(n^{-1/5}\right) \tag{5.261}$$

one obtains an order of convergence of $O(n^{-2/5})$ and has a bias of order $O(h_n^2) = O(n^{-2/5})$. Note also that we cannot choose the stepsize h too small, as then round-off errors might dominate the computations. A practical suggestion to overcome this is given in Jäckel (2003): Given that ϵ is the machine precision of the computer (this can be identified with the smallest possible positive number in the computer) one should choose the stepsize

$$h^* = \sqrt[4]{\epsilon} S\left(t\right). \tag{5.262}$$

This is obtained by a heuristical error analysis and the argument that for this choice bias and round-off error are of the same order. However, to obtain a good overall performance, for given n relation (5.262) only tells us that we should not choose $h_n < h^*$. In total, we would recommend:

To compute the delta of an option with a finite difference method (and a given number n of simulated stock price paths), use the central finite difference

$$\tilde{\Delta}_{cen} = \frac{X\left(t; S\left(t\right) + h, r, \sigma\right) - X\left(t; S\left(t\right) - h, r, \sigma\right)}{2h}, \tag{5.263}$$

with path recycling for a stepsize of $h_n = n^{-1/5}$ (if it is not smaller than h^*).

Computations of other Greeks

As long as we only consider first derivatives, the above described method (and recommendation) stays the same: Simply calculate the option prices for two values θ_1 and θ_2 of the parameter of interest θ with

$$\theta_1 = \theta - h, \quad \theta_2 = \theta + h \tag{5.264}$$

via using path recycling. Then, form the corresponding central difference estimator $\tilde{\theta}_{cen}$. This of course stays the same if we consider an option on many assets where we e.g. have many deltas (i.e. one for each underlying).

If we need a second order derivative such as gamma, then we can again use a central difference. For this we simply use as input the corresponding finite differences as estimates for the first derivatives needed, i.e. we obtain (in simplified terms of the function f):

$$\frac{\frac{f(t+h)-f(t)}{h} - \frac{f(t)-f(t-h)}{h}}{h} = \frac{f(t+h) - 2f(t) + f(t-h)}{h^2} \tag{5.265}$$

Advantages and disadvantages of finite differences

The main strength of the finite difference method is that it is always applicable if the corresponding derivative exists. We only have to be able to calculate the required option prices. There are no further requirements. Also, they can be used in a multiasset setting with obvious modifications (in fact, one only needs to use the notation of partial derivatives, nothing more).

Their main disadvantage is that their use introduces a bias into the calculations, the bias caused by the fact that we approach a derivative by finite difference. Thus, in addition to the Monte Carlo error and the possible discretization error for simulating the stock price paths, a third source of error enters the computations.

5.15.3 The pathwise differentiation method

The pathwise differentiation method is based on the ability to obtain a corresponding derivative with respect to the required parameter for each path of the functional of the stochastic process and on the validity of interchanging this differentiation with the expectation. More precisely, if $B = f(S(T))$ is the payoff of an option and B is differentiable with respect to the initial value $S(0)$ of the stock price, then the delta of this option can be obtained via

$$\frac{\partial}{\partial S(0)}\mathbb{E}(B) = \mathbb{E}\left(f'(S(T))\frac{\partial S(T)}{\partial S(0)}\right) \tag{5.266}$$

if it is allowed to interchange the differentiation of the left-hand side of this equation with the expectation. This is obviously the case if $f(.)$, $f'(S(T))$, and $\frac{\partial S(T)}{\partial S(0)}$ are bounded. At least the last boundedness assumption is rarely satisfied. So, one typically has to use a special justification argument for different options. Further, global differentiability of the option payoff is typically not given. However, we will use the pathwise differentiability method very efficiently in connection with the likelihood ratio method in Section 5.15.5. For the general case, we use the notation

$$\frac{\partial}{\partial \theta}\mathbb{E}(B) = \mathbb{E}\left(f'(S(T))\frac{\partial S(T)}{\partial \theta}\right). \tag{5.267}$$

Note that the corresponding pathwise differentiation Monte Carlo estimator

$$I_{PDE}(\theta;N) := \frac{1}{N}\sum_{i=1}^{N} f'\left(S^{(i)}(T)\right)\frac{\partial S^{(i)}(T)}{\partial\theta} \tag{5.268}$$

is unbiased if the representation (5.267) is valid.

The multidimensional case: The forward and the adjoint method

We now explictly consider the multidimensional case (see Giles and Glasserman [2006]). Think of an exotic option where the pathwise derivatives cannot be easily calculated, and on top of this assume that the price of the underlying follows an SDE that might not be solvable in closed form. Consider the m-dimensional SDE

$$dS(t) = b(S(t))\,dt + \sigma(S(t))\,dW(t) \tag{5.269}$$

where b is $\mathbb{R}^m$-valued, σ an $m \times d$-matrix, and we assumed the SDE to have a unique solution. With the stepsize of $h = T/N$ we introduce the corresponding Euler-Maruyama approximation at time $(n+1)h$ by

$$S^h(n+1) = S^h(n+1) + b\left(S^h(n)\right)h + \sigma\left(S^h(n)\right)\sqrt{h}Z(n+1) \tag{5.270}$$

with $S^h(0) = S(0)$ and $Z(i)$ independent d-dimensional standard Gaussian random vectors. Let $f : \mathbb{R}^m \to \mathbb{R}$ be a function that satisfies conditions such that we can calculate the following delta by the chain rule:

$$\frac{\partial}{\partial S_j(0)}\mathbb{E}(f(S(T))) = \mathbb{E}\left(\sum_{i=1}^{m}\frac{\partial f(S(T))}{\partial S_i(T)}\frac{\partial S_i(T)}{\partial S_j(0)}\right) \tag{5.271}$$

If we have all those derivatives in closed form then we can start the Monte Carlo simulation. If, however, we have to calculate them numerically, then we use the above Euler-Maruyama scheme to approximate the sum in (5.271) by

$$\sum_{i=1}^{m}\frac{\partial f\left(S^h(N)\right)}{\partial S_i^h(N)}\Delta_{ij}(N), \quad \Delta_{ij}(n) = \frac{\partial S_i^h(n)}{\partial S_j^h(0)} \tag{5.272}$$

To obtain $\Delta_{ij}(N)$ – if we do not already have an explicit form – we can simulate it by the Euler-Maruyama scheme above and thus have to simulate

$$\Delta_{ij}(n+1) = \Delta_{ij}(n) + \\ + \sum_{k=1}^{m}\frac{\partial b_i}{\partial S_k^h}\Delta_{kj}(n)\,h + \sum_{l=1}^{d}\sum_{k=1}^{m}\frac{\partial \sigma_{il}}{\partial S_k^h}\Delta_{kj}(n)\sqrt{h}Z_l(n+1) \tag{5.273}$$

with the obvious component notation for the market coefficient functions b, σ both evaluated at $S^h(n)$ and with $\Delta(0) = I_m$ (I_m the m-dimensional identity matrix). By introducing the matrix $D(n)$ via

$$D_{ij}(n) = \delta_{ij} + \frac{\partial b_i}{\partial S_k^h} h + \sum_{l=1}^{d} \frac{\partial \sigma_{il}}{\partial S_k^h} \sqrt{h} Z_l(n+1) \tag{5.274}$$

with $\delta_{ik} = 1$ for $i = k$ and $\delta_{ik} = 0$ else, we obtain the relation

$$\Delta(n+1) = D(n)\Delta(n) = D(n) \cdot \ldots \cdot D(0)\Delta(0)\,. \tag{5.275}$$

From this we get the representation for the vector of all pathwise deltas as

$$\frac{\partial f\left(S^h(N)\right)}{\partial S^h(0)} = \frac{\partial f\left(S^h(N)\right)}{\partial S^h(N)} D(N-1) \cdot \ldots \cdot D(0)\Delta(0)\,. \tag{5.276}$$

There are now two main ways to calculate this vector of deltas.

(a) **The forward method**:
Use relation (5.275) to calculate the right side of representation (5.276). This includes N matrix multiplications which require $O(Nm^3)$ operations.

(b) **The adjoint method**:
Look at the right side of representation (5.276), but start with its left-most term. By introducing the **adjoint relation**

$$V(N) := \left(\frac{\partial f\left(S^h(N)\right)}{\partial S^h(N)}\right)', \quad V(n) = D(n)' V(n+1) \tag{5.277}$$

where $'$ denotes the transposition, there is a second way to calculate

$$\frac{\partial f\left(S^h(N)\right)}{\partial S^h(0)} = V(0)'\Delta(0)\,. \tag{5.278}$$

Note that we have replaced the matrix multiplications by N matrix-vector products which require only $O(Nm^2)$ operations.

Which method to use?

It is clear that both methods deliver exactly the same values. With regard to the necessary operations, if we only want to calculate the deltas of a single option, the adjoint method is much more efficient (compare the orders of operations needed). Indeed, the savings are substantial. However, if we want to calculate the deltas of many options on the same underlyings, then there is an advantage for the forward method as it explicitly determines $\Delta(N-1)$ which can be reused for the delta calculation of all these options. So, in total, the forward method should be used if many options (a whole book) on only some underlyings are considered. If, however, we have only a few options but written on many underlyings, then the adjoint method is preferable.

More Greeks:

The above presented methods can be generalized to the computations of pathwise vegas or pathwise gammas given the option payoff is sufficiently smooth. We refer the reader for the detailed formulae in these cases to Giles and Glasserman (2006).

5.15.4 The likelihood ratio method

While in the pathwise differentiation method the main property was the differentiable dependence of each stock price path from the parameter of interest, in the likelihood ratio method we make use of the differentiability of the density of the stock price with respect to this parameter. More precisely, assume that we have the relation

$$\mathbb{E}\left(f\left(S\left(T\right)\right)\right)=\int f\left(S\right)g\left(S\right)dS \tag{5.279}$$

with $g(.)$ being the density of the stock price at time T. Then, the dependence of the expectation on a parameter θ related to the stock price is summarized in the density $g(.)$. If we could now interchange the differentiation with respect to the parameter θ with the integration, then we obtain

$$\frac{\partial}{\partial\theta}\mathbb{E}\left(f\left(S\left(T\right)\right)\right)=\int f\left(S\right)\frac{\partial}{\partial\theta}g\left(S\right)dS=\int f\left(S\right)\frac{\frac{\partial}{\partial\theta}g\left(S\right)}{g\left(S\right)}g\left(S\right)dS. \tag{5.280}$$

With the definition of the **likelihood ratio** (or score function) as

$$w\left(S;\theta\right):=\frac{\frac{\partial}{\partial\theta}g\left(S\right)}{g\left(S\right)}=\frac{\partial\ln\left(g\left(S\right)\right)}{\partial\theta} \tag{5.281}$$

we can thus – still assuming validity of the interchanging of limits – calculate the partial derivative of the expectation as an expectation with a transformed payoff function:

$$\begin{aligned}\frac{\partial}{\partial\theta}\mathbb{E}\left(f\left(S\left(T\right)\right)\right) &= \int f\left(S\right)w\left(S;\theta\right)g\left(S\right)dS\\ &=: \int\Psi\left(S;\theta\right)g\left(S\right)dS=\mathbb{E}\left(\Psi\left(S\left(T\right);\theta\right)\right).\end{aligned} \tag{5.282}$$

From this representation we can deduce (by the law of large numbers) that the **likelihood ratio estimator**

$$I_{LR}\left(\theta;N\right):=\frac{1}{N}\sum_{i=1}^{N}f\left(S^{(i)}\left(T\right)\right)w\left(S^{(i)}\left(T\right);\theta\right) \tag{5.283}$$

is an unbiased and strongly consistent estimator for the desired derivative. Note one particular advantage of this method: The weight function $w\left(S;\theta\right)$ on the right side of Equation (5.283) is independent of the form of the underlying option! Thus, the weights (at least their functional form) can be reused when the same price sensitivity of a different option shall be calculated. Of course, to obtain Greeks of higher order such as the gamma, one can apply the method in an iterative way.

To illustrate the method, we calculate the weights corresponding to the delta and the rho in the one-dimensional Black-Scholes model. Here, the density of the stock price process at time T is given by

$$g(S) = \frac{1}{S\sqrt{2\pi\sigma^2 T}} \exp\left(-\frac{\left(ln(S/S(0)) - (r - \frac{1}{2}\sigma^2)T\right)^2}{2\sigma^2 T}\right) \tag{5.284}$$

which leads to the weights of

$$w\left(S^{(i)}(T); \Delta\right) = \frac{W^{(i)}(T)}{\sigma S(0) T}, \quad w\left(S^{(i)}(T); \rho\right) = \frac{W^{(i)}(T)}{\sigma} \tag{5.285}$$

where we have used the explicit form of the simulated stock price $S^{(i)}(T)$.

Having derived these formulae, we can see that there are (at least) two good reasons to use the likelihood ratio method for the computation of the Greeks of an option when we compare it to the other two already presented methods:

(i) Comparison to finite difference methods:

Compared to the likelihood ratio method, the discretization error of the derivative is a second source for errors entering the computation of the Greeks in the finite difference method. In particular, it is well-known that numerical differentiation is an ill-posed problem. This is avoided by the likelihood ratio as it does not need a discretization of a derivative.

(ii) Comparison to the pathwise differentiation method:

An obvious advantage of the likelihood ratio method compared to the pathwise differentiation method is the fact that no differentiability or continuity assumption for the payoff function of the option is needed. This allows consideration of even discontinuous payoffs such as digital options.

5.15.5 Combining the pathwise differentiation and the likelihood ratio methods by localization

By having a close look at the likelihood ratio estimator for the delta of a simple European call in the case of the Black-Scholes model,

$$I_{LR}(\Delta; N) = \frac{1}{N} \sum_{i=1}^{N} \left(S^{(i)} - K\right)^+ w\left(S^{(i)}; \theta\right) \tag{5.286}$$

one can realize that the presence of the weight in the estimator results in an increase of the variance compared to the crude Monte Carlo estimator for the call price. Indeed, the weight attains big values exactly when the stock price estimator attains big values, thus amplifying the variance of the call payoff. On the other hand, the call payoff is very smooth in all points different from the strike, and its derivative is much simpler than the original payoff. We thus use a combination of the likelihood ratio and the pathwise differentiation method a **localization** procedure. We highlight it in the simple European call example: we decompose the payoff function as

$$(S(T) - K)^+ = \left[(S(T) - K)^+ - \phi_\delta(S(T))\right] + \phi_\delta(S(T)) \tag{5.287}$$

where $\phi_\delta(S(T))$ is a function that is differentiable and coincides with the call payoff outside an interval of the form $[K-\delta, K+\delta]$. A possible choice of such a smoothing function would be

$$\phi_K(S(T)) = \frac{1}{4K} S(T)^2 \, 1_{\{S(T)\in[0,2K]\}} + (S(T) - K)^+ \, 1_{\{S(T)\notin[0,2K]\}}. \tag{5.288}$$

We then only apply the likelihood ratio method to the first term in the representation (5.287) (i.e. only locally around the strike K) and use the pathwise differentiation method to value the (artificially introduced) option given by the final payoff $\phi_\delta(S(T))$. This yields the following representation of the delta of the call:

$$\begin{aligned}\Delta &= e^{-rT} \frac{\partial}{\partial S(0)} \left[\mathbb{E}\left((S(T) - K)^+ - \phi_\delta(S(T))\right) + \mathbb{E}\left(\phi_\delta(S(T))\right)\right] \\ &= e^{-rT}\mathbb{E}\left(\left[(S(T) - K)^+ - \phi_\delta(S(T))\right] \frac{W(T)}{S(0)\sigma T}\right) + \\ &\qquad + e^{-rT}\mathbb{E}\left(\phi_\delta'(S(T)) \frac{\partial S(T)}{\partial S(0)}\right). \end{aligned} \tag{5.289}$$

Note that in this way we avoid big products in the first expectation. For the special choice of ϕ_K above we obtain

$$\begin{aligned}\Delta &= e^{-rT}\mathbb{E}\left(\left[(S(T) - K)^+ - \frac{1}{4K} S(T)^2\right] \frac{W(T)}{S(0)\sigma T} 1_{\{S(T)\le 2K\}}\right) + \\ &\qquad + e^{-rT}\mathbb{E}\left(\frac{\partial S(t)}{\partial S(0)} \left[\frac{1}{2K} S(T)\, 1_{\{S(T)\le 2K\}} + 1_{\{S(T)>2K\}}\right]\right) \end{aligned} \tag{5.290}$$

which can easily be simulated in the usual way. It should also be noted that the application of the localization approach gets more involved in higher dimensions, as then the construction of smoothing functions is more involved.

REMARK 5.44 A technically very advanced method for computing representations of the Greeks similar to the ones obtained by the likelihood ratio method is based on the application of the so-called **Malliavin calculus** (see the excellent survey paper of Fournié et al. [1999]). However, as long as there is an explicit form of the density of S there is no real need to introduce the relevant technical machinery. If this is not the case, then it is typically very hard to explicitly compute Malliavin derivatives that are needed to obtain representations of the Greeks. We therefore omit the presentation of the Malliavin calculus approach for the Greeks in this book. □

5.15.6 Numerical testing in the Black-Scholes setting

To test the efficiency of the methods of the previous sections we consider some examples in a Black-Scholes setting where analytical solutions serve as a benchmark. We will compare finite differences, the likelihood ratio method, and the combined localization method when it is applicable.

Example 1: European calls

With the test data of $S(0) = 80, K = 100, r = 0.1, \sigma = 0.3, T = 1$ we computed the delta, the gamma, and the vega of a European call by the likelihood ratio method, the corresponding localized likelihood ratio method with smoothing function ϕ_K as in Equation (5.288). The speed of the convergence of the different methods is illustrated by the results in Table 5.8 where we have chosen the values of 100, 10,000 and 100,000 for N.

Greek	Method N	100	10,000	100,000	Exact value
Δ	Likelihood ratio	0.3937	0.4111	0.3929	
Δ	Loc. Likel. ratio	0.3960	0.4000	0.3967	0.3972
Δ	Finite Difference	0.4131	0.4031	0.3950	
Γ	Likelihood ratio	0.0143	0.0167	0.0156	
Γ	Loc. Likel. ratio	0.0167	0.0158	0.0161	0.0161
Γ	Finite Difference	0.0090	0.0146	0.0165	
vega	Likelihood ratio	27.38	32.14	29.97	
vega	Loc. Likel. ratio	31.98	30.89	30.95	30.85
vega	Finite Difference	30.58	31.37	30.18	

Table 5.8: Estimates for Δ, Γ, and $Vega$ of a European Call

In all cases of the estimates for the Greeks of the European call, the localized likelihood ratio method performed best, followed by the finite difference method. As could be expected, the finite difference method has the biggest problem with the calculation of gamma.

Example 2: Digitals

For digital options given by the final payment of

$$B = 1_{S(T) \geq K} \tag{5.291}$$

the advantages of the localization method is becoming even clearer as we have a noncontinuous payoff function. However, here a second order polynomial is not enough as the smoothing function $\phi_\delta(S(T))$. To smooth the payoff function we now need a third-order polynomial connecting the possible values 0 and 1 of the digital payments. Note in particular that for digitals the pathwise differentiation method is definitely not available.

Greek	Method N	100	10,000	100,000	Exact value
Γ	Likelihood ratio	1.5*E-4	1.3*E-4	1.4*E-4	
Γ	Loc. Likel. ratio	1.2*E-4	1.4*E-4	1.4*E-4	1.4*E-4
Γ	Finite Difference	0.0000	-7.2*E-4	3.9*E-4	
vega	Likelihood ratio	0.2854	0.2927	0.2631	
vega	Loc. Likel. ratio	0.2368	0.2628	0.2589	0.2679
vega	Finite Difference	0.2805	0.2950	0.3153	

Table 5.9: Estimates for Γ and *vega* of a European Digital Call

Our results computed in this situation (see Table 5.9 with identical test data as in the call example above) clearly underline the comments given in the European call example. Here, both likelihood ratio methods outperformed the finite difference method. A particular critical behaviour was often shown by the finite difference method for the calculation of gamma. The method needed a long time to get at least the sign correct and then it still was unstable.

REMARK 5.45 For more details on the numerical performance of the different methods for calculating the Greeks we refer the reader to Fournié et al. (1999). □

Computation of the Greeks for Bermudan options

In the above presentations, we implicitly restricted ourselves to European options. For Bermudan options we can make use of the results of Piterbarg (2004) for Bermudan interest rate derivatives. He examines the application of the likelihood ratio method and of the pathwise derivative method for Bermudan options. One of his results states that the application of the likelihood ratio method for the computation of the Greeks can be done similarly to the European setting. In particular, after an (approximately) optimal exercise strategy has been determined, then one should identify the Bermudan option with a European one with a payment structure given by the optimal exercise strategy which is now known. The application of the pathwise delta method requires more technical results. We refer the reader to Piterbarg (2004).

5.16 Basics of interest rate modelling

So far in all financial models the interest rate was either assumed to be constant or not explicitly modelled. However, there is a big need to have a suitable model for the interest rate as:

- Interest rate-related trades (also called **fixed income** trades) have a much higher volume than stock-related trades.
- Interest rates are not as volatile as stock prices, but a look at their empirical dynamics shows that they are far from being constant.

It would now be tempting to use a suitably modified stock price model for modelling interest rates, but there are many empirical and economical reasons for not doing so, as such:

- Stock prices are (in tendency) increasing with time, interest rates tend to move around some specific level.
- There are many interest rates around (for different maturities), not just *the* interest rate.
- Interest rates are not traded, only derivatives on the interest rate are.
- It is not obvious which process to model.
- The value of a bond at maturity is already fixed, its value at any date before maturity is random.
- There are bonds running up to 50 years.
- There is a huge variety of interest rate products.

Mainly, there are three approaches to interest rate modelling in the literature (we often follow Björk [2004] and Brigo and Mercurio [2001] in the remaining sections of this chapter).

1. **The short rate approach** where the evolution of the interest rate for loans that only last an infinitesimal time span is modelled.
2. **The forward rate approach** (or Heath-Jarrow-Morton [**HJM**] framework) where the evolution of a whole interest rate curve over time is modelled.

3. **The market model approach** where the evolution of a finite set of simple market interest rates is modelled.

We will consider all three approaches in detail and in particular will comment on the use of Monte Carlo methods.

5.16.1 Different notions of interest rates

The basic object in interest modelling is the so-called **zero bond**.

DEFINITION 5.46
A zero bond consists of a payment of 1 at maturity T. The set of all current zero bond prices $P(t,T)$ with $t \leq T$ as a function of their maturities T is called the discount curve or the **term structure of bond prices**.

The discount curve has to be distinguished from the price paths of a particular zero bond $P(s,T)$ for $s \leq T$ with a fixed maturity T. The discount curve has to be decreasing, as money today has a higher value than receiving the same amount tomorrow when the interest rate is positive. Empirical discount curves contain nearly no visual information, which is the main reason for looking at a transformation of the discount curve, the (simple) yield curve.

DEFINITION 5.47
(a) The **simple yield** $r(0,T)$ *of a zero coupon bond with maturity T is defined as the equivalent constant interest rate that yields the actual zero bond, i.e. it is given by*

$$P(0,T) = (1 + r(0,T))^{-T} . \tag{5.292}$$

(b) The **simple forward rate** $f(0;T_1,T_2)$ *on the interval $[T_1,T_2]$ is today's simple yield for this period and is defined via*

$$\frac{P(0,T_2)}{P(0,T_1)} = (1 + f(0;T_1,T_2))^{-(T_2-T_1)} . \tag{5.293}$$

The curve of all simple yields shows the market's expectations of the behaviour of the interest rate in the future.

Some standard forms of it are presented in Figure 5.6. A flat term structure of the yields means that there is no time preference. This is often a convenient assumption for theoretical considerations but rarely observed in real life. The normal structure is based on the idea that someone giving money away for a longer time has to be compensated for this with a higher interest rate. With an inverse structure the market is assuming that the interest rate today is too high and that it will decrease in the future. A humped curve is a mixture of a normal and an inverse situation. Also, combinations of these standard forms can appear.

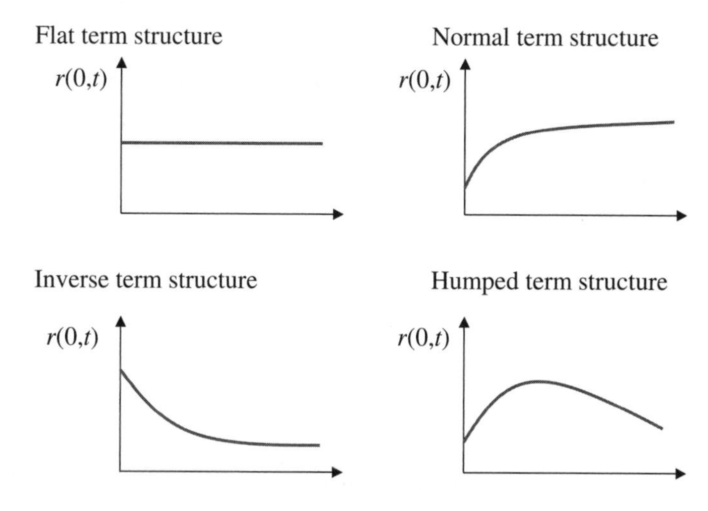

FIGURE 5.6: Standard Forms of Yield Curves

One can deduce more notions of interest rates from the prices of zero bonds.

DEFINITION 5.48
(a) The (infinitesimal) forward rate $f(t,s)$ at time t for the future time s is defined as

$$f(t,s) = -\frac{\partial}{\partial s} \ln (P(t,s)) \text{ for } s \geq t. \tag{5.294}$$

(b) The (infinitesimal) short rate $r(t)$ at time t is defined by

$$r(t) = f(t,t). \tag{5.295}$$

5.16.2 Some popular interest rate products

Besides zero bonds there is a vast amount of products on the interest rate market. We will only present some popular ones.

DEFINITION 5.49
A **coupon bond** *consists of a final payment of 1 at time T and coupon payments c at the times $0 \leq t_1 < t_2 < ... < t_n \leq T$.*

As a coupon bond is a portfolio of a zero bond with maturity T and c zero bonds with maturities t_i, $i = 1, ..., n$, we obtain its $P^c(0,T)$ as

$$P^c(0,T) = P(0,T) + c \sum_{i=1}^{n} P(0,t_i). \tag{5.296}$$

DEFINITION 5.50
In a forward contract on a zero coupon bond party A agrees to deliver B at time T_0 a zero coupon bond with maturity $T > T_0$. B in return pays the fixed price of B_F for this bond. B_F will be chosen such that the initial value of the forward contract equals zero. The value B_F is then called the (T_0-) **forward price** *of the T-bond.*

The forward price is calculated via a replication argument:

- A has to buy the T-bond today for a price of $P(0,T)$ to be able to deliver a T-bond at time T_0.
- To be able to pay exactly B_F at time T_0, party B has to buy today B_F zero bonds maturing at time T_0. This costs $B_F P(0,T_0)$ at time $t = 0$.

The forward contract has zero initial value if both these strategies have the same value which leads to:

$$B_F = \frac{P(0,T)}{P(0,T_0)}. \tag{5.297}$$

It should be emphasized that the value of a forward contract equals zero only at the start of its life. As then the zero bond prices change randomly the value of the above replication strategy also changes. Thus, the forward price at time t has also changed to (use the same replication argument as above!):

$$B_F(t) = \frac{P(t,T)}{P(t,T_0)}. \tag{5.298}$$

Exchange traded forward contracts are called **futures contracts**. They have special features such as daily payments into the so-called margin account that should prevent losses from the default of party A or B.

Besides fixed coupon payments, we also consider floating rate payments. They are typically linked to the evolution of a market rate such as the **3-month LIBOR** (London interbank offer rate) rate. This rate is the interest rate for a 3-month loan between banks at the London stock exchange. Similar rates exist for different maturities (6 months, 9 months, 12 months, ...) or places (the EURIBOR corresponds to the same deals at the Deutsche Börse in Frankfurt). They are changing continuously over time which is why they are termed **floating rates**. Of course, they have to be related to zero bond prices to prevent arbitrage opportunities:

DEFINITION 5.51
(a) The simply compounded spot rate at time t with maturity T (the so-called $(T-t)$ **spot-LIBOR rate***) is given by*

$$L(t,T) = \frac{1 - P(t,T)}{(T-t)\,P(t,T)}. \tag{5.299}$$

(b) The simply compounded forward rate at time t for the time interval $[T, S]$ with $t \leq T < S$ (the so-called $(T-S)$ **forward-LIBOR rate***) is given by*

$$L(t;T,S) = \frac{P(t,T) - P(t,S)}{(S-T)\,P(t,S)}. \tag{5.300}$$

Thus, the spot-LIBOR rate is just the equivalent simply compounded constant interest rate that yields the market price of the T-bond at time t.

DEFINITION 5.52

A **floating rate note** *with payment times $k \cdot a$, $k = 1, ..., n$ pays its owner coupons $C_k = a \cdot L((k-1)a, k \cdot a)$ at times $k \cdot a$ and 1 unit of money at maturity $T = na$.*

A surprising result for the value of such a note follows by a simple replication argument that uses the definition of the LIBOR rates.

THEOREM 5.53

The price $P_f(0,T)$ of a floating rate note with maturity T and coupon payments as defined above is given by $P_f(0,T) = 1$. Further, the value of a floating rate note equals 1 directly after each floating payment.

Interest rate swaps are among the most liquid interest rate products.

DEFINITION 5.54

In a (plain-vanilla) **interest rate swap** *party A pays to party B fixed coupon payments at rate p at times $t_1, ..., t_n$ and receives in return from B floating rate payments of a loan of duration α at times $s_1, s_1 + \alpha, ..., s_1 + k\alpha$. At the final time T, both parties (formally) exchange the face value of 1.*

Swaps can occur in much more complicated forms and can contain any kind of option-like payments and conditions. We only consider the above simple interest rate swap. The main tasks we have to deal with are:

- Determine the **swap rate**, i.e. the fixed rate p such that the value of the swap at its start equals 0.
- Determine the **swap value**, i.e. the value of the swap for a fixed rate p.

THEOREM 5.55

(a) The value $S(0)$ of an interest rate swap – seen from the point of view of the fixed rate payer – with a coupon rate of p at time 0 is given by

$$S(0) = 1 - p \cdot \alpha \sum_{i=1}^{n} P(0, i\alpha) - P(0,T). \tag{5.301}$$

(b) The swap rate of an interest rate swap is given by

$$p_{swap} = \frac{1 - P(0,T)}{\alpha \sum_{i=1}^{n} P(0, i\alpha)}. \tag{5.302}$$

The first claim is valid as the value of a swap is the difference between the value of a floating rate note (which initially equals 1) and a coupon bond with coupon rate p. The second one follows from setting this value equal to 0.

Also, options on swaps, so-called **swaptions**, are popular securities. They are options to enter a swap at some future time for an already fixed swap rate. We will look at them in detail when dealing with market models. Besides swaps the most popular class of interest rate products are the interest rate options such as caps and floors.

DEFINITION 5.56

Let $L(t_{i-1}, t_i)$ be the $(t_i - t_{i-1})$-spot-LIBOR rates at times $t_0 < \dots < t_{n-1}$.

(a) A **cap** *at interest rate level L with face value V delivers at each time t_i, $i = 1, \dots, n$, the payment of*

$$V \cdot (L(t_{i-1}, t_i) - L)^{+} (t_i - t_{i-1}). \tag{5.303}$$

A contract delivering only one such payment is called a **caplet**.

(b) A **floor** *at interest rate level L with value V delivers at each time t_i, $i = 1, \dots, n$, the payment of*

$$V \cdot (L - L(t_{i-1}, t_i))^{+} (t_i - t_{i-1}). \tag{5.304}$$

A contract delivering only one such payment is called a **floorlet**.

The popularity of caps and floors stems from the fact that they are providing insurance against rising interest rate payments and falling interest rate income at times t_i, respectively. They are typically used as an insurance component in a floating rate deal.

In Black (1976) the methodology of the Black-Scholes formula is used to suggest a pricing formula for caps and floors, the so-called Black formula. It is widely used in interest rate markets and will be justified when we look at LIBOR market modelling.

Black formula: The price of a cap with payment times $t_1 < \dots < t_n$, face value V, and level L is given by

$$Cap(0, V, L, \sigma) = Cap_{Black}(0, V, L, \sigma) =$$

$$V \sum_{i=1}^{n} (t_i - t_{i-1}) P(0, t_i) \left[L(0; t_{i-1}, t_i) \Phi(d_1(t_i)) - K \Phi(d_2(t_i)) \right] \tag{5.305}$$

$$d_1(t_i) = \frac{\ln\left(L(0;t_{i-1},t_i)/L\right) + \frac{1}{2}\sigma^2 t_{i-1}}{\sigma\sqrt{t_{i-1}}}, \quad d_2(t_i) = d_1(t_i) - \sigma\sqrt{t_{i-1}} \tag{5.306}$$

where σ is the common volatility of the corresponding forward-LIBOR rates.

To use the Black formula, only the volatility parameter σ has to be estimated. As with the Black-Scholes formula, the Black formula can be used to obtain implied cap-volatilities from quoted market prices of caps. A similar formula is used for floors (with the obvious modifications for the change from call to put option types).

Another approach to pricing caps/floors is to use their relation to bond options. We therefore introduce a call/put with strike K and maturity T on an S-zero bond with $S \geq T$ by the payments of

$$Call(T,S;K) = \left(P(T,S) - K\right)^+, Put(T,S;K) = \left(K - P(T,S)\right)^+ \tag{5.307}$$

at time T. Let $\delta_i = t_i - t_{i-1}$. As the payments of a caplet/floorlet at time t_i are already fixed at time t_{i-1}, they then have a value of

$$Cap_i(t_{i-1};V,L) = P(t_{i-1},t_i)\,V\cdot\delta_i\left(L(t_{i-1},t_i) - L\right)^+, \tag{5.308}$$

$$Floor_i(t_{i-1};V,L) = P(t_{i-1},t_i)\,V\cdot\delta_i\left(L - L(t_{i-1},t_i)\right)^+. \tag{5.309}$$

We can thus directly verify (using the definition of LIBOR rates) that we have

$$Cap_i(t_{i-1};V,L) = V\cdot\delta_i L\cdot Put\left(t_{i-1},t_i;\frac{1}{1+\delta_i L}\right), \tag{5.310}$$

$$Floor_i(t_{i-1};V,L) = V\cdot\delta_i L\cdot Call\left(t_{i-1},t_i;\frac{1}{1+\delta_i L}\right). \tag{5.311}$$

Hence, the values of the caplets/floorlets and the (corresponding multiples of the) bond puts/calls coincide at any time before t_{i-1}, too. Thus, in every interest rate model, we can price caps/floors if we can price bond puts/calls.

5.17 The short rate approach to interest rate modelling

The short rate approach focuses on the modelling of the instantaneous short rate $r(t)$, i.e. the yield for a loan that starts at time t (=now) and ends immediately after that. A question that directly springs to one's mind then is if this is indeed enough to price interest rate products which are of course living longer than just an instant! However, a zero coupon bond can be interpreted as an option with a final payment of 1. From general option pricing theory we know that its price is given as the discounted expectation of one unit of money paid at time T where $\mathbb{Q}$ is an equivalent martingale measure, i.e. as

$$P(0,T) = \mathbb{E}_{\mathbb{Q}}\left(\exp\left(-\int_0^T r(s)\,ds\right)1\right), \tag{5.312}$$

where $r(s)$ is the instantaneous short rate. Consequently, for purposes of zero bond pricing it is enough to model the short rate under an equivalent martingale measure. Thus, the main ingredient of this approach is to model the short rate as the solution of the SDE

$$dr(t) = \mu(t, r(t))\,dt + \sigma(t, r(t))\,dW(t) \tag{5.313}$$

with $W(t)$ a one-dimensional Brownian motion. This is the reason why such a model is called a **one-factor model**. We will also comment on multifactor models. First, we collect some desirable features of a short rate model:

- The short rate should be nonnegative.
- The short rate should be mean-reverting, i.e. there is a certain *true* or *natural* level of the short rate and whenever the actual short rate differs from it the short rate process should have a tendency toward this level.
- The model should allow deriving price formulae for bonds and (simple) derivatives (such as bond options, caps/floors, or swaptions).
- Derived model prices of today should coincide with actually observable market prices ("perfect fit of initial term structure").

5.17.1 Change of numeraire and option pricing: The forward measure

Compared to stock option pricing, a major complication arises in computing contingent claim prices of the form

$$X(0) = \mathbb{E}_{\mathbb{Q}}\left(\exp\left(-\int_0^T r(s)\,ds\right)X\right) \tag{5.314}$$

(with X denoting the terminal payoff) in the stochastic interest rate framework by the presence of the random discount factor. If e.g. X is of the form $X = g(r(T))$, then the discount factor is harder to simulate than X itself!

As the discount factor is positive, one can formally eliminate it by normalization and then performing a suitable Girsanov type change of measure. More precisely, we use the representation (5.312) of a T-zero bond price and the fact that $P(T,T) = 1$ to obtain:

$$X(0) = \mathbb{E}_{\mathbb{Q}}\left(\exp\left(-\int_0^T r(s)\,ds\right)\frac{P(0,T)}{P(0,T)}P(T,T)\,X\right)$$

$$= P(0,T)\,\mathbb{E}_{\mathbb{Q}}\left(X\frac{\exp\left(-\int_0^T r(s)ds\right)P(T,T)}{P(0,T)}\right) =: P(0,T)\,\mathbb{E}_{\mathbb{Q}_T}(X) \quad (5.315)$$

Here, the new probability measure $\mathbb{Q}_T$ is defined by

$$d\mathbb{Q}_T = Z(T)\,d\mathbb{Q} \quad (5.316)$$

with

$$Z(t) = \exp\left(-\int_0^t r(s)\,ds\right)\frac{P(t,T)}{P(0,T)} \quad (5.317)$$

(see also Theorem 4.44, Girsanov's theorem). Note that for the above considerations to be valid, we need $Z(t)$ to be a $\mathbb{Q}$-martingale, a fact that we have always assumed (and will continue to assume in this section), and that actually depends on the coefficients of the short rate equation. By Girsanov's theorem we can then introduce a $\mathbb{Q}_T$-Brownian motion W_T by

$$W_T(t) = W(t) + \int_0^t \beta(s,T)\,\sigma(s,r(s))\,ds \quad (5.318)$$

where $\beta(t,T)$ has its origin in the SDE representation

$$dP(t,T) = P(t,T)\left[r(t)\,dt - \beta(t,T)\sigma(t,r(t))\,dW(t)\right] \quad (5.319)$$

of the T-bond price under $\mathbb{Q}$ (note that we can assume this representation as $P(t,T)$ is positive and must have the drift rate $r(t)$ under $\mathbb{Q}$). Further, by using the W_T-representation for S-zero bond prices with $S \leq T$,

$$dP(t,S) = P(t,S)\left[\left(r(t) + \beta(t,S)\,\beta(t,T)\,\sigma^2(t,r(t))\right)dt\right] - P(t,S)\,\beta(t,S)\sigma(t,r(t))\,dW_T(t), \quad (5.320)$$

application of Itô's formula to the quotient $P(t,S)/P(t,T)$ yields that it actually is a $\mathbb{Q}_T$-martingale. As thus the use of $\mathbb{Q}_T$ as a pricing measure implies that we have to change to the numeraire $P(0,T)$, it is called the T-**forward measure**. Further, we have developed a simple method to calculate contingent claim prices in the stochastic interest rate setting:

1. Introduce the T-forward measure $\mathbb{Q}_T$ by representation (5.316).
2. Calculate the contingent claim price as $X(0) = P(0,T)\mathbb{E}_{Q_T}(X)$.

This procedure also yields an unbiased Monte Carlo estimator to calculate an interest rate option price when we have the T-zero bond price at hand,

$$\tilde{X}_N(0) := P(0,T)\frac{1}{N}\sum_{i=1}^{N} X^{(i)} \tag{5.321}$$

with $X^{(i)}$ the result of simulation run i. Note that here we have to simulate under $\mathbb{Q}_T$, which especially means that the SDE for $r(t)$ is based on $W_T(.)$ and has the form

$$dr(t) = \left(\mu(t, r(t)) - \beta(t,T)\sigma^2(t, r(t))\right)dt + \sigma(t, r(t))\, dW_T(t). \tag{5.322}$$

Of course, the efficiency of this method depends on the exact form of the final payoff X. Finally, note that for a deterministic short rate, $\mathbb{Q}$ and $\mathbb{Q}_T$ obviously coincide as we have $\beta(t,T) = 0$.

5.17.2 The Vasicek model

The earliest, still well known and applied short rate model is the Vasicek model (see Vasicek [1977]). Here, the short rate equation reads as

$$dr(t) = \kappa(\theta - r(t))\, dt + \sigma dW(t) \tag{5.323}$$

with real, positive constants κ, θ, σ. This SDE admits the explicit solution

$$r(t) = r_0 e^{-\kappa t} + \theta\left(1 - e^{-\kappa t}\right) + \sigma\int_0^t e^{-\kappa(t-u)} dW(u). \tag{5.324}$$

Consequently, we have

$$r(t) \sim N\left(r_0 e^{-\kappa t} + \theta\left(1 - e^{-\kappa t}\right), \frac{\sigma^2}{2\kappa}\left(1 - e^{-2\kappa t}\right)\right).$$

Thus, the short rate is mean-reverting around the level θ as its drift is always negative if $r(t)$ is above θ and positive if it is below. As θ equals the asymptotic mean in the above normal distribution, it can be seen as the long term limit of the short rate. The normal distribution of the short rate has computational advantages (see the explicit pricing formulae below), but also has the disadvantage that the short rate can become negative. Further, with only three free parameters at hand, it is in general not possible to perfectly explain an initial term structure of bond prices.

THEOREM 5.57 Bond and option prices in the Vasicek model
In the Vasicek model given by Equation (5.323) *we have:*
(a) T-zero bond prices of the form

$$P(t,T) = e^{-B(t,T)r(t) + A(t,T)} \tag{5.325}$$

with A and B given by

$$B(t,T) = \frac{1}{\kappa}\left(1 - e^{-\kappa(T-t)}\right), \tag{5.326}$$

$$A(t,T) = \left(\theta - \frac{\sigma^2}{2\kappa^2}\right)(B(t,T) - T + t) - \frac{\sigma^2}{4\kappa}B(t,T). \tag{5.327}$$

(b) Bond call and put option prices of the form

$$Call(t,T,S,K) = P(t,S)\,\Phi(d_1(t)) - KP(t,T)\,\Phi(d_2(t)), \tag{5.328}$$

$$Put(t,T,S,K) = KP(t,T)\,\Phi(-d_2(t)) - P(t,S)\,\Phi(-d_1(t)) \tag{5.329}$$

with

$$d_{1/2}(t) = \frac{\ln\left(\frac{P(t,S)}{P(t,T)K}\right) \pm \frac{1}{2}\bar{\sigma}^2(t)}{\bar{\sigma}(t)}, \quad \bar{\sigma}(t) = \sigma\sqrt{\frac{1-e^{-2\kappa(T-t)}}{2\kappa}}B(T,S) \tag{5.330}$$

where K denotes the strike and T the maturity of the options, and $S \geq T$ is the maturity of the underlying zero bond.

(c) Prices for caps with face value V, level L, payment times $t_1 < ... < t_n$

$$Cap(t;V,L,\sigma) = V\sum_{i=1}^{n}\left(P(t,t_{i-1})\,\Phi\left(\tilde{d}_{1,i}(t)\right) - (1+\delta_i L)\,P(t,t_i)\,\Phi\left(\tilde{d}_{2,i}(t)\right)\right) \tag{5.331}$$

for $t < t_0 < t_1$ with

$$\tilde{d}_{1/2,i}(t) = \frac{1}{\bar{\sigma}_i(t)}\ln\left(\frac{P(t,t_{i-1})}{(1+\delta_i L)\,P(t,t_i)}\right) \pm \tfrac{1}{2}\bar{\sigma}_i(t), \tag{5.332}$$

$$\bar{\sigma}_i(t) = \sigma\sqrt{\frac{1-e^{-2\kappa(t_{i-1}-t)}}{2\kappa}}B(t_{i-1},t_i), \quad \delta_i = t_i - t_{i-1}. \tag{5.333}$$

We obtain the log-normal zero bond price formula by computing the expected value in Equation (5.312) using the joint distribution of the short rate $r(t)$ and the integral $\int_0^t r(s)ds$ (see below). The form of the call option price is then a direct application of the log-normal valuation formula, Proposition 5.1.

Of course, these pricing formulae allow for an efficient calibration of the Vasicek model. Note also that today's short rate $r(0)$ is not observable. One therefore has to calibrate it together with the model parameters θ, κ, σ. It is important to point out that a calibration should always contain some nonlinear product such as caps or bond options. If one only uses zero coupon bond prices for calibration, then one might obtain an excellent fit with unreasonable model parameters as the model might not be uniquely determined.

Aspects of Monte Carlo simulation in the Vasicek model

There are various aspects in the Vasicek model that make Monte Carlo simulation convenient. First, the normality of the short rate allows us to derive the joint distribution of its final value and its integral as

$$\left(r(t), \int_0^t r(s)\,ds\right) \sim N\left(m(t), \Sigma(t)\right) \tag{5.334}$$

with

$$m_1(t) = \theta + (r_0 - \theta)e^{-\kappa t}, \quad m_2(t) = \theta t + (r_0 - \theta)\tfrac{1-e^{-\kappa t}}{\kappa}, \tag{5.335}$$

$$\Sigma_{11}(t) = \tfrac{\sigma^2}{2\kappa^2}\left(1 - e^{-2\kappa t}\right), \; \Sigma_{12}(t) = \tfrac{\sigma^2}{2\kappa^2}\left(1 + e^{-2\kappa t} - 2e^{-\kappa t}\right), \tag{5.336}$$

$$\Sigma_{22}(t) = \tfrac{\sigma^2}{\kappa^2}\left(t + \tfrac{1}{2\kappa}\left(1 - e^{-2\kappa t}\right) - \tfrac{2}{\kappa}\left(1 - e^{-\kappa t}\right)\right). \tag{5.337}$$

So to value a contingent claim with a payoff that depends only on $r(T)$, one can simply use the crude Monte Carlo method and simulate both $r(T)$ and (the logarithm of) the discount factor from a two-dimensional normal distribution as in representation (5.334). This in fact is a very efficient method and even avoids the change to the forward measure.

If in contrast we want to avoid the simulation of the discount factor, then we can simulate under the forward measure $\mathbb{Q}_T$ introduced above. For this, note that the bond prices in the Vasicek model satisfy the SDE

$$dP(t,T) = P(t,T)\left[r(t)\,dt - B(t,T)\,\sigma(t, r(t))\,dW(t)\right]. \tag{5.338}$$

This can be seen by applying Itô's formula to the zero bond price formula in Equation (5.325). One can thus simulate the payoff under $\mathbb{Q}_T$ with $\beta(t,T) = B(t,T)$. Note that with Equation (5.322) the short rate SDE under $\mathbb{Q}_T$ has the form

$$dr(t) = \kappa\left(\theta - B(t,T)\,\sigma^2 - r(t)\right)dt + \sigma dW_T(t). \tag{5.339}$$

Hence, the short rate is still normally distributed. This method can also be used for valuing path-dependent options that only depend on a finite number of values $r(t_j)$, $j = 1, .., d$. It is then enough to simulate just these values of the path of the short rate under $\mathbb{Q}_T$. For the Monte Carlo calculation of the price of general path-dependent exotic options we have to simulate (a suitably discretized version of) the whole path $r(t), t \in [0, T]$.

Multifactor generalizations

One can generalize the Vasicek model to a multifactor model by letting the short rate be a linear combination of (possibly correlated) Ornstein-Uhlenbeck processes of the form

$$dx_i(t) = -a_i x_i(t)\,dt + \sigma_i dW_i(t) \tag{5.340}$$

as suggested in Brigo and Mercurio (2001). By the properties of the normal distribution, one still has a normally distributed short rate and similar results for bond prices and bond option prices. The gain of the introduction of the additional parameters are better calibration results; the drawback is a higher computational complexity (although the explicit pricing formulae retain their principle form).

5.17.3 The Cox-Ingersoll-Ross (CIR) model

While keeping the mean-reversion property of the Vasicek model, the main aim of the **C**ox-**I**ngersoll-**R**oss model is to guarantee a nonnegative short rate. In Cox et al. (1985) it is therefore suggested to use a square-root process for the short rate. Thus the short rate satisfies the SDE

$$dr(t) = \kappa(\theta - r(t))\,dt + \sigma\sqrt{r(t)}dW(t) \tag{5.341}$$

for given positive constants κ, θ, σ. We already know there is no explicit solution to the above equation (see Chapter 4). However, by Theorem 4.52 the distribution of $e^{\kappa t}g(t)r(t)$ is known to be a noncentral chi-square distribution with noncentrality parameter λ and d degrees of freedom with

$$g(t) = \frac{4\kappa e^{-\kappa t}}{\sigma^2(1 - e^{-\kappa t})},\ \lambda = r_0 g(t),\ d = 4\kappa\theta/\sigma^2. \tag{5.342}$$

Further, we have

$$\lim_{t\to\infty} E(r(t)) = \theta\ ,\ \lim_{t\to\infty} Var(r(t)) = \frac{\theta\sigma^2}{2\kappa} \tag{5.343}$$

and the short rate in the CIR model stays strictly positive if we have

$$2\kappa\theta > \sigma^2. \tag{5.344}$$

If this is not the case then the origin is attainable by the short rate process. However, if the short rate process reaches the origin it will be reflected. Again, a perfect fit of an arbitrary initial term structure is in general not possible.

Despite the nonexisting explicit solution of the short rate equation, we have explicit formulae for the price of zero bonds and bond calls.

THEOREM 5.58 Bond and option prices in the CIR model
Under the assumption of a short rate that follows the CIR model we have:
(a) T-zero bond prices of the form

$$P(t,T) = e^{-B(t,T)r(t)+A(t,T)} \tag{5.345}$$

with

$$B(t,T) = \frac{2\left[\exp\left((T-t)h\right)-1\right]}{2h+(\kappa+h)\left[\exp\left((T-t)h\right)-1\right]}, \tag{5.346}$$

$$A(t,T) = \ln\left(\left[\frac{2h\exp\left((T-t)(\kappa+h)/2\right)}{2h+(\kappa+h)\left[\exp\left((T-t)h\right)-1\right]}\right]^{2\kappa\theta/\sigma^2}\right), \tag{5.347}$$

$$h = \sqrt{\kappa^2+2\sigma^2}. \tag{5.348}$$

(b) bond call option prices of the form

$$C(t,T,S,K) = P(t,S)\chi^2(a_1;d,\lambda_1) - KP(t,T)\chi^2(a_1;d,\lambda_1) \tag{5.349}$$

with $\chi^2(x;d,\lambda)$ *the distribution function of a noncentral chi-square distribution with* d *degrees of freedom and noncentrality parameter* λ*. Here,* K *is the strike of the call,* T *is its maturity, and* $S \geq T$ *is the maturity of the underlying zero bond. Further, we have*

$$d = 4\kappa\theta/\sigma^2,\ a_1 = 2\bar{r}\left(\rho+\psi+B(T,S)\right),\ a_2 = a_1 - 2\bar{r}B(T,S), \tag{5.350}$$

$$\bar{r} = \frac{\ln(A(T,S)/K)}{B(T,S)},\ \psi = \frac{\kappa+h}{\sigma^2},\ \rho = \frac{2h}{\sigma^2(\exp(h(T-t))-1)}, \tag{5.351}$$

$$\lambda_1 = \frac{2\rho^2 r(t)\exp(h(T-t))}{\rho+\psi+B(T,S)},\ \lambda_2 = \frac{2\rho^2 r(t)\exp(h(T-t))}{\rho+\psi}. \tag{5.352}$$

Aspects of Monte Carlo simulation in the CIR model

In contrast to the Vasicek model, the joint distribution of the final value and the integral of the short rate process does not admit an easy explicit form. However, the Laplace transform of this distribution is explicitly known. In principle one could therefore numerically invert it to simulate from the joint distribution, but this is a comparatively slow procedure.

It is still possible to change to the T-forward measure $\mathbb{Q}_T$ for calculating option prices with a payoff that only depends on a finite number of values $r(t_j)$, $j = 1,...,d$. Again, one can verify that we have

$$\beta(t,T) = B(t,T) \tag{5.353}$$

which is given explicitly in the above theorem. Then, using Equation (5.322) the short rate SDE has the form

$$dr(t) = \left(\kappa\theta - \left(\kappa + B(t,T)\sigma^2\right)r(t)\right)dt + \sigma\sqrt{r(t)}dW_T(t). \tag{5.354}$$

Note that the SDE now contains a time-dependent coefficient of $r(t)$. For pricing exotic options, this emphasizes what had been said in the Heston model. In particular, when simulating a path of the short rate process, it is in general much more efficient to use a numerical discretization scheme such as the Euler scheme with full truncation instead of simulating from the exact distribution (see Section 5.11 on the Heston model).

5.17.4 Affine linear short rate models

There is an obvious generalization of the two presented short rate models which is the class of affine linear models given by

$$dr(t) = (\nu(t) r(t) + \eta(t)) dt + \sqrt{\gamma(t) r(t) + \delta(t)} dW(t) \tag{5.355}$$

with suitable deterministic functions $\nu(t)$, $\eta(t)$, $\gamma(t)$, and $\delta(t)$.
This class has the following properties.

- Vasicek and CIR are particular affine linear models.
- An explicit solution of the above equation is not always possible.
- Depending on the coefficient functions, the short rate is positive or not.
- Four deterministic coefficient functions allow the possibility of a perfect fit of the initial term structure.

Further, there is a general representation of the bond prices.

THEOREM 5.59
In the affine linear short rate class, T-zero bond prices have the form

$$P(t,T) = e^{-B(t,T)r(t)+A(t,T)} \tag{5.356}$$

where $A(t,T)$ and $B(t,T)$ are the unique solutions of the system

$$B_t(t,T) + \nu(t) B(t,T) - \tfrac{1}{2}\gamma(t) B(t,T)^2 + 1 = 0, \quad B(T,T) = 0, \tag{5.357}$$
$$A_t(t,T) - \eta(t) B(t,T) + \tfrac{1}{2}\delta(t) B(t,T)^2 = 0, \quad A(T,T) = 0. \tag{5.358}$$

It is a question of taste to define the class of affine linear models by the form of the short rate equation or by the form of the zero bond prices. Both characterizations are indeed equivalent (see Björk [2004]).

5.17.5 Perfect calibration: Deterministic shifts and the Hull-White approach

The Hull-White approach

We present the Hull-White approach for choosing coefficient functions of an affine linear model to obtain perfect fit of an initial term structure (see Hull and White [1990]). We limit ourselves to the Hull-White version of the Vasicek model which is determined by the short rate equation

$$dr(t) = (\delta(t) - ar(t)) dt + \sigma dW(t), \ a > 0. \tag{5.359}$$

It has a slightly different form as the original Vasicek short rate equation, but can easily be transformed into it when $\delta(t)$ is a constant. The explicit solution to this SDE is given by

$$r(t) = r_0 e^{-at} + \int_0^t e^{-a(t-s)} \delta(s)\, ds + \sigma \int_0^t e^{-a(t-u)} dW(u) \tag{5.360}$$

where the function $\delta(t)$ is chosen such that the zero bond prices in the model, $P(0,T)$, coincide with the actually observed market prices $P^M(0,T)$ (see Theorem 5.61 for its form). Then, the (model) bond prices can be obtained from the general theorem in the section on affine linear models, but will be given explicitly together with the bond call option prices below

THEOREM 5.60 Prices in the Hull-White model
Under the assumption of a short rate that follows the Hull-White Vasicek variant we have:
(a) T-zero bond prices of the form

$$P(t,T) = e^{-B(t,T)r(t)+A(t,T)}, \tag{5.361}$$

$$B(t,T) = \frac{1}{a}\left(1 - e^{-a(T-t)}\right), \tag{5.362}$$

$$A(t,T) = \ln\left(\frac{P^M(0,T)}{P^M(0,t)}\right) + f^M(0,t)\, B(t,T) - \frac{\sigma^2}{4a}\left(1 - e^{-2at}\right) B(t,T)^2 \tag{5.363}$$

where $f^M(0,t)$ denotes today's market forward rate at time t, $P^M(0,t)$ today's market price of a t-bond.
(b) Bond call option prices of the form

$$C(t,T,S,K) = P(t,S)\,\Phi(d_1(t)) - KP(t,T)\,\Phi(d_2(t)), \tag{5.364}$$

$$d_{1/2}(t) = \frac{\ln\left(\frac{P(t,S)}{P(t,T)K}\right) \pm 1/2\bar{\sigma}^2(t)}{\bar{\sigma}(t)}, \quad \bar{\sigma}(t) = \sigma\sqrt{\frac{1-e^{-2a(T-t)}}{2a}}B(T,S) \tag{5.365}$$

where K is the strike of the call, T is its maturity, and $S \geq T$ is the maturity of the underlying zero bond.

Note that we still have log-normal bond prices and thus obtain a Black-Scholes type formula for the price of the bond call option. So, we have obtained a model that is analytically as tractable as the Vasicek model, but has the additional feature of a perfect fit of the initial term structure if we choose $\delta(t)$ appropriately (see Hull and White [1990]).

THEOREM 5.61

Let $f^M(0,T) := -\frac{\partial \ln P^M(0,T)}{\partial T}$ be today's forward rate at time T obtained from today's zero bond market prices. With the choice of

$$\delta(t) = \frac{\partial f^M(0,t)}{\partial T} + af^M(0,t) + \frac{\sigma^2}{2a}\left(1 - e^{-2at}\right) \tag{5.366}$$

the theoretical bond prices $P(0,T)$ in the Hull-White variant of the Vasicek model coincide with today's zero bond market prices $P^M(0,T)$.

Monte Carlo, calibration, and conceptual issues

1. There is one conceptual drawback of the Hull-White model. As the function $\delta(t)$ is introduced to obtain a perfect fit of the initial term structure, the mean-reversion property is no longer explicitly present in the model as there is no real long-term limit of the short rate.

2. Note that the calibration algorithm for the Vasicek model has to be significantly modified. As we have already obtained perfect fit between model and market bond prices, we can no longer use the zero bond prices for calibrating the model coefficients $r(0)$, σ, and θ. For this, we now have to use theoretical and market cap and bond option prices.

3. With regard to the Monte Carlo calculation of the prices of (exotic) options, the remarks made in the Vasicek case remain valid here, too. Of course, the joint distribution of the terminal value and the integral of the short rate is slightly more complicated as the product $\kappa\theta$ is now replaced by the deterministic function $\delta(t)$. However, one can still show that they admit a joint normal distribution. The simulation method with the help of the forward measure $\mathbb{Q}_T$ can be applied without change. The main difference compared to the Vasicek model is that we now have to use either numerical integration or a discretization method to simulate the short rate paths as the time-dependent function $\delta(t)$ enters Equation (5.360).

Perfect initial fit by deterministic shifts

A second method to obtain a perfect fit of the initial term structure is to introduce a so-called deterministic shift of the short rate. More precisely, given a short rate process $r(t)$ and a deterministic function $h(t)$, then the shifted version $r_h(t)$ is obtained as the sum of both,

$$r_h(t) = r(t) + h(t). \tag{5.367}$$

This method goes back to Dybvig (1997), Avellaneda and Newman (1998), and Brigo and Mercurio (2001). It is applicable to general short rate models that do not necessarily belong to the affine linear class. To introduce the convenient choice of the shift function, we have to introduce the notation $P(t,T;r)$ for the T-zero bond price in the **intrinsic short rate model** given by $r(.)$ when at time t we have $r(t) = r$. Similarly, we introduce the notation

$C(t,T,S,K;r)$ for the price of bond call option with strike K, maturity T on an S-zero bond at time t with $r(t) = r$. And finally, let

$$f(0,t;r) = -\frac{\partial ln(P(0,t;r))}{\partial t} \tag{5.368}$$

be the corresponding instantaneous forward rate at time t. We then obtain the following key result (see Brigo and Mercurio [2001]).

THEOREM 5.62
Let the short rate process be given by $r_h(t) = r(t) + h(t)$. We then have:
(a) The price $P(t,T)$ of a T-zero bond at time t is given by

$$P(t,T) = \exp\left(-\int_t^T h(s)\,ds\right) P(t,T;r_h(t) - h(t)). \tag{5.369}$$

(b) With $f^M(0,t) = \partial ln(P^M(0,t))/\partial t$ denoting the instantaneous market forward rates at time t, the choice of

$$h(t) = f^M(0,t) - f(0,t;r(0)) \tag{5.370}$$

is the unique choice of the shift function such that we have a perfect agreement between the market and the model prices of zero bonds at time 0.
3. The price at time t of a European call with maturity T and strike K on an S-zero bond (with $S > T$) is given by

$$C(t,T,S,K) = e^{-\int_t^S h(s)ds} C\left(t,T,S,Ke^{\int_T^S h(s)ds}; r_h(t) - h(t)\right). \tag{5.371}$$

REMARK 5.63 1. Note that the construction of the deterministic shift $h(t)$ ensures a perfect initial fit for **every** parameter constellation that determines the underlying intrinsic short rate model $r(t)$. Thus, we have to calibrate those parameters from other products such as cap prices.

2. If the shift function is differentiable then the shifted short rate obeys the following SDE

$$\begin{aligned} dr_h(t) &= (h_t(t) + \mu(t, r_h(t) - h(t)))\,dt + \sigma(t, r_h(t) - h(t))\,dW(t) \\ &=: \mu_h(t, r_h(t))\,dt + \sigma_h(t, r_h(t))\,dW(t) \end{aligned} \tag{5.372}$$

with $\mu(.,.)$, $\sigma(.,.)$ the coefficient functions of the intrinsic short rate process $r(t)$. One can further show that the shifted bond price satisfies the SDE

$$dP(t,T) = P(t,T)(r_h(t)\,dt - B(t,T)\,\sigma_h(t, r_h(t))\,dW(t)) \tag{5.373}$$

for the $B(t,T)$ of the intrinsic short rate model $r(t)$. Hence, the Brownian motion $W_T(.)$ under T-forward measure $\mathbb{Q}_T$ stays the same as in the intrinsic short rate model and is given by

$$W_T(t) = W(t) + \int_0^t B(s,T)\,\sigma_h(s, r_h(t))\,ds. \tag{5.374}$$

The short rate SDE under $\mathbb{Q}_T$ thus has the form

$$dr_h(t) = \left(\mu_h(t, r_h(t)) - B(s,T)\,\sigma_h^2(t, r_h(t))\right) dt + \sigma_h(t, r_h(t))\,dW(t). \tag{5.375}$$

It can be used to simulate paths of the shifted short rate by a suitable discretization procedure. Depending on the explicit form of $B(t,T)\sigma_h^2(t, r_h(t))$, simulating a path under $\mathbb{Q}_T$ can be comparable to simulating under the martingale measure $\mathbb{Q}$.

3. A new conceptual drawback of the introduction of a deterministic shift is that it does not necessarily preserve positivity of the intrinsic short rate process. This can be of particular importance in the case of a shifted CIR model as in Brigo and Mercurio (2001). □

5.17.6 Log-normal models and further short rate models

A class of models that automatically yield a positive short rate process are the log-normal models. The most popular such model is the Black-Karasinski model (see Black and Karasinski [1991]) where $r(t)$ is given by

$$r(t) = \exp(\tilde{r}(t)), \tag{5.376}$$

$$d\tilde{r}(t) = \kappa(t)\left(\ln(\theta(t)) - \tilde{r}(t)\right) dt + \sigma(t)\,dW(t) \tag{5.377}$$

where $\kappa(t)$, $\theta(t)$, $\sigma(t)$ are deterministic functions chosen to match the initial bond prices, the volatility of the bond price yields, and the cap curve (see Black and Karasinski [1991] for the exact specification of these curves). As there are no explicit price formulae for zero bonds, options, or caps, this has to be done numerically. Black and Karasinski therefore describe a tree procedure. With this procedure they achieve an excellent fit to market data, a fact that is the main reason for the application of this model in real life. However, there is one serious conceptual drawback with the Black-Karasinski model (inherent in all log-normal short rate models): the expected value of the money market account equals infinity, i.e. we have

$$\mathbb{E}_{\mathbb{Q}}\left(e^{\int_0^t r(s)ds}\right) = +\infty. \tag{5.378}$$

This can easily be seen by approximating the integral by a finite sum and using the fact that $\mathbb{E}(\exp(\exp(Z))) = \infty$ for a normally distributed Z. However, as one has to do the relevant computations with the help of discretization procedures, the exploding average of the bank account does not cause problems there. If one is only interested in fitting the initial bond prices, then a shifted version of the exponential Vasicek model given by

$$r_h(t) = r(t) + h(t), \tag{5.379}$$
$$r(t) = \exp(y(t)), \tag{5.380}$$
$$dy(t) = \kappa(\theta - y(t))\,dt + \sigma dW(t) \tag{5.381}$$

is a good alternative to the Black-Karasinski model in its full generality, although it cannot guarantee the positivity of the short rate anymore. To calculate the required forward rates in Equation (5.370) that determine the shift function $h(t)$, one has to resort to a tree procedure similar to the one in Black and Karasinski (1991).

Further positive short rate models

More recent models that guarantee a nonnegative short rate are the model by Flesaker and Hughston (1996) and the potential approach by Rogers (1997). We do not go into detail here, but remark that the potential approach is a framework that in particular allows a multicurrency market modelling in a sparse way.

5.18 The forward rate approach to interest rate modelling

The forward rate approach pioneered by Heath et al. (1992) is based on the relation

$$P(t,T) = \exp\left(-\int_t^T f(t,s)\,ds\right) \tag{5.382}$$

between zero bond prices and forward rates. It implies that it is equivalent to model the evolution of zero bond prices and of forward rates. Note that these two modelling tasks are much more involved than the one of the short rate models. There, we only modelled one (!) particular interest rate, the short rate, evolving over time as a stochastic process. Here, we have to model the evolution of a whole curve (i.e. an uncountable number of points!) through time, no matter if we take the forward rate curve or the bond price curve.

In the HJM framework one decides to model the evolution of the forward rate curve $f(t,T)$, $t \geq 0$, $T \geq t$ by a family of stochastic processes. Before we do this, we point out one very appealing feature of this framework: it easily allows for a perfect calibration of the initial bond prices via choosing the initial forward rates equal to the ones observed at the market, i.e.

$$f(0,t) = f^M(0,t) \;\; \forall t \geq 0 \tag{5.383}$$

lead to agreement between the model and the market zero bond prices

$$P(0,T) = P^M(0,T) \ \forall T \geq 0. \tag{5.384}$$

However, we are heavily restricted with our choice of a forward rate process by the so-called **HJM drift condition**. It is derived from the fact that we must have equality between the two representations of a zero bond price,

$$\exp\left(-\int_t^T f(t,s)\,ds\right) = P(t,T) = \mathbb{E}_{\mathbb{Q}}\left(\exp\left(-\int_t^T r(s)\,ds\right)\right) \tag{5.385}$$

for a suitable martingale measure to prevent arbitrage opportunities. From this relation, Heath et al. (1992) deduced that if we specify a forward rate model as a stochastic process of the form

$$df(t,T) = \mu_f(t,T)\,dt + \sigma_f(t,T)\,dW(t) \tag{5.386}$$

for a d-dimensional Brownian motion $W(.)$ and suitable stochastic processes μ_f, σ_f, then we must have

$$\mu_f(t,T) = \sigma_f(t,T)\int_t^T \sigma_f(t,s)\,ds \quad \textbf{HJM drift condition} \tag{5.387}$$

under $\mathbb{Q}$. This of course is a serious modelling restriction as it can be paraphrased that we are only allowed to choose the volatility structure of the forward rate curve freely. On the other hand, we still have a lot of freedom with the choice of the volatility function.

5.18.1 The continuous-time Ho-Lee model

The historically first model in the HJM framework is a continuous-time version of the Ho and Lee model (1986) that appeared as an example in Heath et al. (1992). There the evolution of the forward rate is modelled as

$$f(t,T) = f^M(0,T) + \sigma W(t) + \sigma^2 t\left(T - \frac{1}{2}t\right) \tag{5.388}$$

for a constant σ and a one-dimensional Brownian motion $W(t)$. From this, one directly obtains explicit formulae for the short rate and zero bond prices

$$r(t) = f(t,t) = f^M(0,t) + \sigma W(t) + \frac{1}{2}\sigma^2 t^2, \tag{5.389}$$

$$P(t,T) = P^M(0,T). \tag{5.390}$$

As we have a normally distributed short rate and log-normal zero bond prices, it is no surprise that we also have a Black-Scholes type formula for bond call options which we do not present here, as the model is too simple to be applied in reality. Note in particular that all forward rates for a given t but varying maturity T are perfectly correlated! However, the model still serves as a good simple introduction to the HJM framework.

5.18.2 The Cheyette model

A practically relevant model that combines tractability with flexibility is the Cheyette (1992) model. Before we present it, we would like to point out one particular problem with an arbitrary specification of the forward rate volatility. Assume for simplicity that we model the forward rate based on a one-dimensional Brownian motion. Then the short rate is given by

$$\begin{aligned} r(t) &= f(t,t) \\ &= f^M(0,t) + \int_0^t \sigma_f(s,t) \int_s^t \sigma_f(s,u)\, du ds + \int_0^t \sigma_f(s,t)\, dW(s). \end{aligned} \quad (5.391)$$

As the integrand of the stochastic integral may depend on t in a general way, the short rate $r(t)$ may no longer be a Markov process. This, however, has serious numerical consequences for the computation of option prices. Updating the drift might require the recomputation of a whole set of volatilities, while the whole path of the past short rate might be needed to calculate an option price. An easy criterion to guarantee that the short rate process has the Markov property is given in Carverhill (1994), who assumes that the volatility is given as the product of two deterministic functions,

$$\sigma_f(t,T) = g(T)\, h(t). \quad (5.392)$$

This is also true in a multifactor forward rate model with a d-dimensional Brownian motion and a d-dimensional volatility vector with components

$$\sigma_{f,i}(t,T) = g_i(T)\, h_i(t), \;\; i = 1, ..., d. \quad (5.393)$$

However, it is possible to have a more general specification of such a product form at the cost that the short rate is no longer a Markov process, but its dependence on the past can be described by just a two-dimensional state process. This insight led to the Cheyette (1992) model which has also been independently developed in Ritchken and Sankarasubramanian (1995), who show that the form

$$\sigma_f(t,T) = \sigma_r(t) \exp\left(- \int_t^T \kappa(x)\, dx \right) \quad (5.394)$$

for a deterministic function $\kappa(x)$ and a suitable adapted stochastic process $\sigma_r(t)$ is an equivalent condition for the term structure of interest rates being determined by a two-dimensional Markov process. Moreover, they show that the zero bond prices are given by

$$\begin{aligned} &P(t,T) \\ &= \frac{P(0,T)}{P(0,t)} \exp\left(-\frac{1}{2}\beta^2(t,T)\,\phi(t) + \beta(t,T)\left(f^M(0,t) - r(t)\right)\right) \end{aligned} \quad (5.395)$$

where we have used the abbreviations

$$\beta(t,T) = \int_t^T e^{-\int_t^u \kappa(x)dx} du, \quad \phi(t) = \int_0^t \sigma_f^2(s,t)\, ds. \tag{5.396}$$

Note that by Equation (5.395), the term structure in the Cheyette model depends only on the two **state processes** $(r(t), \phi(t))$, the short rate and the integrated volatility. Ritchken and Sankarasubramanian (1995) show that they obey the differential representations

$$dr(t) = \mu_r(t)\, dt + \sigma_r(t)\, dW(t), \tag{5.397}$$

$$d\phi(t) = \left(\sigma_r^2(t) - 2\kappa(t)\,\phi(t)\right) dt. \tag{5.398}$$

In particular, differentiation of the short rate process given by Equation (5.391) under the volatility specification (5.394) leads to

$$\mu_r(t) = \kappa(t)\left(f^M(0,t) - r(t)\right) + \phi(t) + \frac{d}{dt} f^M(0,t). \tag{5.399}$$

Note that the short rate is **no** Markov process, as the drift depends also on past values of the volatility via $\phi(t)$. However, the pair $(r(t), \phi(t))$ constitutes a Markov process.

A generalization of the above approach is given in Cheyette (1995) that leads to a Markov process with a higher number of state variables.

PROPOSITION 5.64 (Cheyette [1995])

Assume that the forward rate volatility process can be written in the form of

$$\sigma_f(t,T) = \sum_{i=1}^{N} \beta_i(t) \frac{\alpha_i(T)}{\alpha_i(t)} \tag{5.400}$$

for deterministic functions $\alpha_i(t)$ and adapted processes $\beta_i(t)$. If we then in the risk-neutral world define $N(N+3)/2$ state variables x_i, V_{ij} by

$$x_i(t) = \int_0^t \left(\sum_{k=1}^{N} \beta_k(s) \frac{A_k(t) - A_k(s)}{\alpha_k(s)} \right) \beta_i(s) \frac{\alpha_i(t)}{\alpha_i(s)} ds + \int_0^t \beta_i(s) \frac{\alpha_i(t)}{\alpha_i(s)} dW(s), \tag{5.401}$$

$$V_{ij}(t) = V_{ji}(t) = \int_0^t \beta_i(s)\,\beta_j(s) \frac{\alpha_i(t)\,\alpha_j(t)}{\alpha_i(s)\,\alpha_j(s)} ds \tag{5.402}$$

with $A_k(t) = \int_0^t \alpha_k(s)ds$, the forward rate equation can be expressed as

$$f(t,T) = f(0,T) + \sum_{j=1}^{N} \frac{\alpha_j(T)}{\alpha_j(t)} \left(x_j(t) + \sum_{i=1}^{N} \frac{A_i(t) - A_i(s)}{\alpha_i(s)} V_{ij}(t) \right). \tag{5.403}$$

Further, the state variables x_i, V_{ij} form a joint Markov process and admit the differential representations

$$dx_i(t) = \left(x_i(t) \frac{d}{dt} \ln(\alpha_i(t)) + \sum_{j=1}^{N} V_{ij}(t) \right) dt + \beta_i(t)\, dW(t), \quad (5.404)$$

$$dV_{ij}(t) = \left(\beta_i(t) \beta_j(t) + V_{ij}(t) \frac{d}{dt} (\ln(\alpha_i(t) \alpha_j(t))) \right) dt. \quad (5.405)$$

In particular, we obtain

$$r(t) = f(0,T) + \sum_{j=1}^{N} x_j(t), \quad (5.406)$$

$$P(t,T) = \frac{P(0,T)}{P(0,t)} \exp\left(-\sum_{i=1}^{N} \frac{A_i(T) - A_i(t)}{\alpha_i(t)} x_i(t) \right)$$
$$\exp\left(-\sum_{i,j=1}^{N} \frac{(A_i(T) - A_i(t))(A_j(T) - A_j(t))}{2\alpha_i(t)\alpha_j(t)} V_{ij}(t) \right). \quad (5.407)$$

Also, variations with a multidimensional Brownian motion are considered in Cheyette (1995), together with empirical applications to the U.S. Treasury bond market.

Volatility specifications and aspects of Monte Carlo simulation

Equipped with the above representations, one can now consider specifications of the volatility function. In the case of $N = 1$, a popular choice for a flexible model is a CEV type volatility process, i.e.

$$\sigma_r(t) = \sigma \cdot r(t)^{\gamma} \quad (5.408)$$

for some $\gamma \in [0,1]$ and positive constants σ, κ. One can also imagine other choices of the volatility function such as a displaced diffusion or a Heston type process. It is important to note the advantage of the sparse representation of the Cheyette model. We only have to update a two-dimensional process to obtain the evolution of the whole term structure. Compare this to the general case of the HJM model when one has to update a whole forward rate curve!

To compute the prices of various (European) options on bonds or interest rates we have to simulate the two state variable processes $r(t)$, $\phi(t)$ and the money market account $B(t) = \exp(\int_0^t r(s)ds)$ in a suitable discretized way. This is demonstrated in Algorithm 5.20.

The option considered in the algorithm is implicitly assumed to be written on a path of the short rate. It can also be an option on a path of a zero bond.

Algorithm 5.20 Option pricing in the Cheyette model

Let an initial forward rate curve $f(0,t)$ and an initial term structure $P(0,t)$ be given. Let further $\Delta = T/n$ be a given stepsize.

For $i = 1$ to N do

1. $r^{(i)}(0) = f(0,0), \quad \phi^{(i)}(0) = 0, \quad B^{(i)}(0) = 1.$
2. For $j = 1$ to n do
   (a) $\phi^{(i)}\left((j+1)\Delta\right) = e^{-2\kappa\Delta}\phi\left(j\Delta\right) + \sigma^2 r^{(i)}\left(j\Delta\right)^{2\gamma}\frac{1-\exp(-2\kappa\Delta)}{2\kappa}.$
   (b) Generate a random number $Y^{ij} \sim N(0,1)$.
   (c) $r^{(i)}\left((j+1)\Delta\right) = r^{(i)}\left(j\Delta\right) + \sigma r^{(i)}\left(j\Delta\right)^{\gamma}\sqrt{\Delta}Y^{ij}$
   $+ \left(\kappa\left(f\left(0,j\Delta\right) - r^{(i)}\left(j\Delta\right)\right) + \phi^{(i)}\left(j\Delta\right) + \frac{f(0,(j+1)\Delta)-f(0,j\Delta)}{\Delta}\right)\Delta.$
   (d) $B^{(i)}\left((j+1)\Delta\right) = B^{(i)}\left(j\Delta\right)e^{r^{(i)}(j\Delta)\Delta}.$
3. Compute the discounted payoff $Z^{(i)} = f\left(r\left(t\right), t \in [0,T]\right)/B^{(i)}\left(T\right)$ of an option given by f along path i of $(r(.), \phi(.))$.

Compute the Monte Carlo estimate of the option price via

$$I_{Z,N} = \frac{1}{N}\sum_{i=1}^{N} Z^{(i)}.$$

It might in this case also be useful to update the zero bond price directly via the representation (5.395).

As with the CEV model, a possible variance reduction method is to use a suitable short rate model with explicit valuation formulae for the option under consideration. Possible candidates for such a control variate model are the Vasicek model or the Hull-White model. Numerical examples are given in Ritchken and Sankarasubramanian (1995) and Cheyette (1992).

5.19 LIBOR market models

The so-called **LIBOR market models** are nowadays an industry standard in the interest rate market. There are at least two good reasons for that. One is that they deliver a rigorous derivation ("the story behind ...") of the Black formula for pricing caplets and thus justify the use of a standard market rule. The other reason is that the basic objects that are modelled are directly observable market interest rates such as the 3-month LIBOR rate. This is in particular in contrast to the (instantaneous) short rate and forward rate models where artificial instantaneous rates are modelled. The modelling framework has been introduced in Miltersen et al. (1997), Brace et al. (1997), and in Jamshidian (1997), who all contributed to different aspects of the theory.

5.19.1 Log-normal forward-LIBOR modelling

To present the modelling framework, we introduce some notation. Assume that there is a given **tenor structure** $t = t_0 < t_1 < .. < t_N$, and that zero bonds maturing at the dates t_i are traded. We recall the following definition.

DEFINITION 5.65
The δ_i-forward-LIBOR rate $L_i(t)$ is the simple yield for the time interval $[t_{i-1}, t_i]$, i.e. with $\delta_i = t_i - t_{i-1}$ we define

$$L_i(t) = L(t; t_{i-1}, t_i) = \frac{1}{\delta_i} \frac{P(t, t_{i-1}) - P(t, t_i)}{P(t, t_i)}. \tag{5.409}$$

By the definition of the t_i-forward measure $\mathbb{Q}_i := \mathbb{Q}_{t_i}$, Equation (5.409) implies that $L_i(t)$ is a $\mathbb{Q}_i$-martingale. Thus, it is an immediate consequence that if we want to model log-normal forward-LIBOR rates in a diffusion setting, we have to choose the following dynamics under $\mathbb{Q}_i$:

$$dL_i(t) = L_i(t)\,\sigma_i(t)\,dW_i(t) \tag{5.410}$$

Here, $W_i(.)$ is a (for the moment one-dimensional) $\mathbb{Q}_i$-Brownian motion, $\sigma_i(t)$ a bounded and deterministic function. We further assume that we have modelled all forward-LIBOR rates $L_j(t)$, $j = 1, ..., N$ in analogy to $L_i(t)$ under the corresponding t_j-forward measures $\mathbb{Q}_j$. Then, this log-normal modelling of the forward-LIBOR rates supports the Black formula:

THEOREM 5.66 Cap pricing and the Black formula
Assume that for $i = 1, ..., N$ the δ_i forward-LIBOR rates satisfy

$$dL_i(t) = L_i(t)\,\sigma_i(t)\,dW_i(t)\,, t < t_i. \tag{5.411}$$

(a) Then today's price $C_i(t, \sigma_i(t))$ of a caplet maturing at time t_i with a payment of $\delta_i \cdot (L_i(t_i) - L)^+$ is given by

$$C_i(t, \sigma_i(t)) = \delta_i P(t, t_i)\left[L_i(t)\,\Phi(d_1(t)) - L\Phi(d_2(t))\right], \tag{5.412}$$

$$d_1(t) = \frac{\ln\left(\frac{L_i(t)}{L}\right) + \frac{1}{2}\bar{\sigma}_i^2(t)}{\bar{\sigma}_i(t)},\ d_2(t) = d_1(t) - \bar{\sigma}_i(t), \tag{5.413}$$

$$\bar{\sigma}_i^2(t) = \int_t^{t_{i-1}} \sigma^2(s)\,ds. \tag{5.414}$$

(b) Today's price of a cap in the forward-LIBOR model $Cap_{FL}(t; V, L)$ *with payment times* $t_1 < ... < t_N$ *and level* L *is given by*

$$Cap_{FL}(t; V, L) = V \cdot \sum_{i=1}^{N} C_i\left(t, \sigma_i\left(t\right)\right). \tag{5.415}$$

In particular, if all volatility processes satisfy $\sigma_i(t) = \sigma$ *for some positive constant* σ *then we have*

$$Cap_{FL}(t; V, L) = Cap_{Black}(t, V, L, \sigma), \tag{5.416}$$

i.e. the price of the cap equals the one obtained with the Black formula.

REMARK 5.67 So far we have not specified the exact form of the volatility functions $\sigma_i(t)$ of the forward-LIBOR rates. Actually, for agreement of a single caplet price with one obtained from the Black formula we only need to use the average (squared) volatility $\bar{\sigma}_i^2(t)$ as input for the Black formula. This on the other hand tells us that from a single market caplet price we can only calibrate this average volatility. However, if we have a set of market cap prices and a suitable parameterization of the volatility functions, then we can use a kind of bootstrapping procedure.

Specifying the variance and covariance structure between different forward rates such that it allows for an easy calibration is one of the most important topics of the actual application of LIBOR models in practice. We do not go into detail here, but remark that there is a whole industry of different parameterizations of the covariance structure between the LIBOR rates (see Schoenmakers [2007] for a deep survey of practical aspects and approaches). As an example one can always imagine that the correlations are assumed to be piecewise constant between two tenor times t_{i-1} and t_i (see also Chapter 6 of Brigo and Mercurio [2001] for some examples of covariance structure specifications used in the industry). □

REMARK 5.68 As a cap has an additive payoff structure (although the single payments are nonlinear functions of the underlying floating rate), its caplets could be valued independently. However, it is clear that the forward-LIBOR rates are in general not independent. Therefore, the Brownian motions that drive them should be correlated. We model this by introducing the N-dimensional Brownian motions $W^{(k)}(t)$ under $\mathbb{Q}_k$ that have a correlation matrix of ρ via

$$W^{(k)}\left(t\right) = \left(W_1^{(k)}\left(t\right), ..., W_N^{(k)}\left(t\right)\right)' \sim N\left(0, t \cdot \rho\right). \tag{5.417}$$

Actually, so far we only needed component k of this vector for modelling $L_k(t)$. For simplicity, we will continue to denote it by $W_k(t) := W_k^{(k)}(t)$. However, the introduction of the Brownian vectors allows us to apply Girsanov's theorem when we need the representation of a particular component $W_k(t)$ of the Brownian motion under – say – the measure $\mathbb{Q}_j$. This is the case when we have to price a more complicated derivative than a cap (see the examples in Section 5.19.3 below) where we have to use one probability measure that is responsible for the joint distribution of the LIBOR rates. It is therefore necessary to derive the (joint) dynamics of the LIBOR rate under such a measure. We will first present the dynamics of $L_k(t)$ under a forward measure $\mathbb{Q}_i$ with $i \neq k$. For deriving those dynamics we apply Girsanov's theorem and use the facts that we have

$$\frac{P(t, t_{k-j})}{P(t, t_k)} = \prod_{i=k-j+1}^{k} (1 + \delta_i L_i(t_i)) \tag{5.418}$$

which is a $\mathbb{Q}_k$-martingale, and that under $\mathbb{Q}_k$ we have the representation

$$W_j^{(k)}(t) = \rho_{jk} W_k(t) + \sqrt{1 - \rho_{jk}^2}\, \bar{W}_j(t) \tag{5.419}$$

for a one-dimensional $\mathbb{Q}_k$-Brownian motion $\bar{W}_j(t)$ independent of $W_k(t)$. □

PROPOSITION 5.69

Under the forward measure $\mathbb{Q}_i$ for $t < t_0$ the log-normal forward-LIBOR rate $L_k(t)$ has the following dynamics in the cases $i < k$, $i = k$, and $i > k$:

$$dL_k(t) = \sigma_k(t) L_k(t) \left(dW_k(t) + \sum_{j=i+1}^{k} \frac{\delta_j \rho_{jk} \sigma_j(t) L_j(t)}{1 + \delta_j L_j(t)} dt \right), \tag{5.420}$$

$$dL_k(t) = \sigma_k(t) L_k(t) \, dW_k(t), \tag{5.421}$$

$$dL_k(t) = \sigma_k(t) L_k(t) \left(dW_k(t) - \sum_{j=k+1}^{i} \frac{\delta_j \rho_{jk} \sigma_j(t) L_j(t)}{1 + \delta_j L_j(t)} dt \right). \tag{5.422}$$

Here, $W_k(t) = W_k^{(i)}(t)$ denotes a one-dimensional $\mathbb{Q}_i$-Brownian motion.

The philosophy behind the proposition is to choose one particular forward-LIBOR rate $L_k(t)$ as a reference rate and to express the dynamics of the others in terms of the corresponding forward measure $\mathbb{Q}_k$. A natural alternative for a convenient pricing measure which is more balanced is obtained by introducing the so-called **discrete bank account** as numeraire. This is the wealth process starting with one unit of money corresponding to a roll-over strategy in the zero bond that matures at the next tenor time t_j followed by a corresponding reinvestment. It has a representation of

$$B_{disc}(t) = P\left(t, t_{\beta(t)-1}\right) \prod_{j=1}^{\beta(t)-1} (1 + \delta_j L_j(t_{j-1})) \tag{5.423}$$

where we have $t_{\beta(t)-2} < t \leq t_{\beta(t)-1}$. Let $\mathbb{Q}_{disc}$ be the probability measure such that all traded assets are martingales when B_{disc} is used as numeraire. This measure is called the **spot-LIBOR measure**. Then we have:

PROPOSITION 5.70

The dynamics of the forward-LIBOR rates under the spot-LIBOR measure are given by

$$dL_i(t) = \sigma_i(t) L_i(t) \left(dW_i^{disc}(t) + \sum_{j=\beta(t)}^{i} \frac{\delta_j \rho_{j,k} \sigma_j(t) L_j(t)}{1 + \delta_j L_j(t)} dt \right) \tag{5.424}$$

where $W^{disc}(t) = (W_1^{disc}(t), \ldots, W_N^{disc}(t))'$ *is a* $\mathbb{Q}_{disc}$*-Brownian motion.*

By comparing the way the different LIBOR rates enter the distribution of a generic rate $L_k(t)$ one realizes that choosing a fixed forward measure $\mathbb{Q}_i$ will result in a one-sided influence from the other rates, i.e. either **only earlier** or **only later rates** enter the drift term. This is not the case when the spot-LIBOR measure is used as the underlying pricing measure. Here, there seems to be a more balanced mutual influence of the bias caused by each of the simulated forward-LIBOR rates. It is therefore often advised to simulate under the spot-LIBOR measure than under one of the forward measures.

5.19.2 Relation between the swaptions and the cap market

Besides caps, it is also a market practice to price options on interest rate swaps (also called swaptions) by a suitably adapted Black formula. It is based on the assumption of log-normally distributed forward swap rates. To set the basis for our considerations, we consider an option to enter an interest rate swap. We assume that payment times $t_1 < ... < t_N$ are given and that the floating rates are set at times $t_0 < ... < t_{N-1}$. We look at a payer swap, i.e. at the payment times we have to pay $\delta_i \cdot p$ and receive the floating payments $\delta_i \cdot L(t_{i-1}; t_{i-1}, t_i)$. Today's value of this (forward) swap contract is given by

$$\begin{aligned} S_{(t_1,\ldots,t_N)}(t) &= \sum_{i=1}^{N} P(t, t_i) \delta_i (L_i(t) - p) \\ &= \sum_{i=1}^{N} (P(t, t_{i-1}) - (1 + \delta_i p) P(t, t_i)). \end{aligned} \tag{5.425}$$

Setting this value to zero yields the **forward swap rate** $p_{forward}(t; t_1, ..., t_N)$

$$p_{fsr}(t) := p_{fsr}(t; t_1, ..., t_N) = \frac{P(t, t_0) - P(t, t_N)}{\sum_{i=1}^{N} \delta_i P(t, t_i)}. \tag{5.426}$$

As a **swaption** is the right to enter the above swap contract at time t_0 it yields a payment of

$$B_{swaption} = (p_{fsr}(t_0) - p)^+ \sum_{i=1}^{N} \delta_i P(t_0, t_i) \tag{5.427}$$

at time t_0. It is now easy to put the industry practice of using the Black formula into a rigorous framework:

- Use the numeraire $\sum_{i=1}^{N} \delta_i P(t, t_i)$ to construct the corresponding pricing measure and call it $\mathbb{Q}_{1,N}$.
- Under this measure $\mathbb{Q}_{1,N}$, the forward swap measure, the forward swap rate is a martingale.
- Model log-normal dynamics for the forward swap rate under $\mathbb{Q}_{1,N}$ by

  $$dp_{fsr}(t) = \sigma^{(1,N)}(t)\, p_{fsr}(t)\, dW^{(1,N)}(t) \tag{5.428}$$

  for a bounded, deterministic volatility function and a $\mathbb{Q}_{1,N}$-Brownian motion $W^{(1,N)}(t)$.

Under this assumption, the same argument as in the log-normal forward-LIBOR rate model leads to a Black type formula for swaption prices:

THEOREM 5.71 Swaption pricing with Black's formula
Under the assumption of log-normal forward swap rate dynamics as in Equation (5.428), the price at time t ($< t_0$) of a swaption with payment $B_{swaption}$ given by (5.427) at time t_0 is represented by

$$Swapt(t; p, t_1, ..., t_N; t_0) = \beta(t)\left[p_{fsr}(t)\,\Phi(d_1(t)) - p\Phi(d_2(t))\right] \tag{5.429}$$

$$\beta(t) = \textstyle\sum_{i=1}^{N} \delta_i P(t, t_i) \tag{5.430}$$

$$d_1(t) = \tfrac{\ln(p_{fsr}(t)/p) + \frac{1}{2}\bar{\sigma}_i^2(t)}{\bar{\sigma}_i(t)}, \quad d_2(t) = d_1(t) - \bar{\sigma}_i(t), \tag{5.431}$$

$$\bar{\sigma}_i^2(t) = \textstyle\int_t^{t_0} \sigma^{(1,N)}(s)^2\, ds. \tag{5.432}$$

As forward-LIBOR rates and forward swap rates are connected via zero bond prices, it is a natural question to ask how one can be expressed in terms of the other. Indeed, a comparison between the representation of the forward-LIBOR rates (5.409) and the forward swap rate representation (5.426) yields

$$p_{fsr}(t) = \frac{1 - \prod_{j=1}^{N} \frac{1}{1+\delta_j L_j(t)}}{\sum_{i=1}^{N} \delta_i \prod_{j=1}^{i} \frac{1}{1+\delta_j L_j(t)}}. \tag{5.433}$$

To obtain this equation the relation

$$\frac{P(t, t_i)}{P(t, t_0)} = \prod_{j=1}^{i} \frac{1}{1 + \delta_j L_j(t)}, \quad t < t_0 \tag{5.434}$$

is very useful. A second representation between the two types of rates is

$$p_{fsr}(t) = \sum_{i=1}^{N} \frac{\delta_i P(t,t_i)}{\sum_{j=1}^{N} \delta_j P(t,t_j)} L_i(t) =: \sum_{i=1}^{N} \omega_i(t) L_i(t), \quad t < t_0 \tag{5.435}$$

which directly follows from setting the first representation for the value of a forward contract in Equation (5.425) equal to 0. In applications, the approximation

$$p_{fsr}(t) \approx \sum_{i=1}^{N} \omega_i(0) L_i(t), \quad t < t_0 \tag{5.436}$$

is often used and seems to be supported by empirical evidence. The two representations (5.434) and (5.435) have two important consequences:

- It is enough to have a model for the dynamics of the forward-LIBOR rates $L_i(t)$ for also pricing forward swap rate derivatives.
- Using a log-normal diffusion model for the forward-LIBOR rates as e.g. in Propositions 5.69 or 5.70 does not produce a log-normal forward swap rate model and vice versa (see Brigo and Mercurio [2001], Section 6.8).

The usual way of dealing with the second fact is to

- assume log-normal forward-LIBOR rates and
- price swaptions (and related swap derivatives) numerically under the log-normal forward-LIBOR rate assumption.

5.19.3 Aspects of Monte Carlo path simulations of forward-LIBOR rates and derivative pricing

Propositions 5.69 or 5.70 on the dynamics of the forward-LIBOR rates have one striking thing in common. Both SDEs contain in general highly nonlinear drift terms. Further, the distribution of the resulting forward-LIBOR rate $L_i(t)$ is in general only known under the forward measure $\mathbb{Q}_i$ but not under any other of the measures presented above. It is therefore clear that the only way to simulate $L_i(T)$ for some fixed future time is to use an approximation obtained by a suitable discretization scheme such as the Euler-Maruyama or the Milstein one. Note in particular that valuing derivatives on LIBOR rates is in a natural way a multidimensional problem, a fact which – together with the nonlinearity of the drift term in the LIBOR rate dynamics – makes Monte Carlo simulation a natural, sometimes the only possible calculation method.

Besides the nonlinearity of the forward-LIBOR rate dynamics, we also have to take care for the fact that the different rates are correlated. This in particular implies that they have to be simulated together in one go per path. To do so we take the logarithm of the forward-LIBOR rates and obtain its SDE under – say – the spot measure

$$d\ln\left(L_i(t)\right) = \sigma_i(t)\, dW_i^{disc}(t) - \frac{1}{2}\sigma_i^2(t)\, dt + \\ + \sum_{j=\beta(t)}^{i} \frac{\delta_j \rho_{i,j} \sigma_j(t) L_j(t)}{1+\delta_j L_j(t)} dt \quad \text{for } t_{\beta(t)-2} < t \le t_{\beta(t)-1}. \tag{5.437}$$

The use of the logarithm has the particular advantage that the diffusion coefficient of the SDE is deterministic and thus the Euler-Maruyama and the Milstein schemes coincide. We give an algorithmic description of the simulation of forward-LIBOR paths in Algorithm 5.21.

Algorithm 5.21 Simulation of paths of forward-LIBOR rates under the spot-LIBOR measure

Consider the discretization grid $0 = s_0 < s_1 < \ldots < s_n = T$ with $T < t_0$. Let $L_i(0) = L(0, t_i)$, $i = 1, \ldots, N$ be today's LIBOR rates for a given tenor structure. Set $Z_i(0) = \ln(L_i(0))$.

For $k = 1$ to n

1. Simulate $Y^{(k)} \sim N(0, \rho)$.
2. For $i = 1$ to N set

$$\begin{aligned} Z_i(s_k) &= Z_i(s_{k-1}) + \sigma_i(s_{k-1}) \sqrt{s_k - s_{k-1}}\, Y_i^{(k)} - \\ &- \left(\frac{1}{2}\, \sigma_i^2(s_{k-1}) - \sum_{j=1}^{i} \frac{\delta_j \rho_{i,j} \sigma_j(s_{k-1}) L_j(s_{k-1})}{1+\delta_j L_j(s_{k-1})} \right) (s_k - s_{k-1}), \\ L_i(s_k) &= \exp\left(Z_i(s_k)\right). \end{aligned}$$

REMARK 5.72 1. As $\sigma_i(t)$ is deterministic, we know that we have

$$\int_t^s \sigma_i(r)\, dW_i(r) \sim N\left(0, \int_t^s \sigma_i^2(r)\, dr\right). \tag{5.438}$$

As we also have a deterministic correlation between the stochastic integrals, we know that the increments of the stochastic integrals in Algorithm 5.21 are jointly distributed according to $N(0, \Sigma(s-t))$ with

$$\Sigma(s-t)_{ik} = \int_t^s \sigma_i(r)\, \sigma_k(r)\, \rho_{ik} dr, \ i, k = 1, \ldots, N. \tag{5.439}$$

With this knowledge, if one can easily calculate the corresponding integrals, one can modify Algorithm 5.21 by simulating $Y^{(k)} \sim N(0, \Sigma(s_k - s_{k-1}))$ in Step 1. Then, in Step 2, we can use this $Y_i^{(k)}$ and can drop its multiplying factor $\sqrt{s_k - s_{k-1}}$.

2. For simplicity, we have assumed $s_k < t_0$. However, the algorithm can easily be modified to allow for $s_k > t_0$ for some k. In this case, at time s_k only the LIBOR rates that are not yet determined have to be simulated and the sum of the drift then has to start with index $\beta(s_k)$.

3. Of course, the algorithm can easily be modified for the simulation under a different, forward measure $\mathbb{Q}_i$. Then, we have to discretize the dynamics of the logarithm of the forward-LIBOR rates given by Proposition 5.69.

4. If we simulate the (logarithm of the) LIBOR paths in the above form, we can assume that we are doing this on a sufficiently fine grid such that linear interpolation does not cause too big deviations if we need values in between the discretization points.

5. **Dimension of the underlying Brownian motion**: In Algorithm 5.21 we have assumed that the dimension of the driving Brownian motion underlying the different forward-LIBOR rates equals the number of the different forward-LIBOR rates. If, however, we think that the rates are so strongly dependent that their movements can be explained by a Brownian motion of a lower dimension, then it is easy to make the suitable modifications. □

Drift approximations and upspeeding LIBOR simulations

Algorithm 5.21 is a standard method of simulating paths of forward-LIBOR rates. However, as one typically has to simulate a lot of those rates simultaneously, the computational effort will be huge when one is actually using small time steps in the Euler-Maruyama discretization. As it is the drift term that actually does not allow an exact simulation of the final value of the forward-LIBOR rates in just one large time step, there is a lot of focus on developing approximation methods for this term. A particularly popular method is to **freeze the drift**, i.e. to approximate the drift term (no matter if we consider the logarithm under the spot measure or any other measure) by keeping the forward-LIBOR rates that enter the drift terms at their initial values $L_i(0)$ and calculating the remaining deterministic parts of the drift integral exactly. That is, we use the approximation

$$\int_S^T \frac{\delta_j \rho_{j,k} \sigma_j(t) L_j(t)}{1 + \delta_j L_j(t)} dt \approx \frac{L_j(S)}{1 + \delta_j L_j(S)} \int_S^T \delta_j \rho_{j,k} \sigma_j(t)\, dt \tag{5.440}$$

which actually corresponds to an Euler-Maruyama step over the time interval $[S, T]$ with S, T being two generic times. Note that with this approximation, we can exactly simulate the increment of the forward-LIBOR rates over $[S, T]$ as the stochastic integrals entering the representations of the rates are all normally distributed with the already stated variance-covariance structure (see part 1 of Remark 5.72). Hull and White (2000) tested such an approximation where S is set equal to the initial time and T equal to the maturity of a European swaption and reported a very good performance for the usual interest rates and volatility values usually encountered in North America and Europe.

Of course, this simulation with just one long step can still be improved. Also, it might be necessary to incorporate intermediate times as the payment of an exotic option might depend on the values of the forward-LIBOR rates at those intermediate times. A popular way to do this is the use of a **predictor-corrector method** (see Hunter et al. [2001]) where one is first calculating approximate values $\hat{L}_i(T)$ by an Euler-Maruyama step above. Then, one is using these values in approximating the drift terms with the help of

$$\int_S^T \frac{\delta_j \rho_{j,k} \sigma_j(t) L_j(t)}{1+\delta_j L_j(t)} dt \approx \frac{1}{2}\left(\frac{L_j(S)}{1+\delta_j L_j(S)} + \frac{\hat{L}_j(T)}{1+\delta_j \hat{L}_j(T)}\right) \int_S^T \delta_j \rho_{j,k} \sigma_j(t)\, dt. \quad (5.441)$$

In Joshi and Stacey (2008) the performance of this and various other methods of drift approximation is examined. The authors suggest a simple variant of the predictor-corrector method that clearly performs better, the so-called **iterative predictor-corrector method**. It makes use of the special structure of the forward-LIBOR dynamics when the final forward measure $\mathbb{Q}_N$ is used. Then, one can simulate one rate after the other in starting with $L_N(T)$ which is driftless under $\mathbb{Q}_N$. So, one obtains

$$\ln\left(\hat{L}_N(T)\right) = \ln\left(\hat{L}_N(S)\right) + Y_N(T) - \frac{1}{2}\int_S^T \sigma_N^2(t)\, dt \quad (5.442)$$

with

$$Y(T) \sim N\left(0, \Sigma^{(S,T)}\right), \quad \Sigma_{ij}^{(S,T)} = \int_T^S \sigma_i(t)\sigma_j(t)\rho_{ij} dt. \quad (5.443)$$

This estimator yields a predictor-corrector estimator for the drift term $\hat{\mu}_{N-1}$ of $L_{N-1}(T)$ via

$$\hat{\mu}_{N-1} \approx -\frac{1}{2}\left(\frac{L_N(S)}{1+\delta_N L_N(S)} + \frac{\hat{L}_N(T)}{1+\delta_N \hat{L}_N(T)}\right) \int_S^T \delta_N \rho_{N-1,N} \sigma_N(t)\, dt. \quad (5.444)$$

Repeating this for all the other forward-LIBOR rates with

$$\hat{\mu}_i \approx -\frac{1}{2}\sum_{j=i+1}^{N}\left(\frac{L_j(S)}{1+\delta_j L_j(S)} + \frac{\hat{L}_j(T)}{1+\delta_j \hat{L}_j(T)}\right) \int_S^T \delta_j \rho_{i,j} \sigma_j(t)\, dt. \quad (5.445)$$

constitutes the complete algorithm. This iterative predictor-corrector method performed well in the numerical analysis in Joshi and Stacey (2008). Depending on the type of option, it is applicable as a **long-stepping method** (i.e. just one predictor-corrector step until maturity) or as a discretization method with step lengths given by the structure of the option payments.

For more refined variants of this method that also considers the correlations between the different rates we refer again to Joshi and Stacey (2008).

Monte Carlo pricing of some popular LIBOR derivatives

As we now have algorithms to simulate paths of the forward-LIBOR rates, pricing derivatives by Monte Carlo (MC) simulation is in principle exactly the same task as in all other applications considered so far. We will therefore concentrate on giving an example of popular LIBOR derivatives and comment on some particularities for their valuation.

MC pricing of an auto-cap

An **auto-cap** with j payments consists of n caplets ($j \leq n$) with strike K on the forward-LIBOR rates $L_i, i = 1, \ldots, n$ for the tenor structure $t_1, \ldots, t_n$. The caplets are always exercised if they are in the money and as long as not more than j caplets have already been exercised. Thus, the payment at time t_i is given by

$$B_{auto-cap,i} = (t_i - t_{i-1})\,(L_i\,(t_{i-1}) - K)^+ \cdot 1_{A(i)} \tag{5.446}$$

with $A(i)$ = "at most $(j-1)$ of $L_m(t_{m-1})$, $m = 1, \ldots, i-1$ have been positive". Note that this condition is always satisfied for $i \leq j$. The price of an auto-cap is therefore given by (with $\delta_i = t_i - t_{i-1}$)

$$\begin{aligned} p_{auto-cap} &= \mathbb{E}\left(\sum_{i=1}^{n} e^{-\int_0^{t_i} r(s)ds}\delta_i\,(L_i\,(t_{i-1}) - K)^+ \cdot 1_{A(i)}\right) \\ &= p_{cap}\,(t_1, ..., t_j; K) + \qquad\qquad (5.447) \\ &+ P\,(0, t_n)\,\mathbb{E}_{\mathbb{Q}_n}\left(\sum_{i=j+1}^{n} \frac{\delta_i}{P\,(t_i, t_n)}\,(L_i\,(t_{i-1}) - K)^+ \cdot 1_{A(i)}\right) \end{aligned}$$

where $p_{cap}\,(t_1, ..., t_j; K)$ denotes the price of a cap with strike K and the j first caplets of the auto-cap. To obtain an auto-cap price by Monte Carlo simulation we have to simulate all the relevant forward-LIBOR rates until their fixing times t_{i-1}, together with the corresponding bond prices $P\,(t_i, t_n)$, and in addition keep track of the number of positive payments, which is the only path-dependent feature of an auto-cap. Repeating this for sufficiently many paths and averaging over all the obtained (discounted) payments yields the Monte Carlo estimate for the auto-cap price.

MC pricing of a target redemption note

A **target redemption note** guarantees a fixed total coupon p_{sum} on an underlying face value during a maximum running time N. For ease of exposition, we assume a face value of 1, annual coupon payments and (actual) 12-month LIBOR rates as the underlying. Typically, the first coupon p_1 at time 1 is fixed. The next coupons are of the form $(p_i - L_i(i-1))^+$ if the sum of the already paid coupons does not exceed p_{sum} and also the sum including this last payment does not exceed p_{sum}. Further, this last payment plus the already paid sum are capped by p_{sum}. If the sum does not exceed p_{sum} until maturity, then at the last payment the remaining coupon is added. So, the only uncertainty is the timing of the coupon and the timing of the repayment of the face value. Formally, we have coupon payments of the form

$$B_{trn,1} = p_1, \tag{5.448}$$

$$B_{trn,N} = \left(p_{sum} - p_1 - \sum_{j=2}^{N-1} \left(p_j - L_j\left(j-1\right)\right)^+ \right)^+ \tag{5.449}$$

at the first and at the final payment times and

$$B_{trn,i} = \min\Bigg(\left(p_i - L_i\left(i-1\right)\right)^+ , \left(p_{sum} - p_1 - \sum_{j=2}^{i-1} \left(p_j - L_j\left(j-1\right)\right)^+ \right)^+ \Bigg), \quad i = 2, \ldots, N-1 \tag{5.450}$$

at the times i in between. In addition, there is the possible repayment of the face value at time i given by

$$B^{red}_{trn,i} = 1_{A_i}, \; A_i = \left\{ B_{trn,i} = p_{sum} - p_1 - \sum_{j=2}^{i-1} \left(p_j - L_j\left(j-1\right)\right)^+ \right\}. \tag{5.451}$$

To value such a target redemption we only have to simulate paths of the corresponding forward-LIBOR rates (as long as they are still alive!) and can value the payments $B_{trn,i} + B^{red}_{trn,i}$ at times i in the usual Monte Carlo way, i.e. by discounting and averaging.

More LIBOR derivatives

There is an enormous diversity of further interest rate derivatives with payments that can be expressed as functions of forward-LIBOR or forward swap rates. They have a lot of different features and should sometimes create opposite effects than the current yield curve admits. Among them are so-called **inverse floater** or **steepeners**. There are other derivatives consisting of a sequence of caplets or floorlets where the caps/floors change in a random way depending on already realized forward-LIBOR rates. Examples are so-called **ratchets** and **snowballs**. Another popular derivative is a constant maturity swap which consists of an exchange of fixed for floating rate payments with the floating rate being (a multiple of) a swap rate for a constant maturity always settled at the payment times. All these derivatives come along with different additional features. The only pricing method that can be used for all of them right away is Monte Carlo simulation of the underlying forward-LIBOR rates.

5.19.4 Monte Carlo pricing of Bermudan swaptions with a parametric exercise boundary and further comments

A Bermudan swaption is the Bermudan variant of a European swaption, i.e. it consists of a set of swaptions of maturity times t_i, $i = 1, \ldots, n$. The holder of the Bermudan swaption can now choose one of these exercise times t_i and exercise the swaption or can choose not to exercise any swaption at all. We are thus in the same situation as with Bermudan options on a stock.

Bermudan swaption pricing with the Longstaff-Schwartz algorithm

We have already presented the Longstaff-Schwartz algorithm for pricing Bermudan options with the help of regression methods in Section 5.14. Note that it does not depend on the actual form of the dynamics of the underlying(s). The only requirements for it to work are

- that we are able to simulate paths of the underlying that connect the different exercise times of the Bermudan option,
- that we can determine the intrinsic value of the option which is given by the payment resulting from immediate exercise at the times t_i.

If these two requirements are fulfilled we can start the backward algorithm of the Longstaff-Schwartz algorithm in the usual way. As the payments of the Bermudan swaption typically depend on all forward-LIBOR rates that enter the underlying swap in the future, we are automatically faced with a multidimensional underlying stochastic process. On the positive side, the dimension of the underlying process is always reduced by one with every possible exercise passed as then one of the forward-LIBOR rates is now fixed. However, while in the Black-Scholes model we could easily use exact simulation to get from one payment time to another (and thus only needed one simulation step per payment time), the forward-LIBOR rates cannot be simulated exactly. One can now simulate the required forward-LIBOR paths by either a discretization scheme or with the help of a drift approximation method, both of which are discussed in Section 5.19.3. The accuracy of the method used should always be examined by the use of a dual upper bound method such as the one of Andersen and Broadie (2004) or the variant of Belomestny et al. (2009).

We do not go into more details here, but will instead present a simple alternative that is more suitable in the LIBOR rate setting than in the multiasset stock option world.

Bermudan swaption pricing with a parametric exercise boundary

The method of pricing Bermudan swaptions by using a **parametric exercise boundary** has been introduced in Andersen (1999). By exercise boundary, we simply mean the boundary of the region of the underlying's prices where it is optimal to exercise the Bermudan swaption. To approximate it, we suggest to use a parametric family for the form of the optimal exercise boundary and then try to determine the best-suited parameter(s). Here, the best parameter(s) are those that deliver the highest estimate of the option price on the basis of a sufficiently large number of simulated forward-LIBOR paths. This is based on the fact that every strategy that we are determining can only be suboptimal and therefore the higher the option price estimate, the closer – we believe – is the parametric boundary to the real exercise boundary.

However, to obtain a lower bound for the Bermudan swaption price, we have to use an additional set of simulation runs to obtain a Monte Carlo estimate for the option price on the basis of the just determined (suboptimal) exercising strategy. Otherwise, the maximization over the first set of simulated paths could have led to a higher price estimate as we have solved our maximization of the option value on the basis of the simulated paths which introduces a high bias into the price estimation.

To demonstrate how this method works, we take up an example of Andersen (1999) where a Bermudan swaption is considered. To formalize this, we assume that the Bermudan swaption is given by

- its possible exercise times $T_1 < ... < T_N$,
- N swaps with identical fixed for floating payments and maturity T_s but which start at the different times T_i,$i = 1, \ldots, N$, $N < s$.

The decision problem for the holder of the Bermudan swaption at time t_i is to exercise the swaption and enter the corresponding swap starting at T_i or to wait until a possibly better exercise time. A parametric form of the exercise boundary might now be a deterministic function of the still alive forward-LIBOR rates at time T_i. Indeed, a simple example proposed in Andersen (1999) is the following suggestion for the exercise time τ^*:

$$\tau^* = inf\left\{t \in \{T_1, \ldots, T_N\} : I(t) = 1\right\}, \tag{5.452}$$

$$I(T_i) = \begin{cases} 1 \; if \; S_{i,E}(T_i) > H(T_i) \\ 0 \; else \end{cases} \tag{5.453}$$

where $S_{i,E}(t)$ is the value of the European swaption with maturity T_i and $H(t)$ is a deterministic function. Note in the example above that the holder of the Bermudan swaption exercises it at time T_i in case the intrinsic value of the European swaption maturing at time T_i is larger than the value $H(T_i)$. So what remains is to determine the values $H(T_i)$, $i = 1, ..., N$, and then the parametric exercise strategy is fully determined.

For this, we are looking for those values that maximize the value of the Bermudan swaption with exercise strategy $H(.)$ as given above along a set of K simulated forward-LIBOR rate paths. We also have to decide which numeraire we are actually using (such as i.e. the discrete money market account or some particular zero bond). Let $B(T_i, k)$ be the value of this numeraire at time T_i in the simulated path k.

We determine the optimal parametric exercise values $H(T_i)$ in a backward fashion. Note that at time T_N we must have

$$H\left(T_N\right) = 0 \tag{5.454}$$

as it is the last exercise time, and a swap should be entered then if and only if it has a positive value. At time $H(T_{N-1})$, we know the intrinsic values of the European swaption maturing at T_{N-1} for each path. Also, along those paths we can compute the discounted value according to the strategy if we would not exercise as we have already determined $H(T_N) = 0$. $H(T_{N-1})$ is then determined in such a way that maximizes

$$V_{N-1}\left(H\left(T_{N-1}\right)\right) := \sum_{k=1}^{K} \left(S_{N-1,E}^{(k)}\left(T_{N-1}\right) \cdot 1_{S_{N-1,E}^{(k)}\left(T_{N-1}\right) > H\left(T_i\right)} + \right.$$
$$\left. + \frac{B\left(T_{N-1}, k\right)}{B\left(T_N, k\right)} S_{N,E}^{(k)}\left(T_N\right) \cdot 1_{S_{N,E}^{(k)}\left(T_{N-1}\right) > 0,\ S_{N-1,E}^{(k)}\left(T_{N-1}\right) \leq H\left(T_i\right)} \right). \tag{5.455}$$

We then continue in the same way at the exercise times $T_{N-2}, \ldots, T_1$ and thus obtain all the values $H(T_i)$. As already indicated above, to obtain a lower bound for the Bermudan swaption price by the crude Monte Carlo method, we then have to simulate a new set of corresponding forward-LIBOR rates and estimate the Bermudan swaption price in the usual way by discounting each particular payment with its pathwise discount factor, followed by an averaging over all these discounted payments.

Note that there are many more forms of an exercise rule that one can think of besides the one given in Equation (5.452). As in Andersen (1999), one can consider including the actual values of all the other European swaptions that are still alive. This will, however, greatly increase the computational effort. Further, one can specify particular parametric forms of the function $H(.)$ that depends on less than N parameters, and therefore leads to an even smaller dimensional optimization problem than the iterated one solved above.

In Andersen (1999), good numerical performance is demonstrated for simple piecewise linear functionals. On the down side, let us mention again that we only obtain a lower bound for the price, that we have no real convergence theory (opposed to the Longstaff-Schwartz framework). We are relying on good numerical engineering and experience in guessing a possible form of the exercise boundary.

5.19.5 Alternatives to log-normal forward-LIBOR models

As with nearly all models for price processes on a financial market that started with a log-normal distribution, it has been realized after some time that there are empirical deviations of the prices from the log-normality hypotheses. Of course, one can consider the usual suspects when it comes to **more realistic modelling** that takes care for volatility smiles or skews (see Piterbarg [2003] for a survey). A general class introduced in Andersen and Andreasen (2000) is given by

$$dL_i(t) = f(L_i(t))\,\sigma_i(t)\,dW_i(t), t < t_i. \tag{5.456}$$

for a general function f. Popular choices include:

$$f(x) = ax + b \qquad \text{“displaced diffusion model,”} \tag{5.457}$$

$$f(x) = x^{\gamma} \qquad \text{“CEV type model.”} \tag{5.458}$$

Further extensions include a stochastic volatility component into the forward-LIBOR dynamics. A popular model among practitioners is the so-called SABR-model (stochastic alpha beta rho) introduced by Hagan et al. (2002). Giving a complete and detailed survey on the motivation and technical details of these models is, however, beyond the scope of this book.

Chapter 6

Continuous-Time Stochastic Processes: Discontinuous Paths

6.1 Introduction

Thus far, the continuous-time stochastic processes that we have presented for stock price and interest rate modelling have all had continuous paths. Hence, if we monitor such a process closely, we cannot be caught by surprise by an *exceptionally big* move. However, there are many reasons such as catastrophes or surprising news (unexpected political changes, an economic scandal, and so forth) that can make a real-life process jump. More so, due to discreteness of measurement, it can also be argued that processes modelling the real world should allow for discontinuities in their paths, at least. Finally, in the insurance business the evolution of the absolute number of insurance cases is a fundamental issue that can only be modelled by a counting process. This might be the most natural example of a continuous-time stochastic process with piecewise constant paths that only increases by jumps.

We will now mainly concentrate on the class of Lévy processes as an obvious generalization of the Brownian motion concept. Although there are more general jump processes than Lévy processes, they form a rich class for modelling purposes. Their application and their numerical treatment is a very active area of research where the last words are by far not yet spoken. For the theoretical background on Lévy processes we recommend the monograph by Applebaum (2004). Applications of financial modelling are treated in Cont and Tankov (2003) or in Schoutens (2003).

This chapter begins with an important example of a Lévy process, the Poisson process. As it is possible to use Poisson processes for modelling actuarial and financial data without the need of the full technicalities of Lévy processes, we treat them separately. Combining Poisson processes with Itô processes will give rise to the class of jump-diffusions, which is the second example of jump type processes that we will examine. Finally, we will turn to the general class of Lévy processes presenting the theoretical background, and collect examples of such processes and methods to simulate them.

6.2 Poisson processes and Poisson random measures: Definition and simulation

Poisson processes are among the most simple jump processes. They can easily be understood and simulated. In their simplest form, they are a fundamental building block of the class of Lévy processes. Also, they play a fundamental role in insurance mathematics that is similar to that of the Brownian motion in financial mathematics.

DEFINITION 6.1
A stochastic process $N(t)$ with $N(0) = 0$ is called an inhomogeneous **Poisson process** *with parameter process $\Lambda_{s,t}$ if $N(t)$ has independent increments with*

$$N(t) - N(s) \sim Pn(\Lambda_{s,t}), \; t > s > 0 \tag{6.1}$$

where $Pn(\lambda)$ denotes a Poisson distribution with parameter $\lambda > 0$. If we have

$$\Lambda_{s,t} = \lambda \cdot (t - s), \; t > s > 0 \tag{6.2}$$

for a positive constant λ then we speak of a **homogeneous Poisson process** *with intensity λ. If not stated otherwise, we mean the homogeneous case when we simply speak of a Poisson process.*

REMARK 6.2 1. The paths of a Poisson process are monotonically increasing by jumps of size 1.

2. One can show that in the case of a homogeneous Poisson process with intensity $\lambda > 0$ we have:

- The time between two jumps of a Poisson process is exponentially distributed with parameter λ.
- $\mathbb{E}(N(t)) = \lambda t$, i.e. the number of jumps of a Poisson process is proportional to time, and λ is the average number of jumps per time unit.
- $\mathbb{V}ar(N(t)) = \lambda t$, i.e. the variance of a Poisson process is also proportional to time, a property that it shares with the Brownian motion.

3. We can and will always assume that the paths of a Poisson process are right-continuous with left limits. This in particular means that the value of the jump of the Poisson process actually is already added to the process at the jump time. This is reasonable for applications in finance and insurance. To see this, note that if instead we would have it required to be left-continuous, then this would have allowed us to observe the jump and react on it before

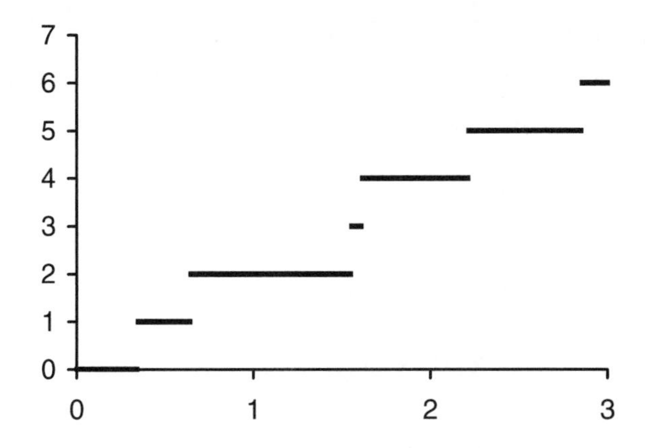

FIGURE 6.1: Poisson process with intensity $\lambda = 2$.

the consequence of it had already hit us. This clearly is not what we had in mind when introducing jumps to our modelling framework. □

Figure 6.1 shows a path of a Poisson process with a parameter of $\lambda = 2$. An immediate generalization of the Poisson process is to allow arbitrary jump heights, which leads to compound Poisson processes.

DEFINITION 6.3
Let $N(t)$ be a Poisson process with parameter λ, let Z_i, $i = 1, 2, \ldots$ be a family of independent, identically distributed random variables which are also independent of the Poisson process $N(t)$. Then the process $X(t)$ defined by

$$X(t) := \sum_{i=1}^{N(t)} Z_i \tag{6.3}$$

is called a **compound Poisson process**.

Note in particular that as negative jump heights are also allowed, a compound Poisson process no longer needs to be an increasing process.

To simulate a compound Poisson process on a finite interval $[0, T]$, we simulate the jump times of a Poisson process N_t and then the jump heights $Z_1, \ldots Z_{N_t}$. In between two jumps the Poisson process is constant. These simple facts yield Algorithm 6.1 to simulate a compound Poisson process.

REMARK 6.4 1. **Simulation of a Poisson process**: By replacing the random number Z by the constant 1 in Algorithm 6.1, we simulate a plain Poisson process. A drift a can be included by replacing $X(R) := X(R-) + Z$

Algorithm 6.1 Simulation of a compound Poisson process

Let L be the distribution of the jump heights, λ be the jump intensity.

1. Set $X(0) := 0$, $R := 0$, $R- := 0$.
2. As long as we have $R < T$:
   (a) Simulate $S \sim Exp(\lambda)$
   (b) Set $R- := R$, $R := R + S$
   (c) If $R > T$
   then $R := T$; $X(R) := X(R-)$
   else
   - Simulate a random number $Z \sim L$
   - Set $X(R) := X(R-) + Z$

   (d) Set $X(t) := X(R-)$ for $t \in (R-, R)$

---

by $X(R) := X(R-) + Z + a \cdot (R - R-)$. Further, we then obtain points between jumps via $X(t) := X(R-) + a \cdot (t - R-)$.

2. **Alternative jump time simulation**: As it is known that given the number $N(T) = k$ of jumps of a (compound) Poisson process on $[0, T]$ the jump times are given as the ordered outcome of the simulation of k independent random variables θ_i which are uniformly distributed on $[0, T]$, one can also simulate the jump times t_i by the following procedure:

1. Simulate the number of jumps $N(T) \sim Pn(\lambda T)$.
2. Simulate $N(T)$ independent random numbers $\theta_i \sim \mathcal{U}[0, T]$.
3. Set $t_i = \theta_{(i)}$ where the subscript (i) denotes the i-th order statistic, i.e. we have $\theta_{(1)} \leq \theta_{(2)} \leq .. \leq \theta_{(N(T))}$.

Then, at t_i one adds the jump height Z_i to the compound Poisson process.

3. A popular model in risk theory (see Section 8.5.2) is the combination of a compound Poisson process with a drift,

$$Y(t) = y + c \cdot t + X(t) \quad . \tag{6.4}$$

□

6.2.1 Stochastic integrals with respect to Poisson processes

As in the Brownian case we will also define an integral with respect to a compound Poisson process to introduce a class of processes similar to Itô processes. For this, it will be convenient to introduce so-called Poisson random measures. However, before doing so, we start with the simple Poisson process to give an idea what we have to take care of. For a general real-valued stochastic process Y we can introduce the stochastic integral of Y with respect to a Poisson process N as

$$\int_0^t Y(s)dN_s := \sum_{i=1}^{N(t)} Y(t_i), \tag{6.5}$$

$t_1, ..., t_{N(t)}$ denoting the jump times of the Poisson process. Thus, the stochastic integral equals the sum of the values of the Y process at the jump times. If the integrand is a compound Poisson process, then the terms in this sum have to be multiplied by the jump heights of the Poisson process to get

$$\int_0^t Y(s)dX(s) := \sum_{i=1}^{N(t)} Y(t_i) \cdot Z_i \tag{6.6}$$

where, in addition to the above example, Z_i denotes the jump height of the compound Poisson process at t_i.

Obviously, the so-defined stochastic integral is in general not a martingale (simply take X as a constant process, integrate with respect to a Poisson process, and obtain a multiple of an increasing process). To obtain this, we replace the integrator by a suitable martingale and also require that the integrand is progressively measurable with respect to a filtration F_t that includes the one generated by the integrator.

DEFINITION 6.5
Let $(F_t)_{t\geq 0}$ be a given filtration. A stochastic process $(X(t), F_t)_{t\geq 0}$ is called **predictable** *(with respect to the filtration F) if the mapping*

$$X : [0, \infty) \times \Omega \longrightarrow \mathbb{R}^d, \quad (t, \omega) \longmapsto X(t, \omega) \tag{6.7}$$

is measurable with respect to the smallest σ-algebra generated by the left-continuous processes $(Y(t), F_t)_{t\geq 0}$.

Predictability thus supports the idea that the value of the integrand should be fixed at time t **before** one knows if there is a jump at time t. This is in particular important for the interpretation of the integrand as a trading strategy in finance. In the case of the Poisson process we obtain the required martingale as an integrator by subtracting its expectation process. We thus introduce the so-called *compensated Poisson process* $\tilde{N}(t)$ via

$$\tilde{N}(t) := N(t) - \lambda t\ . \tag{6.8}$$

Then, one can show that if the integrand X is predictable (with respect to a filtration F that contains the one generated by $N(.)$) and in $L^1([0, T])$ with respect to the Lebesgue measure, then the stochastic integral with respect to the compensated Poisson process

$$\int_0^t X(s)\, d\tilde{N}(s) := \int_0^t X(s)\, dN(s) - \int_0^t X(s)\, \lambda ds \tag{6.9}$$

is well defined and is a martingale on $[0, T]$ (with respect to F).

Below, we will introduce a more general setting that will prove to be convenient in the Lévy process framework and allow us to introduce some notation.

A short excursion: Marked point processes

The idea of a **marked point process** is simply to identify a jump process with the sequence of pairs (t_i, Y_i) where the random times t_i are the jump times and the random variables Y_i characterize the jump heights at time t_i of the process. This allows an easy representation of a compound Poisson process, but also allows us to define a multivariate jump process

$$N(t) := \left(N^{(1)}(t), ..., N^{(m)}(t)\right).$$

For this we use the interpretation that – as before – the t_i-sequence determines the jump times of the process, but the Y_i-sequence identifies **which** of the m components of the process actually jumps at time t_i. More precisely, $Y_i = k$ means that the k-th component $N^{(k)}$ of the jump process increases by 1 at time t_i while the others all remain constant. This can be formulated as

$$N^{(k)}(t) = \sum_{i \geq 1} 1_{t_i \leq t} 1_{Y_i = k} \,. \tag{6.10}$$

Note that the jump process does not necessarily need to be a Poisson process. Both the above interpretations are special cases of the following definition.

DEFINITION 6.6

Let $(E, \mathcal{E})$ be a measurable space with $E \subseteq \mathbb{R}$. Let (t_i, Y_i) be a sequence of pairs of

- *nonnegative random variables $0 < t_1 < t_2 < ...$ and*
- *random variables Y_i taking values in E.*

Then this sequence $(t_n, Y_n)_{n \in \mathbb{N}}$ is called an **E-marked point process**.

Using this newly introduced terminology of an E-marked point process will be the key to introduce stochastic integrals over point processes. Note that for each set $A \subseteq E$ with $A \in \mathcal{E}$ one can construct a new associated marked point process via

$$N(t, A) = \sum_{i \geq 1} 1_{t_i \leq t} 1_{Y_i \in A} \tag{6.11}$$

out of the original marked point process $(t_n, Y_n)_{n\in\mathbb{N}}$. By identifying N as a two parameter family, we can define the filtration

$$F_t^N := \sigma\left\{N(s,A)\,|0 \le s \le t, A \in \mathcal{E}\right\}, \tag{6.12}$$

and an associated **random measure**

$$p\left((0,t], A\right) := N(t,A), 0 \le t, A \in \mathcal{E}. \tag{6.13}$$

This random measure simply counts the "jumps" (with the possible interpretation as choices of indices in the multivariate case) of sizes in A until time t. By assuming that the stopping times t_i do not accumulate before any finite time T, we can now introduce the integral notation

$$\int_0^t \int_E H(s,y)\,p(ds,dy) = \sum_{i\ge 1} H(t_i,Y_i)\,1_{t_i\le t} = \sum_{i=1}^{N(t,E)} H(t_i,Y_i) \tag{6.14}$$

for a given predictable process $H(t,x)$ that can also depend on the value of the jump height Y_i (the "mark") at the jump time.

In our setting of a compound Poisson process the above random measure $p((0,t],A)$ simply counts all jumps of the Poisson process $N(t)$ on $(0,t]$ with jump height values in A. This is modelled by assuming independence between the jump height distribution given by the probability measure $m(dy)$ and the Poisson process with a constant jump intensity λ. The corresponding random measure $p((0,t],A)$ is a **Poisson random measure**, i.e. $p((0,t],A)$ is a Poisson-distributed random variable for all $t \ge 0$, $A \in E$. We then define the **compensated Poisson random measure**

$$q\left((0,t], A\right) := \tilde{N}_t(A) := p\left((0,t], A\right) - \lambda \cdot t \cdot m(A), 0 \le t,\ A \in \mathcal{R}. \tag{6.15}$$

Given then that the integrand $H(t,x)$ is integrable with respect to the compensated Poisson measure and that the jump height distribution has a support of E and an expected value of $\mathbb{E}(Y)$, we have that

$$\int_0^t \int_E H(s,y)\,q(ds,dy) = \sum_{i\ge 1} H(t_i,Y_i)\,1_{t_i\le t} - \lambda t \mathbb{E}(Y) \tag{6.16}$$

is indeed a martingale. As the integral with respect to a compound Poisson process is still essentially only a sum, its simulation is straightforward as described in Algorithm 6.2.

6.3 Jump-diffusions: Basics, properties, and simulation

Roughly speaking, a jump-diffusion is the generalization of the sum of a Poisson process and a Brownian motion, or more precisely the combination

---

Algorithm 6.2 Simulation of a stochastic integral with respect to a compound Poisson process

---

Let $Y(t)$ be a predictable process, $X(t)$ a compound Poisson process with jump height distribution L and intensity λ.

1. Set $I(0) := 0$.
2. Simulate a compound Poisson process X_t with jump times $0 < t_1 < ... < t_{N(T)} \leq T$ and jump heights $Z_1, ..., Z_{N(T)}$.
3. Set $I(t_i) := I(t_{i-1}) + Y(t_i) \cdot Z_i$, $i = 1, ..., N(T)$.
4. Set $I(t) := I(t_{i-1})$ for $t \in [t_{i-1}, t_i)$, $i = 1, ..., N(T+1)$, $t_{N(T+1)} := T$.

---

of a stochastic integral with respect to a (compound) Poisson process (or a Poisson random measure) and a diffusion process. While the sum of a Poisson process and a Brownian motion is a Lévy process, the second one is in general not. Thus, jump-diffusions form no subclass of Lévy processes and are of interest on their own.

Let us consider a probability space $(\Omega, F, \mathbb{P})$ on which both a (d-dimensional) Brownian motion and a (compound) Poisson process are defined.

DEFINITION 6.7
Let $\{(X(t), F_t)\}_{t\in[0,T]}$ be a stochastic process that can be represented as

$$X(t) = X(0) + \int_0^t f(s)\,ds + \int_0^t g(s)\,dW(s) + \int_0^t \int_E h(s,y)\,p(ds,dy) \quad (6.17)$$

with $W(t)$ a one-dimensional Brownian motion independent of the Poisson random measure $p(.,.)$ that corresponds to an underlying (compound) Poisson process $N(t)$. E contains the support of the jump height distribution. The integrands $f(s)$, $g(s)$ are assumed to be progressively measurable, $h(s,y)$ to be a predictable process that all satisfy integrability conditions such that all integrals are defined. Then, we call $X(t)$ a **jump-diffusion process**.

Note that the jump integral in the definition above can be written as a sum

$$\int_0^t \int_E h(s,y)\,p(ds,dy) = \sum_{i=1}^{N(t)} h(t_i, Y_i) \quad (6.18)$$

with the usual interpretation of t_i as the jump times and Y_i as the jump heights of the compound Poisson process N_t, and thus also of the jump-diffusion X_t. As in the diffusion setting, we use the differential notation for the integral representation of the jump-diffusion:

$$dX(t) = f(t)\,dt + g(t)\,dW(t) + \int_E h(t,y)\,p(dt,dy)\,. \tag{6.19}$$

Examples 6.8 Jump-diffusions

1. The simplest, nontrivial example of a jump-diffusion is given by

$$X(t) = W(t) + N(t) \tag{6.20}$$

with $N(t)$ a Poisson process with intensity λ, $W(t)$ a Brownian motion, both being independent. Conditioning on the number of jumps $N(t)$ at time t yields the distribution of $X(t)$ as a Poisson mixture of normal distributions,

$$\mathbb{P}(X(t) \leq x) = \sum_{k=0}^{\infty} e^{-\lambda t}\frac{(\lambda t)^k}{k!}\Phi\left(\frac{x-k}{\sqrt{t}}\right). \tag{6.21}$$

2. A simple jump-diffusion which has no stationary increments (and is thus no Lévy process (see the exact definition in the next section)) is given by

$$X_t = x + \int_0^t s dW_s + N_t. \tag{6.22}$$

Before we turn to the task of simulation we state the Itô formula for jump-diffusion processes as the fundamental tool for working with them.

THEOREM 6.9 Itô formula for jump-diffusions

Let $X(t)$ be a jump-diffusion process that admits a representation given by (6.17), let $F : [0,\infty) \times \mathbb{R} \to \mathbb{R}$ be a $C^{1,2}$-function. Then, we have

$$\begin{aligned} F(t,X(t)) &= F(0,X(0)) + \int_0^t F_x(s,X(s))\,g(s)\,dW(s) \\ &+ \int_0^t \left(F_t(s,X(s)) + F_x(s,X(s))\,f(s) + \frac{1}{2}F_{xx}(s,X(s))\,g^2(s)\right) ds \\ &+ \sum_{i=1}^{N(t)} \left(F(t_i, X(t_i-)(1+h(t_i,Y_i))) - F(t_i,X(t_i-))\right). \end{aligned} \tag{6.23}$$

In the above version we apply the Itô formula for Itô processes between the jumps of $X(t)$ and add the correcting differences at the jump times.

6.3.1 Simulating Gauss-Poisson jump-diffusions

In this section we will concentrate on the simulation of a simple class of jump-diffusions, the Gauss-Poisson jump-diffusion processes, which are a sum of a Brownian motion with drift and a compound Poisson process,

$$X(t) = x + \mu t + \sigma W(t) + \sum_{i=1}^{N(t)} Y_i \tag{6.24}$$

with $W(t)$ a Brownian motion, $N(t)$ a Poisson process with parameter $\lambda > 0$, and all Y_i being independent, identically distributed (i.i.d.) real-valued random variables. The advantage of these processes is that we can exactly simulate their increments. This directly leads to a first simple simulation procedure in Algorithm 6.3.

Algorithm 6.3 Simulation of a Gauss-Poisson jump-diffusion process with a fixed time discretization

---

Let $0 = t_0 < t_1 < ... < t_n = T$ be a fixed time discretization of $[0, T]$, let L be a given jump height distribution, and $\mu, \sigma \in \mathbb{R}$.

1. Set $X(0) := x$.
2. For $i = 1$ to n do:
   - Generate a random number $P \sim Pn(\lambda \cdot (t_i - t_{i-1}))$.
   - If $P > 0$ then for $j = 1$ to P do:
     - Simulate P random numbers $Z_j \sim L$.
     - Set $X(t_i) := X(t_{i-1}) + \sum_{j=1}^{P} Z_j$.
   - Else: $X(t_i) := X(t_{i-1})$.
   - Generate a random number $Z \sim \mathcal{N}(0, 1)$.
   - Set $X(t_i) := X(t_i) + \mu \cdot (t_i - t_{i-1}) + \sigma \cdot \sqrt{t_i - t_{i-1}} Z$.

REMARK 6.10 1. Note that we do not perform a linear interpolation between the time grid points in Algorithm 6.3. This would only make sense if no jump had happened between two grid points. In all other cases, we could neither identify the continuous pieces of the path nor the exact location of the jumps. Thus, a linear interpolation would be misleading.

2. For some applications in finance and insurance it is indeed important to simulate the jumps exactly at the jump times. To achieve this, one first simulates the jump times, then the jump heights, and finally the Brownian motion part on a time grid that is adapted to the jump times. As the times between two jumps can vary quite a lot, we suggest using a given grid for simulation of the Brownian motion and then simply add the jump times of

the Poisson process as grid points as soon as they are known.

3. As in the case of a geometric Brownian motion, we can use the above algorithm also to simulate a process that can be represented as a function $f(X(t))$ of a Gauss-Poisson process $X(t)$ by simulating $X(t)$ first and then evaluating the function at those simulated values.

4. One can generalize the above algorithm to simulate **multidimensional** Gauss-Poisson jump-diffusions. For this, consider a d-dimensional Brownian motion $W(t)$ and k independent compound Poisson processes with different jump height distributions. One can then simulate Gauss-Poisson jump-diffusions of the form

$$X_i(t) = x_i + \mu_i t + \sum_{j=1}^{d} \sigma_{ij} W_j(t) + \sum_{j=1}^{k} \sum_{m=1}^{N_j(t)} Y_m^j, \; i = 1, ..., n \tag{6.25}$$

on a fixed or on a suitably adapted time grid.

5. Note that as we can simulate the exact distribution of (the increments of) the Gauss-Poisson jump-diffusions, there is no additional discretization error when it comes to estimate an expectation of a functional that only depends on the values of the process at finitely many time points,

$$\mathbb{E}\left(g\left(X\left(t_1\right), ..., X\left(t_k\right)\right)\right) \approx \frac{1}{N} \sum_{i=1}^{N} g\left(X^{(i)}\left(t_1\right), ..., X^{(i)}\left(t_k\right)\right). \tag{6.26}$$

Here, the upper index (i) simply indicates the relevant values of the X-process in its i-th simulated path. □

6.3.2 Euler-Maruyama scheme for jump-diffusions

As in the case of Itô processes, in general we do not know the exact distribution of jump-diffusion processes. Therefore, when simulating a path we have to rely on discretization methods. As before, the simplest one is the Euler-Maruyama method. We consider the one-dimensional jump-diffusion $X(t)$ given by the differential representation

$$dX(t) = a(X(t))\,dt + b(X(t))\,dW(s) + c(X(t-))\,dJ(t) \tag{6.27}$$

with $J(t)$ a compound Poisson process with intensity λ and jump height distribution L. Here, we assume that this SDE with jumps admits a unique solution. Examples for this will be seen in the next chapter among the applications in finance.

The Euler-Maruyama described in Algorithm 6.4 has a weak convergence order of 1, a strong convergence order of 1/2 (see Bruti-Liberati and Platen [2007]). A multidimensional extension is straightforward in the sense of Remark 4.66. As in the Gauss-Poisson jump-diffusion case, it is often useful

---

Algorithm 6.4 Euler-Maruyama scheme for jump-diffusions

Let $\Delta t := T/N$ for a given N. Simulate an approximate path $Y_N(t)$ of the jump-diffusion process $X(t)$ given by the SDE with jumps (6.27) via:

1. Set $Y_N(0) = X(0) = x$.
2. For $j = 0$ to $N-1$ do
   (a) Simulate a standard normally distributed random number Z_j.
   (b) Simulate a random variable $\Xi_j \sim Pn(\lambda\Delta t)$.
   (c) Simulate a random variable $\Lambda \sim L$.
   (d) Set $\Delta W(j\Delta t) = \sqrt{\Delta t} Z_j$ and

$$\begin{aligned} Y_N((j+1)\Delta t) &= Y_N(j\Delta t) + a(Y_N(j\Delta t))\Delta t \\ &\quad + \sigma(Y_N(j\Delta t))\Delta W(j\Delta t) + c(Y_N(j\Delta t))\Lambda_j\Xi_j. \end{aligned}$$

---

to have an adapted time grid. This can be obtained in a similar way here. We first have to simulate the jump times τ_i and the corresponding jump heights Λ_i, $i = 1, ..., N(T)$. Then, the jump times are included in the time discretization of $[0, T]$. One can then use the following two step procedure in the **adapted Euler-Maruyama scheme** where the time grid t_i, $i = 1, ..., N$ already includes the jump times:

$$Y_N(t_{j+1}-) = Y_N(t_j) + a(Y_N(t_j))(t_{j+1} - t_j) + \sigma(Y_N(t_j))\Delta W(t_j), \quad (6.28)$$

$$Y_N(t_{j+1}) = Y_N(t_{j+1}-) + c(Y_N(t_{j+1}-))\Delta J(t_{j+1}) \quad (6.29)$$

where $\Delta J(t_{j+1}) = J(t_{j+1}) - J(t_{j+1}-)$ equals 0 at times t_{j+1} where no jump happens. This algorithm might have advantages when the jump intensity λ is small. If, however, λ is large then there might not be big differences between the simple, regular Euler scheme and the jump time-adapted one.

More one-dimensional schemes (such as predictor-corrector schemes in the weak and strong sense) are presented and compared in Bruti-Liberati and Platen (2007).

6.4 Lévy processes: Properties and examples

6.4.1 Definition and properties of Lévy processes

A Lévy process is the natural generalization of both the Brownian motion and the Poisson process.

DEFINITION 6.11
A stochastic process $\{X(t), F_t\}_{t\geq 0}$ *with* $X(0) = 0$ *and independent and stationary increments* $X(t) - X(s)$ *for* $t > s$ *is called a* **Lévy process**.

One may always assume that a Lévy process has right-continuous paths with existing left-hand limits (see Applebaum [2004]). For reasons already explained in the case of a Poisson process, we will therefore in the following always do this. Obviously, (multidimenisonal) Brownian motion and compound Poisson processes are Lévy processes. Further, linear combinations of independent Lévy processes are again Lévy processes.

The main technical problems when dealing with Lévy processes are caused by the small jumps of the process. For instance, there do exist Lévy processes $X(t)$ that only change their value by jumps, but for which we have

$$X(t) \neq X(0) + \sum_{s\leq t} \Delta X(s) \tag{6.30}$$

with the jump process of $X(t)$ being defined by

$$\Delta X(t) = X(t) - X(t-). \tag{6.31}$$

The reason for this behaviour is that the number of jumps is infinite and the sum does not converge.

We recall the notion of a Poisson random measure from the Poisson process section. For $A \in \mathcal{B}(\mathbb{R}^d - \{0\})$ and $t \geq 0$ we introduce

$$N(t, A) := \#\{0 \leq s \leq t; \Delta X(s) \in A\} \tag{6.32}$$

which counts the numbers of jumps of size in A that occur up to time t. One can show (see Applebaum [2004]) that $N(t, A)$ is a Poisson process if A is bounded below (i.e. its closure does not contain 0). We further introduce the following.

DEFINITION 6.12
Let X *be a Lévy process with corresponding counting measure* $N(t, A)$ *as defined in* (6.32). *Then, the* **Lévy measure** ν *of* X *is a measure on* $(\mathbb{R}^d, \mathcal{B}(\mathbb{R}^d))$ *defined by* $\nu(\{0\}) = 0$ *and by*

$$\nu(A) := \mathbb{E}(N(1, A)), \ A \in \mathcal{B}(\mathbb{R}^d - \{0\}). \tag{6.33}$$

Note that by the properties of a Poisson random measure (in particular that the Poisson distribution has a finite second moment) we have $\nu(A) = \mathbb{E}(N(t, A)) < \infty$ for all A which are bounded below. Further, for a fixed A, the process $N(t, A)$ is a compound Poisson process with intensity λ and jump height distribution L given by

$$\lambda = \nu(A), \quad L(dx) = \frac{\nu(dx)}{\nu(A)}. \tag{6.34}$$

It can then be shown that we also have

$$\int_{\mathbb{R}^d} \min\left\{|x|^2, 1\right\} \nu(dx) < \infty \tag{6.35}$$

which is the usual defining equation for a Lévy measure in the literature. From this it follows that we have

$$\nu\left(\mathbb{R}^d - \{x : |x| < 1\}\right) < \infty, \tag{6.36}$$

but that we might have $\nu(\{x : |x| \leq 1\}) = \infty$. This leads to the following definition.

DEFINITION 6.13
Let $X(t)$ be a Lévy process with Lévy measure ν.

(a) We say that $X(t)$ has **finite activity** *if we have*

$$\nu\left(\mathbb{R}^d\right) < \infty. \tag{6.37}$$

Then, each path of $X(t)$ has finitely many jumps on an interval $[0, T]$.

(b) We say that $X(t)$ has **infinite activity** *if we have*

$$\nu\left(\mathbb{R}^d\right) = \infty. \tag{6.38}$$

Then, for $T > 0$ each path of $X(t)$ has infinitely many jumps on $[0, T]$.

The definition highlights that the small jumps can cause the main problems when dealing with a Lévy process. The choice of norm 1 for classifying a jump as small is arbitrary. One could choose any positive number ϵ for this.

The fundamental result on the form of a path of a Lévy process is the Lévy-Itô decomposition. It is the essential tool to understand the simulation of Lévy processes, but also highlights their technical difficulties.

THEOREM 6.14 Lévy-Itô decomposition
Each d-dimensional Lévy process $X(t)$ with a Lévy measure ν admits a decomposition of the form

$$X(t) = \gamma t + \sigma W(t) + \int_{|x| \geq 1} x N(t, dx) + \int_{|x| < 1} x\left(N(t, dx) - t\nu(dx)\right), \tag{6.39}$$

with $\gamma \in \mathbb{R}^d$, $W(t)$ a d-dimensional Brownian motion, $\sigma \in \mathbb{R}^{d,d}$, N a Poisson random measure on $[0, \infty) \times (\mathbb{R}^d - \{0\})$ determined by the Lévy measure ν as in Equation (6.34). Further, N and W are independent.

This decomposition directly shows that a Lévy process is a sum of

- a linear deterministic component γt,
- a Brownian motion with covariance matrix $\sigma\sigma'$,
- a jump process with jumps of absolute value bigger or equal to 1, and
- a compensated jump process with respect to the compensated Poisson random measure $\tilde{N}(t, A) = N(t, A) - t\nu(A)$ with jumps of absolute value smaller than 1.

Note that by the properties of a Poisson jump measure, the two jump processes are also independent. Further, the fourth part has to be considered separately as we might have $\nu(\{x : |x| \leq 1\}) = \infty$. Then, the sum of the small jump would not converge. However, one can show that the compensated sum does.

Thus, when simulating a path of X, one can independently simulate the Brownian motion and the two jump processes and then finally add them. While the simulation of the Brownian motion causes no problem, our considerations on the Lévy measure so far already indicate that simulating the small jumps might become delicate.

We give a definition and some further path properties of a Lévy process (see Cont and Tankov [2003]):

DEFINITION 6.15
Let X be a Lévy process that admits a decompositon (6.39). *Then, the triplet* $(\gamma, \sigma\sigma', \nu)$ *is called the* **Lévy triplet** *of* X.

PROPOSITION 6.16
Let $X(t)$ be a one-dimensional Lévy process with Lévy triple (γ, σ^2, ν).

(a) $X(t)$ has paths of finite variation if and only if we have

$$(\gamma, \sigma^2, \nu) = (\gamma, 0, \nu) \text{ with } \int_{|x|\leq 1} |x|\, \nu(dx) < \infty. \tag{6.40}$$

(b) In the case of (a) with $b = \gamma - \int_{|x|\leq 1} x\, \nu(dx)$ *we have the representation*

$$X(t) = bt + \sum_{s\in[0,t]\,:\,\Delta X(s)\neq 0} \Delta X(s). \tag{6.41}$$

(c) We further have:

$$\mathbb{E}(|X(t)|^n) < \infty \Longleftrightarrow \int_{|x|\geq 1} |x|^n\, \nu(dx) < \infty\ . \tag{6.42}$$

In this particular case we have the explicit representations:

$$\mathbb{E}(X(t)) = \gamma t + t \int_{|x|\geq 1} x\nu(dx),\ \mathbb{V}ar(X(t)) = \sigma^2 t + t \int_{\mathbb{R}} x^2 \nu(dx). \tag{6.43}$$

For later use we also state a suitable version of Itô's formula for Lévy processes (see Applebaum [2004]):

THEOREM 6.17 Itô's formula for Lévy processes
Let $X(t)$ be a one-dimensional Lévy process with representation (6.39). *Let further $f : \mathbb{R} \to \mathbb{R}$ be a C^2-function. Then we have:*

$$f(X(t)) = f(x) + \int_0^t f'(X(s))\, dX(s) + \frac{1}{2}\int_0^t f''(X(s))\, \sigma^2 ds$$
$$+ \sum_{0 \le s \le t} \left[f(X(s)) - f(X(s-)) - f'(X(s-))\, \Delta X(s) \right]. \quad (6.44)$$

Note that the $dX(s)$-integral are indeed four integrals due to Equation (6.39).

6.4.2 Examples of Lévy processes

Of course, there are examples of Lévy processes that are different from Brownian motion and Poisson processes. Some of them will be given below; additional ones are presented in the next chapter as applications in finance.

Example 6.18 Lévy processes from infinitely divisible distributions

This is no particular example, but a general principle how to construct a Lévy process. For this, we need the notion of an infinitely divisible distribution.

DEFINITION 6.19
A distribution L is called **infinitely divisible** *if there exists an $\mathbb{R}^d$-valued random variable X with distribution L such that for every $n \in \mathbb{N}$ there exist i.i.d. random variables $Y_1^{(n)}, ..., Y_n^{(n)}$ with*

$$X \stackrel{D}{=} Y_1^{(n)} + ... + Y_n^{(n)} . \quad (6.45)$$

By the definition above, one only has to check if for all $n \in \mathbb{N}$ the n-th root of the characteristic function $\Phi_\mu(.)$ of the distribution L is again a characteristic function of a distribution. Popular examples of infinitely divisible distributions where this check is straightforward are the multidimensional Gaussian distribution (where the $Y_i^{(n)}$ are again Gaussian with mean and variance scaled by $1/n$) and the Poisson distribution with parameter λ (where the $Y_i^{(n)}$ are again Poisson-distributed but now with parameter λ/n).

If the result above can be extended from the exponent $1/n$ to a general power $t > 0$, then we have a construction procedure for a Lévy process:

1. Let $X(1)$ have an infinitely divisible distribution with characteristic function $\Phi(u)$.

2. Obtain the distribution of $X(t)$ for general t by choosing the one corresponding to the characteristic function $\Phi(u)^t$.

Hence, the property of independent and stationary increments follows by construction. The existence of a Lévy process with the distributional properties above can also be proved (see e.g. Applebaum [2004], p. 62). Of course, the two straightforward examples resulting from the above procedure are the Brownian motion and the Poisson process.

Indeed, this way is the generic one to construct a Lévy process. This can be seen from the famous Lévy-Khinchine formula (see Applebaum [2004]) where we use $\langle x, y\rangle = \sum_{i=1}^{d} x_i y_i$ to denote the scalar product:

THEOREM 6.20 Lévy-Khinchine formula and Lévy processes
(a) The characteristic function $\Phi_L(u)$ of an infinitely divisible distribution L on $\mathbb{R}^d$ is given by

$$\Phi_L(u) = \exp\Bigg(i\langle\gamma, u\rangle - \frac{1}{2}\langle u, Au\rangle + \int\limits_{\mathbb{R}^d-\{0\}} \left(\exp\left(i\langle u, y\rangle\right) - 1 - i\langle u, y\rangle 1_{\{|x|<1\}}\right)\nu(dy)\Bigg), \quad u \in \mathbb{R}^d \qquad (6.46)$$

for some $\gamma \in \mathbb{R}^d$, a positive definite symmetric $A \in \mathbb{R}^{d,d}$, and a Lévy measure ν (i.e. a measure on $\mathbb{R}^d$ that satisfying condition (6.35)). Further, each such mapping $\Phi_L(u)$ is the characteristic function of an infinitely divisible distribution.

(b) The characteristic function of a Lévy process $X(t)$ with triplet $(\gamma, \sigma\sigma', \nu)$

$$\Phi(u;t) := \mathbb{E}\left(exp\left(i\langle u, X(t)\rangle\right)\right) \qquad (6.47)$$

has the form

$$\Phi(u;t) = \left(\Phi_L(u)\right)^t, \quad t \geq 0 \qquad (6.48)$$

with $\Phi_L(u)$ as given in representation (6.46) for $A = \sigma\sigma'$.

Example 6.21 The gamma process
A Lévy process with $X(t) \sim Gamma\left(\mu^2 t/\upsilon, \upsilon/\mu\right)$ with $\mu > 0$, $\upsilon > 0$ is called a **gamma process**. We here use the somewhat strange choice of the scale and the shape parameter a and θ (see Section 2.5.4 for the introduction of the gamma distribution) as this is the one used in the literature on the variance gamma model popular in finance (see Section 7.3.3). Its distribution at time t is given by the density function

$$f_{\mu,\upsilon;t}(x) = \frac{(\mu/\upsilon)^{\mu^2 t/\upsilon}}{\Gamma\left(\mu^2 t/\upsilon\right)} x^{\mu^2 t/\upsilon - 1} e^{-x\mu/\upsilon}, \; x > 0 \qquad (6.49)$$

while its characteristic function has the form

$$\Phi_{\mu,\upsilon}(u;t) = (1 - iu\upsilon/\mu)^{-\mu^2 t/\upsilon}. \tag{6.50}$$

Thus, the Lévy triplet equals

$$(\gamma, \sigma^2, \nu) = \left(\mu\left(1 - e^{-\mu/\upsilon}\right), 0, \tfrac{\mu^2}{\upsilon} x^{-1} e^{-\mu x/\upsilon} 1_{\{x>0\}} dx\right). \tag{6.51}$$

As the gamma distribution is concentrated on the positive half-line, $X(t)$ is an increasing process (its increments are gamma-distributed!). It thus has paths of finite variation. Further, we have explicit expressions for the mean, variance, skewness, and kurtosis of $X(t)$:

$$\mathbb{E}(X(t)) = \mu t \quad , \quad \mathbb{V}ar(X(t)) = \upsilon t\,, \tag{6.52}$$

$$Skew(X(t)) = \frac{2}{\sqrt{\mu^2 t/\upsilon}} \quad , \quad Kurt(X(t)) = 3\left(1 + \frac{2}{\mu^2 t/\upsilon}\right). \tag{6.53}$$

As we have $\nu(\mathbb{R}) = \infty$, the gamma process has infinite activity and thus does not have piecewise constant paths. A simulated path of a gamma process is given in Figure 6.2.

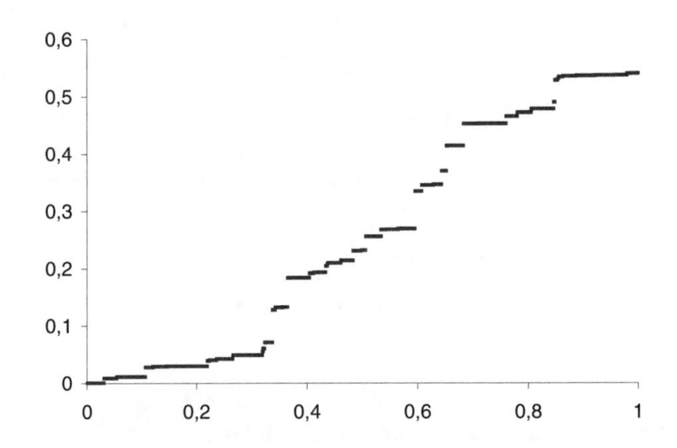

FIGURE 6.2: A simulated path of a gamma process.

Example 6.22 The inverse Gaussian process
A Lévy process is called an inverse Gaussian (IG) process if we have $X(t) \sim IG(\delta t, \gamma)$ with $\delta > 0$, $\gamma > 0$ and where $IG(\delta t, \gamma)$ is the inverse Gaussian distribution with parameters $\delta t, \gamma$. It is given by its probability density

$$f_{IG(\delta t,\gamma)}(x) = \frac{\delta t}{\sqrt{2\pi}} x^{-\frac{3}{2}} e^{\delta t \gamma} e^{-\frac{1}{2}\left((\delta t)^2 x^{-1} + \gamma^2 x\right)}, \; x > 0\,. \tag{6.54}$$

The name of the process stems from the fact that the inverse Gaussian distribution $IG(\delta, \gamma)$ is the distribution of the first passage time of a Brownian motion with drift γ through the level δ.

Again, the process is increasing, as the density of its increments is only concentrated on the positive half-line. We have explicit forms for the characteristic function of $X(1)$ and the Lévy triple given by

$$\Phi_{IG(\delta,\gamma)}(u;1) = \exp\left(-\delta\sqrt{2iu+\gamma^2}-\gamma\right), \tag{6.55}$$

$$\left(\gamma, \sigma^2, \nu\right) = \left(\frac{\delta}{\gamma}\left(2N(\gamma)-1\right), 0, \frac{1}{\sqrt{2\pi}}\delta x^{-\frac{3}{2}}e^{-\frac{1}{2}\gamma^2 x}1_{\{x>0\}}dx\right). \tag{6.56}$$

The mean, variance, skewness, and kurtosis of $X(t)$ read as:

$$\mathbb{E}(X(t)) = \frac{\delta t}{\gamma}, \quad \mathbb{V}ar(X(t)) = \frac{\delta t}{\gamma^3}, \tag{6.57}$$

$$Skew(X(t)) = \frac{3}{\sqrt{\delta t \gamma}}, \quad Kurt(X(t)) = 3\left(1+\frac{5}{\delta t\gamma}\right). \tag{6.58}$$

As its Lévy measure of $\mathbb{R}$ is infinite, the IG process does not have piecewise constant paths and the expected number of jumps in each nonempty time interval is infinite. A simulated path of an IG process is given in Figure 6.3.

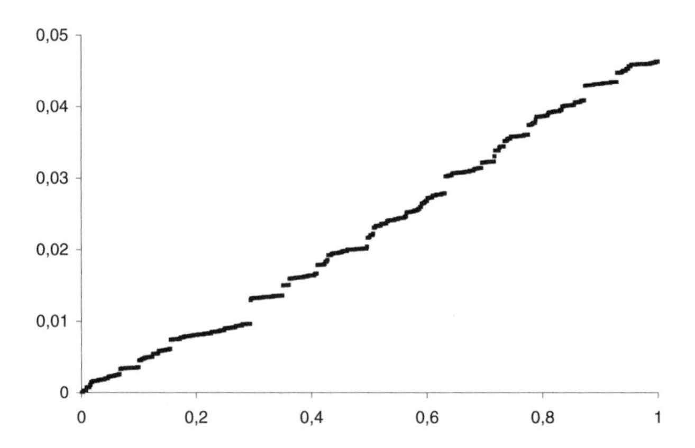

FIGURE 6.3: A simulated path of an IG process.

If we compare this picture with the simulated path of a gamma process in the Figure 6.2, then one observes more small jumps in the IG process path. This can be explained by the fact that we have

$$\int_{\{x:|x|\leq 1\}} |x|\,\nu_{gamma}(dx) < \infty = \int_{\{x:|x|\leq 1\}} |x|\,\nu_{IG}(dx) \tag{6.59}$$

with ν_{IG} and ν_{gamma} being the Lévy measures of the IG and the gamma process, respectively.

A generalization of both the gamma and the IG process is given in the following definition and will play an important role as a stochastic clock in the financial applications of the next chapter.

DEFINITION 6.23
A Lévy process $X(t)$ with increasing paths is called a **subordinator**.

Another class of ditributions that are used to construct Lévy processes are the stable distribution.

DEFINITION 6.24
A distribution L on $\mathbb{R}$ is called **stable** *if for arbitrary $n \in \mathbb{N}$, the sum of n independent random variables X_i all with distribution L satisfy*

$$X_1 + ... + X_n \sim c_n X + d_n \tag{6.60}$$

for some $X \sim L$, $c_n, d_n \in \mathbb{R}$. It is called **strictly stable** *if we have $d_n = 0$ for all $n \in \mathbb{N}$.*

One can even show that c_n must be of the form

$$c_n = \sigma n^{1/\alpha} \text{ for } \sigma > 0, 0 < \alpha \leq 2. \tag{6.61}$$

A further characterization of stable distributions is given in (see Sato [1999]).

THEOREM 6.25
A real-valued random variable X has a stable distribution if and only if there exist $\sigma > 0$, $\mu \in \mathbb{R}$, $-1 \leq \beta \leq 1$ such that for all $u \in \mathbb{R}$ the characteristic function of X is of one of the following forms:

$$\Phi_X(u) = \exp\left(i\mu u - \tfrac{1}{2}\sigma^2 u^2\right), \qquad \alpha = 2, \tag{6.62}$$

$$\Phi_X(u) = \exp\left(i\mu u - \sigma^\alpha |u|^\alpha \left(1 - i\beta sgn(u) \tan\left(\tfrac{\pi\alpha}{2}\right)\right)\right), \alpha \in (1,2), \tag{6.63}$$

$$\Phi_X(u) = \exp\left(i\mu u - \sigma |u| \left(1 + i\beta \tfrac{2}{\pi} sgn(u) \ln(|u|)\right)\right), \qquad \alpha = 1. \tag{6.64}$$

Popular examples are the normal distribution $\mathcal{N}(\mu, \sigma^2)$ for $\alpha = 2$ and the Cauchy distribution for $\alpha = 1, \beta = 0$, and density $f(x) = \sigma/(\pi((x-\mu^2)+\sigma^2))$.
One can then construct a **stable Lévy process** $X(t)$ with distributions given by the characteristic functions

$$\Phi(u;t) = \Phi_X(u)^t \tag{6.65}$$

where $\Phi_X(u)$ is of one of the three types given in Theorem 6.25.

6.5 Simulation of Lévy processes

As already indicated, the main problems of simulating Lévy processes are caused by the presence of the jump part. To be more precise, they are caused by the small jumps in the case of an infinite Lévy measure. As a result, we then have an infinite number of jumps in each interval, and exact simulation of the jump part as in the jump-diffusion case is no longer possible. We will present four different approaches to cope with this situation:

1. A time discretization method where the Lévy process is simulated exactly on a given time grid.

2. An Euler-Maruyama method for Lévy processes.

3. A series representation approach where we use the fact that a Lévy process can be represented as an infinite series of simpler random variables.

4. An approach where the small jumps of the Lévy process are approximated by a simpler process.

6.5.1 Exact simulation and time discretization

As always, the basic case is given when the exact distribution $L\left(\frac{1}{n}\right)$ of the increments $\Delta_i X := X(t_i) - X(t_{i-1})$ is known and can be simulated for a given equidistant partition $t_i := iT/n$, $i = 0, 1, ..., n$, of $[0, T]$. We can simulate the process exactly at the times t_i and leave it constant in between them. This directly leads to Algorithm 6.5, which is rather simple.

Algorithm 6.5 Exact simulation of discretized Lévy processes

1. Set $X(0) := 0$.

2. For $i = 1$ to n

   - Simulate a random number $\Delta_i X$ with distribution $L\left(\frac{1}{n}\right)$ independent on all previous such random numbers.
   - Set $X(t_i) := X(t_{i-1}) + \Delta_i X$.
   - Set $X(t) := X(t_{i-1})\ \forall t \in (t_{i-1}, t_i)$.

However, two aspects should be taken into account for Algorithm 6.5:

- The time discretization smoothes out the paths of the process. This might be particularly problematic for the pricing of path-dependent options such as barrier options.
- As the distribution of the increments sometimes has very complicated forms (involving special functions such as Bessel functions), it might be more efficient to use a different simulation method, although the discretized exact simulation is conceptionally very easy.

Example 6.26
A tractable example different from a compound Poisson process is the simulation of a gamma process. The increments $\Delta_i X$ can be simulated according to an acceptance-rejection method as given in Chapter 2. See Figure 6.2 for a gamma path simulated by the exact method.

Example 6.27
An example where the exact simulation is inefficient is the normal inverse Gaussian process that we will consider in more detail in Section 7.3.4. There, we will use the representation of such a process as a subordinated Brownian motion which allows an easy way of simulating it.

6.5.2 The Euler-Maruyama scheme for Lévy processes

The Euler-Maruyama scheme for Lévy processes can be formulated in its most convenient way for stochastic differential equations (SDEs) which are driven by a (d-dimensional) Lévy process $Z(t)$ instead of a Brownian motion in the usual SDE setting. More precisely, we consider the (Lévy) SDE

$$dX(t) = \mu(X(t))\,dt + \sigma(X(t))\,dZ(t) \tag{6.66}$$

with $Z(t)$ a Lévy process. More details such as conditions for the unique solvability of this equation are given in Protter (2004). To formulate the Euler-Maruyama scheme of Algorithm 6.6, it is necessary that we are able to simulate an increment of the driving Lévy process $Z(t)$ for a given time grid,

$$\Delta Z_i = Z(t_{i+1}) - Z(t_i)\,. \tag{6.67}$$

Note that although the form of the SDE (6.66) looks simple, it is much more complicated than that of a Brownian motion-driven SDE, as $Z(t)$ can be any kind of Lévy process with even very irregular jump behaviour. It is thus not surprising that the convergence behaviour of the Euler-Maruyama scheme depends on the structure of the jumps of $Z(t)$. Therefore, results on weak and strong convergence rates of the Euler-Maruyama scheme for Lévy processes are manifold. The main assertion of Protter P. (1997) is that if the jumps of $X(t)$ are *well-behaved* (e.g. they are bounded) then the weak

Algorithm 6.6 Euler-Maruyama scheme for Lévy processes

Let $\Delta t := T/N$ for a given N. Simulate an approximate path $Y_N(t)$ of the Lévy process $X(t)$ given by the SDE (6.66) on the time grid $t_i = i\Delta t$ via:

1. Set $Y_N(0) = X(0) = x$.
2. For $j = 0$ to $N-1$ do
   Simulate the increment ΔZ_j as given in Equation (6.67) and set

$$Y_N((j+1)\Delta t) = Y_N(j\Delta t) + a(Y_N(j\Delta t))\Delta t + \sigma(Y_N(j\Delta t))\Delta Z_j$$

convergence order of 1 is preserved. However, there are also results that rely in a subtle way on the jump characteristics (see Jacod [2004]).

6.5.3 Small jump approximation

The idea of this method is to use the Lévy-Itô decomposition

$$X(t) = X^{(1)}(t) + X^{(2)}(t) + X^{(3)}(t) + X^{(4)}(t) \tag{6.68}$$

of a Lévy process with triplet $(\gamma, \sigma\sigma', \nu)$ and thus

$$X^{(1)}(t) = \gamma t,\ X^{(2)}(t) = \sigma W(t),\ X^{(3)}(t) = \sum_{s\le t} \Delta X(s) 1_{\{|\Delta X(s)|\ge 1\}} \tag{6.69}$$

$$X^{(4)}(t) = \sum_{s\le t} \Delta X(s) 1_{\{|\Delta X(s)|<1\}} - t\int_{|x|<1} x\nu(dx). \tag{6.70}$$

The first three processes can easily be simulated as a deterministic part, a Brownian motion, and a compound Poisson process with the finite Lévy measure $\nu^{(1)}$ obtained from ν by restricting it to the set $\mathbb{R}^d_I := \mathbb{R}^d - \{x : |x| \ge 1\}$. In particular, we have that the compound Poisson process $X^{(3)}$ has an intensity λ and a jump distribution L given by

$$\lambda = \nu^{(1)}(\mathbb{R}^d) = \nu(\mathbb{R}^d_I),\quad L(dy) = \frac{1}{\lambda}\nu^{(1)}(dy). \tag{6.71}$$

The fourth component $X^{(4)}$ may cause problems, as the Lévy measure of the small jumps might explode. We present two approximation methods:

1. Cutting away the small jumps

Here, the idea is to choose a small number ϵ and to cut away jumps of size $|\Delta X(s)| < \epsilon$ and thus use the approximation

$$X^{(4)}(t) \approx X^{(4,\epsilon)}(t) := \sum_{s\le t} \Delta X(s) 1_{\{\epsilon\le|\Delta X(s)|<1\}} - t\int_{\epsilon\le|x|<1} x\nu(dx). \tag{6.72}$$

Let $B_\epsilon = \{x \in \mathbb{R}^d : \epsilon \leq |x| < 1\}$, ν^ϵ be the restriction of ν to B_ϵ. Again, the process $X^{(4,\epsilon)}(t)$ can be simulated exactly as the difference of a compound Poisson process with intensity λ_ϵ and jump height distribution L_ϵ given by

$$\lambda = \nu^\epsilon(B_\epsilon), \quad L(dy) = \frac{1}{\lambda}\nu^\epsilon(dy) \tag{6.73}$$

and of a deterministic integral. As $X^{(4)}(t)$ is a martingale with expectation zero, so is the compensated process of all the small jumps we have now cut away. Thus, ignoring this compensated small jump process simply means that we are approximating this process by its mean.

2. Approximating the small jumps by a Brownian motion

Replacing the (compensated) small jumps by their mean can at most be justified if the Lévy measure of the corresponding set $B_\epsilon(0) = \{x \in \mathbb{R}^d : |x| < \epsilon\}$ is small. If however we have $\nu(B_\epsilon(0)) = \infty$, then ignoring the influence of small jumps is a too crude approximation. In the univariate case $d = 1$ an approximation of the small jumps by a Brownian motion with a suitable volatility is suggested by Asmussen and Rosiński (2001). With

$$\sigma_\epsilon^2 := \int_{|x|<\epsilon} x^2 \nu(dx), \tag{6.74}$$

a naturally suggested approximation would be

$$X^{(4)}(t) \approx X^{(4,\epsilon)}(t) + \sigma_\epsilon Z(t) \tag{6.75}$$

with $Z(t)$ a Brownian motion that is independent of all the other random sources in $X(t)$. An easy to check condition for the validity of this approximation is given by Asmussen and Rosiński (2001).

THEOREM 6.28
If we have

$$lim_{\epsilon\to 0}\frac{\sigma_\epsilon}{\epsilon} = \infty \tag{6.76}$$

then we obtain

$$\frac{X^{(4)} - X^{(4,\epsilon)}}{\sigma_\epsilon} \xrightarrow{D} Z \text{ for } \epsilon \longrightarrow 0, \tag{6.77}$$

i.e. the normed process of the small jumps converges in distribution towards a Brownian motion $Z(t)$ (in $D[0,T]$).

REMARK 6.29 1. Asmussen and Rosiński (2001) even give an equivalence characterization of the above convergence of the normed small jumps towards a Brownian motion for the condition

$$lim_{\epsilon\to 0}\frac{\sigma_{\min\{c\sigma_\epsilon,\epsilon\}}}{\sigma_\epsilon} = 0 \; \forall c > 0. \tag{6.78}$$

They also show that if ν has no atoms in the neighbourhood of 0, then this condition is equivalent to Equation (6.76).

2. One can easily check that the gamma process does not satisfy condition (6.76) while for the IG process the approximation of the small jumps by a Brownian motion as above is valid as condition (6.76) is satisfied. □

6.5.4 Simulation via series representation

When approximating the small jumps of a Lévy process, we have used a suitable compound Poisson process as the main approximation tool. Note that a compound Poisson process can be represented as an infinite series

$$X(t) = \sum_{j=1}^{\infty} Y_j \cdot 1_{\{t_j \le t\}} \tag{6.79}$$

with Y_j denoting the jump height of the compound Poisson process at the jump time t_j. While in this representation, all terms are of comparably the same size and differ mainly by their time of appearance, we will below introduce another series representation for subordinators. It will have the crucial feature that the terms are ordered by the size of their corresponding jumps, but are on the other hand uniformly distributed in time. We consider the one-dimensional setting and need that the Lévy measure possesses a density, i.e. $\nu(dx) = h(x)dx$. Its tail integral is denoted by

$$\bar{\nu}(x) = \int_x^{\infty} h(y)\, dy, \; x > 0. \tag{6.80}$$

We can then state the following proposition (see Asmussen and Glynn [2007]).

PROPOSITION 6.30
Let the subordinator $X(t), t \in [0,1]$ have a Lévy triplet $(0,0,\nu)$ with $\nu(dx) = h(x)dx$, $h(x) > 0$ for $x > 0$ and let $\tau_1, \tau_2, ...$ be the jump times of a Poisson process with intensity $\lambda = 1$. Let further $U_1, U_2, ...$ be a series of independent random variables that are uniformly distributed on $[0,1]$ and that are independent from the jump times τ_i. Then we have equality in distribution between $X(t)$ and an almost surely converging series:

$$X(t) \overset{D}{=} \sum_{j=1}^{\infty} \bar{\nu}^{-1}(\tau_j)\, 1_{\{U_j \le t\}}, \;\; 0 \le t \le 1. \tag{6.81}$$

REMARK 6.31 1. The above series representation can be used to set up a simulation algorithm for $X(t)$. For this, we first have to decide on a truncation criterion. The natural one is to set $N{=}N(\epsilon)$ with

$$N(\epsilon) = inf\left\{j \in \mathbb{N} : \bar{\nu}^{-1}(\tau_j) < \epsilon\right\}. \tag{6.82}$$

Then, for each path of $X(t)$ we simulate jump times τ_j and calculate the values $\bar{\nu}^{-1}(\tau_j)$ up to $N(\epsilon)$. Finally, we use the approximation

$$X(t) \approx \sum_{j=1}^{N(\epsilon)} \bar{\nu}^{-1}(\tau_j) 1_{\{U_j \leq t\}}, \; 0 \leq t \leq 1. \tag{6.83}$$

Instead of the random limiting sum index $N(\epsilon)$ one could also use a fixed number N. By the law of large numbers we could choose $N = \nu(\epsilon)$ to obtain the same order as the random choice $N(\epsilon)$ in the mean.

2. Note that the approximation method described in the preceding remark is nothing more than cutting away the jumps smaller than ϵ. It is therefore typically not very efficient. There are many more sophisticated series approximations that are conceptually more involved. As we do not need them for our chosen applications in financial and actuarial mathematics, we refer the reader to the survey paper by Rosiński (2001). □

Chapter 7

Simulating Financial Models: Discontinuous Paths

7.1 Introduction

The limitations of diffusion models in explaining observed stock price movements and option prices have led to the introduction of various nondiffusion stock price models. Among them are:

- Jump-diffusion models to explain the smile observed in option prices and the leptokurtic behaviour of stock price returns.
- Lévy processes to fit observed leptokurtic behaviour of stock price returns and skewness in option prices.
- Special subordinated Lévy processes to model an internal clock that accounts for the influence of the speed and volume of trading.

Here, leptokurtic behaviour means that in real financial markets we observe a more spiky and also more heavy-tailed behaviour of log-returns of stock prices than those which can be explained by the normal distribution.

We will look at various nondiffusion models that fall into the above classes. Again, we take the approach to consider some specific examples separately. Our main arguments for this are historical reasons (such as in the Merton jump-diffusion model) and the fact that often presentation of these special cases does not need the full complexity of the general class of Lévy models.

7.2 Merton's jump-diffusion model and stochastic volatility models with jumps

7.2.1 Merton's jump-diffusion setting

Already in Merton (1976), a model that allowed for sudden jumps is considered. The proposed stock price differential equation has the form

$$dS\left(t\right)=S\left(t-\right)\left(\left(\mu-\lambda\kappa\right)dt+\sigma dW\left(t\right)+\left(Y\left(t\right)-1\right)dN\left(t\right)\right). \tag{7.1}$$

Here, μ is the drift and σ the diffusion part of the volatility of the stock price. The jump variables $Y(t)$ are a family of independent random variables all having the same log-normal distribution with

$$\mathbb{E}\left(Y\left(t\right)-1\right)=\kappa \tag{7.2}$$

where λ is the intensity of the Poisson process $N(t)$ and $W(t)$ is a Brownian motion. Note that using $Y(t)-1$ in Equation (7.2) makes it easier to model jump losses. The assumption of a log-normally distributed $Y(t)$ on the one hand ensures analytical tractability, and on the other hand guarantees

$$Y(t)-1>-1, \tag{7.3}$$

which rules out bankruptcy of the company as the stock price never jumps down to a nonpositive value. Indeed, by Itô's formula for jump-diffusions, the explicit solution of the stochastic differential equation (SDE) is given by

$$S\left(t\right)=S_0\exp\left(\left(\mu-\lambda\kappa-\frac{1}{2}\sigma^2\right)t+\sigma W\left(t\right)\right)\cdot\prod_{i=1}^{N(t)}\tilde{Y}_i. \tag{7.4}$$

Here, $\tilde{Y}_i=Y(t_i)$ simply denotes the value of the jump at the i-th jump time. We also introduce the expected value and the variance of the jumps in the log-return, $\ln(S(t_i)/S(t_i-))=\ln(\tilde{Y}_i)$ as

$$\mathbb{E}\left(\ln\left(\tilde{Y}_i\right)\right)=\mu_J-\frac{1}{2}\sigma_J^2,\quad \mathbb{V}ar\left(\ln\left(\tilde{Y}_i\right)\right)=\sigma_J^2. \tag{7.5}$$

Note that this in particular implies the relation

$$\kappa=\exp\left(\mu_J\right)-1. \tag{7.6}$$

The multiplicative separation between the continuous and the jump parts in the explicit solution formula allows for a very easy way to simulate the paths of the stock price in the Merton jump-diffusion setting as in Algorithm 7.1.

Note that in this algorithm one still has to decide how to actually choose the discretization. Depending on the final task, a fixed discrete time grid or a jump-adapted one as explained in the jump-diffusion part of the previous chapter can be chosen.

Option pricing in the Merton jump-diffusion model: Merton's approach

As the number of jumps is unbounded in Merton's model, the corresponding market model is incomplete (see Merton [1976]). Therefore, pure arbitrage considerations do not lead to a unique option price. In Merton (1976) it is assumed that the jumps of the stock prices are diversifiable (via holding a suitable portfolio of shares) and therefore should not be priced at all. Merton

Algorithm 7.1 A path in the Merton jump-diffusion model

1. Simulate the path $B(t) = \left(\mu - \lambda\kappa - \frac{1}{2}\sigma^2\right)t + \sigma W(t)$ of a Brownian motion with drift.

2. Simulate the path of a Poisson process $N(t), t \in [0,T]$.

3. Simulate $N(T)$ independent standard normally distributed random variables $Z_1, ..., Z_{N_T}$ and set

$$\tilde{Y}_i = \exp\left(\mu_J - \frac{1}{2}\sigma_J^2 + \sigma_J^2 Z_i\right).$$

4. Obtain $J(t) = \prod_{i=1}^{N(t)} \tilde{Y}_i, t \in [0,T]$ and set

$$S(t) = S_0 \exp(B(t)) \cdot J(t), t \in [0,T].$$

---

thus suggests a hedging strategy that hedges the diffusion risk completely but does not care for the jump risk. Expressed in terms of the equivalent martingale that he uses for calculating the price of an option, this means that he uses the same transformation as in the simple Black-Scholes setting, i.e. the diffusion drift μ is changed to r, and all other parameters remain unchanged. With this choice, one can then simply calculate the price of a European call option by conditioning on the number and on the heights of the jumps, then make use of the fact that the remaining parts of the stock price process follow an appropriate log-normal distribution. More precisely, by introducing the Black-Scholes call price operator as

$$C\left(S, K, r, \sigma^2; T\right) := S\Phi(d_1) - Ke^{(-rT)}\Phi(d_2) \tag{7.7}$$

$$d_1 = \frac{\ln(S/K) + \left(r + \frac{1}{2}\sigma^2\right)T}{\sigma\sqrt{T}}, \quad d_2 = d_1 - \sigma\sqrt{T}, \tag{7.8}$$

one obtains the following representation of the Merton price for a European call in the above jump-diffusion setting:

$$C_{Merton}(0, S_0) = \sum_{n=0}^{\infty} e^{-\tilde{\lambda}T}\frac{1}{n!}\left(\tilde{\lambda}T\right)^n C\left(S, K, r_n, \sigma_n^2; T\right), \tag{7.9}$$

$$\tilde{\lambda} = \lambda(k+1), \; r_n = r - \lambda k + n\frac{\mu_J}{T}, \; \sigma_n^2 = \sigma^2 + n\frac{\sigma_J^2}{T}. \tag{7.10}$$

With the help of formula (7.9), one can now calibrate the model in the usual way, i.e. one determines the unknown parameters λ, σ^2, μ_J, σ_J^2 by fitting the model to market prices of a set of traded call options. Having obtained the parameters, one is then able to price more complicated exotic options by the Monte Carlo method. More precisely, one uses Algorithm 7.1 to simulate paths in the Merton jump-diffusion models and then obtains from these the price of an exotic option by averaging over the resulting discounted payoffs. Of course, depending on the special type of options, variance reduction methods and special modifications are possible.

REMARK 7.1 1. **The pricing formula:** To derive the pricing formula (7.9), note that conditional on the number of jumps n, the stock price $S(T)$ is log-normal with additional terms (compared to the Black-Scholes case) of

$$\left(-\lambda\kappa + n\frac{\mu_J}{T}\right)T + \sum_{i=1}^{n}\left(\sigma_J Z_i - \frac{1}{2}\sigma_J^2 T\right). \tag{7.11}$$

Here, the Z_i are independent standard normally distributed random variables. Having this in mind, similar calculations as in the Black-Scholes case lead to the pricing formula if one additionally takes into account that we have

$$\mathbb{Q}\left(N\left(T\right)=n\right)=e^{-\lambda T}\frac{1}{n!}\left(\lambda T\right)^{n}. \tag{7.12}$$

It should be noted that the (relative) jump drift μ_J and the mean jump size κ enter the option pricing formula. As these parameters are subjective parameters of the investor, we can no longer talk of an objective valuation.

2. **Why jumps anyway?** A main reason for considering jump type models in option pricing is that diffusion-based continuous-path models cannot explain the very skewed behaviour of the implied volatility of options that are close to maturity. Indeed, there might only be the fear of a sudden jump of the stock price close to maturity of the option that could be a reasonable explanation for the observed behaviour of the implied volatility.

3. **Further valuation approaches**: In Grünewald (1998) the above valuation approach is compared to various other types of option pricing and hedging approaches in the Merton jump-diffusion model. Among them are the local risk-minimizing approach by Schweizer (1991) and an equilibrium approach by Bates (1996). They lead to similar results but differ in the way the jump risk is priced, and therefore yield different option prices.

4. **Multiasset models**: A multiasset formulation of the model is also possible. For this, one can directly consider multidimensional Gauss-Poisson processes as given in Equation (6.25). A particular case where the underlying market model is even complete is given in Jeanblanc-Picqué and Pontier (1990). There, the dynamics of n stock prices are modelled via

$$dS_i\left(t\right)=$$
$$S_i\left(t-\right)\left(\left(\mu_i-\sum_{j=1}^{n-d}\lambda_j\kappa_{ij}\right)dt+\sum_{j=1}^{d}\sigma_{ij}dW_j\left(t\right)+\sum_{j=1}^{n-d}\kappa_{ij}dN_j\left(t\right)\right) \tag{7.13}$$

where the (relative) jump heights of the stock prices are constant. Although this model has the nice feature of leading to a complete market, the assumption of such constant jump heights is not realistic. □

7.2.2 Jump-diffusion with double exponential jumps

Another jump-diffusion model that allows explicit computations is the double exponential model suggested by Kou (2002). It has also been used by Acar (2006) in the context of the modelling of optimal capital structure. The main difference to the Merton jump-diffusion model consists in the assumption that instead of log-normal jump heights, it is assumed that the logarithm of the jump height $V = \ln(Y)$ has a double exponential distribution, i.e. V possesses the density function

$$f(v) = p \cdot \eta_1 e^{-\eta_1 v} 1_{\{v \geq 0\}} + (1-p) \cdot \eta_2 e^{\eta_2 v} 1_{\{v<0\}} \tag{7.14}$$

with $\eta_1 > 1$, $\eta_2 > 0$, $p \in [0,1]$ leading to

$$\mathbb{E}(Y) = \mathbb{E}\left(e^V\right) = p\frac{\eta_1}{\eta_1 - 1} + (1-p)\frac{\eta_2}{\eta_2 + 1}. \tag{7.15}$$

This motivates the surprising condition $\eta_1 > 1$, which in particular guarantees an average upward jump less than 1, indeed a reasonable assumption.

REMARK 7.2 For path simulation, we need a double exponential random variable. It can be generated by inversion of the distribution function. Another very intuitive method is to first simulate a uniform random variable $U \sim \mathcal{U}[0,1)$, then to decide in a zero-one experiment with success probability p if U should be transformed in an exponential distribution with parameter η_1 (in case of "1") or with parameter η_2 (in case of "0"). From this one obtains a double exponential random variable Z_i. With this simple change of Step 3 of the Algorithm 7.1 for a path of the Merton jump-diffusion model, we obtain a path of the double exponential jump-diffusion model, too. □

Option pricing in the double exponential model

Building upon work by Naik and Lee (1990), Kou (2002) used the framework of a rational expectations equilibrium to obtain an option price in an incomplete market setting. There, the utility of a representative investor is maximized and the option price is then determined such that it is optimal for this investor to hold a zero position in the option. We do not go into detail here, but the consequences of this derivation are stated as follows:

- The pricing formula can be interpreted as being obtained in a market equipped with an equivalent martingale measure $\mathbb{Q}$ such that the drift and the jump characteristics of the stock price are suitably transformed.

However, the transformed price process has a drift of r but otherwise the same form as the original stock price process with different jump parameters. For purposes of option pricing, we can adopt our usual approach to assume that we directly model in this transformed market. This in particular means that a simple Monte Carlo pricing can be based on the path simulation algorithm indicated above.

- Kou (2002) derived an explicit pricing formula for European calls (among others) that resembles the Black-Scholes formula. There, the analogues to $\Phi(d_i(t))$ are defined in terms of so-called *Hn*-functions that appear due to the fact that the distribution of sums of normal and double exponential random variables have to be calculated. Kou also gives a detailed algorithm to compute the *Hn*-functions.

7.2.3 Stochastic volatility models with jumps

As mentioned previously, even stochastic volatility models of diffusion type cannot explain the very skewed behaviour of the implied volatility of options that are close to maturity. Therefore, in Bates (1996), the Merton jump-diffusion model is combined with the Heston stochastic volatility model. Thus, for the purpose of option pricing, the stock price and volatility equations are given by

$$dS(t) = S(t-)\left((r - \lambda\kappa)\,dt + \sigma(t)\,dW(t) + (Y(t) - 1)\,dN(t)\right), \quad (7.16)$$

$$d\sigma(t) = \theta(\sigma_\infty - \sigma(t))\,dt + \nu\sqrt{\sigma(t)}d\tilde{W}(t) \quad (7.17)$$

with $r, \lambda, \kappa, \theta, \nu, \sigma_\infty$ suitable positive constants. The two Brownian motions are independent from the Poisson process. However, they have a correlation of $\rho \in [-1, 1]$,

$$Corr\left(W_t, \tilde{W}_t\right) = \rho. \quad (7.18)$$

Due to the independence between the Brownian parts and the Poisson process, one can use any method to simulate the Heston-like part of the stock price (including the volatility process!) and then combine it with a jump part as in the jump-diffusion setting. For this, we refer to the simulation algorithms for the Heston stock price model and the Merton jump-diffusion model.

7.3 Special Lévy models and their simulation

The introduction of Lévy models into finance started in the mid 1990s with a series of papers by Barndorff-Nielsen (1997), Eberlein and Keller (1995), and Küchler et al. (1999), just to mention a few. All those papers were centred around the class of hyperbolic distributions that had already been introduced by Barndorff-Nielsen (1977) in the context of turbulence.

In these models often the logarithm of stock prices is modelled by a Lévy process to ensure a nonnegative stock price process. Further, all the models that we are going to present below lead to incomplete financial markets. Thus, there are typically infinitely many equivalent martingale measures that can serve as pricing measures. While in the jump-diffusion type models the pricing measure used was satisfied by equilibrium considerations, in general Lévy process models it is often quite involved to find at least one equivalent martingale measure (EMM) that is analytically tractable. A popular way of constructing such an EMM is the use of the so-called Esscher transform.

7.3.1 The Esscher transform

Using the Esscher method to construct an EMM is a convenient choice which is mainly justified by its simplicity. To explain it, let us assume that the stock price model we are considering is of the form

$$S(t) = S_0 e^{Z(t)} \tag{7.19}$$

where $Z(t)$ is a Lévy process that admits a density function $f(x;t)$. Then, the basic principle is a multiplication of the density by an exponential factor $e^{\theta x}$ with $\theta \in \mathbb{R}$ yielding the new density function

$$f(x;t,\theta) := \frac{e^{\theta x} f(x;t)}{\int_{-\infty}^{\infty} e^{\theta y} f(y;t)\, dy} \tag{7.20}$$

where the denominator ensures that the integral of the new function is indeed equal to one. By this transformation we can introduce a new probability measure via

$$dP_t^{\theta}(x) := \frac{e^{\theta x}}{\int e^{\theta x} dP_1(x)} dP_t(x) \tag{7.21}$$

with P_t being the original probability measure with the density $f(x;t)$. The constant θ is now determined such that the probability measure P_t^{θ} is a martingale measure for $S(t)$, i.e. $e^{-rt}S_t$ has to be a P_t^{θ}-martingale. For this, we look at the moment generating function $M(u;t)$ of $Z(t)$,

$$M(u;t) = \mathbb{E}\left(e^{uZ(t)}\right) \tag{7.22}$$

and at the moment generating function under P_t^{θ} given by

$$\begin{aligned} M(u;t,\theta) &= \int_{-\infty}^{\infty} e^{ux} f(x;u,\theta)\, dx \\ &= \frac{\int_{-\infty}^{\infty} e^{ux} e^{\theta x} f(x;u)\, dx}{\int_{-\infty}^{\infty} e^{\theta y} f(y;u)\, dy} = \frac{M(u+\theta;t)}{M(\theta;t)}. \end{aligned} \tag{7.23}$$

The martingale requirement of

$$S_0 = e^{-rt}\mathbb{E}^{\theta}\left(S\left(t\right)\right) = e^{-rt}\mathbb{E}^{\theta}\left(e^{Z(t)}\right) = S_0 e^{-rt}\frac{M\left(1+\theta;t\right)}{M\left(\theta;t\right)} \tag{7.24}$$

thus leads to an implicit equation for θ,

$$M\left(\theta;t\right) = e^{-rt}M\left(1+\theta;t\right) \tag{7.25}$$

which due to the Lévy-Khinchine formula is equivalent to

$$M\left(\theta;1\right) = e^{-r}M\left(1+\theta;1\right). \tag{7.26}$$

If there is a solution θ^* to Equation (7.26) then the so-determined probability measure P^{θ^*}, the *Esscher measure*, is indeed an EMM. It can then be used to determine arbitrage-free prices of options. Besides convenience, there are also equilibrium-based arguments for its use (see e.g. Gerber and Shiu [1994]).

7.3.2 The hyperbolic Lévy model

Although not the most general one, we will start the section of Lévy process models by presenting the hyperbolic model which has been introduced by Eberlein and Keller (1995) and by Küchler et al. (1999). Both works were motivated by remarks by Barndorff-Nielsen (1977) who proposed the use of the hyperbolic distribution in relation with turbulence and sand flow. In Eberlein and Keller (1995) two possible ways of generalizing the geometric Brownian motion are given, the replacement of the Brownian motion by a hyperbolic Lévy motion in either the exponent of the geometric Brownian motion or as the driving process of a linear SDE. Here, we will only consider the model of the form

$$S_t = S_0 e^{Z^{\zeta,\delta}(t)}. \tag{7.27}$$

$Z^{\zeta,\delta}(t)$ is the Lévy process generated by a symmetric form of the hyperbolic distribution. It is uniquely determined by its density at time $t = 1$,

$$h_{\zeta,\delta}\left(x\right) = \frac{1}{2\delta K_1\left(\zeta\right)}e^{-\zeta\sqrt{1+(x/\delta)^2}}. \tag{7.28}$$

Here, $K_1\left(x\right)$ denotes the modified Bessel function of the third kind, i.e.

$$K_1\left(x\right) = \frac{1}{2}\int_0^{\infty} e^{-\frac{1}{2}x\left(u+\frac{1}{u}\right)}du, \tag{7.29}$$

for real constants ζ, δ. As Bessel functions play an important role in engineering applications, they are implemented in many mathematical software packages. A drawback of the model is that the hyperbolic distribution is infinitely divisible but not closed under convolution. This in particular means that we have in general only exactly hyperbolic log-returns of the stock at times t which are integers. As a consequence of this, (simple) exact simulation with the help of the hyperbolic density is only possible at these times, which makes the use of this simulation method a very limited one.

The characteristic function of $Z^{\zeta,\delta}(1)$ has the form of

$$\phi_1(u) = \frac{\zeta}{K_1(\zeta)} \frac{K_1\left(\sqrt{\zeta^2+\delta^2 u^2}\right)}{\sqrt{\zeta^2+\delta^2 u^2}} \tag{7.30}$$

(see Eberlein and Keller [1995]) from which we obtain $\phi_t(u)$ via

$$\phi_t(u) = \phi_1(u)^t \tag{7.31}$$

and the moment generating function as

$$M^{\zeta,\delta}(u,1) = E\left(e^{uZ_1^{\zeta,\delta}}\right) = \frac{\zeta}{K_1(\zeta)} \frac{K_1\left(\sqrt{\zeta^2-\delta^2 u^2}\right)}{\sqrt{\zeta^2-\delta^2 u^2}} \tag{7.32}$$

for $|u| < \zeta/\delta$. With this explicit form we can determine the parameter θ^* which is needed to obtain the Esscher measure P^* by an Esscher transformation as the unique solution of the consequence of Equation (7.26),

$$r = ln\left(\frac{K_1\left(\sqrt{\zeta^2-\delta^2(\theta+1)^2}\right)}{K_1\left(\sqrt{\zeta^2-\delta^2\theta^2}\right)}\right) - \frac{1}{2} ln\left(\frac{\zeta^2-\delta^2(\theta+1)^2}{\zeta^2-\delta^2\theta^2}\right). \tag{7.33}$$

This equation can be solved numerically for θ for given values of r, δ, and ζ.

Thus, having obtained a parameter θ^* that yields the Esscher measure $P^* = P^{\theta^*}$, one can calculate option prices with the help of the P^*-density $f^{\zeta,\delta}(x;t,\theta^*)$. In the case of a European call with strike K, Eberlein and Keller (1995) obtain

$$E^*\left(e^{-rT}(S(T)-K)^+\right) =$$
$$S_0 e^{-rT}\int_c^\infty e^x f^{\zeta,\delta}(x;T,\theta^*)\,dx - e^{-rT}K\int_c^\infty f_T^{\zeta,\delta}(x;T\theta^*)\,dx \tag{7.34}$$

with $c = ln(K/S_0)$. Here, the density functions have no explicit representation, but have to be calculated numerically via inversion of the characteristic function for P^{θ^*} via the Fourier inversion formula

$$f^{\zeta,\delta}(x;t,\theta) = \frac{1}{\pi}\int_0^\infty cos(ux)\,\phi_t^*(u)\,du \tag{7.35}$$

from the characteristic function.

Although the hyperbolic model shows a very good fit to data, it has drawbacks with respect to tractability. As the hyperbolic family is not closed under convolution, we typically know the density of the price increments only implicitly which leads to the necessity of heavy numerical computations. Also, generating hyperbolic random numbers is not an easy issue (see Hartinger and Predota [2003] for some applications). Therefore, more tractable Lévy models will be presented below.

7.3.3 The variance gamma model

If one looks at time series of stock price returns, then one often observes phases of high frequency price changes followed by phases where the intensity of price changes is comparably low. This is often referred to as **volatility clustering**. To model this, one introduces a different clock where time moves in velocity proportional to the trading activity. When there are a lot of trades, the internal time of the process runs faster and thus formally enlarges the observation interval. It runs slower if nearly nothing happens. In such a way it is hoped to care for the main weakness of the Brownian motion-based Black-Scholes model which moves in a too uniform way over time (by this, we mean that the log-returns behave too similar over time). With this random time change, a Brownian motion can again be used as the basic building block for modelling the uncertainty of the future stock price evolution.

The appropriate class of Lévy processes to model such a random time evolution are the subordinators introduced in Definition 6.23. There, we have already seen some examples of subordinators: each compound Poisson process with nonnegative jump heights, the gamma process, and the inverse Gaussian process are subordinators. A simple recipe for generating a new Lévy process is to use a subordinator as a model for the evolution of time and plugging it into a Brownian motion. Adding a suitable drift term and correcting for the quadratic variation of the stochastic part then yields a model for the log-return of a stock.

We start to consider the variance gamma model introduced by Madan and Seneta (1990), by Madan and Milne (1991), and generalized in Madan et al. (1998). There, the so-called variance gamma (VG) process is used to model the (log-) returns of the stock, i.e. we have

$$S(t) = S_0 e^{\mu t + X(t) + \omega t} \tag{7.36}$$

with $X(t)$ a VG process, μ the stock price drift, and the *compensation term*

$$\omega t = \ln\left(1 - \theta\nu - \frac{\sigma^2 \nu}{2}\right)\frac{t}{\nu} \tag{7.37}$$

is introduced to ensure that we have

$$\mathbb{E}\left(S\left(t\right)\right) = S_0 e^{\mu t}. \tag{7.38}$$

We implicitly assumed the validity of the assumption

$$1 > \left(\theta + \frac{\sigma^2}{2}\right)\nu \tag{7.39}$$

and will do so for the rest of this section. As before, we are mostly interested in option pricing and therefore assume that we are working in a risk-neutral setting. We thus always assume

$$\mu = r \tag{7.40}$$

when option pricing problems are considered. However, we would like to point out that if we want to switch to a risk-neutral setting starting from our original model via a measure transformation, we also have to change the coefficients σ, θ, and ν in a suitable way (see e.g. Madan et al. [1998] for details). As we will never need this explicit change of measure here, we can however simply assume that the parameters σ, θ, and ν are already the transformed ones when we assume $\mu = r$.

Before continuing with the application and properties of this stock price model, we will first devote some time to the introduction of the VG process. By doing this, we directly choose the three parameter variant of Madan et al. (1998) that includes the earlier two parameter variant of Madan and Seneta (1990) as a special case, the so-called symmetric VG process.

DEFINITION 7.3
Let $B(t;\theta,\sigma) := \theta t + \sigma W_t$ be a one-dimensional Brownian motion with drift θ and volatility σ. Let also $\gamma(t;\nu)$ be a gamma process with parameter ν and Gamma $(t/\nu, 1)$ distribution at time t. Then, the process

$$X(t) := X(t;\sigma,\nu,\theta) := B(\gamma(t;\nu);\theta,\sigma) \tag{7.41}$$

is called a **variance gamma process** *with parameters σ, ν, and θ.*

REMARK 7.4 1. In addition to the usual volatility parameter σ, the parameter θ is introduced to control the skew while the kurtosis of the stock prices is cared for by ν. This can be seen by looking at the relevant moments (see Madan et al. [1998]). While mean and variance have the simple form of

$$\mathbb{E}(X(t)) = \theta t, \quad \mathbb{V}ar(X(t)) = \left(\theta^2\nu + \sigma^2\right)t, \tag{7.42}$$

this is only the case for the skewness and the kurtosis if we assume $\theta = 0$:

$$Skew(X(t)) = 0, \; Kurt(X(t)) = 3\left(1 + \frac{\nu}{t}\right) \quad \text{for } \theta = 0. \tag{7.43}$$

2. The characteristic function of the VG process is given by

$$\Phi_{X(t)}(u) = \left(\frac{1}{1 - i\theta u\nu + (\sigma^2\nu/2)\,u^2}\right)^{t/\nu}. \tag{7.44}$$

3. The name *variance gamma* process is directly derived from the above representation of the process as a (subordinated) Brownian motion with drift where a gamma process describes the evolution of the variance over time. □

This parameterization of the VG process now allows a very easy possibility for simulation by subordination as in Algorithm 7.2.

Algorithm 7.2 Simulating a variance gamma process by subordination

1. Set $X(0) = 0$.
2. Choose a time discretization $0 = t_0 < t_1 < ... < t_n = T$ of $[0, T]$.
3. For $i = 1$ to n
   - Simulate a random number $G_i \sim Gamma\left(\left(t_i - t_{i-1}\right)/\nu, 1\right)$ independent of all other yet simulated random numbers.
   - Simulate a standard normally distributed random number Y_i.
   - Set $X(t_i) := X(t_{i-1}) + \sqrt{G_i}Y_i + G_i\theta$.

As the gamma process is strictly increasing by jumps and the Brownian motion is continuous, the VG process only changes by jumps under the above transformation. Figure 7.1 shows a simulated path of a VG process. Here, we have drawn the paths as a continuous function for better visualization. Actually, a VG process is a pure jump process with paths of infinite activity but finite variation.

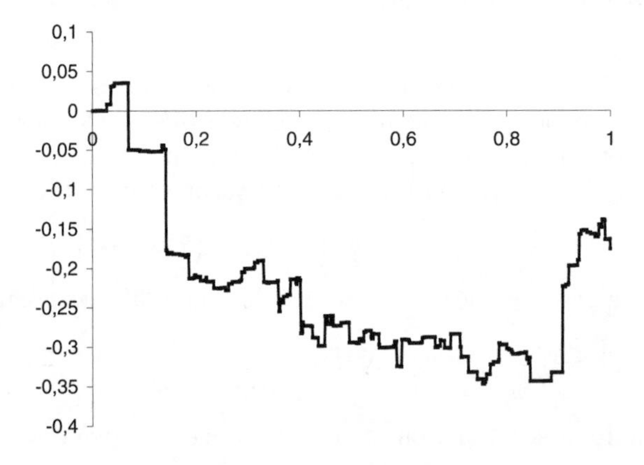

FIGURE 7.1: A simulated path of a VG process.

This can be seen from a second representation of a VG process which will be found by using the following product decomposition of the characteristic function of the VG process

$$\Phi_X(u;t) = \left(\frac{1}{1 - i\theta u\nu + (\sigma^2\nu/2)\,u^2}\right)^{t/\nu}$$
$$= \left(\frac{1}{1 - i\eta_p u}\right)^{t/\nu} \left(\frac{1}{1 + i\eta_n u}\right)^{t/\nu} = \Phi_{\gamma^+}(u;t) \cdot \Phi_{\gamma^-}(u;t) \quad (7.45)$$

where we have $\eta_p - \eta_n = \theta\nu$, $\eta_p\eta_n = \frac{1}{2}\sigma^2\nu$. Thus, we obtain

$$\eta_p = \frac{\theta\nu}{2} + \sqrt{\frac{\theta^2\nu^2}{4} + \frac{\sigma^2\nu}{2}}, \quad \eta_n = -\frac{\theta\nu}{2} + \sqrt{\frac{\theta^2\nu^2}{4} + \frac{\sigma^2\nu}{2}}\,. \quad (7.46)$$

Consequently, we have shown that the above VG process can be represented as the difference of two independent gamma processes $\gamma^+(t) := \gamma(t;\mu_p,\nu_p)$ and $\gamma^-(t) := \gamma(t;\mu_n,\nu_n)$ with

$$\mu_p = \eta_p/\nu, \quad \mu_n = \eta_n/\nu, \quad \nu_p = \mu_p^2\nu, \quad \nu_n = \mu_n^2\nu. \quad (7.47)$$

Consequently, we have

$$\gamma^+(t) \sim Gamma\,(t/\nu, \mu_p\nu)\,, \quad \gamma^-(t) \sim Gamma\,(t/\nu, \mu_n\nu) \quad (7.48)$$

and arrive at the **difference of gamma representation**

$$X(t) = \gamma^+(t) - \gamma^-(t) \quad (7.49)$$

of the VG process. With this representation, Algorithm 7.3 is straightforward and represents another possibility for simulating paths of the VG process.

Algorithm 7.3 Variance gamma path by differences

1. $X(0) = 0$.
2. For $i = 1$ to n
   - Generate the independent gamma processes $\gamma_i^+(t), \gamma_i^-(t)$.
   - $X(t_i) = \gamma_i^+(t) - \gamma_i^-(t)$.
   - $X(t) = X(t_{i-1})$ for all $t \in (t_{i-1}, t_i)$.

REMARK 7.5 1. **Option pricing in the VG model**: As always, the first ingredient for option pricing is the choice of an equivalent martingale measure. As the market for a VG model is incomplete, Madan et al. (1998)

used equilibrium arguments to figure out a particular risk measure. They then derived an explicit pricing formula for European calls that resembles the Black-Scholes formula. However, the computation of the probabilities (i.e. the analogues to $\Phi(d_i(t))$ in the Black-Scholes formula) is quite involved. Therefore, we do not present it here.

2. **Conventional Monte Carlo option pricing in the VG model**. As we have two possibilities for sampling VG paths, the Monte Carlo method for option pricing is straightforward: sample a sufficiently big number of stock price paths of the VG process to obtain VG stock price paths under the risk-neutral measure, calculate the corresponding option payoffs, and estimate the option price by the discounted average over these payoffs. Note in particular that for path-independent options it is enough to generate the final value $X(T)$ of the VG process and not the whole path. For computing the prices of various exotic options, a combination of bridge sampling and of using the difference of gamma representation to bound the option prices is, however, far more efficient. We will present it in more detail below. □

In Carr et al. (2002) the VG model is generalized to the so-called CGMY model, which we do not consider here.

The difference of gamma bridge sampling method

We have already seen how to sample a path of a Brownian motion by the Brownian bridge method in Section 4.4.3. In the case of the VG process a similar bridge method is presented in Avramidis and L'Ecuyer (2006). There, in addition, the authors make use of explicit properties of the VG process to obtain very efficient (quasi-) Monte Carlo algorithms to calculate the price of popular exotic options. They consider the difference of gamma representations

$$X\left(t\right)=\gamma^{+}\left(t\right)-\gamma^{-}\left(t\right) \tag{7.50}$$

of the VG process and sample both gamma processes by a bridge sampling algorithm (as explained below). An important ingredient of their method is the observation that for a fixed path ω of the VG process we always have

$$\gamma^{+}\left(t_1,\omega\right)-\gamma^{-}\left(t_2,\omega\right)\leq X\left(t,\omega\right)\leq\gamma^{+}\left(t_2,\omega\right)-\gamma^{-}\left(t_1,\omega\right) \tag{7.51}$$

for $t_1 \leq t \leq t_2$ due to the fact that both gamma processes are increasing. It will later be used to obtain bounds for option payoffs along simulated paths.

For a bridge sampling algorithm we need conditional distributions of the process in between two given times. Let therefore $\gamma(t)$ be a gamma process with parameters (μ,ν) as described in Example 6.21. Then, for $0\leq\tau_1<t<\tau_2$ the conditional distribution of $\gamma(t)$ given $\gamma(\tau_1)$ and $\gamma(\tau_2)$ equals that of $\gamma(\tau_1)+(\gamma(\tau_2)-\gamma(\tau_1))Y$ with

$$Y\sim Beta\left(\left(t-\tau_1\right)\mu^2/\nu,\left(\tau_2-t\right)\mu^2/\nu\right). \tag{7.52}$$

Here, $Beta\,(\alpha,\beta)$ denotes the beta distribution with parameters (α,β). It is defined on the unit interval $(0,1)$ and given by the density

$$f(x) = \frac{x^{\alpha-1}(1-x)^{\beta-1}}{\int_0^1 y^{\alpha-1}(1-y)^{\beta-1}\,dy} = \frac{x^{\alpha-1}(1-x)^{\beta-1}}{B(\alpha,\beta)}\,. \tag{7.53}$$

Given the values of a gamma process for two endpoints of a time interval, one can thus generate the value of this gamma process at any time inside the interval. An algorithm for a gamma bridge sampling on $[0,T]$ now starts with $\gamma(0)=0$ and by simulating $\gamma(T)$ and then fills the gaps inside the interval in a suitable way by conditional sampling. The most convenient such way is a dyadic partition, i.e. a successive halving of all the intervals that appear during the bridge sampling process. It yields Algorithm 7.4, the difference of gamma bridge sampling (DGBS) (see Avramidis and L'Ecuyer [2006]).

Algorithm 7.4 A VG path by the DGBS algorithm

Let $N\in\mathbb{N}$ with $N=2^K$, $\nu,\nu_p,\nu_n,\mu_p,\mu_n$ be given as in Equation (7.47). Set $h=T$, $\gamma^+(0)=\gamma^-(0)=0$ and

$$\gamma^+(T)\sim\gamma\left(T/\nu,\nu_p/\mu_p\right),\quad \gamma^-(T)\sim\gamma\left(T/\nu,\nu_n/\mu_n\right)$$

with the corresponding gamma-distributed random variables.
For $k=1$ to K do

1. Set $h=h/2$.
2. For $j=1$ to 2^{k-1} do
   (a) Generate independent random numbers $Z_1,Z_2\sim Beta(h/\nu,h/\nu)$ and set

   $$\gamma^+((2j-1)h)=\gamma^+((2j-2)h)+\left(\gamma^+(2jh)-\gamma^+((2j-2)h)\right)Z_1,$$

   $$\gamma^-((2j-1)h)=\gamma^-((2j-2)h)+\left(\gamma^-(2jh)-\gamma^-((2j-2)h)\right)Z_2.$$

   (b) $X((2j-1)h)=\gamma^+((2j-1)h)-\gamma^-((2j-1)h)$.

One can also set up the above algorithm for a nondyadic partition of $[0,T]$ a more involved notation. As in the Brownian setting, the backward formulation of the bridge algorithm has advantages when, instead of Monte Carlo methods, the corresponding quasi-Monte Carlo methods are used.

The truncated difference of gamma bridge option pricing method

We will now show how to construct an efficient option pricing algorithm which makes use of payoff bounds that rely on the relation (7.51). For this we look again at the stock price equation in the risk-neutral setting

$$S(t) = S_0 e^{rt+\omega t+X(t)} = S_0 e^{\tilde{r}t+\gamma^+(t)-\gamma^-(t)}. \tag{7.54}$$

For a given time discretization $0 = t_{m,0} < t_{m,1} \dots < t_{m,m} = T$, we now introduce lower and upper bounds for the stock price process via

$$L_m(t) := S_0 e^{\tilde{r}t+\gamma^+(t_{m,i-1})-\gamma^-(t_{m,i})}, \quad t_{m,i-1} < t < t_{m,i}, \tag{7.55}$$

$$U_m(t) := S_0 e^{\tilde{r}t+\gamma^+(t_{m,i})-\gamma^-(t_{m,i-i})}, \quad t_{m,i-1} < t < t_{m,i}, \tag{7.56}$$

$$L_m(t_{m,i}) := U_m(t_{m,i}) := S(t_{m,i}), \qquad i = 0, 1, \dots, m. \tag{7.57}$$

Obviously, we then have

$$L_m(t) \le S(t) \le U_m(t), \quad \forall\, t \in [0, T] \tag{7.58}$$

by the difference of a gamma process representation of a VG process. Such an estimation would not be possible for a process with a Brownian motion component. We can make use of this relation if the payoff functional of an option admits a monotonicity property in the stock price paths. Therefore let

$$C = e^{-rT} f(S(t), t \in [0, T]) \tag{7.59}$$

be the discounted option payoff while $C_{U,m}$ and $C_{L,m}$ be the counterparts when S is replaced by U_m and L_m, respectively. Let further

$$F_m = \left(t_{m,1}, \gamma^+(t_{m,1}), \gamma^-(t_{m,1}), \dots, t_{m,m}, \gamma^+(t_{m,m}), \gamma^-(t_{m,m})\right) \tag{7.60}$$

be the (simulated) parts of the components of the VG process along the given time discretization. Then we have the following (see Avramidis and L'Ecuyer [2006]).

PROPOSITION 7.6

Suppose that conditional on F_m C is a monotone nondecreasing function of $S(t)$, $t \notin \{t_{m,0}, \dots, t_{m,m}\}$. Then we have

$$C_{L,m} \le C \le C_{U,m}. \tag{7.61}$$

The inverse relation holds if C is nonincreasing instead.

Of course, the applicability of the proposition hinges critically on the monotonicity assumption of the option payoff. However, it is easily verified that many examples of traded options share this property. Among them are Asian options, lookbacks, and barrier options.

Example 7.7 Pricing an up-and-in call
For an up-and-in call with the final payoff

$$C_B = e^{-rT} \left(S(T) - K\right)^+ \cdot 1_{\sup_{0 \le t \le T} S(t) > b} \tag{7.62}$$

we obtain lower and upper bounds as

$$C_{L,m} = e^{-rT} \left(S(T) - K\right)^+ \cdot 1_{\max_{1 \le i \le m} S(t_{m,i}) > b}, \tag{7.63}$$

$$C_{U,m} = e^{-rT} \left(S(T) - K\right)^+ \cdot 1_{\max_{1 \le i \le m} U_{m,i} > b} \tag{7.64}$$

where we have used the fact that the lower bounds for $S(t)$ are attained at the values $t_{m,i}$ and that the upper bounds are attained for

$$U_{m,i} = \sup_{t_{m,i-1} < t < t_{m,i}} U_m(t). \tag{7.65}$$

It is not hard to see that the lower and the upper bounds for the payoff C_B coincide in the cases of $S(T) \le K$ and when we have

$$\max_{1 \le i \le m} U_{m,i} \le b \quad \text{or} \quad \max_{1 \le i \le m} S(t_{m,i}) > b. \tag{7.66}$$

While in the first two cases the final payoff vanishes, the option is "in" in both the discrete and the continuous settings. For each path there is a significantly big value of M such that the lower and the upper bounds $C_{L,M}$, $C_{U,M}$ coincide. In such a case they also coincide with the payoff, but M is a priori unknown and thus a random variable. Further, increasing the number m above M will not change the option payoff for that given sample path. One therefore fixes an upper bound m^* and simulates paths of the VG process by the difference of the gamma bridge sampling up to a fineness given by the minimum of M and m^*.

One can then set up an example for a barrier option pricing algorithm by the so-called **truncated** difference of gamma bridge sampling method (truncated DGBS method) as in Algorithm 7.5.

Avramidis and L'Ecuyer (2006) calculate the prices of an up-and-in call (and further options such as lookback or Asian options) by the truncated DGBS method with both Monte Carlo and quasi-Monte Carlo methods (where the random numbers are replaced by suitable quasirandom numbers). As the usual dominating error in barrier option pricing is given by $O(\sqrt{1/m})$, they also used extrapolation estimators via

$$C_{B,extra}(N) = \frac{2^{0.5} C_B^{m^*}(N) - C_B^{m^*/2}(N)}{2^{0.5} - 1}. \tag{7.67}$$

where the superscript indicates the fineness of the discretization of the sampled stock price paths. These extrapolation estimators in particular showed an excellent performance.

Algorithm 7.5 Pricing up-and-in calls by the DGBS method

Let N be the number of path replications and m^* be the number of discretization points for the DGBS method. Set

$$C_B(N) = C_{low} = C_{up} = 0.$$

For $i = 1$ to N do

1. Simulate a VG model path by the DGBS method with fineness m^*:
   - At each step $M \in \{1, ..., m^*\}$ calculate $C_{L,M}^{(i)}$ and $C_{U,M}^{(i)}$ as in (7.64).
   - If there is an $M < m^*$ with $C_{L,M} = C_{U,M}$ then stop and set

$$C_{L,m^*}^{(i)} = C_{U,m^*}^{(i)} = C_{L,M}^{(i)}.$$

2. Set $$C_{low} = C_{low} + C_{L,m^*}^{(i)}, \quad C_{up} = C_{up} + C_{U,m^*}^{(i)}.$$

Obtain the Monte Carlo estimates for the lower bound, the upper bound, and the option price:

$$C_{low} = \frac{1}{N} C_{low}, \quad C_{up} = \frac{1}{N} C_{up}, \quad C_B(N) = \frac{1}{2}\left(C_{low} + C_{up}\right).$$

REMARK 7.8 1. **More barrier options**: With obvious changes and modifications, the above way of applying the truncated DGBS method also goes through for the other one-sided barrier option types (such as down-and-out call/put, up-and-out call/put, ...).

2. The choice $C_B(0) = 1/2(C_{low} + C_{up})$ is an obvious choice for an estimate of the option price. However, it is not the only one. Alternatives would be to use the upper and lower bounds on the stock price process to obtain an estimate of the stock price, either as an arithmetic or a geometric mean. This can then be used to obtain an estimate for the (pathwise) option payoff which then has to be updated after each simulated path. □

7.3.4 Normal inverse Gaussian processes

Another popular model from the class of subordinated Lévy processes is the normal inverse Gaussian model (NIG model). It is defined similarly as the VG model but is based on the inverse Gaussian process as the subordinator. More precisely, we look at the stock price model as suggested by Barndorff-Nielsen (1997), where we assume that we are already working under an equivalent martingale measure,

$$S(t) = S(0)\frac{\exp(rt + \sigma X(t))}{\mathbb{E}(\exp(\sigma X(t)))} \tag{7.68}$$

where $X(t)$ is an NIG process defined as

$$X(t) = W(G(t)) + \beta G(t) \tag{7.69}$$

for a Brownian motion $W(t)$ and an inverse Gaussian (IG) process $G(t) \sim IG(\delta t, \gamma)$. We then say that $X(t)$ has an $NIG(\alpha, \beta, \delta t)$-distribution with $\alpha = \sqrt{\beta^2 + \gamma^2}$ where it is required that we have

$$\alpha > 0, \quad \delta > 0, \quad \alpha > |\beta|. \tag{7.70}$$

The $NIG(\alpha, \beta, \delta)$-distribution has a probability density of the form

$$f_{NIG(\alpha,\beta,\delta)}(x) = \frac{\alpha\delta}{\pi} e^{\delta\sqrt{\alpha^2-\beta^2}+\beta x} \frac{K_1\left(\alpha\sqrt{\delta^2 + x^2}\right)}{\sqrt{\delta^2 + x^2}}, \quad x > 0 \tag{7.71}$$

where again $K_1(x)$ is the Bessel function of third kind with index 1 as given in Equation (7.29). Its characteristic function and its first two moments are given by

$$\Phi_{NIG(\alpha,\beta,\delta)}(u) = \exp\left(-\delta\sqrt{\alpha^2 - (\beta + iu)^2} - \sqrt{\alpha^2 - \beta^2}\right), \tag{7.72}$$

$$\mathbb{E}(X(t)) = \frac{\beta\delta t}{\sqrt{\alpha^2-\beta^2}}, \quad \mathbb{V}ar(X(t)) = \frac{\delta t}{\alpha\sqrt{\left(1-(\beta/\alpha)^2\right)^3}}, \tag{7.73}$$

$$Skew(X(t)) = 3\frac{\beta}{\alpha\sqrt{\delta\gamma}}, \quad Kurt(X(t)) = 3\frac{\alpha^2+4\beta^2}{\delta\alpha^2\gamma} \tag{7.74}$$

(see Ribeiro and Webber [2003]). From these considerations we in particular obtain that we have

$$\mathbb{E}(X(t)) = \exp\left(\left(\delta\gamma - \delta\sqrt{\alpha^2 - (1+\beta)^2}\right)t\right) \tag{7.75}$$

which is needed to give the price model of Equation (7.68) an explicit form.

A property that makes the NIG-distribution suitable for a log-return model is the fact that it is much more flexible than the normal distribution. Also, it can produce a higher peak in the centre and at the same time more heavy tails than the normal distribution while having the same mean and variance.

Of course, we could try to simulate the NIG process directly. However, as its distribution is not easy to invert, we recommend in Algorithm 7.6 to use the method via subordination (see also Rydberg [1997] for a detailed treatment of the aspects of simulating the NIG process).

A path of an NIG process is simulated in Figure 7.2. Note that for various parts it looks like a "Brownian motion with holes" and has an appearance different from a VG process.

Algorithm 7.6 Simulation of an NIG path

Let $\alpha, \beta, \gamma, \delta$ satisfy requirement (7.70). Further, consider a time discretization $0 = t_0 < ... < t_N = T$. Set $X(0) = 0$.
For $i = 1$ to N do

1. Simulate $G_i \sim IG(\delta(t_i - t_{i-1}), \gamma)$.
2. Simulate a standard normally distributed random number Y_i.
3. Set $X(t_i) := X(t_{i-1}) + \sqrt{G_i}Y_i + \beta G_i(t_i - t_{i-1})$.
4. Set $X(t) := X(t_{i-1}),\ t \in (t_{i-1}, t_i)$.

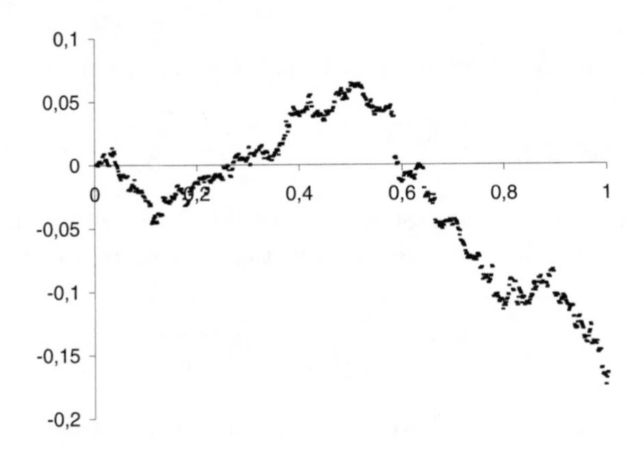

FIGURE 7.2: A simulated path of an NIG process.

Putting the just simulated path of an NIG process into the stock price representation (7.68) yields a path of the corresponding stock price. This simulation method yields a crude Monte Carlo algorithm for option pricing.

More sophisticated alternatives to the above method that deliver promising results can be found in Ribeiro and Webber (2003), who give a bridge sampling algorithm similar to the one presented in the VG process framework, and in Benth et al. (2006) who consider a quasi-Monte Carlo approach.

7.3.5 Further aspects of Lévy type models

Multivariate models

Thus far, different from jump-diffusions, we considered only univariate Lévy processes. One could of course easily think about a direct generalization of the subordinator idea: Just replace the one-dimensional Brownian motion by a d-dimensional one. However, as pointed out by Cont and Tankov (2003) the use of the same subordinator for all stock price processes automatically creates a certain dependence structure. So it is important to have independent (nontrivial) assets in such a setting. A possibility out of this dilemma would be the use of a factor model for the time evolution. More precisely, we need more than one subordinator that determines the actual clock inherent in the movement of asset prices.

Another alternative is the use of so-called Lévy copulas to model multiasset markets. Again, we refer to Cont and Tankov (2003) for a survey on this issue.

More models

Another class of processes that has become popular recently is the class of Meixner processes which was introduced by Schoutens and Teugels (1998). Its construction is intimately related to orthogonal polynomials. As it does not seem to be favourable when compared to the VG process or the NIG process, we do not present it here in detail, but refer the interested reader to Schoutens (2000).

A criticism of the Lévy models considered thus far is that they do not include a stochastic volatility term. Due to the stationarity of the increments, Lévy processes are too similar over time. Therefore, Barndorff-Nielsen and Shephard (2001) introduced a Lévy model with an additional mean-reverting volatility parameter. More precisely, they assumed a log-price $Z(t)$ of the form

$$dZ\left(t\right) = \left(r - \lambda\kappa\left(-\rho\right) - \sigma\left(t\right)/2\right)dt + \sqrt{\sigma\left(t\right)}dW\left(t\right) + \rho dL\left(\lambda t\right) \tag{7.76}$$

where the volatility process $\sigma(t)$ is given as

$$d\sigma(t) = -\lambda\sigma\left(t\right)dt + \lambda dL\left(\lambda t\right). \tag{7.77}$$

Here, $L(t)$ is a subordinator, $\lambda > 0$, $-\kappa(u) = \ln(\mathbb{E}(\exp(-uL(1))))$, and $W(t)$ is a Brownian motion independent of $L(t)$. Note that the form of the volatility specification has a tendency to decrease to zero slowly with time while sudden jumps increase it again. The parameter ρ models a correlation effect between the stock price and the volatility.

Lévy models with stochastic volatility are further discussed in Schoutens (2003).

Chapter 8

Simulating Actuarial Models

8.1 Introduction

In the preceding chapters we have been interested in pricing isolated financial contracts. There, the main principle to deal with market risk, the risk of losses due to unfavourable price moves, is to switch to an equivalent market martingale measure and calculate the present value of a financial contract. This approach is based on the assumption that assets underlying these contracts can be traded to reduce or even eliminate the inherent risks.

In insurance mathematics, we look at the risks arising from insurance contracts (such as in life insurance or car insurance). However, they cannot be traded and the arbitrage argument often plays no role in their valuation. As these contracts are often sold in high numbers, suitable variants of the law of large numbers suggest the expected present value of the future payments as an indicator for the value of a contract. To be on the safe side, safety loadings are included in the premium calculations. Also, dependencies can play an extremely important role in judging the risk arising from the whole portfolio of sold contracts. We will therefore consider the two important topics of premium principles and of dependence modelling. Further important types of risks that we will explicitly look at are the risk of rare events in nonlife insurance and the longevity risk in life insurance. In both cases Monte Carlo methods are suitable tools. The concepts of copulas or quantiles that we are dealing with in this chapter also have applications in finance.

As insurance mathematics is a classical subject, there are various monographs on different aspects and types of insurance. Among them are standard and recent texts such as Bühlmann (2005), Gerber (1997), Mikosch (2004), and Møller and Steffensen (2007), just to name a few.

8.2 Premium principles and risk measures

The **premium** of an insurance contract is that part of its price that should be sufficient to cover the risk that the insurance company takes over with this contract. The actual price of the contract also contains parts that are needed for covering administrational expenses and further costs. This is sometimes called **gross premium**. We will not consider administrational costs and only look at the premium as described above. To calculate this premium a so-called **premium principle** is used.

As premium principles are used to judge the risk inherent in an insurance contract, it is reasonable to present along with them so-called (financial) **risk measures** that are developed for judging and managing financial risks.

We will introduce both concepts, comment on the aspects of their Monte Carlo simulation, and comment on the relationship between risk measures and premium principles.

8.2.1 Properties and examples of premium principles

To introduce a premium principle, we first have to introduce the notion of a **risk** X as a random variable for which we always assume suitable integrability properties when they are needed. We further assume that the considered insurance contract starts right away, which implies that the risk is present immediately. We will consider a different case in life insurance when contracts might start in the future. For this, we then have to introduce a suitable discounting. Formal definitions are given below.

DEFINITION 8.1
Let $(\Omega, \mathcal{F}, \mathbb{P})$ be a probability space.

(a) A **risk** *X is a nonnegative random variable on $(\Omega, \mathcal{F}, \mathbb{P})$.*

(b) A functional $p(.)$ on the space of risks $\mathcal{X}$ is called a **premium principle**.

As there are many suggestions around for different premium principles to use, there are also properties in the literature by which the suitability of a premium principle should be judged. We state four of them (see Sundt [1993]) but refer the reader to an impressive collection of nearly 20 properties listed in Laeven and Goovaerts (2008):

DEFINITION 8.2
Let X, Y be two risks. Then, some reasonable properties of a premium principle $p(.)$ are:

1. $$p(X+Y) \le p(X) + p(Y)$$ **subadditivity**

2. $$p(X) \le p(X+Y))$$ **monotonicity**

3. $$p(X) \ge \mathbb{E}(X)$$ **nonnegative safety loading**

4. $$p(X) \le \sup_{\omega \in \Omega} X(\omega)$$ **no ripoff**

REMARK 8.3 The interpretations of the above properties are indicated by their names:

1. Property 1 requires that it should not be profitable to split the risk $X + Y$ and sign two contracts, one for X and the other one for Y. However, this property is at debate of being reasonable without further assumptions on the dependence between X and Y.

2. Property 2 is a monotonicity requirement: Additional risk needs additional premium.

3. Property 3 is motivated by the law of large numbers: If the company would charge less than the expected loss $\mathbb{E}(X)$ ("fair premium"), it would for sure go bankrupt given the number of sold contracts is large.

4. Property 4 is reasonable as no customer would sign a contract that costs more than the highest possible claim size.

With the choice of presenting these properties we do not claim that they are the most important ones. They simply serve as popular examples. □

We present some popular premium principles and check to see if they share the above properties.

DEFINITION 8.4
The **expectation principle** $p_{exp}(X)$ *for a claim* X *and a constant* $\mu > 0$ *is given by*

$$p_{exp}(X) = (1 + \mu)\mathbb{E}(X). \tag{8.1}$$

The expectation principle satisfies Properties 1, 2, and 3 of Definition 8.2. However, it obviously violates Property 4 for constant claims or for claims with a maximum smaller than $(1 + \mu) \cdot \mathbb{E}(X)$. However, as in both cases such a contract could not be sold, it is clear that such a value for μ would not be used in practice. A further weakness seems to be that the fluctuations of the risk X plays no role for its premium.

Principles that explicitly take the fluctuations of the claim sizes into account are the variance and the standard deviation principles.

DEFINITION 8.5
Let $\mu > 0$ *be a given constant.*

1. The **variance principle** $p_{var}(X)$ *for a claim* X *is given by*

$$p_{var}(X) = \mathbb{E}(X) + \mu \cdot \mathbb{V}ar(X). \tag{8.2}$$

2. The **standard deviation principle** $p_{sd}(X)$ *for a claim* X *is given by*

$$p_{sd}(X) = \mathbb{E}(X) + \mu \cdot \sqrt{\mathbb{V}ar(X)}. \tag{8.3}$$

Although at first sight more sophisticated than the expectation principle, both these premium principles might violate Property 2 of Definition 8.2 which is a serious defect. The reason for this is that expectation and variance/standard deviation are related in a nonlinear and nonmonotonic way. To demonstrate this, we assume that we have a probability q for a claim X to arise in the next period and that the height of the claim is $\Gamma(1, 500)$-distributed. The premium arising for this configuration for the standard deviation principle with $\mu = 1$ is displayed in Figure 8.1.

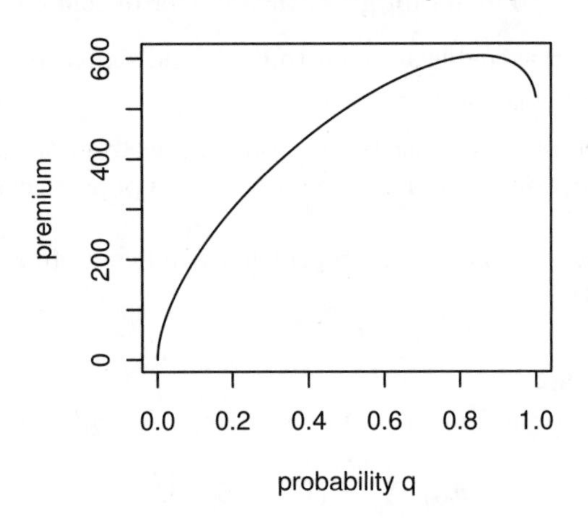

FIGURE 8.1: Premium from standard deviation principle.

Note that the resulting premium for $q = 1$, i.e. that a claim occurs for sure, is lower than for one which only occurs with probability $q = 0.856$. This violates the monotonicity requirement and is not acceptable. So one should take great care when using one of those premium principles. On top of that the variance principle also violates the subadditivity property.

A premium principle that avoids these problems (see Fischer [2003]) is the semistandard deviation principle which incorporates only deviations from the mean by high claims.

DEFINITION 8.6

The **semistandard deviation principle** $p_{ssd}(X)$ *for a claim* X *and a con-*

stant $0 \leq \mu \leq 1$ *is given by*

$$p_{ssd}(X) = \mathbb{E}(X) + \mu \cdot \sqrt{\mathbb{E}\left\{[\max(0,\ X - \mathbb{E}(X))]^2\right\}}. \tag{8.4}$$

All premium principles that we have considered so far were closely related to the strong law of large numbers and the central limit theorem (which motivates the use of standard deviation or variance as a measure for judging the risk of deviation from the expectation of a claim) and explicitly added some safety loadings. The following principle – which is known as the expected utility principle – incorporates the attitude towards risk of the insurer by the introduction of a utility function.

DEFINITION 8.7
Let $U(x)$ be a utility function (i.e. a concave, strictly increasing function). A premium $p_{eu}(X)$ for a claim X is said to be calculated by the **expected utility principle** *if we have*

$$U(c) = \mathbb{E}\left(U\left(c - X + p_{eu}(X)\right)\right) \tag{8.5}$$

where c is a (positive) constant, e.g. the wealth of the insurer.

REMARK 8.8 1. The premium is such that the utility from signing the new contract is equal to the utility of not signing it.

2. Property 1 of a premium principle is not fulfilled. However, this is desired, because higher risk should lead to overproportionally increasing premiums. The reasoning behind this can be seen when comparing the risks corresponding to n customers of the same age having identical life insurance contracts with the risk of a single customer insured on n-times the sum payable at death and annuity rate. In the second case, the longevity risk and the early death risk are much higher than n-times the risks in the first case, because in the first case the longevity and early death risks average out over the n customers.

3. Instead of the constant c one should insert the random variable C representing the whole portfolio of claims and replace the left-hand side by $\mathbb{E}\left(U\left(C\right)\right)$. This leads to a high premium for a claim being highly correlated to C and a low premium for a claim leading to a diversification in the portfolio.

4. Note that for the particular choice of the utility function

$$U(x) = \frac{1}{\alpha}\left(1 - e^{-\alpha x}\right), \text{ for a fixed } \alpha > 0 \tag{8.6}$$

the premium calculated by the expected utility principle is independent of c and is explicitly given by

$$p_{eu}(X) = \frac{1}{\alpha}\ln\left(\mathbb{E}\left(e^{\alpha X}\right)\right) \tag{8.7}$$

(see Laeven and Goovaerts [2008] for this and also for more premium principles). This principle only yields a finite premium for exponentially bounded risks. □

8.2.2 Monte Carlo simulation of premium principles

After the choice of a premium principle has been made, there remains the task of explicitly calculating the premium of an insurance contract. If this cannot be done explicitly, Monte Carlo simulation is a possible method of choice. This is straightforward for the expectation principle (of course, besides the fact that contracts with a complicated insurance payment structure might require methods as sophisticated as for exotic options) and also for the expected utility principle in case of the exponential utility function when the explicit expression of Equation (8.7) is used.

However, there is a new aspect introduced by the premium principles that include the variance in any form as an ingredient. To calculate the variance we already need the expectation. Of course, for large values of N (the number of Monte Carlo runs) one can use the Monte Carlo estimate of the expectation, the arithmetic mean. As we can calculate the variance as

$$\mathbb{V}ar\,(X) = \mathbb{E}\left(X^2\right) - \left(\mathbb{E}\,(X)\right)^2 \tag{8.8}$$

the Monte Carlo estimation of $\mathbb{E}(X)$ and of $\mathbb{E}(X^2)$ can be done simultaneously. However, for estimating the semivariance $\mathbb{E}([\max(0,\ X - \mathbb{E}(X))]^2)$ such a decomposition is not available. Thus, one could perform a two-step procedure:

1. Estimate the mean $\mathbb{E}(X)$ by $\bar{X}_{N_1}$ based on N_1 simulation runs.

2. Estimate the semivariance $\mathbb{E}([\max(0,\ X - \mathbb{E}(X))]^2)$ based on N_2 new simulation runs by

$$\frac{1}{N_2}\sum_{i=1}^{N_2}\left(\max\left(0,\ X - \bar{X}_{N_1}\right)\right)^2.$$

8.2.3 Properties and examples of risk measures

A risk measure is related to a financial position $\tilde{X}$ and a time horizon T. Here, the position $\tilde{X}$ can be both positive or negative. In contrast to insurance claims, $\tilde{X} > 0$ describes a profit.

Föllmer and Schied (2002) state the requirement on a risk measure clearly: "...a risk measure is viewed as a capital requirement: We are looking for the minimal amount of capital which, if added to the position and invested in a risk-free manner, makes the position acceptable."

DEFINITION 8.9
A risk measure ρ is a real-valued mapping defined on the space of random variables.

As this definition is fairly weak, we present some requirements on a risk measure that are popular in the literature.

DEFINITION 8.10
Let $\tilde{X}, \tilde{Y}$ be two financial positions. Some reasonable properties of a risk measure $\rho(.)$ are:

1. $\rho(\tilde{X} + m) = \rho(\tilde{X}) - m \ \forall m \in \mathbb{R}$ **translation invariance**
2. $\tilde{X} \geq \tilde{Y} \ a.s. \Rightarrow \rho(\tilde{X}) \leq \rho(\tilde{Y})$ **monotonicity**
3. $\rho\left[\lambda\tilde{X} + (1-\lambda)\tilde{Y}\right] \leq \lambda\rho(\tilde{X}) + (1-\lambda)\rho(\tilde{Y}) \text{ for } \lambda \in [0,1]$ **convexity**
4. $\rho(\lambda\tilde{X}) = \lambda\rho(\tilde{X}) \text{ for } \lambda \geq 0$ **positive homogeneity**

REMARK 8.11 The meaning of the properties of risk measures can already be understood by their names:

1. Translation invariance means that riskless money changes the risk of a position by exactly the same amount. In particular, we observe $\rho(\tilde{X} + \rho(\tilde{X})) = 0$, i.e. if we invest the *risk premium* $\rho(X)$ in a risk-free manner then there is no risk anymore.
2. Monotonicity simply says that less risk requires less money set aside.
3. Convexity of the risk measure favours diversification.
4. Positive homogeneity implies that risk increases linearly in the units owned of a particular risky good. This property is heavily discussed in the literature as it totally ignores liquidity risk. From an insurance point of view, it also means that insuring 10 high towers, in for example San Francisco, against earthquakes bears the same risk as insuring 10 high towers in 10 different places. With regard to extreme risk, this is not reasonable as an earthquake in San Francisco will likely damage all high towers, whereas it is rather unrealistic that an earthquake happens at all the 10 places at the same time.

To normalize the range of the risk measure one can also require

$$\rho(0) = 0 \qquad \textbf{normalization}$$

which has the reasonable interpretation that a zero position has no risk. □

In the literature, mainly two types of risk measures are considered.

DEFINITION 8.12
A risk measure is called **convex** *if it satisfies the requirements 1 to 3 of Definition 8.10.*

A risk measure is called **coherent** *if it satisfies the requirements 1 to 4 of Definition 8.10.*

Properties 3 and 4 imply that a coherent risk measure is also subadditive. If it attains a finite value $\rho(0)$ then it also has the normalization property $\rho(0) = 0$.

We will look at some popular risk measures. The one which is mainly used in banks and has become an industry standard is the value-at-risk.

DEFINITION 8.13
The **value-at-risk** *of level α (VaR_α) is the α-quantile of the loss of the financial position $\tilde{X}$ at time T:*

$$VaR_\alpha(\tilde{X}) = -\inf\left\{u \in \mathbb{R} \middle| \mathbb{P}\left(\tilde{X} \geq u\right) \geq 1-\alpha\right\} \tag{8.9}$$

where α is a high percentage such as 95% or 99%.

As a quantile, VaR_α is easy to understand and very popular in applications. However, it does not give us an idea about the height of the actual loss above that quantile. Further, it is not convex and so it does not necessarily support diversification. To see this, consider the positions X, Y with:

$$\tilde{X} = \tilde{Y} = \begin{cases} 100 & \text{with probability } 0.901 \\ 0 & \text{with probability } 0.009 \\ -200 & \text{with probability } 0.09, \end{cases}$$

then for $\alpha = 90\%$, we obtain

$$VaR\left(\frac{1}{2}\tilde{X} + \frac{1}{2}\tilde{Y}\right) = 50,$$
$$\frac{1}{2}VaR(\tilde{X}) + \frac{1}{2}VaR(\tilde{Y}) = -100.$$

DEFINITION 8.14
The conditional value-at-risk (or average value-at-risk) is defined as

$$CVaR_\alpha(\tilde{X}) = \frac{1}{1-\alpha}\int_\alpha^1 VaR_\gamma(\tilde{X})\, d\gamma. \tag{8.10}$$

The $CVaR_\alpha$ coincides with the **expected shortfall** or **tail conditional expectation** defined by

$$TCE_\alpha(\tilde{X}) = -\mathbb{E}\left(\tilde{X} \middle| \tilde{X} \leq VaR_\alpha\right) \tag{8.11}$$

if the probability distribution of $\tilde{X}$ has no atoms. $CVaR_\alpha$ is indeed a coherent risk measure (see Acerbi and Tasche [2002]).

As for premium principles there is a risk measure based on expected utility:

DEFINITION 8.15

Let $U : \mathbb{R} \to \mathbb{R}$ be a utility function (i.e. strictly increasing and concave). Then, the risk measure based on utility of a financial position $\tilde{X}$ is given by

$$\rho_{utility}(\tilde{X}) = \inf\left\{m \in \mathbb{R} \middle| \mathbb{E}\left[U(\tilde{X}+m)\right] \geq U(0)\right\}. \tag{8.12}$$

It can be shown that the just defined risk measure based on a utility function is a convex risk measure (see Föllmer and Schied [2002]).

8.2.4 Connection between premium principles and risk measures

As both concepts are used to judge risks, they should have many features in common (indeed, already in Deprez and Gerber [1985] convex premium principles were discussed). However, before we comment on parallels one should also keep in mind that a premium principle is closer to a pricing principle as it is focused on a single contract. But the main concept behind this pricing approach is the strong law of large numbers and not the arbitrage principle of finance. Therefore, the classical premium principles such as the expectation (see Definition 8.4) or the variance principle as introduced in Definition 8.5 are not directly related to ideas of risk measures that have a tendency to concentrate on valuing the extreme risks. Examples for this point of view are the VaR, presented in Definition 8.13, or the CVaR, introduced in Definition 8.14. However, the semistandard deviation premium principle from Definition 8.6 also concentrates on the high claims.

An approach that connects both concepts is the one based on utility, the expected utility approach for premium principles and risk measures as in Definitions 8.7 and 8.15.

One can define a premium principle p out of a given risk measure ρ via the requirement of

$$\rho(p(X) - X) \stackrel{!}{=} 0 \tag{8.13}$$

for each claim X. Since $p(X)$ is riskless, this leads to the identification

$$p(X) = \rho(-X) \tag{8.14}$$

which would also allow us to extend the definition of a premium principle to general random variables.

Further, we can then compare the conditions imposed on premium principles and on risk measures. Given that the risk measure is convex and normalized, then it follows directly that the above defined premium principle is monotonic and also condition 4 of a premium principle is implied. Moreover, the subadditivity of a coherent risk measure implies the subadditivity of the premium principle. Condition 3 of a premium principle cannot be directly verified, as a risk measure is a priori defined without reference to a probability measure. However, for special choices of ρ such as CVaR and expected utility, this condition can be explicitly verified.

For convex risk measures which are not coherent, the above premium principle will also typically fail to be subadditive.

8.2.5 Monte Carlo simulation of risk measures

Quantile estimation and VaR

The first ingredient of estimating risk measures is the estimation of a quantile. The natural Monte Carlo estimator for an α-quantile $q_\alpha = F^{-1}(\alpha)$ of a random variable X with distribution function F is obtained by generating N realizations of X and then using the α-quantile $\hat{q}_{\alpha,N}$ of the empirical distribution $F_N(x)$.

Algorithm 8.1 Crude Monte Carlo simulation of the α-quantile

Let F be a given distribution function, $\alpha \in [0,1]$.

1. Simulate N independent random numbers $X_1, \ldots, X_N$, $X_i \sim F$.

2. Compute the empirical distribution function

$$F_N(x) = \frac{1}{N} \sum_{i=1}^{N} \mathbf{1}_{\{X_i \le x\}}.$$

3. Estimate the quantile q_α by

$$\hat{q}_{\alpha,N} = F_N^{-1}(\alpha).$$

REMARK 8.16 1. Of course, if F is explicitly known, then one would calculate the quantile via numerically solving $F(x) = \alpha$. So the crude Monte Carlo method of Algorithm 8.1 is only used when F is not available or hard to compute. We have already seen such examples in Sections 5.6.1 and 5.6.2 in the case of basket or Asian options, where F has been the distribution function of sums of log-normals which is not known explicitly. The above inversion of the empirical distribution function is of course done in a simple way: Order the simulated values by their size and then pick out the one at position $k = \min\{n \in \{1, ..., N\} \,|n/N \geq \alpha\}$.

2. To obtain a confidence interval for q_α, one can use the central limit theorem for quantiles (see Glynn [1996]) that states

$$\sqrt{N}\left(\hat{q}_{\alpha,N} - q_\alpha\right) \xrightarrow{D} \mathcal{N}(0, \sigma^2) \text{ (as } N \to \infty), \quad \sigma^2 = \frac{\alpha\,(1-\alpha)}{f\,(q_\alpha)} \tag{8.15}$$

where $f(.)$ denotes the density of the distribution function F. This directly gives us a 95%-confidence interval for q_α as

$$\left[\hat{q}_{\alpha,N} - 1.96\frac{\alpha\,(1-\alpha)}{f\,(q_\alpha)\sqrt{N}},\ \hat{q}_{\alpha,N} + 1.96\frac{\alpha\,(1-\alpha)}{f\,(q_\alpha)\sqrt{N}}\right]. \tag{8.16}$$

Note that we have the usual $1/\sqrt{N}$-convergence. However, the value $f\,(q_\alpha)$ is not under our control. It can become arbitrarily small, and typically is also very hard to estimate, in particular for large or small values of α. As an example, consider the simulation of a 0.995-quantile of a standard normally distributed random variable which is given by $q_{0.995} = 3.2905$. To obtain a 95%-confidence interval of length 0.001 we already need approximately N=1,215,492 if (!) the density value at $q_{0.995}$ is known. □

Note that to estimate the desired quantile accurately, it is necessary to have many observations close to it. However, as quantiles like value-at-risk are typically extreme quantiles, a crude Monte Carlo simulation leaves us with exactly the opposite situation: We generate a lot of observations far away from the quantile and only a few close to it. This directly calls for an application of importance sampling (as presented in Section 3.3.5) to reduce the variance of the quantile (and thus also to reduce the length of the confidence interval).

However, we then face a second problem. As we do not know the quantile and typically also not the form of the distribution function F, we need at least an approximation for F and f around the quantile. Here, a large deviations result (see Bucklew [1990] or Glynn [1996]) of the form

$$\mathbb{P}\,(X > x) \approx \exp\left(-x\theta_x + C\,(\theta_x)\right), \quad x >> \mathbb{E}\,(X), \tag{8.17}$$

with $C(.)$ the cumulant generating function of X and θ_x is given by $C'(\theta_x) = x$, comes to our help. If we want to estimate a high quantile q_α, $\alpha \approx 1$ then (in the case of a continuous distribution) we can simply solve the equation

$$1 - \alpha = \mathbb{P}\,(X > q_\alpha) = \exp\left(-q_\alpha\,(C')^{-1}\,(q_\alpha) + C\left(\left(C'\right)^{-1}(q_\alpha)\right)\right). \tag{8.18}$$

Note that there can be situations where the cumulant generating function $C(.)$ is easier to compute than the distribution function. It might therefore be possible that the root of Equation (8.18) can be computed while it is nearly impossible to solve the equation $F(q_\alpha) = \alpha$, indeed the only case where a Monte Carlo simulation is useful.

If the large deviations result would be exact then the root of this equation would equal the quantile q_α. However, it is typically only a crude approximation $\tilde{q}_\alpha$. But as it is an approximation, it gives us an idea to where we should shift the distribution F to obtain a new distribution function $\tilde{F}$ that is (more) concentrated in the neighbourhood of the quantile.

Indeed, for this we can use the method of exponential twisting as presented in Section 3.3.5. We change the original distribution from F to $\tilde{F}$ via

$$\tilde{F}(dx) = \exp\left(-x\theta_{\tilde{q}_\alpha} + C\left(\theta_{\tilde{q}_\alpha}\right)\right) F(dx). \tag{8.19}$$

This new distribution has a mean of $\tilde{q}_\alpha$ which is close to q_α, and thus we can use the modified quantile estimator

$$\hat{q}_{\alpha,N}^{imp,1} = F_{N,imp,1}^{-1}(\alpha), \tag{8.20}$$

$$F_{N,imp,1}(x) = \frac{1}{N}\sum_{i=1}^{N} L(X_i)\, 1_{X_i \le x}, \tag{8.21}$$

$$L(x) = \exp\left(x\theta_{\tilde{q}_\alpha} - C\left(\theta_{\tilde{q}_\alpha}\right)\right) \tag{8.22}$$

and where all the random variables X_i are generated under the distribution function $\tilde{F}$ of Equation (8.19).

An obvious alternative is the estimator which considers only the high values:

$$\hat{q}_{\alpha,N}^{imp,2} = F_{N,imp,2}^{-1}(\alpha), \tag{8.23}$$

$$F_{N,imp,2}(x) = 1 - \frac{1}{N}\sum_{i=1}^{N} L(X_i)\, 1_{X_i > x}. \tag{8.24}$$

Note that in the nontransformed setting the two quantile estimators coincide. As pointed out in Glynn (1996), $\hat{q}_{\alpha,N}^{imp,1}$ should be used for estimating low quantiles (i.e. for small values of α), while $\hat{q}_{\alpha,N}^{imp,2}$ is the choice for estimating high quantiles. Further, there is a generalized central limit theorem for this transformed quantile estimator. However, we do not give it here as it does not directly lead to an easy-to-compute confidence interval. We collect all the above work in formulating Algorithm 8.2.

Example 8.17

We look at an artificial example where we assume that we have $X \sim \mathcal{N}(0,1)$ and we want to compute $VaR_{0.999}(X) \approx 3.0902$ by the two Monte Carlo variants given above. For reasons of comparison we also give the difference between the values that are five positions higher and lower than the quantile estimator. To do the explicit calculation we can use

$$C(\theta) = \frac{1}{2}\theta^2 \tag{8.25}$$

Algorithm 8.2 Importance sampling for quantiles

Let the distribution function F, the level $\alpha \in (0,1)$, and $N \in \mathbb{N}$ be given.

1. Compute an approximate quantile $\tilde{q}_\alpha$ via solving Equation (8.18).
2. Generate random numbers X_i according to the distribution $\tilde{F}$ as in Equation (8.19).
3. If $\alpha < 0.5$ then use the estimator $\hat{q}_{\alpha,N}^{imp,1}$ according to Equation (8.20).
4. If $\alpha \geq 0.5$ then use the estimator $\hat{q}_{\alpha,N}^{imp,2}$ according to Equation (8.23).

for the cumulant generating function leading to the approximate value of

$$\tilde{q}_{0.999} = \sqrt{-2\ln(1-0.999)} = 3{,}7169. \tag{8.26}$$

For the choices of N= 1,000 and 10,000 we obtain the results of Table 8.1.

Method N	1,000	10,000
Crude MC quantile	3.03815	3.03146
Importance sampling	3.09744	3.09044

Table 8.1: $VaR_{0.999}$ Estimates for $\mathcal{N}(0,1)$ (True Value = 3.0902)

Note that these choices are actually very risky ones for the use of the crude method, as there the VaR is determined in its exact form by only a few numbers while for the importance sampling case the VaR lies close to the middle of the range of the distribution and is much more stable against outliers. This argument and the numerical results clearly demonstrate the advantages of the importance sampling method.

REMARK 8.18 1. The importance sampling method for calculating quantile performs very well if we can indeed compute all the necessary ingredients, in particular the cumulant generating function C. This is often a nontrivial task and in particular when we are not able to calculate the distribution function F. On the other side, if F is explicitly given then Monte Carlo methods are not needed! Therefore, the direct application of this method for calculating risk measures is quite limited.

2. As a second drawback, the calculation of an (asymptotic) confidence interval requires the calculation of the density function f at the quantile q_α itself. One can replace the quantile by its estimate $\hat{q}_{\alpha,N}^{imp,i}$, $i = 1, 2$, whichever

is more suitable. However, we still have the problem to estimate the density value itself which is a delicate task when it comes to accuracy. □

VaR via importance sampling and delta-gamma approximation

As we have seen in the previous section, the calculation of a quantile is not possible if the cumulant generating function of the underlying distribution is not known. This, however, is the typical case for a portfolio of a bank or an insurance company that also contains complicated products such as derivatives.

On the other hand, banks and insurance companies have to calculate risk measures such as VaR for their portfolio (sometimes) daily. We will present a workable way out of this that has been developed in a series of papers by Glasserman et al. (1999, 2000a, 2000b, 2001). This method rests on:

1. Calculating the probability of losses above a given value x instead of calculating VaR_{α}.
2. Use of importance sampling to reduce computational effort.
3. Repeating Step 1 with varying levels until the obtained loss probability is close to α.

To present it, we assume that the **loss** L over a given time horizon t is a function of a vector of underlying **risk factors** $(W_1, \ldots, W_n)$,

$$L = f(0, \ldots, 0) - f(t, W_1, \ldots, W_n). \tag{8.27}$$

Such a function typically is the sum of many functions $h^{(i)}(t, W_1, \ldots, W_n)$ that describe the prices of the different securities (stocks, derivatives, ...), that make up the portfolio of an investor, a bank, or an insurance company, between times 0 and t. One is now interested in estimating the probability to suffer a loss above a certain given value x,

$$\mathbb{P}(L > x) = \mathbb{E}(1_{L>x}). \tag{8.28}$$

As we have a representation of an expectation, this is a standard problem that can be dealt with by Monte Carlo simulation. Of course, there are two particular problems. One problem is that we consider x to be large, which means that with a crude Monte Carlo approach we will indeed not have a lot of observations that help us to estimate the probability accurately.

The second problem is that the evaluation of the loss L will take a lot of time if the underlying portfolio is large and contains a lot of functions $h^{(i)}$ which are nonlinear in the risk factors. In particular, these functions can be prices of exotic options and might themselves need separate Monte Carlo simulations to obtain these prices. As the loss probabilities have to be calculated by banks and insurance companies daily, approximations of L are popular. They typically rest on the assumptions that W has a multivariate normal distribution and that the loss function $f(.)$ is approximated by a Taylor polynomial. The so-called **delta approximation** rests on a linear approximation and is too coarse for typical portfolios containing derivatives. The **delta-gamma approximation** uses a second-order Taylor approximation

$$L = f(0,\ldots,0) - f(t,W) \approx -f_t(0)\,t - \nabla f(0)\,W - \frac{1}{2}W' Hess_f(0)\,W, \quad (8.29)$$

and is popular in applications. This is the starting point for the Monte Carlo analysis in the above mentioned series of papers by Glasserman et al. (1999, 2000a, 2000b, 2001). Note that the gradient ∇f contains all the partial derivatives $\partial f/\partial W_i$ (the *deltas*) and the Hesse matrix $Hess_f$ contains the second order partial derivatives $\partial^2 f/(\partial W_i \partial W_j)$ (the *gammas*). As both the deltas and gammas are the sum of the deltas and gammas of the members of the portfolio, they are usually calculated anyway by the traders and so do not create additional effort. The same is typically true for the time derivative, the *theta*.

The main idea of Glasserman et al. (1999, 2000a, 2000b, 2001) now is to use the delta-gamma approximation (8.29) as a substitute for L, assume $W \sim \mathcal{N}(0,\Sigma)$, and then apply an importance sampling step by exponential twisting (see Section 3.3.5) based on Equation (8.29). More technically, we first introduce a matrix B with

$$\Sigma = BB' \quad \text{and} \quad -\frac{1}{2}B' Hess_f(0)\,B = D \quad (8.30)$$

with D a diagonal matrix containing all the eigenvalues of $-1/2B'Hess_f(0)B$. Such a matrix exists as one can use a decomposition $\Sigma = AA'$, then diagonalize $-1/2AHess_f(0)A' = UDU'$, use the property that U is orthogonal, define $B = AU$, and use $D = -1/2B'Hess_f(0)B$. We can further assume (by permuting the indices of W_i if necessary) that the eigenvalues satisfy

$$d_1 \geq \ldots \geq d_n. \quad (8.31)$$

Thus, with the definition of $W = BX$ for some $X \sim \mathcal{N}(0,I)$ we can write

$$\begin{aligned} L &\approx -f_t(0)\,t - (B'\nabla f(0))'\,X - \tfrac{1}{2}X'B' Hess_f(0)\,BX \\ &=: f^{(0)} + b'X + X'DX = f^{(0)} + \sum_{i=1}^{n}\left(b_i X_i + d_i X_i^2\right) \\ &=: f^{(0)} + Q(X). \quad (8.32) \end{aligned}$$

We then transform the distribution of $Q(X)$ such that its mean under the new importance sampling distribution equals $x - f^{(0)}$. There might be many possibilities to achieve this. We will use the exponential twisting method as in the quantile estimation case, i.e. we use the importance sampling density

$$\ell(X) = \exp(-\theta Q(X) + C(\theta)) \tag{8.33}$$

with $C(.)$ the cumulant generating function of $Q(X)$.

As $Q(X)$ is a quadratic form in independent standard normals X_i, $C(.)$ is explicitly known and given by (see e.g. Imhof [1961] or Baldessari [1967])

$$C(\theta) = \sum_{i=1}^{n} \frac{1}{2}\left(\frac{(\theta b_i)^2}{1-2\theta d_i} - \ln(1-2\theta d_i)\right) = \sum_{i=1}^{n} C^{(i)}(\theta). \tag{8.34}$$

Under the importance sampling distribution the X_i remain independent, normally distributed, but with means $\mu_i(\theta)$ and variances $\sigma_i^2(\theta)$ given by

$$\sigma_i^2(\theta) = \frac{1}{1-2\theta d_i}, \quad \mu_i(\theta) = \theta b_i \sigma_i^2(\theta). \tag{8.35}$$

Note that we have an increase (decrease) of the variance for those X_i with $d_i > 0$ ($d_i < 0$). For the μ_i we have a similar effect for the signs of the b_i, but this can get mixed with the variance effect. It remains to choose the parameter θ as in the quantile estimation case, i.e. as the (unique) root of

$$C'(\theta_x) = x - f^{(0)} \tag{8.36}$$

which has to be done numerically and which ensures that we have

$$\mathbb{E}(\ell(X;\theta_x) Q(X)) = C'(\theta_x) = x - f^{(0)}. \tag{8.37}$$

Putting all our considerations together, we arrive at Algorithm 8.3.

REMARK 8.19 1. An approximate 95%-confidence interval for $\mathbb{P}(L > x)$ can be obtained via the usual formula of $\hat{p}_N^{\theta_x} \pm \frac{s_N}{\sqrt{N}}$ with s_N^2 being the sample variance of the $\ell(X^{(j)}) 1_{L^{(j)} > x}$. This should always be compared with the 95%-confidence interval corresponding to the crude Monte Carlo estimator $\hat{p}_N^0$ given by

$$\left[\hat{p}_N^0 - 1.96\frac{\sqrt{\hat{p}_N^0(1-\hat{p}_N^0)}}{\sqrt{N}},\ \hat{p}_N^0 + 1.96\frac{\sqrt{\hat{p}_N^0(1-\hat{p}_N^0)}}{\sqrt{N}}\right]. \tag{8.38}$$

To illustrate the size of the N that is typically needed, assume that we know $p = 0.001$ exactly. Of course a confidence interval should have a smaller order than the estimate itself. So to obtain a confidence interval of length 0.0001, we need $N \approx 10^6$. This is particularly high as each evaluation of the portfolio loss function is very costly in terms of computing time.

Algorithm 8.3 Value-at-Risk via importance sampling of the delta-gamma approximation

Let the loss function $L = f(W)$, $W \sim \mathcal{N}(0, \Sigma)$ and the loss level x be given. Assume further that $f(0)$, $\nabla f(0)$, $Hess_f\,(0)$ are given together with the (formal) delta-gamma approximation (8.29).

1. Preparation:
   (a) Determine B, D, b and $Q(X)$ as given in Equations (8.30) and (8.32).
   (b) Determine θ_x via solving Equation (8.36).
2. Simulation:
   For $j = 1$ to N
   (a) Simulate $X^{(j)} = (X_1^{(j)}, \ldots, X_n^{(j)})$ where the $X_i^{(j)}$ are independent with $X_i^{(j)} \sim \mathcal{N}(\mu_i(\theta_x), \sigma_i^2(\theta_x))$ as given in Equation (8.35).
   (b) Obtain $L^{(j)} = f(0) - f(t, BX^{(j)})$ and $\ell(X^{(j)})$ according to Equation (8.33).

   Obtain the loss probability estimator by: $\hat{p}_N^{\theta_x} = \frac{1}{N}\sum_{j=1}^{N} \ell(X^{(j)}) 1_{L^{(j)}>x}$.

2. Remember that our original intention was the calculation of a risk measure such as VaR_α. For this problem we need a good initial guess of x, i.e. we should have $\mathbb{P}(L > x) \approx \alpha$. Then, one should iteratively change x until the corresponding loss probability is close enough to α.

In particular, one should always be on the safe side, i.e. the 95%-confidence interval for $\hat{p}_N^{\theta_x}$ should be above α. To start the iteration, we can again use our knowledge on the delta-gamma approximation. As we have its cumulant generating function $C(\theta)$, we obtain explicit forms for its mean and variance:

$$\mathbb{E}\left(Q(X)\right) = C'\left(0\right) = \sum_{i=1}^{n} d_i, \quad \mathbb{V}ar\left(Q(X)\right) = C''\left(0\right) = \sum_{i=1}^{n}\left(b_i + 2d_i^2\right). \tag{8.39}$$

One can now start with an initial guess of

$$x = \mathbb{E}\left(Q(X)\right) + y\sqrt{\mathbb{V}ar\left(Q(X)\right)} \tag{8.40}$$

and then increase or decrease y depending on the resulting loss probability estimate. A simple choice for a value of y to start with would be the α-quantile of the standard normal distribution $q_\alpha = \Phi^{-1}(\alpha)$.

3. In Glasserman et al. (1999), the authors additionally apply a stratification procedure to reduce the variance of the loss probability estimator

even further. As the main factor of variance reduction is the above described importance sampling, we skip the presentation of the stratification step here.

4. In Dunkel and Weber (2007) a similar Monte Carlo approach is applied to the estimation of utility-based shortfall risk measures. □

Some aspects of the practical applications

We will illustrate the main aspects of the performance of the importance sampling method via simple examples. More detailed examinations of the numerical performance are given in Glasserman et al. (1999) and other papers by the same authors. There, it is shown that one can reduce the variance of the crude Monte Carlo estimator by factors well above 10 and even more.

In contrast to showing the numerical performance of the method for large portfolios, we will highlight some important aspects that one should be aware of before (!) applying the method:

1. **Computing risk measures versus pricing of derivatives.** When we are interested in calculating risk measures for the evolution of a portfolio of securities at a fixed future time T, we have to simulate paths of the underlying factors (such as interest rates, stock prices, ...) under the **real-world** probability measure. This is because we like to get a feeling for the height of possible losses which of course occur in the real world and not in the risk-neutral world. A particular consequence would be that we have to simulate the evolution of a stock price – in say the Black-Scholes setting – with a drift of μ which in general does not coincide with the riskless interest rate of r.

2. **Time horizon and normal approximation.** As the time horizon when risk measures have to be calculated are usually quite small (such as a day or a week), it is common practise to neglect the influence of the drift in stock prices and assume that we have

$$\Delta S := S(t + \Delta t) - S(t) \approx S(t)\,\sigma\,(W(t + \Delta t) - W(t)), \tag{8.41}$$

an approximation obtained by neglecting the deterministic parts in the exponent of the stock price (justified by the fact that Δt is small compared to $\sqrt{\Delta t}$ which is the order of the standard deviation of the random change) and using the first order approximation to the exponential function.

3. **Accuracy of the delta-gamma approximation and its consequences.** It is crucial to understand that the delta-gamma approximation is **not** (!) used as an approximation procedure in our method. It is used as an **orientation** in the search for a good position for our importance sampling transformation. So even if the approximation is poor the method might still work. On the other hand, as we only approximate the (in general unknown) distribution of the evolution of the value of the portfolio, we cannot automatically expect to have a performance of the importance sampling method as when the distribution is known. We highlight this by the following example:

Numerical illustrations: The delta-gamma approximation for a call option and for a portfolio of puts and calls. We start with a simple portfolio of one European call option on a stock with maturity T in a Black-Scholes setting. Our aim is to compute (an approximation of) the loss distribution of the portfolio for a fixed time $t + \Delta t$. Assuming that Δt is small, we use approximation (8.41). Denoting the value of the call at time t by $C(t, S(t))$ we like to compute the distribution function for the (possible) loss

$$L\left(\Delta t, \Delta W\right) = C(t, S(t)) - C(t + \Delta t, S(t)\sigma\Delta W). \tag{8.42}$$

Figure 8.2 shows its real distribution function $F_1(x)$ and the distribution function $F_2(x)$ based on the delta-gamma-approximation

$$\begin{aligned} &C(t, S(t)) - C(t + \Delta t, S(t) + \Delta S) \\ &\quad\approx -\frac{\partial C}{\partial t}(t, S(t))\Delta t - \frac{\partial C}{\partial S}(t, S(t))\Delta S - \frac{1}{2}\frac{\partial^2 C}{(\partial S)^2}(t, S(t))\left(\Delta S\right)^2 \end{aligned} \tag{8.43}$$

with ΔS given by Equation (8.41). Note that the two distribution functions are nearly completely identical. Hence, one expects an importance sampling procedure based on the delta-gamma approximation to be nearly as effective as one based on the original (but in general unknown) distribution. This often good approximation by the delta-gamma distribution is the main reason for the good performance of the method by Glasserman et al. (1999). In our

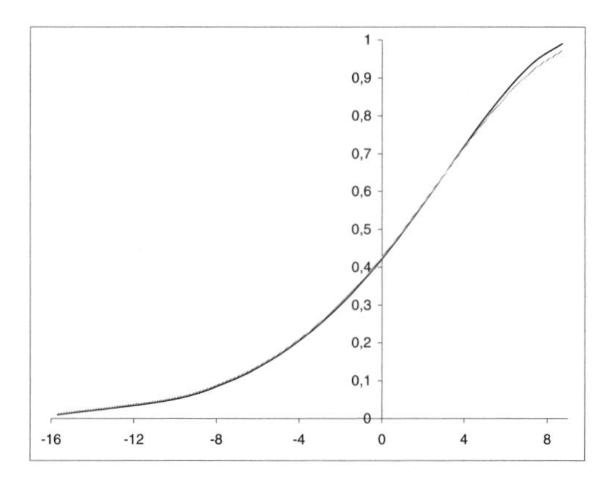

FIGURE 8.2: Exact loss distribution function $F_1(x)$ (black) and approximation $F_2(x)$ (grey) for $\Delta t = 0.1, \sigma = 0.3, T = t{+}0.5$, and strike $K = S(t) = 100$.

example, the importance sampling approach consequently yields a big variance reduction when applied at different loss probability levels. Table 8.2 contains estimates of the loss probabilities for $p = 0.05, 0.01$, and 0.005 for both the crude Monte Carlo method and the importance sampling method based on the delta-gamma approximation. Further, it contains the quotient of the variance of the crude MC method divided by that of the importance sampling approach. All computations are done for $N=$ 100,000 simulation runs. As

p	0.05	0.01	0.005
Crude MC estimator	0.0502	0.0099	0.00489
Delta-gamma IS estimator	0.0502	0.0100	0.00496
Variance ratio	9.65	51.94	109.21

Table 8.2: Crude Monte Carlo and Delta-Gamma-Based Importance Sampling (IS) for Estimating Loss Probabilities p

expected, the variance reduction (as a variance ratio much bigger than one) gets larger for smaller loss probabilities. To illustrate the importance sampling transformation we report that the $N(0,1)$-distributed input for the crude MC method is transformed to an $N(-2.782, 0.1776)$-distribution, i.e. the input is transformed, making losses much more likely by shifting the mean in the appropriate direction which is emphasized by reducing the variance.

We now consider the same characteristics for the loss of a portfolio consisting of 10 calls as above and 5 puts with the same input data. Although the loss of this portfolio is now no longer a monotonic function in the change of the stock price ΔS, the delta-gamma approximation of the loss function is still very accurate. We plot the distribution function of the loss together with the distribution function corresponding to the loss based on the delta-gamma approximation in Figure 8.3.

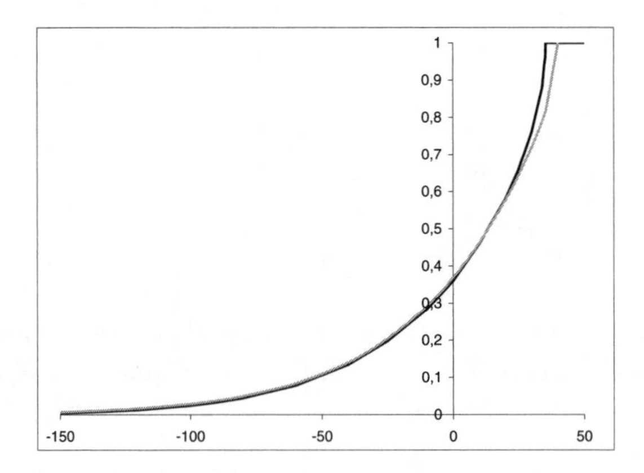

FIGURE 8.3: Exact loss distribution function $F_1(x)$ (black) and approximation $F_2(x)$ (grey) for $\Delta t = 0.1$, $\sigma = 0.3$, $T = t{+}0.5$, and strike $K = S(t) = 100$.

Note that for high quantiles the delta-gamma approximation overestimates the real losses while it is more accurate for low quantiles (i.e. for gains). However, the approximation is on both sides of the distribution good enough such that we can base the importance sampling approach on the delta-gamma approximation. For $N = 100{,}000$ simulation runs we looked at the ruin probabilities for the short version of the above portfolio. In Table 8.3, we compare the performance of the crude Monte Carlo method with the importance sampling approach. Note again the excellent behaviour of the importance sampling

p	0.05	0.01	0.005
Crude MC estimator	0.0494	0.0010	0.00494
Delta-gamma IS estimator	0.0501	0.0100	0.00500
Variance ratio	4.31	16.52	30.18

Table 8.3: Crude Monte Carlo and Delta-Gamma-Based Importance Sampling for Estimating Loss Probabilities p

approach. The main difference to the single-call portfolio case is that for the short call-put portfolio the big losses occur for both very high and very low values of ΔS. Therefore, the importance sampling method goes for a combination of shifting the distribution to the right (i.e. the mean of ΔS is increased) but also increases the variance (for the 0.005 level we have a shifted mean of 2.383 and a new variance of 2.625).

One has to keep in mind that in general we do not know the distribution function of the portfolio loss. In this case, we also have to start the iterative process to get into the region of the desired quantiles. Therefore, the above computations should only serve as an orientation about the performance and accuracy of the importance sampling method.

Of course, there are also situations when the delta-gamma approximation might fail. A typical such case is a portfolio which is both delta and gamma hedged. Then the delta-gamma approximation is simply the current value of the portfolio and therefore does not help as an orientation for the importance sampling distribution.

8.3 Some applications of Monte Carlo methods in life insurance

Life insurance mathematics is a classical subject, maybe one of the oldest applied mathematical subjects with an economic background. Also, nearly everyone has had a contact with life insurance products. They, however, have many variants, far too many to deal with all of them in this book. There is also a big number of monographs on the subject and we mention Gerber (1997) as a standard reference for the classical approach.

The two main sources of uncertainty in life insurance are the duration of the lifetime of the insured and the interest rate risk. Interest rate risk enters the scene as typically every life insurance product consists of uncertain payments in the future where time and/or height of the future payment are not known in advance. Thus, suitable discounting is an essential part of calculating the premium ("the price") the insured customer has to pay. As interest rate modelling has been dealt with in Chapter 5, we will here concentrate on the modelling of the evolution of **the force of mortality** over time.

8.3.1 Mortality: Definitions and classical models

We consider a customer of the age of x today (at time $t = 0$). Then, by D_x we denote the **time of his death** measured in years from now on. As the time of his death is random, we look at its distribution.

DEFINITION 8.20
Let $G_x : [0,\ \infty] \to [0,\ 1]$ be the distribution of D_x, the time of death of a customer of age x today, i.e.

$$G_x(t) = \mathbb{P}(D_x \leq t). \tag{8.44}$$

Its **survival probability** *for the next t years is denoted by*

$${}_tp_x := 1 - G_x(t). \tag{8.45}$$

Further, the **force of mortality** *$\mu_x(t)$ at time t is defined as*

$$\mu_x(t) := -\frac{d}{dt}\ln\left[1 - G_x(t)\right] = -\frac{d}{dt}\ln\left({}_tp_x\right). \tag{8.46}$$

REMARK 8.21 The importance of the force of mortality lies in the following facts. First, for a small time interval $[t,\ t + \Delta_t]$ the conditional probability for a customer of age x to die within it satisfies

$$\mathbb{P}\left(t < D_x \leq t + \Delta_t | D_x \geq t\right) \approx \mu_x(t) \cdot \Delta_t. \tag{8.47}$$

Further, one can show that the distribution G_x of the time of death and its density $g_x(t)$ are determined by the force of mortality as we have

$$g_x(t) = \mu_x(t) \cdot \left[1 - G_x(t)\right], \tag{8.48}$$

$$G_x(t) = 1 - {}_tp_x = 1 - \exp\left(\int_0^t -\mu_x(s) \cdot ds\right). \tag{8.49}$$

Thus, it is convenient to specify a certain form of the force of mortality to determine the distribution of the remaining lifetime of the customer. We will review some of them below. □

De Moivre (1724) introduced a maximal age A_{max} and assumed that the time of death is uniformly distributed on $[0,\ A_{max}]$, which corresponds to a force of mortality of

$$\mu_x^{\text{De Moivre}}(t) = \frac{1}{A_{max} - x - t}, \qquad 0 < t \leq A_{max} - x. \tag{8.50}$$

As the force of mortality usually grows with age, **Gompertz** (1824) dropped the assumption of a maximal age and suggested an exponentially growing force of mortality

$$\mu_x^{\text{Gompertz}}(t) = b \cdot \exp\left(c \cdot (x+t)\right) \quad b, c \text{ positive constants.} \tag{8.51}$$

Makeham (1860) extended the Gompertz model by adding a positive constant a (the so-called **young mortality**) to the force of mortality

$$\mu_x^{\text{Makeham}}(t) = a + b \cdot \exp\left(c \cdot (x+t)\right). \tag{8.52}$$

This model is still very popular in life insurance.

A popular model in material science is given by **Weibull** (1939) who suggests a polynomially growing force of mortality

$$\mu_x^{\text{Weibull}}(t) = a \cdot (x+t)^b, \quad a, b \text{ positive constants.} \tag{8.53}$$

8.3.2 Dynamic mortality models

One of the main current problems of both life insurers and pension funds is the continuing growth of the lifetime of the insured population. Underestimating the mean lifetime leads to too high costs of contracts compared to the premium for which they have once been sold.

One reaction to the change in expected lifetime has been the introduction of so-called generation life tables, i.e. there are different life tables for different generations. This takes care of the fact that a 60-year-old today has a different survival probability for the next year compared to that of a 60-year-old 20 years ago. On the academic side this evolution has caused an increasing interest in developing so-called **dynamic mortality models**.

An easy way to incorporate calendar time is an extrapolation approach (see Pitacco [2003]). It is based on the assumption that the survival probability ${}_tp_x\,(t)$ is a function of calendar time t for fixed age x. The basic idea of the extrapolation approach is to use the realized survival probabilities (i.e. the relative frequency of surviving members of the insured population of an age of x) during recent years as input for setting up an interpolation function such as

Algorithm 8.4 Modelling dynamic mortality by extrapolation

1. Interpret the realized survival probabilities of insured of age x at time t

$$_ip_x(t) = \frac{\text{No. of } (x+i)\text{-year old insured at time } t+i}{\text{No. of } x\text{-year old insured at time } t},$$

$t \in \{-(i+1), -(i+2), ..., -N_i\}$ as a function of calendar time t.

2. Use the realized survival probabilities to approximate this function by an interpolation function such as a spline or a polynomial.

3. Using the just determined interpolation function $f_x^{(i)}(.)$ obtain estimates $_i\hat{p}_x(0)$ for the current survival probabilities $_ip_x(0)$ by

$$_i\hat{p}_x(0) = f_x^{(i)}(0), \quad i = 1, ..., N .$$

---

a spline function or a polynomial. Predictions for future survival probabilities are then simply obtained by extrapolating the so-obtained function.

Below, we present a simulation approach that goes back to Lee and Carter (1992) (see Algorithm 8.5). Its main idea is to take a parametric mortality model such as the Gompertz-Makeham one and introduce uncertainty into it by modelling some of its components as stochastic processes.

Algorithm 8.5 Stochastic dynamic mortality modelling

1. Choose a parametric stochastic form for the mortality model of choice.

2. Determine the realized mortality rates of the past from past data.

3. Calibrate the parameters of the stochastic process underlying the stochastic mortality model of Step 1 to the time series of realized mortality rates from Step 2.

4. Choose the obtained stochastic process for simulating future mortality rates (or for calculating premiums).

Compared to the extrapolation method, the main advantages of this approach are the possibility to obtain error bounds via simulating many runs of the future mortality rates and that we can now use Monte Carlo methods for pricing all kinds of longevity products. Of course, one can also model survival probabilities in this framework instead of mortality rates.

In the literature on longevity models, various models of different complexity that follow the Lee and Carter approach are suggested (see e.g. Cairns et al. [2006]). As a particular example we look at the **stochastic Gompertz model** as given in Korn et al. (2006).

Example 8.22 The stochastic Gompertz model
Here, the suggested dynamic mortality model is given by

$$\mu_x^{SG}(t) = \alpha(t)\, e^{\beta(t)x}, \tag{8.54}$$
$$d\alpha(t) = -\kappa\alpha(t)\, dt, \quad \kappa > 0,\ \alpha(0) = \alpha_0 > 0, \tag{8.55}$$
$$d\beta(t) = \nu dt + \sigma dW(t), \quad \beta(0) = \beta_0 > 0 \tag{8.56}$$

with $W(t)$ a one-dimensional Brownian motion. Note that t now is also related to the calendar time (although it does not necessarily have to equal it).

The reasoning behind the above form of the equations for $\alpha(t), \beta(t)$ are the required positivity for $\alpha(t)$, the decrease of the overall mortality level over calendar time, and the seemingly linear behaviour of $\beta(t)$ as a function of time when calibrated to data (see Korn et al. [2006]). Further, the empirical evidence in the data considered by Korn et al. (2006) suggested that a one-factor model would be sufficient to explain the randomness in the evolution of the mortality rates over time. This is supported by the correlation structure in the Cairns-Blake-Dowd two-factor model (see Cairns et al. [2006]).

To be able to set up a simulation algorithm at time $\bar{t}$, we still have to calibrate the parameters κ, ν, σ, $\alpha(0)$, and $\beta(0)$. They can be obtained in a two step procedure:

1. First, obtain $\alpha(0)$, κ and $\beta(0)$ by fitting a standard Gompertz model with these parameters to the realized mortality rates at past times $t = 0, 1, ..., \bar{t} - 1$.

2. Estimate ν and σ from the time series $\beta(0), ..., \beta(\bar{t} - 1)$.

With all necessary model parameters obtained above, the simulation of the future mortality rates is now straightforward as described in Algorithm 8.6.

We will demonstrate in Section 8.3.4 how Algorithm 8.6 can be used when pricing mortality related insurance products.

REMARK 8.23 1. Note that we simulated the mortality rates on a yearly basis. As our model is a continuous-time one, it can be simulated on a finer scale. We suggest adapting the scale to that of the publishing time of the survival or mortality data to which the model is actually related. We could easily incorporate this in the above algorithm.

2. As we have survival probabilities of the form

$${}_1p_x(t) = \mathbb{E}\left(\exp\left\{-\int_0^1 \mu_{x+s}(t+s)\, ds\right\}\right) \tag{8.57}$$

Algorithm 8.6 Simulating dynamic mortality rates in the stochastic Gompertz model

Let $\alpha(\bar{t}), \beta(\bar{t})$ be given.
For the future times $t = \bar{t}+1, \bar{t}+2, ..., \hat{t}$ and all relevant age groups x do

1. $\alpha(t) = \alpha(t-1)e^{-k}$.
2. Simulate a random number $Z \sim \mathcal{N}(0,1)$.
3. $\beta(t) = \beta(t-1) + \nu + \sigma Z$.
4. $\mu_{\bar{x}}(t) = \alpha(t)e^{\beta(t)\bar{x}}$.
5. For $x = \bar{x}+1, ...\hat{x}$ set

$$\mu_x(t) = \mu_{x-1}(t)e^{\beta(t)}.$$

one could also argue for the need of a continuous-time simulation of the mortality rate. However, this would also require a continuous-time simulation along the age variable x. For this reason, we adopt the method of Korn et al. (2006) to approximate the integral above by the value of the integrand on the left-hand side. Simulating this value N times leads to the following estimate for the 1-year survival probability of

$$_1\hat{p}_x(t) = \frac{1}{N}\sum_{j=1}^{N}\mu_x^{(j)}(t) \tag{8.58}$$

where the upper index (i) denotes the simulated mortality rate from run j. The i-year survival probability will then simply be a product of the 1-year survival probabilities:

$$_i\hat{p}_x(t) = \prod_{j=1}^{i} {_1\hat{p}_{x+j-1}}(t+j-1). \tag{8.59}$$

3. **Further dynamic mortality models**: An example of a very detailed but also technically demanding approach is the Cox-Ingersoll-Ross type approach of Dahl and Møller (2006). It is a natural suggestion that also information about the future development of the life circumstances such as the climate, the medical, and the social treatment should be included in a dynamic mortality model. A suggestion of a corresponding multifactor model is made in Bauer and Russ (2006). □

8.3.3 Life insurance contracts and premium calculation

There is a great variety of life insurance-related products around. We will consider some basic ones and hint how to value more complicated variants.

Examples of simple life insurance contracts

We consider two basic subcases, **payments at death**, where a payment happens after the customer's death, and **payments when alive**, where payments happen (continuously) during the lifetime of the customer. Examples of the first type are:

- **Whole life insurance**: One unit is paid at death of the customer.
- **Term insurance of duration** n: One unit is paid at death of the customer if the death happens in the first n years.
- n **years deferred whole life insurance**: One unit is paid at death of the customer if the customer is still alive after n years since the start of the contract.

Examples of the second type are:

- **Pure endowment of duration** n: One unit is paid after n years if the customer is still alive.
- **Whole life annuity**: The customer obtains an annuity rate of one as long as the customer is alive.
- n **year temporary life annuity**: The customer obtains an annuity rate of one as long as the customer is alive, but for a maximum of the first n years.
- n **years deferred whole life annuity**: If the customer is alive after n years, then the customer obtains a life-long annuity rate starting then.

Of course, any combinations of these types are possible.

Premium calculations

In the following, we use $C(t)$ for a sum payable at the time of death t while we use $c(t)$ to denote an annuity rate (which may depend on t) that is paid as long as the insured is alive. On the pricing side, $\Pi(t)$ stands for a premium that has to be paid at time t while $\pi(t)$ denotes a premium rate at time t. If not otherwise stated, we make the following assumption.

ASSUMPTION 8.24
Interest and mortality rates are deterministic.

So, let all future interest rates be known today and assume that mortality $\mu_x(t)$ is deterministic. We also introduce $r_0(t), t \geq 0$ as today's yield curve, which is determined by the relation

$$P(0,t) = e^{-r_0(t)t} \tag{8.60}$$

where $P(0,t)$ is today's price of a zero bond maturing at time t.

As already indicated in the introduction of this chapter, the arbitrage principle of pricing in finance, which is based on replication or at least on hedging of the relevant payments, can in general not be used to value insurance contracts. The reason is that (at least substantial parts of) the underlying risk cannot be traded. Therefore, premium principles are used. We recall the so-called **net premium principle** in a form suitable for life insurance products.

DEFINITION 8.25

Assume that Assumption 8.24 holds. We consider an insurance contract for a customer of age x today (at time zero) that consists of a payment of $C(D_x)$ at the time of death and an annuity rate of $c(t)$ during the lifetime of the customer. We then say that this contract is valued by the **net premium principle** *if the single premium $\Pi(t)$ payable at time $t \geq 0$ and the premium rate $\pi(s)$ for $s \geq 0$ until death are determined such that we have*

$$\Pi(t) \cdot e^{-r_0(t) \cdot t} \cdot (1 - G_x(t)) + \int_0^\infty \pi(s) \cdot e^{-r_0(s)s} \cdot (1 - G_x(s))\, ds$$
$$= \int_0^\infty C(s)\, e^{-r_0(s)s} dG_x(s) + \int_0^\infty c(s) e^{-r_0(s)s} (1 - G_x(s))\, ds. \tag{8.61}$$

REMARK 8.26 1. The net premium principle is equal to the expectation principle of Definition 8.4 with the risk premium factor $\mu = 0$.

2. By noting that under Assumption 8.24 we have for example

$$\mathbb{E}\left(e^{-r_0(D_x)D_x} C(D_x)\right) = \int_0^\infty C(s) \cdot e^{-r_0(s)s} dG_x(s), \tag{8.62}$$

the net premium principle formulated above simply says that the net present value of the payments have to equal those of the premiums. We can thus directly generalize it also to the cases of dynamic (stochastic) mortality rates and to stochastic interest rates.

3. Of course, all the other premium principles presented in Section 8.2 can be applied to value life insurance products.

To obtain unique premium payments one first has to decide e.g. on the time when the premium should be paid, if it should be paid upfront, or in constant or varying rates. These properties are usually already fixed as part of the insurance contract. Then, valuing such a contract under Assumption 8.24 is straightforward. If Assumption 8.24 is not satisfied then Monte Carlo methods enter the scene. ∎

8.3.4 Pricing longevity products by Monte Carlo simulation

Longevity products are any kind of financial contracts that have payments which are adapted to the survival behaviour of a specified cohort of insured people. One can think of the population of e.g. all insured 65-year-old males in Germany. Two prominent examples are the EIB/BNP-longevity bond offered in 2004 and the launch of the LifeMetrics framework by JP Morgan in 2007.

We will only consider the pricing of the longevity bond, as the forward contracts of the LifeMetrics framework are only linear contracts. The EIB/BNP-longevity bond was launched in November 2004. It had a face value of 540 million Euro, 25-year duration, and its annual coupon payments should be multiplied with the fraction of survivors of the cohort of English and Welsh males aged 65 in 2003. The longevity bond was withdrawn in 2005 due to insufficient interest and due to accounting problems. One reason for the insufficient interest was the structure of the bond that did not supply enough hedging against the longevity risk as only one cohort of 65-year-old males do not cover the whole population of insured of a particular insurer. Also, it offered no protection against longevity risk in the liabilities beyond 25 years.

For the pricing of a longevity bond of the type described above, let us assume that a cohort is fixed and introduce

$$S(i) = \frac{\text{Number of survivors of the cohort at time } i}{\text{Size of cohort at time } 0}, \tag{8.63}$$

the fraction of survivors i years after the start of the longevity bond. Let z be the annual coupon payment of the longevity bond if the whole cohort would survive. If we now take a financial mathematics valuation view then under a suitable pricing measure $\mathbb{Q}$, we would obtain the price of the longevity bond as a discounted expectation. For this we assume independence of the evolution of the mortality rate and the interest rate (under $\mathbb{Q}$). Then the price of a longevity bond with N coupon payment times $1, 2, ..., i$ (with zero denoting the starting time of the bond) is given by

$$\begin{aligned} P^{(L)}(t) &= \mathbb{E}_Q\left(\sum_{i=1}^{N} \exp\left(-\int_t^i r(s)\,ds\right) S(i)\, 1_{\{t\leq i\}} \,|f_t\right) \\ &= \sum_{i=1}^{N} P(t,i)\, \mathbb{E}_Q\left(S(i)\,|f_t\right) 1_{\{t\leq i\}} \end{aligned} \tag{8.64}$$

where $P(t,s)$ is the market price of a zero bond at time t maturing at time $s \geq t$. We thus can benefit from the fact that in principle we only have to value the mortality risk as the fixed income market has already priced the bond components of the longevity bond.

As a possible choice for the *mortality component* of the valuation measure $\mathbb{Q}$, one could use the measure $\mathbb{P}$ that underlies the modelling of the dynamic mortality process which would result in

$$P^{(L)}(t) = \sum_{i=1}^{N} P(t,i)\, \mathbb{E}_{\mathbb{P}}\left(S(i)\,|f_t\right) 1_{(t \leq i)}. \tag{8.65}$$

As, however, there is not yet a market for trading mortality risk, it remains questionable if the application of the financial valuation principle which is based on arbitrage considerations can be justified at all. A hint which points in the direction of this question is the fact that the actual price $P_M^{(L)}(t)$ of the offered longevity bond has been higher than the one computed with the above pricing measure. There are (at least) two possible explanations for this:

- The issuer of the longevity bond has used an actuarial valuation principle. Indeed, the expectation principle would have included a risk premium factor $\mu > 0$ that leads to

$$P_M^{(L)}(t) = (1+\mu)\, P^{(L)}(t) \ . \tag{8.66}$$

- Another possible (but more restricted) explanation is that as mortality cannot be traded, the market where the longevity bond is traded is incomplete. One could therefore argue to replace the *mortality measure* $\mathbb{P}$ by another measure $\mathbb{P}(\mu)$ that is equivalent to $\mathbb{P}$ and that explains the observed market price, i.e. that in our case satisfies

$$P_M^{(L)}(t) = \sum_{i=1}^{N} P(t,i)\, \mathbb{E}_{\mathbb{P}(\mu)}\left(S(i)\,|f_t\right) 1_{(t \leq i)} \ . \tag{8.67}$$

  The determination of μ from this equation is actually the usual calibration principle of financial valuation in incomplete markets.

We consider the longevity bond pricing in the framework of Korn et al. (2006). There, the class of probability measures $\mathbb{P}(\mu)$ is parameterized by a parameter $\mu \in \mathbb{R}$. The change to the new measure $\mathbb{P}(\mu)$ corresponds to a change of the original drift λ of the underlying Brownian motion to $\lambda - \mu\sigma$. We describe how we can obtain the suitable parameter μ^* in Algorithm 8.7.

REMARK 8.27 Note that to obtain the probabilities ${}_ip_x^{\lambda}$ in Algorithm 8.7 at least approximately, we typically have to run a Monte Carlo simulation. For this, we simply generate a large number of paths of the mortality process μ_x^t as described in Algorithm 8.6. Then, we determine (pathwise) estimates of the survival probabilities by Remark 8.23 and average over all of them to obtain our Monte Carlo estimates (for the different times i). As a change to another μ means that the drift in the mortality process is changed by a constant, we only have to generate the mortality process paths once and can then simply change them by correcting for the difference in the drift. □

Algorithm 8.7 Measure calibration in the stochastic Gompertz model with a traded longevity bond

Let $P_M^L(0)$ be the market price of a traded longevity bond.

1. Calibrate the unknown parameters of a stochastic Gompertz model to realized mortality rates.

2. Determine the parameter μ^* such that we have

$$P_M^{(L)}(0) = \sum_{i=1}^{N} P(0,i)\, \mathbb{E}_{\mathbb{P}(\mu^*)}(S(i)) = \sum_{i=1}^{N} P(0,i)\, {}_ip_x^{\mu^*}.$$

REMARK 8.28 More complicated products such as variable annuities or index fund-based contracts with possibly early exercise features can be valued by the appropriately modified algorithms already presented in Chapter 5. We do not go into detail here, but we mention a fundamental difference to the financial market. It is by far not clear that the owners of such products will exercise their contracts in an optimal way. There might be different arguments than the pure value-based one that leads to seemingly unreasonable early exercise. Knowledge of this type of (average) exercise behaviour of the customers is therefore a crucial ingredient in the insurance setting. □

8.3.5 Premium reserves and Thiele's differential equation

As the equality between the net present values of the payments to the customer and of the incoming premiums does not necessarily need to hold after the insurance contract is sold, it is important to know how this possible difference evolves over time. It is in particular important for the insurance company to know the **premium reserve**, i.e. the amount of money it must have invested in order to be able to fulfill the expected future liabilities to the customer. This is the conditional expectation of the future payments, conditioned on the survival of the customer until time u. It is called the prospective premium reserve, because it considers only future payments.

DEFINITION 8.29

The **prospective premium reserve** *$V_x(u)$ at time $u \geq 0$ of an insurance contract of a customer aged x at the start of the contract at time zero and still alive at time u is defined as the conditional expectation of the future payments to the customer minus the conditional expectations of the future premiums (conditioned on the event that the customer is alive at time u).*

Example 8.30 Prospective reserve of a term insurance contract
To calculate the representation of a term insurance contract for n years, we assume that Assumption 8.24 is valid and that the contract consists of the following payments:

- $C(D_x)$, the death sum paid out if the customer dies before the n years,
- C^a, a lump sum paid out upon survival until time n,
- $\pi(t) = \pi$, the constant continuous premium rate paid as long as the insured is alive.

We further assume that the interest rate is given by r and the force of mortality is given by $\mu_x(t)$ and is deterministic. Under these assumptions we obtain the prospective premium reserve as

$$V_x(t) = C^a \cdot e^{-r(n-t)} \cdot {}_{n-t}p_{x+t} + \int_t^n e^{-r(s-t)} \cdot {}_{s-t}p_{x+t} \left(\mu_{x+s} C(s) - \pi\right) ds. \tag{8.68}$$

Additional contractual payments lead to additional terms in this equation.

A celebrated result in life insurance mathematics is Thiele's differential equation that describes the evolution of the prospective reserve over time given that the customer is still alive. Differentiating representation (8.68) with respect to t yields the following theorem.

THEOREM 8.31 Thiele's differential equation
We consider a term insurance contract as in Example 8.30 and assume that the yield curve is flat, i.e. $r_0(t) = r$ for all $t > 0$. Further, we assume that the distribution $G_x(t)$ has a density and thus the mortality process $\mu_x(t)$ is defined. Then, the prospective premium reserve $V_x(t)$ solves

$$\frac{d}{dt} V_x(t) = r \cdot V_x(t) + \pi(t) + \left[V_x(t) - C(t)\right] \mu_x(t) \ \forall t \in [0, n), \tag{8.69}$$

$$V_x(n) = C^a. \tag{8.70}$$

Use of Thiele's differential equation, generalizations, and Monte Carlo simulation

The main advantage of Thiele's differential equation is that given the features of the payment, we can use it to calculate the initial premium $\Pi(0)$ by equating it to $V_x(0)$ using the net premium principle, or to (re-)design the contract via choosing either the premium rate, the lump sum payment, or the death sum. As, however, Thiele's differential equation indeed describes the dynamic evolution of expectations, we can also calculate these expectations directly with the help of Monte Carlo simulation. More precisely, we can simulate the lifetime of the insured, calculate the relevant payment streams, do this N times, average over the results, and obtain an approximation $\hat{V}_x(0)$ of the prospective reserve, and then (re-)design the contract. An important point of this approach is that we **do not** need the simplifying Assumption 8.24. In this more general case and in situations such as equity-linked contracts there exist generalizations of Thiele's differential equation that are derived with the help of a financial mathematics-based pricing approach. Steffensen (2000) is a nice reference for this subject where Thiele's differential equation actually has the form of a partial differential equation which – in view of the Feynman-Kac representation 4.56 – is not surprising.

However, note that in this generalized setting the so-calculated value of the prospective reserves **cannot** be justified by the law of large numbers. Under Assumption 8.24 only the uncertainty about the lifetime of the customer remains for which it could be argued that the payment levels average out over the many different customers due to the law of large numbers. If on the other hand interest rate uncertainty and maybe even stock price uncertainty enter into the future payments, then taking an expectation as a suggestion for the value of the reserves can only be justified if all the payments can be reproduced at the financial market. Thus, it is better to speak of the **mean prospective reserves**. We give a Monte Carlo framework for calculating these mean prospective premium reserves in Algorithm 8.8.

Algorithm 8.8 Simulation of the (mean) prospective reserve of an insurance contract

---

For $i = 1$ to N do

1. Simulate the lifetime of the customer $l_x^{(i)}$.
2. Simulate a path $r^{(i)}(t),\ t \in [0, min\{n, l_x^i\}]$.
3. Calculate all payments of the contract based on the lifetime $l_x^{(i)}$, discount them with the suitable discount factor $\exp(-\int_0^t r(s)\, ds)$, and add them up to obtain $\hat{V}_x^{(i)}(0)$.

Calculate the approximation for the (mean) prospective reserve:

$$\hat{V}_x(0) = \frac{1}{N}\sum_{i=1}^{N} \hat{V}_x^{(i)}(0).$$

There are many more possible applications of Monte Carlo simulation in life insurance problems. However, we only give the above framework as it can easily be applied to all kinds of contracts without the need of a solution of a suitable variant of Thiele's differential equation. One should of course always think carefully if the risks of deviation from the (approximated) means inherited in these computations are **automatically hedged** by the validity of a law of large numbers (due to averaging personal properties, such as life length, over many customers) or not (such as stock return risk).

8.4 Simulating dependent risks with copulas

Apart from the family of normal distributions, there do not seem to be other popular families of distributions which allow a natural multivariate generalization such that one can easily simulate dependent random variables. Often the joint distribution of random variables can only be explicitly computed if the random variables are independent. The concept of copulas is a very useful tool to overcome this problem. It has therefore become popular in recent years in the areas of credit risk modelling and in e.g. nonlife insurance mathematics: In the latter, dependence between different insurance contracts cannot be neglected (think of a hail storm that might affect many car insurance contracts in a given region in a similar way).

In this section, we introduce the concept of copulas together with their main properties and ways to simulate dependent random variables ("dependent risks") with their help (see Embrechts et al. [2003] for most of the proofs of the results given in this section; see McNeil et al. [2005] for further references).

8.4.1 Definition and basic properties

DEFINITION 8.32
A **copula** *C is a distribution function on $[0,\ 1]^n$ for $n \in \mathbb{N}$ with uniformly distributed marginals, i.e.*

$$C(1,\ldots,\ 1,\ x_i,\ 1,\ldots,\ 1) = x_i \quad \forall i \in \{1,\ldots,n\}. \tag{8.71}$$

A main result in the theory of copulas is the theorem by Sklar (1960).

THEOREM 8.33 Sklar's theorem
Let $X_1,\ldots,X_n$ be real-valued random variables with marginal distributions $F_1,\ldots,F_n$ and joint distribution function F. Then, there exists a copula C such that

$$F(x_1,\ldots,x_n) = C\left[F_1(x_1),\ldots,F_n(x_n)\right]. \tag{8.72}$$

This copula is uniquely determined if the marginal distributions are continuous. Otherwise, it is only unique on the range of $(F_1,\ldots,F_n)$.

Conversely, let C be a copula and let $X_1, \ldots, X_n$ be real-valued random variables with distribution functions $F_1, \ldots, F_n$. Then, the function F as defined in Equation (8.72) is an n-dimensional distribution function with marginal distributions given by $F_1, \ldots, F_n$. It is called the **joint distribution function** F *of* $(X_1, \ldots, X_n)$ **generated by the copula** C.

Sklar's theorem has a clear message: The marginal distributions and the dependence structure of an n-dimensional real-valued random vector can be strictly separated. While the marginal distributions are determined by the univariate distribution functions, the dependence structure is determined by the copula.

With the generalized inverse H^{-1} of a univariate distribution function H,

$$H^{-1}(y) = \inf\{x \in \mathbb{R} \,|\, H(x) \geq y\}, \tag{8.73}$$

we can also directly construct the copula from the joint distribution.

PROPOSITION 8.34

Let F be an n-dimensional distribution function with continuous marginals $F_1, \ldots, F_n$ and a copula C according to Sklar's theorem.

Then, we have

$$C(u_1, \ldots, u_n) = F\left(F_1^{-1}(u_1), \ldots, F_n^{-1}(u_n)\right) \text{ for } u \in [0,\ 1]^n. \tag{8.74}$$

We collect some more properties of copulas.

PROPOSITION 8.35

Let $X_1, \ldots, X_n$ be real-valued random variables.

(a) If $X_1, \ldots, X_n$ are independent then the copula generating their joint distribution is given as

$$C(z_1, \ldots, z_n) = \prod_{i=1}^{n} z_i \qquad \text{"independence copula"} \tag{8.75}$$

(b) **Transformation invariance:** *Let the functions $h_1, \ldots, h_n$ all be strictly monotone increasing (decreasing) and denote by $Y_i = h_i(X_i)$ the transformed random variables. Then the copula corresponding to the joint distribution of $X_1, \ldots, X_n$ as described in Sklar's theorem 8.33 coincides with the one for the joint distribution of to $Y_1, \ldots, Y_n$.*

(c) For every n-dimensional copula $C(u)$ we have

$$W^n(u) = \max(u_1 + \ldots + u_n - n + 1, 0) \leq C(u) \leq \min(u_1, \ldots, u_n) = M^n(u). \tag{8.76}$$

$M^n(u)$ is called the **upper Frechet copula**. *For $n = 2$, $W^n(u)$ is called the* **lower Frechet copula**. *For general n, $W^n(u)$ is in general no copula.*

REMARK 8.36 1. The only one-dimensional copula is obviously equal to $C(x) = x$ for $x \in [0,\ 1]$.

2. Due to the translation invariance of copulas we can concentrate on standardized random variables (i.e. random variables with zero expectation and unit variance) in connection with copulas. ▯

Now, we introduce two concepts of dependency that are different from the usual, linear concept of correlation. The first one is a concept of local dependence, the so-called tail dependence.

DEFINITION 8.37

Let (X_1, X_2) be a random vector with marginal distributions F_1, F_2.

(a) The **coefficient of upper tail dependence** *of (X_1, X_2) is defined by*

$$\lambda_U\left(X_1, X_2\right) = lim_{u \uparrow 1} \mathbb{P}\left(X_2 > F_2^{-1}\left(u\right) \middle| X_1 > F_1^{-1}\left(u\right)\right) \tag{8.77}$$

if the limit exists.

(b) The **coefficient of lower tail dependence** *of (X_1, X_2) is defined by*

$$\lambda_L\left(X_1, X_2\right) = lim_{u \downarrow 0} \mathbb{P}\left(X_2 \leq F_2^{-1}\left(u\right) \middle| X_1 \leq F_1^{-1}\left(u\right)\right) \tag{8.78}$$

if the limit exists.

(c) In case of $\lambda_U > 0$ ($\lambda_L > 0$) we say that (X_1, X_2) admits **upper tail dependence** (**lower tail dependence**).

REMARK 8.38 Note that it is possible to have both upper and lower tail dependence. In contrast to the usual correlation that measures linear dependence between two random variables, the tail dependence only concentrates on the dependence in the extreme values of two random variables. It is thus a local measure. Positive tail dependence means that the probability for both random variables to attain very high values simultaneously is of the same order as for the single random variables separately. The same is true for extremely small values in the case of negative tail dependence. ▯

To measure global dependence between two random variables we introduce the following.

DEFINITION 8.39
For the $\mathbb{R}^2$-valued random pair (X,Y) **Kendall's tau** $\tau(X,X)$ *is defined as*

$$\tau(X,Y) = \\ = \mathbb{P}\left(\left(X-\tilde{X}\right)\left(Y-\tilde{Y}\right)>0\right) - \mathbb{P}\left(\left(X-\tilde{X}\right)\left(Y-\tilde{Y}\right)<0\right) \quad (8.79)$$

where $(\tilde{X},\tilde{Y})$ is an independent copy of (X,Y).

REMARK 8.40 Kendall's tau focuses on the **monotone dependence** of X and Y. It checks how the order between X and Y is preserved. It is positive if (in tendency) for randomly drawn pairs $(x,y),(x',y')$ from the distribution of (X,Y), we observe that y is bigger than y' if x has been bigger than x'. The absolute size of the values does not enter Kendall's tau. It obviously attains values in [0, 1]. We further have (see Embrechts et al. [2003])

$$\tau(X,Y) = 1 \Longleftrightarrow C = M^2; \quad \tau(X,Y) = -1 \Longleftrightarrow C = W^2 \quad (8.80)$$

where C is the copula corresponding to (X,Y) and M^2, W^2 are the Frechet copulas. Also, Kendall's tau can be directly related to the copula C via

$$\tau(X,Y) = 4\int\int_{[0,1]^2} C(x,y)\,dC(x,y) - 1 \quad (8.81)$$

(see Embrechts et al. [2003]). □

Example 8.41 Kendall's tau and correlation
An example highlighting the main differences between Kendall's tau and the usual correlation is given by the following setting:
Let $X \sim \mathcal{N}(0,1)$ and $Y = \exp(X)$ and obtain:

$$\mathbb{C}orr(X,Y) = \frac{\mathbb{E}\left(X\cdot\exp(X)\right)}{1\cdot\sqrt{\mathbb{V}ar(Y)}} = \frac{\exp(1/2)}{\sqrt{\exp(2)-\exp(1)}} \approx 0.763 \neq 1 \\ = \mathbb{P}\left(\left(X-\tilde{X}\right)\cdot\left(\exp(X)-\exp(\tilde{X})\right)>0\right) = \tau(X,Y).$$

So, while Kendall's tau directly realized the monotone, nonlinear dependence, only its linear part is indicated by the correlation coefficient.

8.4.2 Examples and simulation of copulas

Gaussian copula

The Gaussian copula is derived from the multidimensional normal distribution which explains the name. To introduce it, it suffices to consider standardized random variables (see Remark 8.36).

DEFINITION 8.42
Let $X_1, \ldots, X_n$ be $\mathcal{N}(0, \Sigma)$-distributed with $\mathbb{V}ar(X_i) = 1$. Let Φ_Σ^n be the corresponding joint n-dimensional normal distribution function, and denote the standard normal marginal distributions by Φ. Then, the **Gaussian copula** *C_{Gauss} (with correlation matrix Σ) is given by*

$$C_{\text{Gauss}}(x_1, \ldots, x_n) = \Phi_\Sigma^n \left[\Phi^{-1}(x_1), \ldots, \Phi^{-1}(x_n)\right]. \tag{8.82}$$

THEOREM 8.43
The density of the n-dimensional Gaussian copula is given by

$$\frac{\partial^n C_{\text{Gauss}}}{\partial x_1 \cdot \ldots \cdot \partial x_n}(x_1, \ldots, x_n) = \frac{\tilde{\varphi}\left[\Phi^{-1}(x_1), \ldots, \Phi^{-1}(x_n)\right]}{\varphi\left[\Phi^{-1}(x_1)\right] \cdot \ldots \cdot \varphi\left[\Phi^{-1}(x_n)\right]} \tag{8.83}$$

where $\tilde{\varphi}$ is the density of the n-dimensional normal distribution Φ_Σ^n and φ is the density of the standard normal distribution.

REMARK 8.44 If Σ is equal to the identity matrix, i.e. the random variables are independent, then the Gaussian copula is the independence copula from Proposition 8.35 and the density coincides with the indicator function of $[0,\ 1]^n$, which coincides with the result above. □

The density of the two-dimensional Gaussian copula for different correlations ρ is given in Figure 8.4. It is important to notice the symmetry in each density and the different heights of the density: the density for a correlation close to -1 or 1 is more peaked. We cannot plot the density for the whole range, because the theoretical value in the corners with the peaks is $+\infty$.

The simulation of random variables with this dependence structure and arbitrary marginal distributions is rather simple as shown in Algorithm 8.9.

REMARK 8.45 Instead of the Cholesky decomposition in Algorithm 8.9, one can also use the square root of Σ obtained by the singular value decomposition. □

If we use normally distributed marginals, then the resulting correlation matrix of the newly constructed random variables equals Σ. However, for e.g. binomial distributions, we obtain different correlations. For illustration, we simulated 500,000 Monte Carlo runs where in each run we simulated two binomially distributed random variables with the number of trials equal to 5 and equal probability of success. For the two-dimensional Gaussian copula we used a correlation of 0.5. The number of trials of the binomial random variables is restricted to five to avoid the validity of the asymptotics of the central limit theorem. The resulting correlation curve for different success probabilities can be seen in Figure 8.5.

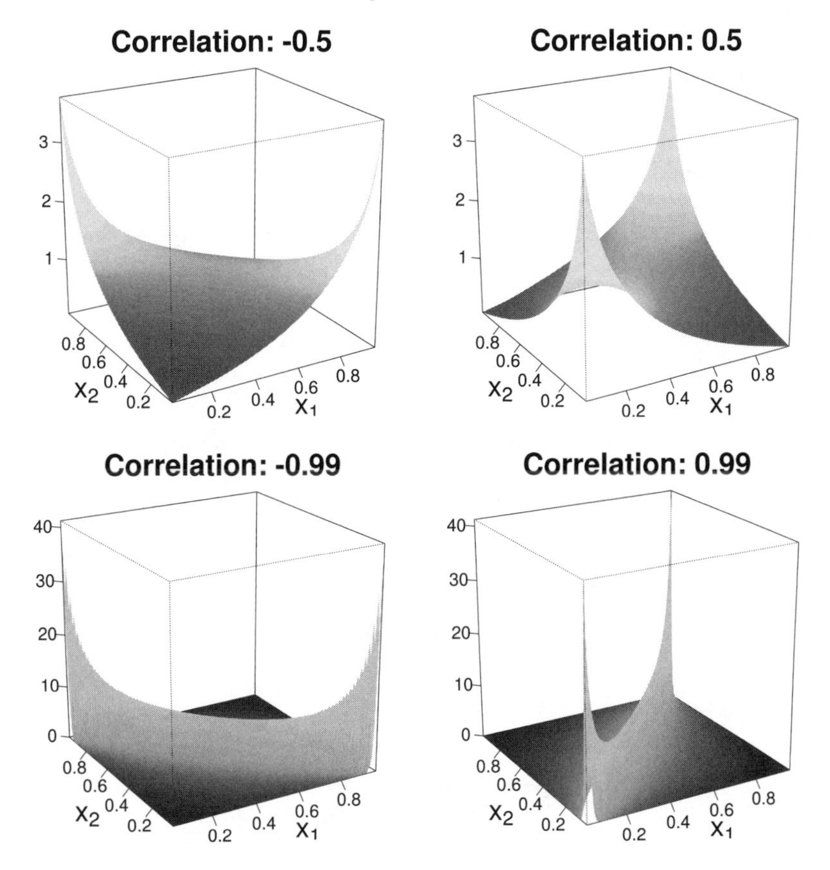

FIGURE 8.4: Density of Gaussian copula for different correlations ρ.

t-copula

The t-copula is closely related to the Gaussian copula.

DEFINITION 8.46
Let $Y_1, \ldots, Y_n$ be standard normally distributed random variables with corresponding correlation matrix Σ. Let Z be a χ^2-distributed random variable with m degrees of freedom. We denote by t_m the distribution function of a t-distribution with m degrees of freedom. Then, the joint distribution function $C_t(x_1, \ldots, x_n)$ of the random variables

$$X_i = t_m\left(\sqrt{m} \cdot \frac{Y_i}{\sqrt{Z}}\right) \tag{8.84}$$

is called **t-copula** *with m degrees of freedom and correlation matrix Σ.*

Algorithm 8.9 Simulating with a Gaussian copula

Let $F_1, \ldots, F_n$ be the desired marginal distributions and Σ the $n \times n$ correlation matrix of the desired Gaussian copula.

1. Compute the Cholesky decomposition of Σ as $\Sigma_{Chol} \cdot \Sigma_{Chol}^t = \Sigma$.
2. Simulate n independent random variables $Y_i \sim \mathcal{N}(0, 1)$.
3. Set
$$\begin{pmatrix} Z_1 \\ \vdots \\ Z_n \end{pmatrix} = \Sigma_{Chol} \cdot \begin{pmatrix} Y_1 \\ \vdots \\ Y_n \end{pmatrix}.$$
4. The random variables X_i with the desired distribution are obtained via
$$X_i = F_i^{-1}\left[\Phi(Z_i)\right]$$
where Φ is the standard normal distribution function and F_i^{-1} is the inverse of the desired marginal distribution function.

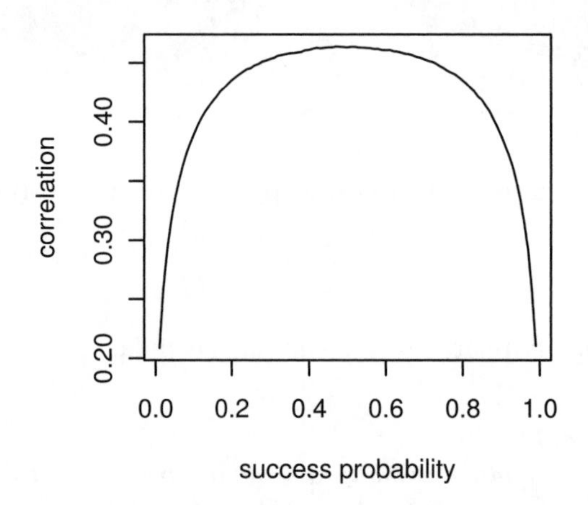

FIGURE 8.5: Correlation between X_1, X_2 with X_1, $X_2 \sim B(5, p)$ and Gaussian copula with correlation 0.5.

To justify calling Σ a correlation matrix, note that if for the marginal distributions we use $F_i = t_m$, then it can be shown that the correlations between the random variables are indeed given by Σ.

THEOREM 8.47
The density of a t-copula with $m \in \mathbb{N}$ degrees of freedom and nonsingular correlation matrix Σ is given by

$$\frac{\partial^n C_t}{\partial x_1 \cdot \ldots \cdot \partial x_n}(x_1, \ldots, x_n) = \frac{\Gamma\left(\frac{m+n}{2}\right) \cdot \left[\Gamma\left(\frac{m}{2}\right)\right]^{n-1}}{\sqrt{\det(\Sigma)} \cdot \left[\Gamma\left(\frac{m+1}{2}\right)\right]^n} \cdot$$

$$\cdot \frac{\sqrt{\left[\frac{\left[t_m^{-1}(x_1)\right]^2}{m} + 1\right] \cdot \ldots \cdot \left[\frac{\left[t_m^{-1}(x_1)\right]^2}{m} + 1\right]}^{\,m+1}}{\sqrt{\frac{1}{m} \cdot \left(t_m^{-1}(x_1); \ldots; t_m^{-1}(x_n)\right) \cdot \Sigma^{-1} \cdot \begin{pmatrix} t_m^{-1}(x_1) \\ \vdots \\ t_m^{-1}(x_n) \end{pmatrix} + 1}^{\,m+n}} \tag{8.85}$$

where t_m^{-1} is the inverse of the distribution function of the t-distribution with m degrees of freedom. $\det(\Sigma)$ *denotes the determinant of* Σ.

A graphical representation of the density of a two-dimensional t-copula for different correlations is shown in Figure 8.6. The same limitations apply here as for the Gaussian copula, i.e. the value in two of the four corners is $+\infty$, i.e. we plot only the range $[0.03; 0.97]$. Note that the density looks similar to the density of the Gaussian copula. The main difference is that it is even more peaked, but still symmetric.

REMARK 8.48 1. If the number of degrees of freedom m approaches infinity, then the t-copula approaches the Gaussian copula. Further, from the formula for the density we can observe that for Σ being the identity matrix, the density does **not** become the density of the independence copula. In fact, we can never obtain the independence copula when using the t-copula.

2. Both the t- and the Gaussian copulas are symmetric and very fast to simulate. There is one striking difference between the two. While the Gaussian copula has no tail dependence, the t-copula has both upper and lower tail dependence. As it is symmetric we only state the upper one. For $\sigma_{12} = \mathbb{C}orr(X,Y)$ we have

$$\lambda_U = 2 \cdot \left(1 - t_{m+1}\left(\frac{\sqrt{(m+1)(1-\sigma_{12})}}{\sqrt{1+\sigma_{12}}}\right)\right) \tag{8.86}$$

where $t_m(x)$ denotes the distribution function of a t-distribution with m degrees of freedom. □

From the definition of a t-copula it is easy to simulate random variables with given marginal distributions (see Algorithm 8.10).

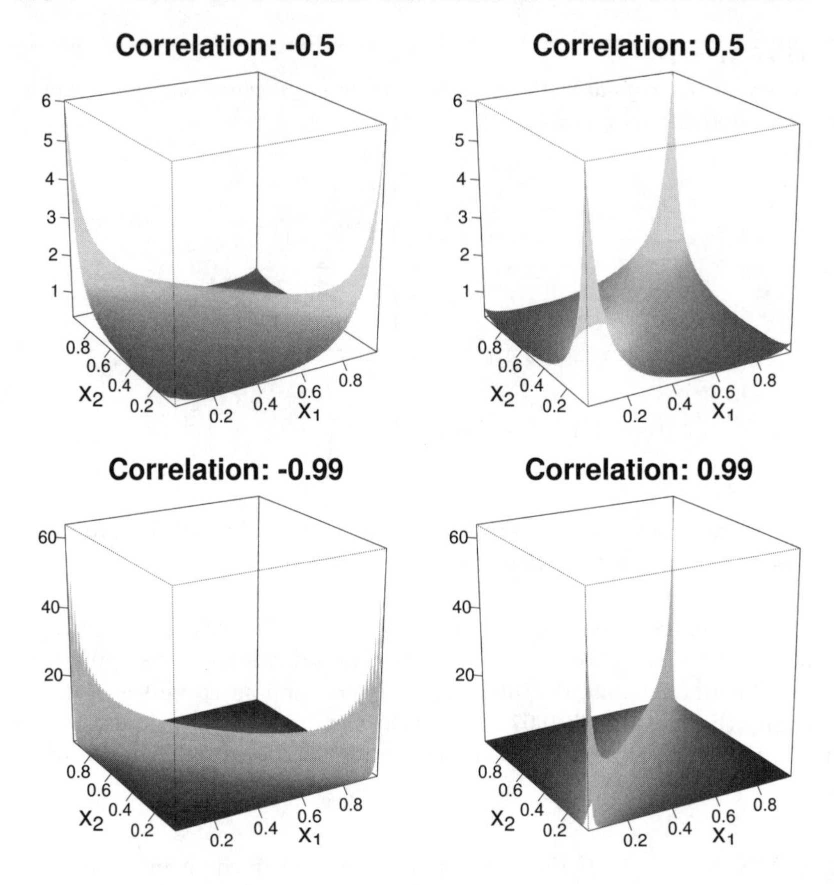

FIGURE 8.6: Density of t-copula with 3 degrees of freedom for different correlations ρ.

To illustrate the difference between a t-copula and a Gaussian copula, as in Figure 8.5, we simulate again two binomially distributed random variables. The resulting correlation curve is shown in Figure 8.7.

Archimedean copulas

An Archimedean copula is characterized by a single one-dimensional distribution. To introduce it, we recapitulate the notion of the Laplace transform in a way that is suitable for us.

DEFINITION 8.49

Let $Z : \Omega \to \mathbb{R}_+$ be a nonnegative real-valued random variable with distribution function F where $F(0) = 0$. The **Laplace transform** $\mathsf{L}_Z : \mathbb{R}_+ \to \mathbb{R}$ *of Z is*

Algorithm 8.10 Simulating with a t-copula

Let $F_1, \ldots, F_n$ be the desired marginal distributions and Σ the desired $n \times n$ correlation matrix of the t-copula with m degrees of freedom.

1. Compute the Cholesky decomposition of Σ as $\Sigma_{Chol} \cdot \Sigma_{Chol}^t = \Sigma$.
2. Simulate n independent random variables $\tilde{Y}_i \sim \mathcal{N}(0,1)$.
3. Set
$$\begin{pmatrix} Y_1 \\ \vdots \\ Y_n \end{pmatrix} = \Sigma_{Chol} \cdot \begin{pmatrix} \tilde{Y}_1 \\ \vdots \\ \tilde{Y}_n \end{pmatrix}.$$
4. Simulate m independent random variables $\tilde{Z}_i \sim \mathcal{N}(0,1)$ and set
$$Z = \tilde{Z}_1^2 + \ldots + \tilde{Z}_m^2.$$
5. The random variables X_i with the desired distribution are obtained via
$$X_i = F_i^{-1}\left[t_m\left(\sqrt{m} \cdot \frac{Y_i}{\sqrt{Z}}\right)\right]$$
where t_m is the t–distribution function with m degrees of freedom.

---

given by

$$\mathsf{L}_Z(x) = \mathbb{E}\left[\exp(-x \cdot Z)\right] = \int_{\mathbb{R}} \exp(-xz)dF(z) \tag{8.87}$$

If in addition the random variable Z has a density f, then we can reconstruct it by the so-called inverse Laplace transform.

DEFINITION 8.50

The Laplace transform L_Z can be extended to complex arguments with positive real part. The **inverse Laplace transform** *is given by*

$$\mathsf{L}_Z^{-1}(x) = \frac{1}{2\pi i} \int_{1-i\infty}^{1+i\infty} \exp(sx)\mathsf{L}_Z(s)ds. \tag{8.88}$$

We can now return to the definition of an Archimedean copula.

DEFINITION 8.51

Let F be the distribution function of a one-dimensional random variable Z with $F(0) = 0$. Let L_Z be the Laplace transform of Z. We define $\varphi(u) =$

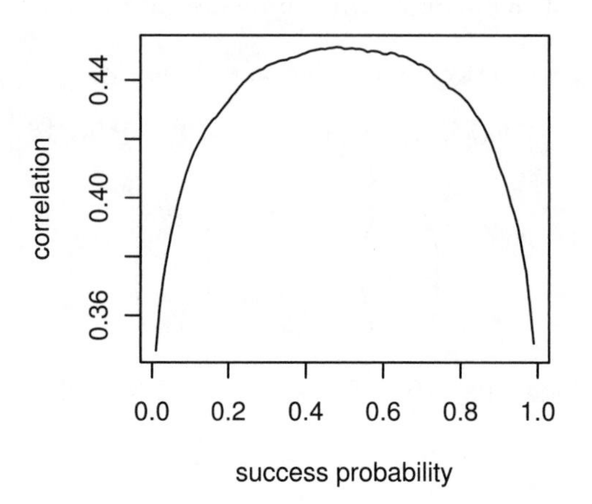

FIGURE 8.7: Correlation between X_1, X_2 with X_1, $X_2 \sim B(5,p)$ and t-copula with correlation 0.5 and 3 degrees of freedom.

$inf\{v | \mathsf{L}_Z(v) \geq u\}$ and call it the **generator** *of the Archimedean copula. The corresponding* **Archimedean copula** *$C_{\text{Archimedean}}$ is then given by*

$$C_{\text{Archimedean}}(x_1, \ldots, x_n) := \varphi^{-1}\left(\sum_{i=1}^{n} \varphi(x_i)\right) \tag{8.89}$$

where φ^{-1} is the inverse function of the generator and coincides with the Laplace transform of Z.

REMARK 8.52 An Archimedean copula has several advantages and disadvantages compared to the t- and the Gaussian copulas:

Advantages: It is possible to obtain a nonsymmetric tail dependence. This is in contrast to the t- and the Gaussian copulas, because the densities of these are symmetric around the point $(0.5, \ldots, 0.5)$.

Disadvantages: The main disadvantages of the Archimedean copula can be seen directly from its construction. First, the number of free parameters is strongly limited, as one usually chooses for F a function with maximal two or three parameters. Second, as the random components enter via a (transformed) sum we have

$$C_{\text{Archimedean}}(x_1, \ldots, x_n) = C_{\text{Archimedean}}\left(x_{\Pi(1)}, \ldots, x_{\Pi(n)}\right) \tag{8.90}$$

for a permutation $\Pi : \{1, \ldots, n\} \to \{1, \ldots, n\}$. Thus, the dependence structure between any finite subset of the random variables is identical. This is different to the Gaussian and the t-copulas where a correlation matrix allows for different dependencies between the components. □

Algorithm 8.11 shows the general framework for simulating with an Archimedean copula (see Marshall and Olkin [1988]).

Algorithm 8.11 Simulating with an Archimedean copula

Let $F_1, \ldots, F_n$ be the desired marginal distributions, φ the generator of the Archimedean copula, and φ^{-1} the inverse function.

1. Simulate n independent random variables $Y_i \sim \mathcal{U}[0,\ 1]$.
2. Simulate another random variable Z independent of $Y_1, \ldots, Y_n$ with Laplace transform equal to φ^{-1}.
3. Define $Z_i = \varphi^{-1}\left[-\frac{1}{Z} \cdot \ln(Y_i)\right]$.
4. The random variables with the desired distribution are obtained by

$$X_i = F_i^{-1}(Z_i).$$

To specialize the above concept, we now introduce some specific examples for Archimedean copulas.

Gumbel copula:

The Gumbel copula for $\alpha \geq 1$ is given by its generator

$$\varphi(t) = [-\ln(t)]^{\alpha}, \quad \varphi^{-1}(u) = \exp\left(-u^{\frac{1}{\alpha}}\right). \tag{8.91}$$

For $n = 2$, the Gumbel copula has upper tail dependence with

$$\lambda_U = 2 - 2^{1/\alpha} \quad (\text{for } \alpha > 1\ ) \tag{8.92}$$

and a Kendall's tau of

$$\tau_\alpha(X, Y) = 1 - 1/\alpha. \tag{8.93}$$

Clayton copula:

The Clayton copula for $\alpha > 0$ is given by its generator

$$\varphi(t) = \frac{1}{\alpha} \cdot \left(t^{-\alpha} - 1\right), \quad \varphi^{-1}(u) = (\alpha \cdot u + 1)^{-\frac{1}{\alpha}}. \tag{8.94}$$

For $n = 2$, the Clayton copula has lower tail dependence with

$$\lambda_L = 2^{-1/\alpha} \quad (\text{for } \alpha > 1\ ) \tag{8.95}$$

and a Kendall's tau of

$$\tau_\alpha(X, Y) = \frac{\alpha}{\alpha + 2}. \tag{8.96}$$

Frank copula:

The Frank copula for $\alpha > 0$ is given by its generator

$$\varphi(t) = -\ln\left(\frac{\exp(-\alpha \cdot t) - 1}{\exp(-\alpha) - 1}\right),$$
$$\varphi^{-1}(u) = -\frac{1}{\alpha} \cdot \ln\left\{\exp(-u) \cdot [\exp(-\alpha) - 1] + 1\right\}. \tag{8.97}$$

The Frank copula has neither lower nor upper tail dependence. An explicit formula for Kendall's tau is available but lengthy (see Embrechts et al. [2003]).

8.4.3 Application in actuarial models

Suppose an insurance company is considering a set of possibly dependent businesses $X_1, \ldots, X_n$ and wants to calculate an expected value of a loss functional of them, i.e.

$$\mathbb{E}\left(f\left(X_1, \ldots, X_n\right)\right) = ? \tag{8.98}$$

The obvious way to do this would be to use a crude Monte Carlo estimate, i.e. to sample N realizations of the random vector $(X_1, \ldots, X_N)$ and take the average of the outcomes of $f(X_1, \ldots, X_n)$ as an estimate. However, for doing this, we need the joint distribution of the businesses. If we do not have it then the approximation procedure of Algorithm 8.12, which is based on using a family of copulas, is a workable approach.

To fill this framework with life, still some questions such as

- Which copula should one use in a particular application?
- How to fit a copula to given data?

have to be answered.

The answer to the first question is not obvious. Besides tractability (which would favour the use of a Gaussian copula) an important point is of course the power of explaining the empirically observed phenomena of the corresponding data. One should therefore check the data for, e.g.

- tail dependence,
- symmetry,
- extreme values.

This should lead to a decision in favour of a particular copula family. Then, the next step is to fit the family of copulas to existing data. As fitting univariate distributions is well understood, we propose the following two step procedure already indicated in the algorithmic framework above.

Algorithm 8.12 Copula framework for dependent risks

1. Estimate the marginal distributions $F_1, \ldots, F_n$ of $X_1, \ldots, X_n$ from past (univariate) data.

2. Decide on a family of copulas $C_\theta, \theta \in \Theta$ according to phenomena observed in empirical data.

3. Determine the parameter θ^* corresponding to the best copula from the parametric family via estimating quantities characterizing the chosen family of copulas (such as Kendall's tau, upper/lower tail dependence, or correlation).

4. Use the marginal distribution and the copula C_{θ^*} to generate N independent random samples $(X_1^{(i)}, \ldots, X_n^{(i)})$ according to the corresponding copula simulation algorithms of the preceding section.

5. Obtain the crude Monte Carlo estimate from the just generated sample for $\mathbb{E}(f(X_1, \ldots, X_n))$ in the usual way.

---

1. **Fit marginal distributions** F_i to the X_i. This could be done by looking at past data and using well-established univariate models for the business under consideration. Often, assumptions about these univariate distributions have already been made by past experience.

2. **Fit the copula to the multivariate data.** This depends on the chosen family. In case of a Gaussian or a t-copula, one has to estimate the correlation matrix. In case of an Archimedean family, one can use a least squares approach by estimating – say – Kendall's tau for all pairs (X_i, X_j) and choose that parameter θ that leads to the smallest sum of quadratic deviations between theoretical and estimated Kendall's taus. Also, lower and/or upper tail dependence can enter the decision.

8.5 Nonlife insurance

In nonlife insurance, the variation of the claims can be much higher than in life insurance. Consequences of catastrophes such as hurricanes or earthquakes often lead to extreme damages. These very rare events typically have a very high impact on the risk of the insured portfolio and therefore also on the result of the whole business of the insurance company. As new modelling ingredients that take care of this situation, heavy-tailed distributions enter the scene.

Also, both occurrence and financial consequences cannot be as easily predicted as in life insurance, simply because a law of large numbers does in general not hold. Therefore, estimation of the probability of ruin of (at least a particular part of) the business is an essential topic of nonlife insurance mathematics. Closely related to this is the determination of the total claim size within a prespecified time interval (normally 1 year). Modelling claim sizes and the arrival process of these claims are central modelling questions. A very recent monograph surveying this area is Mikosch (2004). Further, modelling the impact of relations between different claims is an actual topic with an emphasis on introducing copula models into simulation.

8.5.1 The individual model

In the individual model, the collective set of all insurance policies is investigated by modelling each contract separately.

DEFINITION 8.53
In the **individual (risk) model** *contract i in the insurer's portfolio is identified with the random variable X_i, the claim size of the contract in a given time period (e.g. 1 year). The* **total claim** *S_n of the portfolio is given by*

$$S_n = \sum_{i=1}^{n} X_i. \tag{8.99}$$

Mainly, the insurance company is interested in the distribution of the total claim S_n. To determine it, several simplifying assumptions are introduced:

ASSUMPTION 8.54

1. *The claims $X_1, \ldots, X_n$ of the individual contracts are independent.*
2. *$X_1, \ldots, X_n$ are identically distributed according to an infinite divisible distribution.*

The first assumption yields that the distribution of the total claim is the convolution of the distributions of the individual contracts. The second assumption ensures that the distribution of the total claim is the same, except for the parameters, as the distribution of the individual contracts.

These assumptions for the individual model are indeed very restrictive, because the individual contracts normally have different characteristics. Furthermore, we have to take into account that the claim size of an individual contract may be zero. Thus, if the total claim size should be simulated, we have to specify a probability mass for having a zero (individual) claim size during the period.

Apart from this criticism, Monte Carlo simulation of the total claim size simply amounts to simulating a fixed sum of random variables.

8.5.2 The collective model

In the collective model (also often refered to as the **Cramér-Lundberg model**), we switch from investigating each contract separately to investigating each claim, i.e. X_i denotes the i-th occurring claim and is in general not related to the i-th contract.

DEFINITION 8.55
The **collective (risk) model** *consists of a stochastic process N_t that models the number of claims occurring until time t. The nonnegative random variables X_i, $i = 1, \ldots, N_t$, denote the sizes of these individual claims.*

The following assumptions are imposed in the collective model.

ASSUMPTION 8.56

1. *The number of claims is independent of the claims.*
2. *The individual claims are independent and identically distributed.*

These assumptions allow the derivation of some distributional properties of the total claim, as e.g. the expectation and variance given these quantities are known for the distribution of the number of claims and the individual claims:

THEOREM 8.57
If the individual claim X_i has expectation c and variance σ_c^2 and the number of claims N_t has expectation n and variance σ_n^2, then the total claim S_t until time t has expectation $n \cdot c$ and variance $n \cdot \sigma_c^2 + c^2 \cdot \sigma_n^2$.

In the classical Cramér-Lundberg model (see Lundberg [1903]) the claim occurrences are modelled as a Poisson process with intensity $\lambda > 0$.

We now focus on one of the main tasks within nonlife insurance mathematics, the computation of the ruin probabilities. For this, we need to introduce the **initial reserve** h and the premium rate $\pi(t)$ received at time t.

DEFINITION 8.58
Let s be the initial reserve, $\pi(t)$ be the premium rate received at time t, and $X_1, X_2, \ldots$ be the claims occurring at times $t_1, t_2, \ldots$.

Then, the **ruin probability** *is given by*

$$p_{\text{ruin}} = \mathbb{P}\left(\exists t : h + \int_0^t \pi(s)\, ds - \sum_{i=1}^{\infty} \mathbf{1}_{\{t_i \le t\}} \cdot X_i < 0 \right). \tag{8.100}$$

Its converse $1 - p_{\text{ruin}}$ is called the **survival probability**.

Important results characterizing the collective model are seen below (see Mikosch [2004]).

THEOREM 8.59
(a) If the premium rate $\pi(t)$ is calculated according to the expectation principle (see Definition 8.4), then the ruin probability is one if we choose a safetly loading of $\mu = 0$, i.e. a positive safety loading μ is necessary for the insurance company to survive.
(b) Assume that we have a premium rate of $\pi(t) = c \cdot t$, a i.i.d. sequence of indenpendent, identically distributed (i.i.d.) claim interarrival arrival times W_i with $t_i = W_1 + ... + W_i$ satisfying

$$\mathbb{E}(X_1) - c\mathbb{E}(W_1) < 0.$$

Assume further that for $Z_1 = X_1 - cW_1$ the moment generating function $m_{Z_1}(h)$ exists for all $h \in (-h_0, h_0)$ and some $h_0 > 0$. If then we have a unique positive solution r to the equation

$$m_{Z_1}(r) = \mathbb{E}\left(e^{R(X_1 - cW_1)}\right) = 1, \tag{8.101}$$

then we obtain the **Lundberg bound** *for the ruin probability with an initial reserve of h:*

$$p_{\text{ruin}} \leq \exp(-R \cdot h). \tag{8.102}$$

The number R is called the adjustment coefficient.

For general distributions of the claim size X_i the determination of the ruin probability is nearly impossible. Therefore, one specializes to certain distributions for X_i, derives approximations, or uses Monte Carlo methods (see Mikosch [2004] for a survey on various specifications of the claim size and possible dependencies between the claims) to compute ruin probabilities.

Simulation and ruin probability in the Cramér-Lundberg model

To calculate the probability of ruin in the classical Cramér-Lundberg model up to a fixed time T, one can simply use the algorithms to simulate paths of compound Poisson processes given in Section 6.2. It is easy to include the deterministic drift $\int_0^t \pi(s)\, ds$ and to obtain paths of the reserve process

$$R(t) = h + \int_0^t \pi(s)\, ds - \sum_{i=1}^{\infty} \mathbf{1}_{\{t_i \leq t\}} X_i \text{ for } t \in [0, T].$$

The ruin probability up to time T can then be estimated by

$$\hat{p}_{\text{ruin}}(T) := \frac{1}{M} \sum_{i=1}^{M} \mathbf{1}_{\{R^{(i)}(t) < 0 \text{ for some } t \in [0,T]\}} \tag{8.103}$$

where $R^{(i)}(t)$, $t \in [0, T]$ is the i-th simulated path of the reserve process.

The application of more sophisticated Monte Carlo methods hinges critically on the distributional properties of both the claims arrival process N_t and the claim sizes X_i. We will comment on this in Section 8.5.3.

Path simulation in the Cramér-Lundberg model: Generalizations

There is a number of good reasons why a homogeneous Poisson process can be a quite crude approximation for a real life claims arrival process. Typically, insured populations are formed of subpopulations that can be distinguished but that are homogeneous inside themselves. A simple example are the subpopulations of male and female drivers. For this, the suitable generalization of a homogeneous Poisson process is a mixed Poisson process:

DEFINITION 8.60
Let Λ be a nonnegative random variable with finite first two moments. Then, the stochastic process N_t is called a **mixed Poisson process** *with structure variable Λ if N_t conditional on $\Lambda = \lambda$ is a homogeneous Poisson process with intensity λ.*

Conditioning on Λ and integrating over its possible values leads to

$$\mathbb{E}(N_t) = t \cdot \mathbb{E}(\Lambda), \quad \mathbb{V}ar(N_t) = t \cdot (\mathbb{E}(\Lambda) + \mathbb{V}ar(\Lambda)), \tag{8.104}$$

a fact that in particular allows for a variance of the claims arrival process which is higher than its expectation, indeed a property often empirically observed.

The path simulation of a mixed Poisson process is very easy (see Algorithm 8.13)

Algorithm 8.13 Simulating a path of a mixed Poisson process

1. Simulate a realization λ of the random variable Λ.
2. Simulate a path of a homogeneous Poisson process N_t, $t \in [0, T]$ with an intensity of λ.

If, however, the intensity process varies randomly with time then the notion of a Cox process is the appropriate concept.

DEFINITION 8.61
Let Λ_t be a nonnegative stochastic process satisfying

$$\int_0^T \Lambda_t dt < \infty \quad \forall T > 0. \tag{8.105}$$

Then, the stochastic process N_t (which is adapted to the same filtration as Λ_t) is called a **Cox process** *or a* **doubly stochastic process** *with intensity process Λ_t if N_t conditional on $\Lambda_t = \lambda_t$ is an inhomogeneous Poisson process with intensity funtion λ_t.*

Before we turn to special choices of the intensity process Λ_t, we will comment on the way to simulate paths from it. Indeed, the Cox process is related to an inhomogeneous Poisson process in the same way as a mixed Poisson process is related to a homogeneous one. We therefore first explain how to simulate a path of an inhomogeneous Poisson process (see Section 6.2 for its definition).

There are (at least) two possibilities to simulate an inhomogeneous Poisson process. One relies on knowledge of the exact distribution of the interarrival times between two jumps. As this is rarely the case for nontrivial intensity rate functions, the second variant which is based on the acceptance-rejection method (see Chapter 2) is the preferable one (see Algorithm 8.14).

Algorithm 8.14 Simulating a path of an inhomogeneous Poisson process

---

1. Set $t_0 = 0 = \hat{t}_0$, $\bar{\lambda} = \max\{\lambda_t | 0 \le t \le T\}$
2. While $t_i < T$ do
   (a) Generate: $Z \sim Exp\left(\bar{\lambda}\right), \quad U \sim \mathcal{U}(0,1)$.
   (b) Set $\hat{t} = \hat{t} + Z$.
   (c) If $U \le \lambda\left(\hat{t}\right)/\bar{\lambda}$ then $t_i = \hat{t}$, $i = i + 1$ else go to Step 2a.

As we can now simulate a path of an inhomogeneous Poisson process, the simulation of a path of a Cox process is straightforward (see Algorithm 8.15).

Algorithm 8.15 Simulating a path of a Cox process

---

1. Simulate a realization λ_t of the intensity rate process Λ_t, $t \in [0,T]$.
2. Simulate a path of an inhomogeneous Poisson process N_t, $t \in [0,T]$ with the intensity rate function λ_t, $t \in [0,T]$ by Algorithm 8.14.

One can surely think of many types of an intensity process in the Cox process. A particular one would be a mean-reverting processs which could model an intensity that randomly fluctuates around a typical value and is attracted by it.

Another one that is popular in theory is the **Poisson shot noise** process (see Cox and Isham [1980] and Klüppelberg and Mikosch [1995]). The idea behind the concept of the shot noise process is that we have a mixture between a normal business and extraordinary times, such as a dramatical increase of incoming claims due to some catastrophe. After the catastrophe it takes some time until all corresponding claims are reported and then the intensity level tends back to normal. The same procedure then repeats after the next catastrophe. As a particular example, we introduce the form of the shot noise process introduced in Cox and Isham (1980),

$$\lambda_t = \lambda_0 e^{-\delta t} + \sum_{i=1}^{K_t} y_i e^{-\delta(t-s_i)}. \tag{8.106}$$

Here, λ_0, δ are positive numbers. The random variable y_i is the shot in the intensity caused by catastrophe i at time s_i, which itself is the i-th jump time of the homogeneous Poisson process K_t with intensity ρ. To illustrate the behaviour of a Poisson shot noise process, we refer to Figures 8.8 and 8.9. The first one depicts the intensity process (with a jump height of 10 at the jump time), the second one shows the claim numbers. Note the dramatic increase of the claims frequency after the intensity jump.

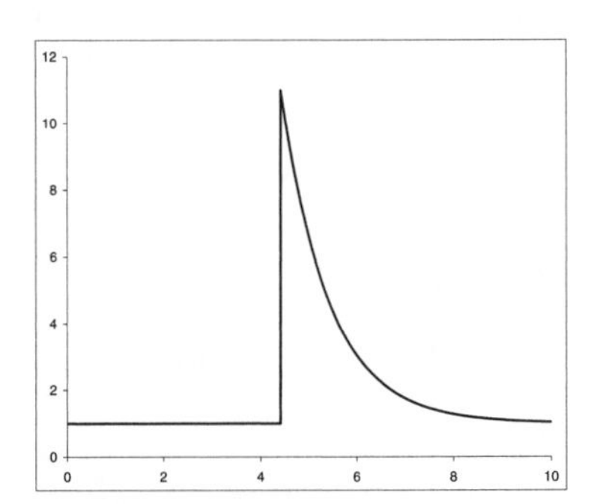

FIGURE 8.8: Shot noise intensity process λ_t.

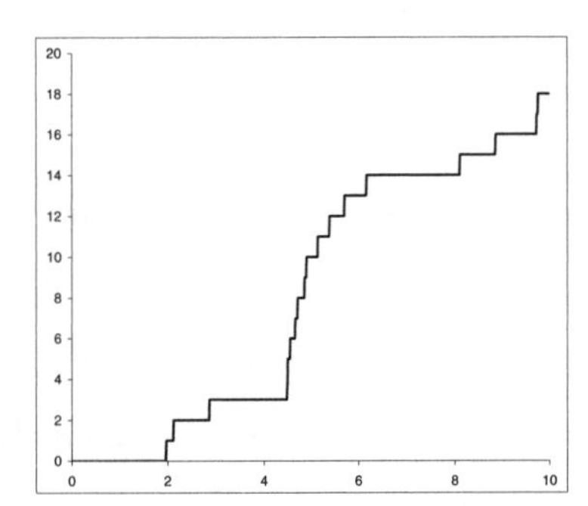

FIGURE 8.9: Claim number process.

More general forms of the shot noise process and also analytical approximations of the ruin probability both for a finite and an infinite time horizon can be found in Albrecher and Asmussen (2006).

8.5.3 Rare event simulation and heavy-tailed distributions

An area which is closely related to the computation of a ruin probability or to risk measures is the simulation of rare events. However, as these events by definition occur very rarely, a crude Monte Carlo simulation of the corresponding experiments followed by an estimation of the probability of the rare event by its relative frequency of occurrence cannot be an efficient method.

As it is also clear that the tail behaviour of the underlying distribution is the decisive characteristic for the occurrence of such a rare event, we give the following classification of distributions.

DEFINITION 8.62
Let X be real-valued random variable with distribution function F. The **tail** *of the distribution is denoted by $\bar{F}(x) := 1 - F(x)$.*

(a) We say that a (univariate) distribution is **light-tailed** *if its tail satisfies*

$$\bar{F}(x) \leq C \cdot \exp(-\alpha x) \text{ for } x > x_0 \tag{8.107}$$

for some x_0 and some positive constants C, α.

(b) We say that a (univariate) distribution is **heavy-tailed** *if its moment generating function $M(u) = \mathbb{E}(\exp(uX))$ does not exist for any $u > 0$.*

REMARK 8.63 1. At first sight, the different forms of the conditions characzerizing light- and heavy-tailed distributions are surprising. However, condition (8.107) implies that the moment generating function of a light-tailed distribution $M(u) = \mathbb{E}(\exp(uX))$ exists for some $u > 0$. Also, the nonexistence of the moment generating function for each $u > 0$ for a heavy-tailed distribution implies that its tail must be heavier than every exponential distribution.

2. Popular examples of light-tailed distributions are the exponential, the gamma, and the normal distribution.

3. Popular examples of heavy-tailed distributions are

- the Pareto distribution with $\bar{F}(x) = \left(\frac{x}{b}\right)^{-\alpha}$ for $\alpha, b > 0$ and $x \geq b$,
- the well-known log-normal distribution, and
- the Weibull distribution with $\bar{F}(x) = \exp(-\gamma x^a)$ for $a, \gamma, x > 0$.

4. A popular class of heavy-tailed distributions are the **subexponential distributions** (see e.g. Klüppelberg [1988]). The distribution F is a subexponential one if for each set of independent identically distributed nonnegative random variables $X_1, ..., X_n$ (all distributed according to F) we have:

$$\lim_{x \to \infty} \frac{\mathbb{P}(\max(X_1, \cdots, X_n) > x)}{\mathbb{P}(X_1 + \cdots + X_n > x)} = 1. \tag{8.108}$$

The interpretation of this property is that a large value for the sum of such subexponentially distributed random variables is typically dominated by a single very large value, a situation sometimes encountered for insured portfolios in nonlife insurance. □

In this section we will only consider the simulation of rare events or, more precisely, the computation of the probabilities of their occurrence in the heavy-tailed situation. For the light-tailed setting, importance sampling with exponential twisting as described in detail in Section 8.2.5 and adapted to the setting of random sums is the method of choice (see Chapter X in Asmussen [2000]). We will therefore not consider this situation in detail here.

In the heavy-tailed setting, exponential twisting of the density function is by definition of a heavy-tailed distribution not possible. Among the approaches considered in the literature, two variants of an approach given in Asmussen and Kroese (2006) show the best performance. We will limit our presentations to these approaches. For general treatment and analytical approximations of ruin probabilities we refer the interested reader to the monographs of Asmussen (2000) and Mikosch (2004).

To judge the performance of a Monte Carlo estimator for small probabilities, we introduce two performance criteria that are based on the relative length of the corresponding confidence interval. They both require that for a sequence of small probabilities, the variance of the estimator should converge faster to zero than the sequence itself.

DEFINITION 8.64

We consider a sequence of events $A(u)$ depending on a parameter u with probabilities

$$\mathbb{P}\left(A\left(u\right)\right) =: z\left(u\right) \longrightarrow 0 \text{ for } \; u \rightarrow \infty. \tag{8.109}$$

An unbiased estimator $Z(u)$ of $z(u)$ is said to have a **bounded relative error** *if it satisfies*

$$\limsup_{u\rightarrow\infty} \frac{\mathbb{V}ar\left(Z\left(u\right)\right)}{Z\left(u\right)^{2}} < \infty. \tag{8.110}$$

An unbiased estimator $Z(u)$ of $z(u)$ is said to be **logarithmically efficient** *(or* **asymptotically efficient***) if it satisfies*

$$\limsup_{u\rightarrow\infty} \frac{\mathbb{V}ar\left(Z\left(u\right)\right)}{Z\left(u\right)^{2-\epsilon}} = 0 \quad \forall\; \epsilon > 0. \tag{8.111}$$

The estimator given below is concerned with estimating the tail probability of a sum of independent and identically distributed random variables $S_n = X_1 + \ldots + X_n$. The Asmussen-Kroese estimator is based on the identity

$$\mathbb{P}\left(S_n > u\right) = n \cdot \mathbb{P}\left(S_n > u, M_n = X_n\right) \tag{8.112}$$

for continuous distributions and where $M_n = \max\{X_1, \ldots, X_n\}$. To further reduce the variance of the Monte Carlo estimator of the right-hand side of this equation, conditional Monte Carlo is invoked by introducing the Asmussen-Kroese estimator as

$$n\mathbb{P}(S_n > u, M_n = X_n | X_1, \ldots, X_{n-1}) = n\bar{F}(\max\{M_{n-1}, u - S_{n-1}\}) \tag{8.113}$$

where $\bar{F} = 1 - F$ and F is the distribution of X_i.

If we now consider an insurance company with initial reserves of u at the beginning of the period and no further premium payments before the end of the period, then

$$z(u) := \mathbb{P}(S_N > u) \tag{8.114}$$

indeed equals its ruin probability by the end of the period if N is the random number of claims during the period. By replacing the fixed n in Equation (8.113) by the random variable N, we obtain the Asmussen-Kroese estimator for a compound sum,

$$Z(u) = N\bar{F}(\max\{M_{N-1}, u - S_{N-1}\}). \tag{8.115}$$

Theoretical results in Asmussen and Kroese (2006) show that the estimator is asymptotically efficient for a Weibull distribution with parameter a satisfying $2^{1+a} < 3$. It is even of bounded relative error for distributions with regularly varying tail. As reported in Asmussen and Kroese (2006), the estimator shows a superior performance compared to other suggested estimators in the area. We describe its simulation in Algorithm 8.16.

Algorithm 8.16 Simulating the Asmussen-Kroese estimator

Let N be a random variable with a given distribution, $X_1, X_2, \ldots$ be i.i.d. random variables with distribution function F, $u > 0$.
For $i = 1$ to K do

1. Simulate a realization $N^{(i)}$ of the random variable N.
2. Simulate $X_1^{(i)}, \ldots, X_{N-1}^{(i)}$ independently according to the distribution F.
3. Calculate $M_{N-1}^{(i)} = \max\left\{X_1^{(i)}, \ldots, X_{N-1}^{(i)}\right\}$, $S_{N-1}^{(i)} = \sum_{j=1}^{N-1} X_j^{(i)}$.
4. Set $Z^{(i)}(u) = N^{(i)}\bar{F}\left(\max\left\{M_{N-1}^{(i)}, u - S_{N-1}^{(i)}\right\}\right)$.

Set $\quad Z_K(u) := \frac{1}{K}\sum_{i=1}^{K} Z^{(i)}(u)$.

To improve the performance of the estimator even further, Asmussen and Kroese (2006) propose the use of a control variate approach. As in the subexponential distribution case where the tail probability of the sum is asymptotically equal to n times the tail probability of each single random variable, $N \cdot \bar{F}(u)$ is suggested as a control variate, leading to the Asmussen-Kroese estimator with control variate of the form

$$Z_{con}(u) = N \cdot \left(\bar{F}\left(\max\left\{M_{N-1}, u - S_{N-1}\right\}\right) - \bar{F}\left(u\right)\right) + \mathbb{E}\left(N\right) \cdot \bar{F}\left(u\right). \quad (8.116)$$

The performance of the Asmussen-Kroese estimator with and without control variate is in detail analyzed in Asmussen and Kroese (2006). There, the authors highlight the superior behaviour of the two estimators with a clear advantage of the version including the control variate as given in Equation (8.116).

8.5.4 Dependent claims: An example with copulas

The individual as well as the collective model are both based on independence assumptions. This is unrealistic for earthquakes where the force and intensity vary over time, because a large earthquake is often followed directly by many small earthquakes. This can be dealt with if we define this as a single claim. But we observe also that a large earthquake in a region is followed in the future by small ones until the tension is high enough again for a large one. While we have already introduced a Poisson shot noise model as a possible framework for this in Section 8.5.2, we will here use a copula approach for modelling the dependence structure between claims.

Example 8.65 Ruin probability due to earthquakes
Let us assume that every tenth earthquake is large and those in between are smaller ones. We model an individual claim X_i by a gamma distribution with shape 0.9 and scale 30. The dependence between different claims is modelled by a Gaussian copula with a correlation matrix having a Toeplitz structure with $-0.1, -0.05$ on the first side diagonals and 0.3 on the tenth side diagonal.

The waiting time between two earthquakes is modelled by an exponential distribution with an intensity of 0.1, i.e. in the mean, an earthquake occurs every tenth year. We also introduce a dependence between the claim size of an earthquake and the waiting time until the next earthquake by a Gaussian copula with correlation 0.7, i.e. a large earthquake increases the waiting time until the next earthquake occurs. The reasoning is that a strong earthquake reduces the tension much more.

We perform this simulation for the next 1,000 years. We set the premium rate $\pi(t) = 1.1 \cdot \frac{\mathbb{E}(X_i)}{10} = 2.97$. For reasons of comparison, we also perform the same simulation under the independence assumption. The results can be seen in Figures 8.10 through 8.12.

In Figure 8.10, we see the evolution of the reserve when starting with no initial reserve. Between two earthquakes, the premiums lead to an increasing reserve and in the event of an earthquake, the reserve drops down by the claim

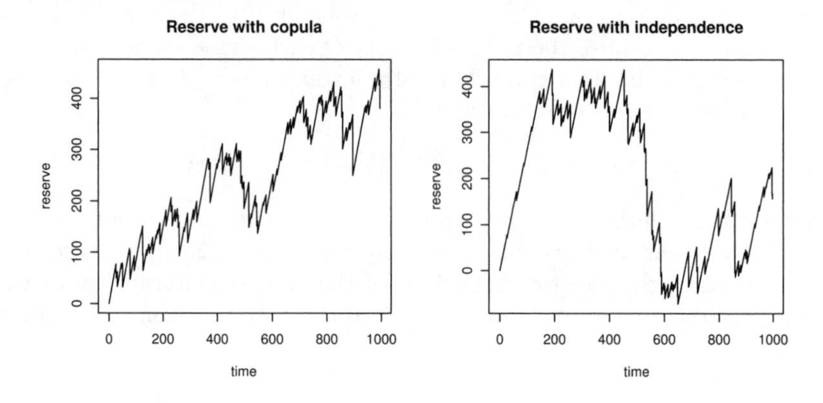

FIGURE 8.10: Paths of the reserve when starting with an initial reserve of zero.

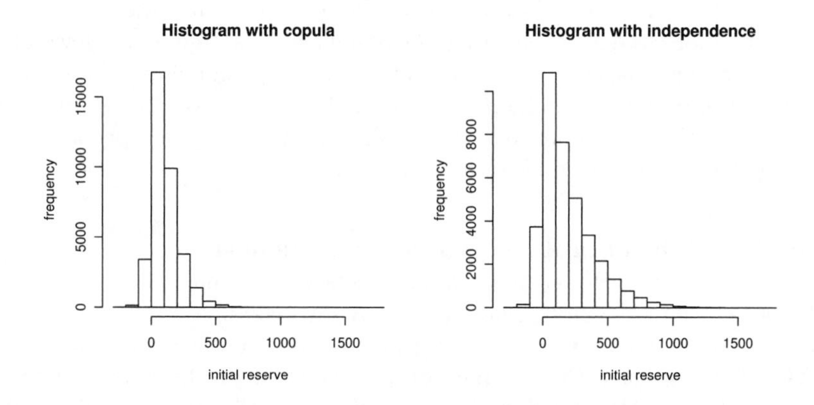

FIGURE 8.11: Histograms of the necessary initial reserves to avoid bankruptcy.

amount. If we compare the paths, then in the dependent case we observe that after a large earthquake the waiting time is longer until the next earthquake occurs, and also the claim size of the second one becomes smaller, which is exactly as desired in our model.

The most interesting result is the difference in the needed initial reserves for avoiding ruin or, vice versa, the ruin probability for a fixed initial reserve. In Figures 8.11 and 8.12 we observe that the dependent case leads to a much smaller needed initial reserve or to a much smaller ruin probability.

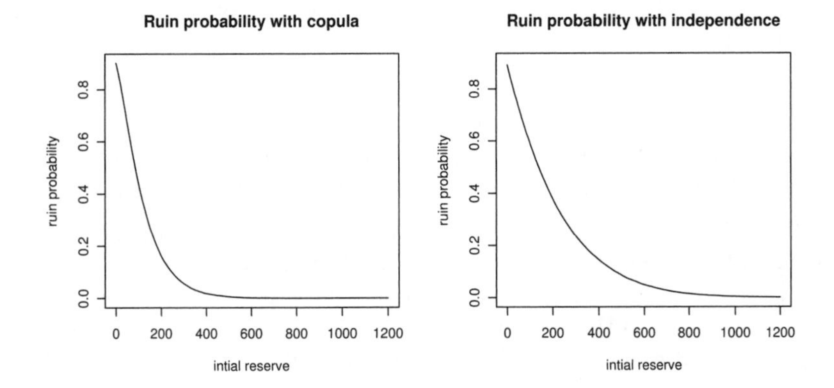

FIGURE 8.12: Ruin probability depending on initial reserve.

8.6 Markov chain Monte Carlo and Bayesian estimation

In this section we introduce the concepts of the Markov chain Monte Carlo method (MCMC) and of Bayesian estimation. MCMC is a basic tool for performing Bayesian estimation methods and for Monte Carlo methods when the simulation of the underlying random numbers is inefficient or impossible by conventional ways. For both situations, numerous applications in actuarial mathematics and in financial mathematics exist.

8.6.1 Basic properties of Markov chains

The term **Markov chain** is used in various ways in the stochastic process literature. This ranges from simply identifying Markov process and Markov chain to the restriction that a Markov chain is a discrete-time Markov process with a countable state space. We use the latter as seen below.

DEFINITION 8.66
Let $\{X(n), n \in \mathbb{N}\}$ be a discrete-time stochastic process such that $X(n)$ only attains values in a countable set $\mathcal{S}$, the so-called **state space**. *For convenience we always identify $\mathcal{S}$ with (a subset of) $\mathbb{N}$. It is called a (discrete-time)* **Markov chain** *if we have*

$$\mathbb{P}(X(n+1) = j | X(n) = i, X(n-1) = i_{n-1}, \ldots, X(0) = i_0)$$
$$= \mathbb{P}(X(n+1) = j | X(n) = i) =: p_{ij}(n) \;\; \forall i, j, i_k \in \mathcal{S}. \quad (8.117)$$

The possibly infinite matrix $(p_{ij}(n))$ is called the **transition matrix** *from time n to time $n+1$.*

REMARK 8.67 1. A customer of a life insurance company is often identified with the "state" he is in, i.e. he can be "healthy," "ill," or "dead". A Markov chain is a model for the evolution of this state process over time.

2. Note that for the discrete state setting we can formulate the Markov property (8.117) element-wise (compare to Definition 4.6 for the general case). Of course, the interpretation stays the same: the future evolution of the process solely depends on the present state and not on the past history. Further, we typically choose $\mathcal{F}(n)$ to be the filtration generated by the elements of the Markov chain up to time n.

3. **Continuous state space**. There is also a well-developed theory for discrete-time Markov chains with a general state space. Indeed, it should be natural that a Markov chain could have transitions from one time instant to the next governed by – say – a normal distribution. In such a situation the role of the transition matrix $(p_{ij}(n))$ is taken over by a transition kernel, i.e. a mapping $p(x, A)$ such that for each fixed $x \in \mathcal{S}$ $p(x, .)$ is a probability distribution with support in $\mathcal{S}$. We do not elaborate on this case, but will occasionally present examples with a noncountable state space such as $\mathbb{R}^d$.

As we are performing all our simulations on a computer with its finite set of numbers, one could also argue that restricting to a finite state space $\mathcal{S}$ would not be a restriction at all. □

Non-Markovian discrete-time stochastic processes that only depend on a fixed part of the past can be turned into a Markov process by enlarging the state space. It should then include all those parts of the past that determine the future evolution of the time series. More precisely, we have the following lemma that can easily be verified.

LEMMA 8.68

Let $\{X(n), n \in \mathbb{N}\}$ be a stochastic process such that $X(n+1)$ can be expressed by $X(n), \ldots, X(n-k)$ via the representation

$$X(n+1) = f_{n+1}(X(n), \ldots, X(n-k), \varepsilon(n+1)) \tag{8.118}$$

with $\varepsilon(n+1)$ a random variable which is independent of $X(l)$ for any $l \leq n$ and f_{n+1} a sequence of deterministic functions.

Then the vector-valued stochastic process $Y(n) = (X(n), \ldots, X(n-k))'$ is a Markov chain.

An important and often practically relevant special case of a Markov chain appears when its transition probabilities do not change over time.

DEFINITION 8.69
A Markov chain $\{X(n)\}_{n\in\mathbb{N}}$ with state space $\mathcal{S}$ is called **homogeneous** *if for any $i, j \in \mathcal{S}$ and any $n \in \mathbb{N}$ we have*

$$\mathbb{P}(X(n+1) = j|X(n) = i) = \mathbb{P}(X(1) = j|X(0) = i) = p_{i,j}. \tag{8.119}$$

REMARK 8.70 It can directly be verified that the n-step transition matrix $p^{(n)}$ from time 0 to time n defined by

$$p_{i,j}^{(n)} := \mathbb{P}\left(X(n) = j|X(0) = i\right)$$

actually equals the n-th power of the transition matrix,

$$p^{(n)} = p^n. \tag{8.120}$$

□

In our life insurance motivation, there are different types of states. Some might only be visited such as "ill" and others can never be left such as "dead." We formalize this in the following definitions.

DEFINITION 8.71
Let $i \in \mathcal{S}$ be some state of the homogeneous Markov chain $\{X(n)\}_{n\in\mathbb{N}}$ and assume that $X(0) = i$. Denote by

$$\tau_i = \min\{n > 0|X(n) = i\} \tag{8.121}$$

the first recurrence time, i.e. the first time of a revisit of the state i by X. The state i is called **transient** *if we have*

$$\mathbb{P}(\tau_i < \infty) < 1. \tag{8.122}$$

i is called **recurrent** *if we have*

$$\mathbb{P}(\tau_i < \infty) = 1. \tag{8.123}$$

If a recurrent state satifies $\mathbb{E}(\tau_i) < \infty$ then we call it **positive recurrent**.

DEFINITION 8.72
Two states $i, j \in \mathcal{S}$ of the homogeneous Markov chain $\{X(n)\}_{n\in\mathbb{N}}$ are called **connected**, *if for some $n_{12}, n_{21} \in \mathbb{N}$ we have*

$$\mathbb{P}\left(X(n+n_{12}) = j|X(n) = i\right) \cdot \mathbb{P}\left(X(n+n_{21}) = i|X(n) = j\right) > 0. \tag{8.124}$$

The Markov chain is called **aperiodic** *if for all states $i \in \mathcal{S}$ we have*

$$1 = g.c.d.\{n \in \mathbb{N}|\mathbb{P}(X(n) = i|X(0) = i) > 0\} \tag{8.125}$$

where g.c.d. denotes the greatest common divisor.

REMARK 8.73 1. It is allowed to have $n_{12} = n_{21} = 0$, i.e. that each state is connected with itself. Further, it can easily be verified that the notion of being connected defines an equivalence relation on the set of states. In particular, we obtain a partition of the state space into different subsets each consisting of either recurrent or transient states.

2. If we are in a recurrent state, then it is impossible to reach a state in any other equivalence class in the future. In order to be able to reach any possible state in a simulation, the homogeneous Markov chain should therefore consist of a single equivalence class. If this is not the case, we must choose the distribution of $X(0)$ carefully to get into the desired class. □

DEFINITION 8.74
A homogeneous Markov chain $\{X(n)\}_{n\in\mathbb{N}}$ with state space $\mathcal{S}$ is called **irreducible** *if it has only one equivalence class of connected states. Otherwise we call it* **reducible**.

If we want to simulate a path of a Markov chain and if we do not know its initial value, then we need a probability distribution $\mu(.)$ on the state space from which we can draw the initial state $X(0)$. A particular such initial distribution that is persistent through time is the following one:

DEFINITION 8.75
A distribution $\pi(.)$ on the state space $\mathcal{S}$ is called a **stationary distribution** *of the homogeneous Markov chain $\{X(n)\}_{n\in\mathbb{N}}$ if all $X(n)$ are distributed according to $\pi(.)$ when the starting value $X(0)$ is distributed according to $\pi(.)$.*

REMARK 8.76 Usually, a stationary distribution π of a Markov chain is defined as a nonnegative solution of the equations

$$\pi p = \pi, \quad \sum \pi(i) = 1. \tag{8.126}$$

Below, we will see that the two definitions are essentially equivalent. Although not every homogeneous Markov chain possesses a stationary distribution, there are many results that guarantee the existence of a (unique) stationary distribution (see e.g. Haeggstrøm [2003] or Durrett [1999]), some even give its explicit form. □

A collection of standard results is summarized below (see Durrett [1999]).

THEOREM 8.77
(a) Let $\{X(n)\}_{n\in\mathbb{N}}$ be a homogeneous, irreducible, and aperiodic Markov chain

with positive recurrent states. Then it has a unique stationary distribution π *that satisfies*

$$\lim_{n\to\infty} p_{i,j}^{(n)} = \pi\left(j\right) \quad \forall i,j \in \mathcal{S} \tag{8.127}$$

and is the unique nonnegative solution to the Equation (8.126).

(b) If $\mathcal{S}$ *is finite and the Markov chain* $\{X(n)\}_{n\in\mathbb{N}}$ *is homogeneous, irreducible, and aperiodic then the distribution of* $X(n)$ *converges exponentially fast towards its unique stationary distribution* π *in the following sense:*

$$\max_{1\leq i\leq N} \max_{1\leq j\leq N} \left| p_{i,j}^{(n)} - \pi\left(j\right)\right| \leq C \cdot e^{-nc} \tag{8.128}$$

for some $c, C > 0$ *where* N *is the size of the state space.*

Besides these convergence results for the distribution of a Markov chain, there is another class of convergence theorems for Markov chains which relate them to Monte Carlo simulation (see Durrett [1999] for a proof).

***THEOREM 8.78* Strong law for Markov chains**

Let $\{X(n)\}_{n\in\mathbb{N}}$ *be a homogeneous and irreducible Markov chain with a unique stationary distribution* π*. Let* f *be a real-valued function such that* $\mathbb{E}(f(X))$ *is defined and finite for* X *distributed according to* π*. We then have*

$$\frac{1}{n}\sum_{k=1}^{n} f\left(X\left(k\right)\right) \overset{n\to\infty}{\longrightarrow} \mathbb{E}\left(f\left(X\right)\right) \quad a.s. \tag{8.129}$$

where X *is distributed according to the stationary distribution* π*.*

Note one particular aspect: We did not (!) need independence between the $X(k)$ for the convergence to hold, a fact that is also crucial for the convergence of Monte Carlo estimates obtained by the MCMC method below.

8.6.2 Simulation of Markov chains

The simulation for a path of a Markov chain $\{X(t), \in \mathbb{N}\}$ is quite easy if the transition probabilities are known in advance. To have a unified framework, we will always draw the initial state $X(0)$ from some distribution D in the algorithm below. If we already know the starting value then D is simply the distribution that puts all mass in this particular value. We assume that we know the transition matrix completely and that we are able to draw a random number from all the discrete distributions that are induced by the rows of the transition matrix. Then, we can give the simple Algorithm 8.17.

The Algorithm 8.17 can be inefficient if the state space is infinite and/or if the transition probabilities have to be calculated from scratch at each state.

Algorithm 8.17 Simulating a path of a homogeneous Markov chain with a precalculated transition matrix

1. Set $n = 0$. Draw a random number $X(0)$ from the distribution D.

2. For $n = 1$ to N draw a random number $X(n)$ from the discrete distribution given by $\{p_{X(n-1),j}, j \in \mathbb{N}\}$.

In such a situation it can be more efficient to use the inversion method for generating random numbers from the relevant discrete distribution in an iterative way, as we will do in Algorithm 8.18.

Algorithm 8.18 Simulation of a Markov chain

1. Set $n = 0$. Draw a random number $X(0)$ from the distribution D.

2. For $n = 1$ to N do

   (a) Draw a random number $u \sim \mathcal{U}(0,1]$ and set $j = 0, sum = 0$.

   (b) Calculate $p_{X(n-1),j}$.

   (c) Set $sum = sum + p_{X(n-1),j}$.

   (d) If $sum \geq u$ then set $X(n) = j$ else set $j = j + 1$ and go to (b).

8.6.3 Markov chain Monte Carlo methods

The basic idea behind Markov chain Monte Carlo methods (MCMC methods) is that one can obtain random numbers that are distributed – at least approximately – according to a given distribution π by simulating a Markov chain that has this distribution as its unique stationary distribution. Here, π can either be a discrete distribution given by its probability function $\pi(.)$ or a continuous one given by its density function $g(.)$. There exist a lot of introductions and survey papers on the properties and applications of MCMC. We mention the monographs by Asmussen and Glynn (2007), Gilks et al. (1996), and Liu (2001), to state a few.

The Metropolis-Hastings algorithm

The most popular MCMC algorithm is the **Metropolis-Hastings algorithm** (MH algorithm), Algorithm 8.19 (see Metropolis et al. [1953] and Hastings [1970]). It constructs a Markov chain that starts in an arbitrary state and has a transition probability $p_{i,j}$ such that the Markov chain is reversible, i.e. it satisfies the **detailed balance equation**

$$\pi(i)p_{i,j} = \pi(j)p_{j,i}. \tag{8.130}$$

Hence, starting with the initial distribution π the probability of getting from state i to j equals that for getting from state j to state i. Summing both sides of Equation (8.130) over j shows that then π is indeed a stationary distribution of the chain (which under suitable conditions is unique).

Algorithm 8.19 Metropolis-Hastings algorithm

Let π be a given probability distribution. Let $q(x, y)$ be a given transition matrix. Set further $X(0) = \bar{x}$ for some value $\bar{x}$ with $\pi(\bar{x}) > 0$.

For $k = 0$ to $N - 1$ do

1. Draw a random number Y according to the transition probability $q(X(k), .)$ and draw a random number $U \sim \mathcal{U}[0, 1]$.
2. Calculate $\alpha(X(k), Y) = \min\left\{1, \frac{\pi(Y)q(Y,X(k))}{\pi(X(k))q(X(k),Y)}\right\}$.
3. If $\alpha(X(k), Y) > U$, then $X(k + 1) = Y$, k=k+1, go to Step 1.
   Else go to Step 1.

REMARK 8.79 Properties/modifications of the MH algorithm

1. The transition probabilities in the MH algorithm are given by

$$p_{i,j} = \begin{cases} q(i,j)\alpha(i,j), & i \neq j, \\ q(i,i)\alpha(i,i) + \sum_{j \in \mathcal{S}} q(i,j)\,(1 - \alpha(i,j)), & i = j. \end{cases} \tag{8.131}$$

2. One can directly imitate the considerations leading to the algorithm for a distribution π with a density. By slight misuse of notation, we denote this density function again by $\pi(x)$. Further, in this case the transition probability $q(x, y)$ is then replaced by a transition density $q(x, y)$, i.e. for each fixed x $q(x, .)$ is a density function. Again, one obtains that the detailed balance equation is satisfied, but now the transition probability of the chain is replaced by a transition kernel $p(x, y)$:

$$\pi(x)p(x, y) = \pi(y)p(y, x). \tag{8.132}$$

Integrating over y yields the stationarity condition in the density case,

$$\pi(x) = \int \pi(y)p(y, x)dy. \tag{8.133}$$

Thus, it should be kept in mind that we also cover the density case below by simply replacing point probabilities by densities. However, for simple notation, we will mainly restrict ourselves to the discrete state space setting.

3. **How to choose the proposal function $q(x, y)$?** There are two popular choices of the transition probability $q(x, y)$. One is that of a symmetric q, i.e. one has

$$q(x,y) = q(y,x) \quad \forall x, y \in \mathcal{S} \tag{8.134}$$

where $\mathcal{S}$ denotes the support of the distribution Π. This choice simplifies the calculation of the acceptance function as we then have

$$\alpha(x,y) = \min\left\{1, \frac{\pi(y)}{\pi(x)}\right\}. \tag{8.135}$$

In particular, for this q a simulated state Y is always accepted if its probability $\pi(Y)$ exceeds $\pi(X(k))$. On the other hand, if we use a proposal function q which attains its highest value at $X(k)$ (such as a normal distribution centered around $X(k)$) then the acceptance probability is always below one.

A second popular choice is the **independence sampler**, i.e. the choice of

$$q(x,y) = g(y) \quad \forall x, y \in \mathcal{S} \tag{8.136}$$

for some probability function (or density) $g(.)$.

However, one should keep the following problem in mind when deciding about which transition probability function $q(x, y)$ to use: if $q(x, y)$ has a tendency to put too much probability mass to the near neighbourhood of x, the chain might always stay very close to its initial state. If $q(x, y)$ has a tendency to put too much probability on large jumps away from x, one faces the danger of ending up in the tails of the distribution, which leads to too many nonrepresentative values of the MH chain.

4. **Convergence properties and stationarity behaviour**. To ensure the convergence of the MH chain toward the desired stationary distribution, we have to ensure that the relevant convergence results from the Markov chain section can be used. In the discrete state space case, we have that the MH chain is irreducible and aperiodic if we have

$$q(i,j) > 0 \quad \forall\, i, j \in \mathcal{S} \text{ and} \tag{8.137}$$

$$p(i,i) > 0 \text{ for at least one } i \in \mathcal{S}. \tag{8.138}$$

In the density case, we need $p(x, x) > 0$ for almost all $x \in \mathcal{S}$ instead of condition (8.138). Note that by our choice of $q(i, j)$ we can keep those conditions under control.

As the chain is also homogeneous, we then have verified the assumptions of both convergence results, the distributional convergence theorem 8.77, and the strong law for Markov chains. We thus have obtained the desired convergence toward the stationary distribution and the convergence of the Monte Carlo estimator based on the MH algorithm.

As a special case, Asmussen and Glynn (2007) show a fast convergence for the choice of the independence sampler if we have

$$A := \sup_x \frac{\pi(y)g(x)}{\pi(x)g(y)} < \infty. \tag{8.139}$$

So, in particular $g(.)$ has to be very similar to $\pi(.)$ to deliver good results.

5. **The burn-in period**. Although the required convergence of the Monte Carlo estimator based on the MH chain can be ensured for the use of the whole chain, this convergence is usually faster if one only includes those members of the chain that are already close to the stationary distribution state. One therefore lets the chain run for a **burn-in period** and only uses the members of the MH chain afterward.

Various theoretical considerations for the optimal size of the burn-in period exist. Some of them are hard to verify, some are only asymptotically valid. A special recent example is Rudolf (2009), where an exact formula for the optimal burn-in period length to obtain a given mean-square error between the Monte Carlo estimator and the desired expected value is given. It is based on the concepts of laziness and conductance of the MCMC chain. Both those technical concepts will not be introduced here.

An empirically justified criterion that indicates a convergence of the chain to the stationary state is the similarity of the generated data, i.e. we should decide on a graphical basis about the data point from which onwards we believe the MCMC chain behaves stationary (see the example below). Note however that this can only be a proof for nonstationarity in case the simulated MCMC chain values look different from what we expect for the desired stationary distribution.

6. **Total chain length**. An aspect that has to be considered to decide upon the total chain length is the variance of the corresponding Monte Carlo estimator based on the MCMC chain. To see this, let $\sigma^2 = \mathbb{V}ar(f(X))$ with X distributed according to π. Let ρ_k be the autocorrelation of order k of the MCMC chain elements. Then we have

$$\mathbb{V}ar\left(\sum_{k=1}^{N} f\left(X\left(k\right)\right)\right) =$$

$$= \sigma^2 \cdot \left(1 + 2\sum_{k=1}^{N} \frac{N-k}{N}\rho_k\right) \overset{n\to\infty}{\longrightarrow} \sigma^2 \cdot \left(1 + 2\sum_{k=1}^{\infty} \rho_k\right). \tag{8.140}$$

It thus depends on the autocorrelation function between the generated sample values if we have to generate more values by an MCMC chain than the usual N to obtain an accuracy of $O(1/N)$. Thus looking at the (sample) autocorrelations is also a part of performing an MCMC simulation.

7. **Modifications**. There exist various modifications of the MH algorithm to improve the quality of the generated random numbers. A particular example is to use only every k-th member of the MH chain to have at least approximately independent random numbers. Of course, a high value of k makes this method quite inefficient. For more variants see Liu (2001).

8. By construction of the MH algorithm, it is only necessary to know the stationary distribution $\pi(.)$ up to a norming constant, as only the quotients $\pi(x)/\pi(y)$ enter the computations. This fact will be important for the application of MCMC methods in Bayesian estimation. □

Example 8.80

To illustrate the behaviour of an MCMC chain we look at the toy example to generate the Poisson distribution with intensity $\lambda = 3$ and use the transition matrix of a simple random walk reflected in $x = 0$. More precisely, we use

$$q(x, y) = \begin{cases} 1/2 \text{ for } y \in \{x-1, x+1\}, \ x > 0, \\ \ 1 \ \text{ for } y = 1, \ x = 0, \\ \ 0 \ \text{ else.} \end{cases}$$

We simulated a MCMC chain of length 10,000 with a start in $X(0)$=1 and used the first 1,000 members of the chain for the burn-in period. Figure 8.13 seems to indicate that the chain becomes stationary quite soon. This is underlined by Figure 8.14, where the simulated frequencies of the different values among the 9,000 MCMC chain members are compared to their theoretical counterparts. The differences are as small as if one would simulate directly from the Poisson distribution. This observation was independent from the starting value, even for very high ones such as $X(0) = 20$.

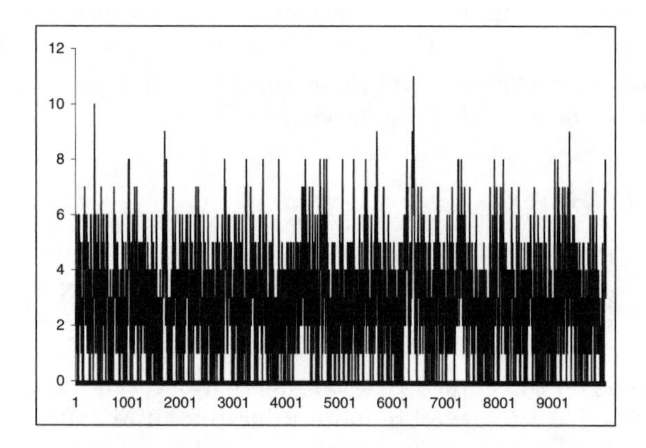

FIGURE 8.13: MCMC chain for a Poisson distribution ($\lambda = 3$).

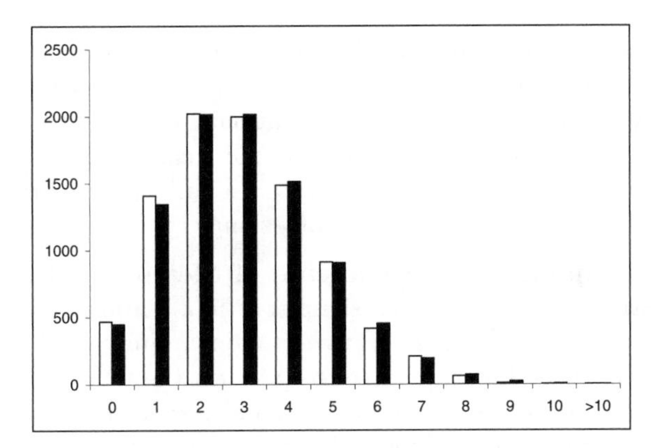

FIGURE 8.14: Actual frequencies (white) of the values of an MCMC chain after burn-in for a Poisson distribution ($\lambda = 3$) and expected theoretical frequencies (black).

The Gibbs sampler

In the case of a desired multivariate limit distribution of the MCMC chain, we often face a situation where the marginal distribution of a component given the remaining components of the state variable is easily available (such as in the situation of a multivariate normal). One then updates the Markov chain created by the MCMC method component by component, always using the already available newly generated component and the remaining old ones, a method which is known as the **Gibbs sampler** (see Algorithm 8.20 and Geman and Geman [1984]).

Algorithm 8.20 Gibbs sampler

Let $\pi(.)$ be a given probability distribution (or a given probability density function). Set $X(0) = (x_1, ..., x_d)$ for some value x with $\pi(x) > 0$.

For $k = 0$ to $N - 1$ do

1. Generate $X(k+1) = (X_1(k+1), ..., X_d(k+1))$ by generating the following random numbers one after each other:
   $X_1(k+1) \sim \pi\left(x \,|X_2(k), ..., X_d(k)\right),$
   $X_2(k+1) \sim \pi\left(x \,|X_1(k+1), X_3(k), ..., X_d(k)\right),$
   ...
   $X_d(k+1) \sim \pi\left(x \,|X_1(k+1), X_2(k+1), ..., X_{d-1}(k+1)\right).$
2. Set k=k+1 and go to Step 1.

REMARK 8.81 1. The above variant of the Gibbs sampler is also called the **systematic-scan** Gibbs sampler. Some authors argue that updating just one randomly chosen component i at iteration k and leaving the remaining ones unchanged (**random-scan** Gibbs sampler) can outperform systematic-scan in terms of the speed of convergence (see Liu [2001] for this and the discussion of further variants of the Gibbs sampler).

2. **Gibbs sampling as a special case of Metropolis-Hastings.** Note that by construction of the Gibbs sampler π is a stationary distribution of the corresponding MCMC chain. However, the marginal conditional distributions $\pi(.|.)$ have to be known exactly to perform the Gibbs sampler. Under these conditions the Gibbs sampler is indeed a special case of the Metropolis-Hastings algorithm, with the choice of $q(x_{-m}, z) = \pi(z|x_{-m})$ as the proposal transition probability in the update step for the chain. Here, x_{-m} denotes the vector x without its m-th component. Let $y = (x_1, ..., x_{m-1}, y_m, x_{m+1}, ...x_n)$. We then obtain

$$\alpha(x,y) = \frac{\pi(y)q(y,x_m)}{\pi(x)q(x,y_m)} = \frac{\pi(y_m|x_{-m})\pi\left(x_{-m}\right)\pi(x_m|x_{-m})}{\pi(x_m|x_{-m})\pi\left(x_{-m}\right)\pi(y_m|x_{-m})} = 1, \qquad (8.141)$$

i.e. the acceptance probability for the proposed value y that only differs from x by its m-th component always equals 1. Also, the convergence considerations of the Metropolis-Hastings algorithm apply for the Gibbs sampler.

3. As in many applications conditional distributions can often be specified while the unconditional ones are not known, the use of the Gibbs sampler is very popular. We will see particular applications in Bayesian statistics and in actuarial mathematics later in this section. □

Example 8.82

This simple application is taken from Liu (2001) where a two-dimensional normal distribution with

$$X \sim N\left(\begin{pmatrix}0\\0\end{pmatrix}, \begin{pmatrix}1 & \rho\\ \rho & 1\end{pmatrix}\right), \qquad -1 < \rho < 1$$

is the target distribution. As in this case, we can represent the two components X_1, X_2 of this normal distribution via

$$X_1 = \sqrt{1-\rho^2}Y_1 + \rho X_2, \quad X_2 = \sqrt{1-\rho^2}Y_2 + \rho X_1$$

for two independent standard normally distributed Y_1, Y_2, the construction of the Gibbs sampler implies

$$\begin{aligned} X_1(k+1)|X_2(k) &\sim N\left(\rho X_2(k), (1-\rho^2)\right), \\ X_2(k+1)|X_1(k+1) &\sim N\left(\rho X_1(k+1), (1-\rho^2)\right). \end{aligned}$$

Hence, starting at $(X_1(0), X_2(0))$, the unconditional distribution at iteration k is given by

$$X(k) \sim N\left(\begin{pmatrix} \rho^{2k-1}X_2(0) \\ \rho^{2k}X_2(0) \end{pmatrix}, \begin{pmatrix} 1-\rho^{4k-2} & \rho-\rho^{4k-1} \\ \rho-\rho^{4k-1} & 1-\rho^{4k} \end{pmatrix}\right)$$

which obviously converges to the desired target distribution.

8.6.4 MCMC methods and Bayesian estimation

The main principle of Bayesian estimation of a parameter $\theta(\in \mathbb{R}^d)$ consists of three ingredients:

1. **Preknowledge** expressed by the fact that we assume that θ is a realization of a distribution (the **prior distribution**):
$$\theta \sim G(\theta).$$

2. **Combination of observations** $X = (X_1, ..., X_n)$ **with preknowledge** expressed by updating our view on the distribution of θ to obtain the conditional distribution (the **posterior distribution**):
$$\theta | X_1, ..., X_n \sim G(\theta | X_1, ..., X_n).$$

3. **Point estimation of** θ as either the mean or the mode of the posterior distribution.

Instead of just a point estimator for the unknown parameter θ, Bayesian estimation yields a whole distribution of it, the posterior distribution. If we choose an uninformative prior distribution (such as a uniform distribution on the possible values for θ) then the mode of the posterior distribution coincides with the familiar maximum likelihood estimator. The main tasks in Bayesian estimation are therefore

- the computation of the posterior distributio and
- the computation of the posterior mode or the posterior mean.

In case of a discrete distribution or of a distribution with density we have

$$G(\theta|x) = \begin{cases} \frac{f(x|\theta)g(\theta)}{\int_y f(x|y)g(y)dy}, & \text{density case} \\ \frac{p(x|\theta)g(\theta)}{\sum_y p(x|y)g(y)}, & \text{discrete case} \end{cases} \quad (8.142)$$

where the likelihood functions $f(x|y)$ and $p(x|y)$ are assumed to be known.

While in both cases the computation of the numerator might be possible, it is the norming constant in the denominator that in general is hard – if not impossible – to compute (note that for a multivariate distribution this requires the computation of a high-dimensional integral!). Thus, in the general case, we need efficient numerical methods to compute the posterior distribution.

To avoid this, in Bayesian statistics a big focus is laid on searching for so-called **conjugate priors**. These are prior distributions $G(\theta)$ such that for a given likelihood function the posterior distribution $G(\theta|x)$ belongs to the same family of distributions as the prior. As an example, one can verify by direct computation that the beta distribution is a conjugate prior for the Bernoulli distribution (see e.g. Lee [1997] for an introduction to this and further aspects of Bayesian statistics). We give an example that highlights the increase the efficiency of MCMC methods for conjugate priors:

Example 8.83 MCMC with conjugate priors

There are indeed two critical aspects in the form of Equation (8.142) that can make MCMC simulation of the posterior distribution really slow. One is the tremendous task of computing the denominator numerically, the second one is the evaluation of the likelihood function in the nominator. Both these tasks can be avoided by a suitable choice of a conjugate prior distribution.

As an example, we consider the estimation problem of the mean θ for a normal distribution with known variance σ^2. Let us also assume that we have a data set $x_1, ..., x_n$ which then leads to a likelihood function of

$$f(x|\theta) = \prod_{i=1}^{n} \left(2\pi\sigma^2\right)^{-n/2} \exp\left(-\frac{1}{2\sigma^2}\sum_{i=1}^{n}(x_i-\theta)^2\right).$$

If we now use the density of a normal distribution with mean ν and variance τ^2 as a prior distribution $g(.)$ for θ, then we obtain a posterior distribution proportional to

$$f(x|\theta)\, g(\theta) = const \cdot \exp\left(-\frac{1}{2\tilde{\sigma}^2}(\theta-\tilde{\mu})^2\right)$$

with

$$\tilde{\mu} = \tilde{\sigma}^2 \cdot \left(\frac{\nu}{\tau^2} + \frac{\sum_{i=1}^{n} x_i}{\sigma^2}\right), \quad \tilde{\sigma}^2 = \left(\frac{1}{\tau^2} + \frac{n}{\sigma^2}\right)^{-1}. \tag{8.143}$$

Hence, the posterior distribution is again a normal distribution, but now with the new parameters $\tilde{\mu}$, $\tilde{\sigma}^2$ as given above. This has two consequences:

1. We know the denominator of the posterior distribution without having explictly computed it.

2. If we run an MCMC chain for the unknown parameter θ, then it is only necessary to simulate the next element $\theta^{(i)}$ at each step. It is not necessary to evaluate the likelihood function, only the estimates for the parameters of the prior have to be updated.

The advantages of conjugate priors can be stressed even more for a multidimensional parameter estimation when the complete conditional distributions together with suitable conjugate priors are available. Then, Bayesian estimation with the Gibbs sampler can be very efficient. Indeed, each updating step only consists of the simulation from the complete conditional distributions followed by an update of the parameters that determine the distribution family of the relevant parameter. There is no need to evaluate the likelihood function, which saves a large amount of computation time.

REMARK 8.84 Bayesian estimation and MCMC methods 1. Numerous applications of MCMC methods for Bayesian estimation of model parameters in financial models are given in the excellent survey by Johannes and Polson (2010). The models covered range from the Black-Scholes model via stochastic volatility models to various interest rate models. As we will present an actuarial application in detail below, we will not give details for the financial applications here.

2. The software package WinBUGS is a highly useful and freely available tool (see www.mrc-bsu.cam.ac.uk/bugs/winbugs/contents.shtml) that is based on the application of Gibbs sampling for use in Bayesian statistics. The basic reference for it is Lunn et al. (2000).

3. When the likelihood function of a particular model has no simple form, the introduction of **latent variables** sometimes improves the MCMC-based Bayesian estimation procedure. A possible example (also covered in Johannes and Polson [2010]) to highlight this would be the introduction of the jump times and jump heights in a Merton jump-diffusion model. Given their knowledge, observed stock prices are log-normally distributed. So if we are able to condition on the jump heights and jump times, we can use the log-normal distribution as a likelihood function. Of course, there is a price to pay for that. As the jump times and heights are not available as observed data (at least, we assume this for the moment), they are added to the unknowns and have to be simulated in the MCMC procedure, too. However, if – by a suitable choice of priors – we are able to generate a situation where the evolution of the likelihood functions could be avoided, then this typically would lead to an upspeeding of the whole computations although the number of variables would be increased. □

8.6.5 Examples of MCMC methods and Bayesian estimation in actuarial mathematics

In the following we will present two specific applications of both MCMC methods and Bayesian estimation in actuarial mathematics. However, this will only be a small selection of such applications. Many more are presented in e.g. Scollnik (2001) and the references given therein.

Calibrating parameters and prediction in total claims models

We consider an example given in Czado (2004). Here, the total claim size in year t is given by

$$S_t = Y_{t,1} + ... + Y_{t,N_t} \tag{8.144}$$

with all single claim heights $Y_{t,i}$ being independent and identically distributed, and N_t the number of claims in year $t \in \{t_1, ..., t_n\}$. We also assume that the single claim heights $Y_{t,i}$ are independent of N_t. Further, the following distributional assumptions are made:

$$Y_{t,i} \sim \text{Pareto}\,(a, 20)\,, \tag{8.145}$$
$$N_t|\lambda \sim \text{Pn}\,(\lambda)\,. \tag{8.146}$$

The use of the Pareto and the Poisson distributions for (single) claim heights and claim frequency are standard assumptions in actuarial mathematics. The data used in Czado (2004) consist of claims above 20 million Danish crowns from a well-documented data set.

We now look at the following two tasks:

1. Bayesian estimation of the parameters a and λ determining claim size and claim frequency.

2. Predicting the claim frequency and claim height distribution of year t_{n+1} given the information contained in the data set.

To deal with the estimation task note that the posterior distribution of λ given the relevant data is

$$G\,(\lambda|N_{t_1}, ..., N_{t_n}) = const \cdot p\,(N_{t_1}, ..., N_{t_n}|\lambda)\, g\,(\lambda)\,. \tag{8.147}$$

As the Poisson likelihood function $p\,(N_{t_1}, ..., N_{t_n}|\lambda)$ possesses the gamma distribution as a conjugate prior, we choose the $Gamma(0.001, 0.001)$ distribution as an (approximately) uniformative prior for the parameter λ.

The conditional independence of the N- and Y-variables leads to the posterior density of

$$G\left(a|Y_{t_1,1}, ..., Y_{t_n,N_{t_n}}\right) = const \cdot f\left(Y_{t_1,1}, ..., Y_{t_n,N_{t_n}}|a\right) g\,(a)\,. \tag{8.148}$$

Again, the choice of the $Gamma(0.001, 0.001)$ distribution as an (approximately) uniformative prior for the parameter λ is also a conjugate prior.

Both corresponding chains are then run to generate 10,000 iterates (with the help of WinBUGS). The estimation results are then based on the last 9,500 observations, which means that a burn-in period of 500 has been chosen. The resulting parameters are summarized in Table 8.4 (see also Czado [2004]).

To predict the total claims characteristics for year t_{n+1} by Bayesian prediction, one computes both the predictive distribution of the number of claims and that of the total claims amount. For the number of claims, we can use

	Mean	St. dev.	2.5%	Mean	97.5%
a	1.810	0.298	1.270	1.796	2.422
λ	3.305	0.577	2.273	3.267	4.550

Table 8.4: Bayesian Estimation for Total Claim Parameters

the just generated MCMC chain $\lambda^{(i)}$, $i = 501, ..., 10{,}000$ to obtain the approximate predictive probabilities given the posterior distribution of λ:

$$\mathbb{P}\left(N_{t_{n+1}} = n\right) \approx \frac{1}{9,500} \sum_{i=501}^{10{,}000} \mathbb{P}\left(N_{t_{n+1}} = n|\lambda^{(i)}\right)$$
$$= \frac{1}{9,500} \sum_{i=501}^{10{,}000} e^{-\lambda^{(i)}} \frac{(\lambda^{(i)})^n}{n!}. \quad (8.149)$$

As for the predictive for the total claims amount we would have to calculate the double integral

$$f\left(S_{t_{n+1}}|data\right) = \int_0^\infty \int_0^\infty f\left(S_{t_{n+1}}|\lambda, a\right) f\left(\lambda, a|data\right) d\lambda da \quad (8.150)$$

that in particular contains the density $f\left(S_{t_{n+1}}|\lambda, a\right)$, which is based on a convolution, but we prefer to use MCMC simulation again. Thus, with the choice of a burn-in period length of 500 and a total chain length of 10,000 we perform:

For $i = 500$ to 10,000:

1. Generate $N_{t_{n+1}}^{(i)} \sim Pn\left(\lambda^{(i)}\right)$.
2. If $N_{t_{n+1}}^{(i)} = 0$ then set $S_{t_{n+1}}^{(i)} = 0$.
3. If $N_{t_{n+1}}^{(i)} > 0$ then
   (a) Generate $Y_{t_{n+1},k}^{(i)} \sim Pareto\left(a^{()}, 20\right)$, $k = 1, .., N_{t_{n+1}}^{(i)}$,
   (b) Set $S_{t_{n+1}}^{(i)} = \sum_{k=1}^{N_{t_{n+1}}^{(i)}} Y_{t_{n+1},k}^{(i)}$.

Finally, estimate the predictive density by its empirical counterpart obtained from just the simulated data.

Credibility and experience rating

Our next application only has a Bayesian component, not an MCMC aspect. It is concerned with the theoretical background of a premium calculation technique called experience rating. The need for experience rating arises from the fact that the whole portfolio of particular insurance contracts consists of different homogeneous subpopulations. Typical examples for this can occur in health insurance (male/female, young/old, ...) or car insurance (type of car, type of driver).

If in such a situation one assigns every contract the same premium, then the "good" risks (i.e. those contracts who would (at least in the mean) not correspond to claims above average) would move to other insurance companies and the "bad" ones would stay (and happily enjoy a low premium). It is thus reasonable to use a weighted average of a portfolio mean and an individual mean of the claim height as a basis for constructing the premium. As the individual component is based on the past experience that the insurance company has had with the customer, this type of setting of a premium is called **experience rating**. We will follow Norberg (2002) in our presentation below.

If one interprets the claims arising from a single insurance contract as a realization $m(\Theta)$ of a function of a random variable Θ (=the customer) and if we have a data vector X (the history) for Θ, then the so-called **total accuracy approach** to experience rating postulates to use that function $\hat{m}(X)$ as a premium based on the individual history of the customer that minimizes

$$MSE\left(\hat{m}(X)\right) := \mathbb{E}\left(\left(m\left(\Theta\right) - \hat{m}(X)\right)^2\right). \tag{8.151}$$

It then follows directly from

$$\begin{aligned}&\mathbb{E}\left(\left(m\left(\Theta\right) - \mathbb{E}\left(m\left(\Theta|X\right)\right) + \mathbb{E}\left(m\left(\Theta|X\right)\right) - \hat{m}(X)\right)^2\right) = \\ &\quad = \mathbb{E}\left(\left(m\left(\Theta\right) - \mathbb{E}\left(m\left(\Theta|X\right)\right)\right)^2\right) + \mathbb{E}\left(\left(\mathbb{E}\left(m\left(\Theta|X\right)\right) - \hat{m}(X)\right)^2\right) \end{aligned}\tag{8.152}$$

that the optimal estimator in the above mean-square error (MSE) sense is

$$\hat{m}(X) = \mathbb{E}\left(m\left(\Theta|X\right)\right), \tag{8.153}$$

the conditional mean of $m(\Theta)$ given the history X. Hence, it only remains to compute this conditional expectation for a specified customer θ (i.e. we condition on $\Theta = \theta$) given his history. To do this, we need the posterior distribution of Θ given X, indeed a task that could be performed by running a suitable MCMC chain. As we have seen, this could be done by choosing efficiently a suitable conjugate prior distribution for Θ.

However, such a choice of prior distribution mainly for reasons of computational convenience is questionable here, as the preknowledge plays the main conceptual part in experience rating. The alternative suggested in Bühlmann (1967, 1969) is to restrict the set of Bayesian estimators to a specific linear class of the form

$$\hat{m}_{lin}(X) = (1-z)a + z\bar{X}_n \tag{8.154}$$

for data $X = (X_1, ..., X_n)$ and where $\bar{X}_n$ denotes the arithmetic mean of the observations. Here, the assumptions are that conditional on Θ the annual claims $X_1, ..., X_n$ are i.i.d. with a mean of $m(\Theta)$ and a variance of $s^2(\Theta)$. As stated in the above papers by Bühlmann (1967, 1969), the MSE-optimal premium in this linear class is given by

$$a = \mathbb{E}\left(m(\Theta)\right), \; z = \frac{\lambda n}{\lambda n + h}, \; \lambda = \mathbb{V}ar\left(m(\Theta)\right), \; h = \mathbb{E}\left(s^2(\Theta)\right). \quad (8.155)$$

The coefficient z is called the **credibility factor**. Note that it will tend to 1 if the number of observations n gets large. So if we have a lot of information about the individual customer, our rating is nearly fully based on our judgement of the customer and not so much on the behaviour of the total portfolio.

To be able to give periods a different weight according to the insured volumes p_j in period j, Bühlmann and Straub (1970) assume that the conditional variances have the form $\mathbb{V}ar\left(X_j|\Theta\right) = s^2(\Theta)/p_j$. This leads to an optimal linear estimator and a credibility factor of the forms

$$\hat{m}_{lin}(X) = (1-z)\mathbb{E}\left(m(\Theta)\right) + z\textstyle\sum_{i=1}^{n} \frac{p_i}{\sum_{j=1}^{n} p_j} X_j, \quad (8.156)$$

$$z = \frac{\sum_{j=1}^{n} p_j \lambda}{\sum_{j=1}^{n} p_j \lambda + h}, \quad \lambda = \mathbb{V}ar\left(m(\Theta)\right), \quad h = \mathbb{E}\left(s^2(\Theta)\right). \quad (8.157)$$

For further generalizations such as multidimensional credibility and aspects of estimation we refer to Norberg (2002).

8.7 Asset-liability management and Solvency II

8.7.1 Solvency II

The key word **Solvency II** stands for the introduction of new regulations for the risk calculations within insurance companies in Europe. The rules of Solvency II are still not in force, but there are already different countries having their own regulations like Switzerland and Great Britain. The main principles of Solvency II are already installed and can be found at the home page for the **C**ommittee of **E**uropean **I**nsurance and **O**ccupational **P**ensions **S**upervisors (CEIOPS) at www.ceiops.eu.

Main principles of Solvency II

The main principles underlying the concept of Solvency II include:

1. The economic value of assets and liabilities should be determined by "mark to market," i.e. the values should be equal to actual market prices if they exist for the corresponding assets and liabilities. If these are not observable, a reasonable model for the determination of the values should be applied.

2. The value of the technical provisions (i.e. the premium and claims reserves arising from unearned premiums and not fully paid out claims) has to be given as the sum of a best estimate and a risk margin where:
   - The best estimate should be based on all cash flows of current contracts during their whole lifetime. The amounts are discounted by the risk-free interest rate yield curve valid at the valuation date.
   - The risk margin must take into account different risks such as operational risk, underwriting risk, and counterparty default risk as well as different risks for each line of business.
   - The risk margins must be calculated separately for each line of business and it is not allowed to assume any diversification.

3. The **solvency capital requirement** should equal the economic capital that n insurance company should hold to ensure that ruin occurs with a maximal probability of 0.05 during 1 year, i.e. the solvency capital equals the Value-at-Risk at the 99.5%-level for a time horizon of 1 year of the capital needed such that the insurance company can meet its obligations to policy holders and beneficiaries.
   - The calculation of the solvency capital requirement is split up into several parts in the standard formula.
   - The exact calculation of each part is exactly specified and reflects the opinion of the commission about the quantiles. The market risk for the yield curve is characterized by a prespecified shock down- or upwards. For equities it is specified by a crash of 32% or 45%, depending on the type of share.
   - The individual parts are related by a prespecified correlation matrix. Hence, some diversification effects are incorporated in the rules.

4. Each insurance company can use an internal model for the calculation of the solvency capital requirement. However, the rules for such internal models are not specified yet.

5. Besides the solvency capital requirement, the insurer must determine the **minimal capital requirement**, too. It represents the minimal capital that is required to transfer the business to another insurance:
   - The minimal capital requirement is not based on any quantile of any risk.
   - Its value is capped and floored by percentages of the solvency capital requirement.

We do not state exact formulae here, but again refer the reader to www.ceiops.eu for more information on the different standard formulae for different risks and businesses.

Consequences of the calculations

The supervisory authority reacts as soon as the capital of the company does not cover the solvency capital requirement. In that case the supervisory authority will demand a plan from the insurance company with the aim to fulfill the solvency capital requirement as soon as possible. This plan must include several milestones and has to be accepted by the supervisory authority. The supervisory authority can force the insurance company to reduce the risk, to stop paying dividends, or to take a credit, too.

If the capital falls below the minimal capital requirement, then the insurance company must present within 1 week a restructuring plan such that within 3 months the minimal capital requirement will be fulfilled again. However, the time span for fulfilling the solvency capital requirement is not increased.

All these regulations will require a number of computations that often can only be performed via extensive Monte Carlo simulations. From a theoretical point of view, one can regard them as a special case of asset-liability management.

8.7.2 Asset-liability management (ALM)

Asset-liability management (ALM) is the main challenge for Monte Carlo simulations in finance and insurance as it can possibly rely on all the types of methods and models we have presented thus far.

Aim of ALM

The aim of ALM is to determine an optimal long term investment strategy for the assets in order to maximize the bonus for the customers (in the case of life insurance) or to minimize the premiums for the customers (in the case of nonlife insurance). One typically considers only constant portfolio strategies such as to invest e.g. 30% of the capital into shares, 60% into bonds and the remaining 10% into real estate. These proportions should then be kept constant over time. Note that to do so at least approximately we have to trade quite often, theoretically even at each time instant.

The determination of the bonus or premiums is based on the investigation of the technical provisions, i.e. the optimal investment strategy thus depends on the evolution of the assets and of the liabilities. Therefore, we must introduce a constraint based on the riskiness of the investment strategy, taking into account the distribution of the asset values and the technical provisions. This can either be to limit the probability of default or to use a risk measure of Section 8.2.3 for the difference of the value of the assets and the technical provisions. Another possibility is to use the regulations of Solvency II as constraints. It is reasonable to maximize the return under the constraint that the company will fail to fulfill the solvency capital requirement at any time in at most $x\%$ of the possible future scenarios.

Connecting assets and liabilities

As ALM considers both the evolution of assets and liabilities, they should be simulated **together** in a joint framework when they are used to determine the optimal investment strategy. This is particularly reasonable as the evolution of the liabilities also influences the capital reserve of the insurance company, and the liabilities often move in parallel to factors influencing the asset side (such as interest rates, inflation, and exchange rates).

Although the connection between the asset and the liability side is obvious, the investment strategy is still often determined solely by the fund managers of an insurance company. The actuaries only require a certain return without taking into account the relation of the performance of the investment strategy and the liabilities for the success of the company in total.

We collect some examples to illustrate the connection between assets and liabilities. In a pension insurance the premiums are often coupled to the income of the insured. In this case, we model the yield curve and the inflation as part of the assets which steer the value of the bonds and real estate and which influence the incomes, as the labor unions typically ask for a real income increase. Since the premiums are directly coupled to the incomes, we have modelled a connection between the assets and liabilities.

A weaker relation holds for life insurance, because the evolutions of bonds and of liabilities directly depend on the yield curve. Therefore, life insurance companies normally follow a rather conservative investment strategy with a high percentage of the capital invested in bonds. The time to maturity of these bonds should then match the so-called **duration** of the liabilities, i.e. the mean time a unit of the premiums stays within the insurance company. This strategy enables the life insurance company to fulfill the liabilities without the risk of failing the solvency capital requirement as long as the relevant bond yield is higher than the guaranteed interest rates of the insurance contracts. However, this strategy fails as soon as the yield curve is below the guaranteed interest rate as happened in Europe in 2002.

In nonlife insurance companies the damages are in general independent of the evolution of the assets, but their absolute height is of course related to inflation, again a possible relation between assets and liabilities.

Challenges and realization of ALM

An ALM study is done by Monte Carlo simulations of both the assets as well as the liabilities. We must specify a time horizon for the ALM study, e.g. 15 years, and a time discretization for the simulation such as e.g. 1 month. We have both challenges on the asset and on the liability side as explained below.

Challenges on the asset side

Many asset classes. Typically, an insurance company has a diversification strategy to invest in many different assets such as e.g. bank accounts, governmental bonds, corporate bonds with default risk, shares, options, real estate, and inflation-linked bonds. Therefore, one has to simulate the evolution of many different assets using a lot of different models. This is typically a high-dimensional problem.

Different countries. To diversify further, insurance companies do not only invest in their home country, but also in other countries worldwide, such as Germany as a home country, and as the other countries the United States, Great Britain, Japan, and emerging markets.

As a consequence, we must model the exchange rates, too. This can be based on the "Purchase-Power-Parity." This principle states that the investment in different countries should lead in the mean to the same return as well as the prices for goods should evolve in the same way in the mean.

Therefore, the exchange rate F for one unit of the foreign currency (indicated with index f) denominated in one unit of the domestic (or "home") currency (indexed h) can be modelled in the risk-neutral world by

$$dF(t) = F(t)t\left\{\left[(r_h(t) - r_f(t)) + (i_h(t) - i_f(t))\right]dt + \sigma dW(t)\right\} \qquad (8.158)$$

where $r_h(t), r_f(t)$ are the short rates and $i_h(t), i_f(t)$ are the inflation rates in the different countries. As we prefer a sparsely parameterized model, we often assume that the inflation rates are already included in the short rates. Thus, the term $i_h(t) - i_f(t)$ disappears in the model of the exchange rate.

With the help of bonds of both the domestic and the foreign country it is then possible to hedge the exchange rate risk for the foreign currency.

Real and risk-neutral world: Which one has to be used? The answer is easy, but we need to be very rigorous in ALM as we need models and simulations in both worlds, the real and the risk-neutral one.

So far, we mainly considered pricing problems for derivatives. To calculate prices, one can always assume to be in a risk-neutral world. However, when we want to gain information about the future evolution of stock prices, interest rates, or other quantities that evolve with time, then we have to do this in the real world. This in particular means that we have to use our own subjective views on the drift of the stock price or the short rate.

More precisely:

1. Whenever the **evolution of price processes over time** should be modelled (may it be for risk calculations or for ALM) then the corresponding price paths have to be simulated based on the real-world model.
2. **Only for price calculations**, paths of the underlying stock prices or interest rates have to be simulated in the risk-neutral world.

As a particular example, in a Black-Scholes framework, we would simulate the stock price paths in the real world by

$$dS(t) = S(t)[\mu dt + \sigma dW(t)], \quad S(0) = s. \tag{8.159}$$

If we want to calculate the price of a European call option at time $t > 0$, then we would insert the simulated stock price $S(t)$ with the real-world drift μ into the standard Black-Scholes formula (see Corollary 5.12), i.e. we would insert the simulated value from the real world into the pricing formula in the risk-neutral world. However, if we want to price at time $t > 0$ an option with issue date t that requires a Monte Carlo simulation itself (e.g. Asian or barrier options as in Section 5.6.2), then we have to perform the Monte Carlo pricing algorithm in the risk-neutral world by using $S(t)$ coming from the simulation in the real world, as price of the stock at time t. If the issue date t_0 of this path-dependent option lies before t, then we even have to use the simulated values $S(\tilde{t})$ from the simulation in the real world in the pricing algorithm for those time points $\tilde{t}$ that fulfill $t_0 \leq \tilde{t} \leq t$.

Which models to choose? This does not seem to be a particular question related to ALM. Of course, one should always use realistic models that are calibrated to market data. However, one also has to consider both the behaviour of the model and its tractability when deciding which one to use.

A special aspect for ALM is that – as pointed out above – we often need to simulate models in both the risk-neutral and the real world. We thus need models where it is easy (and possible!) to switch between both worlds. In the Black-Scholes model this is simply done by a change of drift from μ (the subjective drift) to r, the risk-neutral drift. The most convenient way to switch from the risk-neutral to the real world consists via the introduction of a **market price of risk**. In the general one-dimensional diffusion setting of the risk-neutral model

$$dS(t) = S(t)[r(t)dt + \sigma(t, S(t))dW(t)], \tag{8.160}$$

one can introduce this market price of risk as

$$\lambda(t) = \frac{\mu(t) - r(t)}{\sigma(t, S(t))}, \tag{8.161}$$

use Girsanov's theorem (see Theorem 4.44) to define a Brownian motion $\tilde{W}(t)$ under the real-world measure $\mathbb{P}$ via

$$\tilde{W}(t) = W(t) - \int_0^t \lambda(s)\,ds, \tag{8.162}$$

and obtain the real-world SDE representation with the desired drift $\mu(t)$

$$dS(t) = S(t)\left[\mu(t)dt + \sigma(t, S(t))d\tilde{W}(t)\right], \tag{8.163}$$

where we always assume sufficient integrability of the market price of risk.

By this approach we can show that many popular models stay in the same model class in both worlds. A simple such example is the change of the mean-reversion level of the short rate in the Vasicek model from θ to $\tilde{\theta}$ by introducing

$$\lambda(t) = \frac{\kappa\left(\tilde{\theta} - \theta\right)}{\sigma} \tag{8.164}$$

and introducing the real-world Brownian motion $\tilde{W}$ as in (8.162). Further, the measure changes introduced in Section 5.17.1 show how to introduce additional drift parts for the zero bond prices.

Parameter calibration: Risk-neutral world or historical data? The calibration of the parameters of the corresponding models in the real and risk-neutral worlds highlights a positive (but also conflicting) aspect of the existence of the two alternatives.

We can in principle calibrate all parameters to historical data. To use them for pricing, we then have to take the short rate as drift in the risk-neutral world. However, it is also a viable alternative to calibrate all parameters but the drift to prices of derivatives (such as bonds for the yield curve or options for shares). The (stock price) drifts then still have to be calibrated from real world historical data, or they simply can be parameters that the investor can forecast based on his expectations.

Including crash scenarios. As the relevant time horizon in ALM for insurance companies is often around 15 years, one should also consider the possibility of an economic crisis. Thus, it is important to include some crash or stress scenarios. Examples for such scenarios can be the explicit inclusion of a prespecified crash of a given height of the whole market at a predetermined time, a shift of the complete yield curve, or other catastrophic events.

On the modelling side, one can use a jump-diffusion process to include crash possibilities. One can also model a correlation structure that can change over time. This is reasonable as one often observes that just before a crash, correlations of stocks get close to one followed by a down jump of the market.

Challenges on the liability side

The main challenge on the liability side is to keep the model as representative as possible and on the other hand as small as possible in order to keep the simulation effort small.

Aggregation of contracts. In this chapter, we have already discussed approaches for life and nonlife insurance liabilities where the aggregation is the main trick to keep the simulation effort as small as possible and as accurate as necessary to simulate the liabilities in a reasonable time. However, to aggregate the contracts to a representative one is only admissible if the law of large numbers can be assumed to hold for the contracts. If the risks underlying the different contracts are very heterogeneous, then we must include several representative contracts in our simulations to ensure that our sample cares for all essential types of claim sizes.

New customers or run-off simulation? Another point to take into account is the question whether we investigate the so-called run-off or allow new customers. In the run-off case the investment strategy becomes irrelevant as soon as no claims can occur any more and the value of the assets is positive, i.e. it makes only sense to do a run-off investigation for a short time horizon. If we allow for new customers and contracts, we also have to decide about modelling their occurrence.

Performing an ALM study

After having set up the relevant models that are necessary to simulate all the important evolutions on the asset and liability sides, one actually has to simulate a large number of paths of all these relevant factors. So a Monte Carlo simulation kernel such as the simulation engine ALMSim, developed by the Fraunhofer Institute for Industrial Mathematics ITWM (see www.itwm.fraunhofer.de/en/fm__projects__ALMSim/almsim/), is at the heart of each ALM performance.

On the basis of these simulated paths, one can now evaluate the performance of different investment strategies and thereby figure out the optimal one (of course, one should have also clarified in which sense "optimal" should be understood!). Of course, the consequences of choosing an investment strategy such as the evolution of the total wealth, the assets, the liabilities,... of a company are functionals of the simulated paths, they are also calculated directly when each path is simulated. As usual, one obtains Monte Carlo estimates of the distributions, the means, the variances, and/or the risk measures via a suitable averaging over the different paths.

An ALM study is a tremendous task. It cannot be put together in just some lines. Many considerations besides the ones already raised by us above will enter into the task. Putting together all the details needed and also setting up the simulation concept has to be done by a team and has to be well-organized. We will not give a "toy" example here, as it would be too simplistic.

References

M. Abramowitz and I. A. Stegun. *Handbook of Mathematical Functions.* Dover, New York, USA, 1972.

S. K. Acar. *Aspects of optimal capital structure and default risk.* PhD thesis, University of Kaiserslautern, Germany, 2006.

C. Acerbi and D. Tasche. On the coherence of expected shortfall. *Journal of Banking and Finance*, 26(7):1487–1503, 2002.

H. Albrecher and S. Asmussen. Ruin probabilities and aggregate claims distributions for shot noise Cox processes. *Scandinavian Actuarial Journal*, pages 86–110, 2006.

L. Andersen. Efficient simulation of the Heston stochastic volatility model, 2007. URL `http://papers.ssrn.com/sol3/papers.cfm?abstract_id=946405`.

L. Andersen. A simple approach to the pricing of Bermudan swaptions in the multifactor LIBOR market model. *Journal of Computational Finance*, 3 (2):5–32, 1999.

L. Andersen and J. Andreasen. Volatility skews and extensions of the LIBOR market model. *Applied Mathematical Finance*, 7(1):1–32, 2000.

L. Andersen and M. Broadie. Primal-dual simulation algorithm for pricing multidimensional American options. *Management Science*, 50(9):1222–1234, 2004.

D. Applebaum. *Lévy Processes and Stochastic Calculus.* Cambridge University Press, Cambridge, UK, 2004.

P. Artzner, F. Delbean, J.-M. Eber, and D. Heath. Coherent measures of risk. *Mathematical Finance*, 9(3):203–228, 1999.

S. Asmussen. *Ruin Probability.* World Scientific Publishing Company, Singapore, 2000.

S. Asmussen and P. Glynn. *Stochastic Simulation: Algorithms and Analysis.* Stochastic Modelling and Applied Probability. Springer, Berlin, Germany, 2007.

S. Asmussen and D. P. Kroese. Improved algorithms for rare event simulation with heavy tails. *Advances in Applied Probability*, 38:545–558, 2006.

S. Asmussen and J. Rosiński. Approximations of small jumps of Lévy processes with a view towards simulation. *Journal of Applied Probability*, 38(2):482–493, 2001.

M. Avellaneda and J. Newman. Positive Interest Rates and Non-Linear Term Structure Models. CIMS-NYU Working Paper (unpublished), 1998.

A. Avramidis and P. L'Ecuyer. Efficient Monte Carlo and quasi-Monte Carlo option pricing under the variance gamma model. *Management Science*, 52 (12):1930–1944, 2006.

L. F. Bachelier. Théorie de la spéculation. *Annales Scientifique de l'École Normal Superieure*, 17:21–86, 1900.

B. Baldessari. The distribution of a quadratic form of normal random variables. *Annals of Mathematical Statistics*, 38:1700–1704, 1967.

V. Bally and D. Talay. The law of the Euler scheme for stochastic differential equations. I: Convergence rate of the distribution function. *Probability Theory and Related Fields*, 104(1):43–60, 1996.

V. Bally, G. Pagès, and J. Printems. A quantization tree method for pricing and hedging multidimensional American options. *Mathematical Finance*, 15(1):119–168, 2005.

O. E. Barndorff-Nielsen. Exponentially decreasing distributions for the logarithm of particle size. *Proceedings of the Royal Society of London, Series A*, 353:401–419, 1977.

O. E. Barndorff-Nielsen. Processes of normal inverse Gaussian type. *Finance and Stochastics*, 2(1):41–68, 1997.

O. E. Barndorff-Nielsen and N. Shephard. Non-Gaussian Ornstein-Uhlenbeck-based models and some of their uses in financial economics. *Journal of the Royal Statistical Society, Series B*, 63:167–241, 2001.

D. S. Bates. Jumps and stochastic volatility: Exchange rate processes implicit in Deutsche Mark options. *Review of Financial Studies*, 9(1):69–107, 1996.

D. Bauer and J. Ruß. Pricing longevity bonds using implied survival probabilities, 2006. URL `http://www.ifa-ulm.de/downloads/ImpliedSurv.pdf`.

D. Belomestny, C. Bender, and J. Schoenmakers. True upper bounds for Bermudan products via non-nested Monte Carlo. *Mathematical Finance*, 19(1):53–71, 2009.

F. E. Benth, M. Groth, and P. C. Kettler. A quasi Monte Carlo algorithm for the normal inverse Gaussian distribution and valuation of financial derivatives. *International Journal of Theoretical and Applied Finance*, 9(6):843–867, 2006.

L. Bergomi. Smile dynamics II. *Risk*, October:67–73, 2005.

P. Billingsley. *Convergence of Probability Measures.* Wiley, New York, USA, 1968.

N. H. Bingham and R. Kiesel. *Risk-Neutral Valuation: Pricing and Hedging of Financial Derivatives.* Springer Finance. Springer, Berlin, Germany, 1998.

T. Björk. *Arbitrage Theory in Continuous Time.* Oxford University Press, Oxford, UK, 2nd edition, 2004.

F. Black. The pricing of commodity contracts. *Journal of Financial Economics*, 3:167–179, 1976.

F. Black and P. Karasinski. Bond and option pricing when short rates are log-normal. *Financial Analysts Journal*, 47(4):52–59, 1991.

F. Black and M. Scholes. The pricing of options and corporate liabilities. *Journal of Political Economics*, 81:637–654, 1973.

D. Blake, A. Cairns, and K. Dowd. Living with mortality: Longevity bonds and other mortality-linked securities. *British Actuarial Journal*, 12:153–197, 2006.

P. P. Boyle. Options: A Monte Carlo approach. *Journal of Financial Economics*, 4:323–338, 1977.

P. P. Boyle, M. Broadie, and P. Glasserman. Monte Carlo methods for security pricing. *Journal of Economic Dynamics and Control*, 21:1267–1321, 1997.

A. Brace, D. Gatarek, and M. Musiela. The market model of interest rate dynamics. *Mathematical Finance*, 7:127–155, 1997.

P. Bratley and B. Fox. ALGORITHM 659: Implementing Sobol's quasi random sequence generator. *ACM Transactions on Mathematical Software*, 14 (1):88–100, 1988.

D. Brigo and F. Mercurio. *Interest Rate Models: Theory and Practice.* Springer Finance. Springer, Berlin, Germany, 2001.

M. Broadie and Ö. Kaya. Exact simulation of stochastic volatility and other affine jump diffusion processes. *Operations Research*, 54(2):217–231, 2006.

M. Broadie, P. Glasserman, and S. Kou. A continuity correction for discrete barrier options. *Mathematical Finance*, 7:325–349, 1997.

M. Broadie, P. Glasserman, and S. Kou. Connecting discrete and continuous path-dependent options. *Finance and Stochastics*, 3(1):55–82, 1999.

N. Bruti-Liberati and E. Platen. Approximation of jump diffusions in finance and economics. *Computational Economics*, 29(3):283–312, 2007.

J. A. Bucklew. *Large Deviation Techniques in Decision, Simulation and Es-*

timation. Wiley, New York, USA, 1990.

H. Bühlmann. *Mathematical Methods in Risk Theory*. Grundlehren der mathematischen Wissenschaften. Springer, Berlin, Germany, 2005.

H. Bühlmann. Experience rating and credibility. *ASTIN Bulletin*, 4:199–207, 1967.

H. Bühlmann. Experience rating and credibility. *ASTIN Bulletin*, 5:157–165, 1969.

H. Bühlmann and E. Straub. Glaubwürdigkeit für Schadensätze. *Mitteilungen der Vereinigung Schweizer Versicherungsmathematiker*, 70:111–133, 1970.

A. Cairns, D. Blake, and K. Dowd. A two-factor model for stochastic mortality with parameter uncertainty: Theory and calibration. *Journal of Risk and Insurance*, 73:687–718, 2006.

P. Carr, H. Geman, D. Madan, and M. Yor. The fine structure of asset returns: An empirical investigation. *Journal of Business*, 75:305–332, 2002.

A. Carverhill. When is the short rate Markovian? *Mathematical Finance*, 4 (4):305–312, 1994.

O. Cheyette. Term structure dynamics and mortgage valuation. *Journal of Fixed Income*, March:28–41, 1992.

O. Cheyette. Markov representation of the Heath-Jarrow-Morton model, 1995. URL `http://papers.ssrn.com/sol3/papers.cfm?abstract_id=6073`.

E. Clément, D. Lamberton, and P. Protter. An analysis of a least squares regression method for American option pricing. *Finance and Stochastics*, 6 (4):449–471, 2002.

P. D. Coddington. Random number generators for parallel computers, 1996. URL `http://wotug.ukc.ac.uk/parallel/nhse/NHSEreview/RNG/`.

R. Cont and P. Tankov. *Financial Modelling with Jump Processes*. Financial Mathematics Series. Chapman & Hall, CRC Press, Boca Raton, Florida, USA, 2003.

D. Cox and V. Isham. *Point Processes*. Chapman & Hall, CRC Press, Boca Raton, Florida, USA, 1980.

J. C. Cox, J. E. Ingersoll, and S. A. Ross. A theory of the term structure of interest rates. *Econometrica*, 53:385–407, 1985.

C. Czado. Einführung zu Markov Chain Monte Carlo Verfahren mit Anwendung auf Gesamtschadenmodelle. *Blätter der DGVFM*, 26(3):331–350, 2004.

M. Dahl and T. Møller. Valuation and hedging of life insurance liabilities with systematic mortality risk. *Insurance Mathematics & Economics*, 39 (2):193–217, 2006.

D. Davydov and V. Linetsky. The valuation and hedging of barrier and look-back options under the CEV process. *Management Science*, 47:949–965, 2001.

G. Deelstra and F. Delbaen. Convergence of discretized stochastic (interest rate) processes with stochastic drift term. *Applied Stochastic Models and Data Analysis*, 14(1):77–84, 1998.

F. Delbaen and W. Schachermayer. *The Mathematics of Arbitrage.* Springer Finance. Springer, Berlin, Germany, 2006.

O. Deprez and H. Gerber. On convex principles of premium calculation. *Insurance: Mathematics and Economics*, 4:179–189, 1985.

L. Devroye. *Non-Uniform Random Variate Generation.* Springer, New York, USA, 1986.

L. Devroye. Random variate generation in one line of code. In *1996 Winter Simulation Conference Proceedings*, 265–272, ACM, New York, USA, 1996.

M. D. Donsker. Justification and extension of Doob's heuristic approach to the Kolmogorov-Smirnov theorems. *Annals of the Institute of Statistical Mathematics*, 23:277–281, 1952.

U. L. Dothan. On the term structure of interest rates. *Journal of Financial Economics*, 6:59–69, 1978.

D. Duffie. *Dynamic Asset Pricing Theory.* Princeton Series in Finance. Princeton University Press, Princeton, New Jersey, USA, 3rd edition, 2001.

D. Duffie and P. Glynn. Efficient Monte Carlo simulation of security prices. *Annals of Applied Probability*, 5(4):897–905, 1995.

D. Dufresne. The integrated square-root process. Technical report, University of Montreal, 2001. URL `http://mercury.ecom.unimelb.edu.au/SITE/actwww/wps2001.shtml`.

J. Dunkel and S. Weber. Efficient Monte Carlo methods for convex risk measures in portfolio credit risk models. In S. Henderson, B. Biller, M.-H. Hsieh, J. Shortle, J. Tew, and R. Barton, editors, *Proceedings of the 2007 Winter Simulation Conference*, 958–966, ACM, New York, USA, 2007.

B. Dupire. Pricing and hedging with smiles. In M. A. Dempster and S. R. Pliska, editors, *Mathematics of Derivative Securities*, 103–111, Cambridge University Press, Cambridge, UK, 1997.

R. Durrett. *Essentials of Stochastic Processes.* Springer Texts in Statistics. Springer, Berlin, Germany, 1999.

P. H. Dybvig. Bond and bond option pricing based on the current term structure. In M. A. Dempster and S. R. Pliska, editors, *Mathematics of Derivative Securities*, 271–293, Cambridge University Press, Cambridge, UK, 1997.

E. Eberlein and U. Keller. Hyperbolic distributions in finance. *Bernoulli*, 1: 281–299, 1995.

D. Egloff. Monte Carlo algorithms for optimal stopping and statistical learning. *Annals of Applied Probability*, 15(2):1396–1432, 2005.

J. Eichenauer-Herrmann. Pseudorandom number generation by non-linear methods. *International Statistical Reviews*, 63:247–255, 1995.

J. Eichenauer-Herrmann and J. Lehn. A non-linear congruential pseudo random number generator. *Statistical Papers*, 27:315–326, 1986.

P. Embrechts, F. Lindskog, and A. McNeil. Modelling dependence with copulas and applications to risk management. In S. T. Rachev, editor, *Handbook of Heavy Tailed Distributions in Finance*, 329–384, Handbooks in Finance, Elsevier, Amsterdam, Netherlands, 2003.

K. Entacher. A collection of selected pseudorandom number generators with linear structure. *Technical report 97-1, ACPC-Austrian Center for Parallel Computation*, 1997.

N. Etemadi. An elementary proof of the strong law of large numbers. *Zeitschrift für Wahrscheinlichkeitstheorie und Verwandte Gebiete*, 55:119–122, 1981.

M. Evans and T. Swartz. *Approximating Integrals via Monte Carlo and Deterministic Methods.* Oxford University Press, Oxford, UK, 2000.

W. Feller. Diffusion processes in genetics. In J. Neyman, editor, *Proceedings of the Second Berkeley Symposium on Mathematical Statistics and Probability, 1950*, 227-246, University of California Press, Berkeley, California, USA, 1951.

T. Fischer. Risk capital allocation by coherent risk measures based on one-sided moments. *Insurance: Mathematics and Economics*, 32(1):135–146, 2003.

B. Flesaker and L. Hughston. Positive interest. *Risk*, January:46–49, 1996.

H. Föllmer and A. Schied. *Stochastic Finance: An Introduction in Discrete Time.* de Gruyter, Berlin, Germany, 2002.

E. Fournié, J.-M. Lasry, J. Lebuchoux, P.-L. Lions, and N. Touzi. Applications of Malliavin calculus to Monte Carlo methods in finance. *Finance and Stochastics*, 3(4):391–412, 1999.

M. C. Fu, D. B. Madan, and T. Wang. Pricing continuous Asian options: A comparison of Monte Carlo and Laplace transform inversion method. *Journal of Computational Finance*, 2(2):49–74, 1999.

H. Geman and M. Yor. Bessel processes, Asian options and perpetuities. *Mathematical Finance*, 3(4):349–375, 1993.

S. Geman and D. Geman. Stochastic relaxation, Gibbs distributions and the Bayesian restoration of images. *IEEE Transactions on Pattern Analysis and Machine Intelligence*, 6:721–741, 1984.

H. U. Gerber. *Life Insurance Mathematics*. Springer, Berlin, Germany, 3rd edition, 1997.

H. U. Gerber and E. S. Shiu. Option pricing by Esscher transforms. *Transactions of the Society of Actuaries*, 46:99–191, 1994.

M. Giles. Improved multilevel Monte Carlo convergence using the Milstein scheme. In *Monte Carlo and Quasi-Monte Carlo Methods 2006*, 343–358, Springer, Berlin, Germany, 2007.

M. Giles. Multi-level Monte Carlo path simulation. *Operations Research*, 56 (3):607–617, 2008.

M. Giles and P. Glasserman. Smoking adjoints: Fast Monte Carlo Greeks. *Risk*, January:88–92, 2006.

W. Gilks, S. Richardson, and D. Spiegelhalter. *Markov Chain Monte Carlo in Practice*. Chapman & Hall, CRC Press, Boca Raton, Florida, USA, 1996.

P. Glasserman. *Monte Carlo Methods in Financial Engineering*. Springer, New York, USA, 2004.

P. Glasserman and J. Staum. Conditioning on one-step survival for barrier option simulations. *Operations Research*, 49(6):923–937, 2001.

P. Glasserman and B. Yu. Large sample properties of weighted Monte Carlo estimators. *Operations Research*, 53(2):298–312, 2005.

P. Glasserman, P. Heidelberger, and P. Shahabuddin. Stratification issues in estimating value-at-risk. In P. Farrington, H. Nembhard, D. Sturrock, and G. Evans, editors, *Proceedings of the 1999 Winter Simulation Conference*, IEEE Press, Phoenix, Arizona, USA, 1999.

P. Glasserman, P. Heidelberger, and P. Shahabuddin. Importance sampling and stratification for value-at-risk. In Y. S. Abu-Mostafa, B. LeBaron, A. W. Lo, and A. S. Weigend, editors, *Computational Finance 1999*, MIT Press, Cambridge, Massachusetts, USA, 2000a.

P. Glasserman, P. Heidelberger, and P. Shahabuddin. Variance reduction techniques for estimating value-at-risk. *Management Science*, 46(10):1349–

1364, 2000b.

P. Glasserman, P. Heidelberger, and P. Shahabuddin. Efficient Monte Carlo methods for value-at-risk. In C. Alexander, editor, *Mastering Risk*, Prentice Hall, Upper Saddle River, New Jersey, USA, 2001.

P. W. Glynn. Importance sampling for Monte Carlo estimation of quantiles. Technical report, Publishing House of Saint Petersburg University, 1996. URL `http://citeseer.ist.psu.edu/70000.html`.

E. Gobet. Advanced Monte Carlo methods for barrier and related exotic options. In P. Ciarlet, A. Bensoussan, and Q. Zhang, editors, *Mathematical Modeling and Numerical Methods in Finance, 15*, 497–528, Handbook of Numerical Analysis, Elsevier, Amsterdam, Netherlands, 2009.

E. Gobet and S. Menozzi. Discrete sampling of functionals of Itô processes. In C. Donati-Martin, M. Émery, A. Rouault, and C. Stricker, editors, *Séminaire de Probabilités XL*, 355–374, Lecture Notes in Mathematics, Springer, Berlin, Germany, 2007.

B. Grünewald. *Hedging in unvollständigen Märkten am Beispiel des Sprung-Diffusionsmodells.* PhD thesis, University of Mainz, Germany, 1998.

O. Haeggstrøm. *Finite Markov Chains and Algorithmic Applications.* Number 52 in Student Texts. London Mathematical Society, London, UK, 2003.

P. S. Hagan, D. Kumar, A. S. Lesniewski, and D. E. Woodward. Managing smile risk. *Wilmott Magazine*, September:84–108, 2002.

J. M. Hammersley and D. C. Handscomb. *Monte Carlo Methods.* Chapman & Hall, CRC Press, Boca Raton, Florida, USA, 1964.

H. Haramoto, M. Matsumoto, T. Nishimura, F. Panneton, and P. L'Ecuyer. Efficient jump ahead for $\mathbb{F}_2$-linear random number generators. *Journal on Computing*, 20(3):385–390, 2008.

J. Harrison and S. R. Pliska. Martingales and stochastic integrals in the theory of continuous trading. *Stochastic Processes and Applications*, 11:215–260, 1981.

J. Hartinger and M. Predota. Simulation methods for valuing Asian option prices in a hyperbolic asset price model. *IMA Journal of Management Mathematics*, 14(1):65–81, 2003.

W. K. Hastings. Monte Carlo sampling methods using Markov chains and their applications. *Biometrika*, 57:97–109, 1970.

M. B. Haugh and L. Kogan. Pricing American options: A duality approach. *Operations Research*, 52(2):258–270, 2004.

D. Heath and E. Platen. A variance reduction technique based on integral representations. *Quantitative Finance*, 2(5):362–369, 2002.

D. Heath, R. A. Jarrow, and A. Morton. Bond pricing and the term structure of interest rates: A new methodology for contingent claims valuation. *Econometrica*, 60(1):77–105, 1992.

S. Heinrich. Multilevel Monte Carlo methods. In S. Margenov, J. Wasniewski, and P. Yalamov, editors, *Large-Scale Scientific Computing. 3rd International Conference*, 58–67, Lecture Notes in Computer Science, Springer, Berlin, Germany, 2001.

P. Hellekalek. Inversive pseudorandom number generators: Concept, results and links. In D. Goldsman, C. Alexopoulos, and K. Kang, editors, *Proceedings of the 1995 Winter Simulation Conference*, 252–262, ACM, New York, USA, 1995.

S. L. Heston. A closed-form solution for options with stochastic volatility with applications to bond and currency options. *Review of Financial Studies*, 6 (2):327–343, 1993.

T. Hida. *Brownian Motion.* Applications of Mathematics. Springer, Berlin, Germany, 1980.

D. J. Higham and X. Mao. Convergence of Monte Carlo simulations involving the mean-reverting square root process. *Journal of Computational Finance*, 8(3):35–61, 2005.

D. Hincin. Asymptotische Gesetze der Wahrscheinlichkeitsrechnung. *Ergebnisse der Mathematik*, 2(4), 1933.

T. S. Y. Ho and S.-B. Lee. Term structure and pricing interest rate contingent claims. *Journal of Finance*, 41(5):1011–1029, 1986.

J. Hull and A. White. Forward rate volatilities, swap rate volatilities, and the implementation of the LIBOR market model. *Journal of Fixed Income*, 10 (3):46–62, 2000.

J. Hull and A. White. The pricing of options on assets with stochastic volatilities. *Journal of Finance*, 42(2):281–300, 1987.

J. Hull and A. White. Pricing interest rate derivative securities. *Review of Financial Studies*, 3(4):573–592, 1990.

C. Hunter, P. Jäckel, and M. Joshi. Getting the drift. *Risk*, July:81–84, 2001.

J. Imhof. Computing the distribution of quadratic forms in normal variables. *Biometrika*, 48:419–426, 1961.

N. Imkeller. The multi-level Monte Carlo method with applications in financial mathematics. Master's thesis, University of Kaiserslautern, Germany, 2009.

P. Jäckel. *Monte Carlo Methods in Finance.* Wiley, Chichester, UK, 2003.

J. Jacod. The Euler scheme for Lévy driven stochastic differential equations: Limit theorems. *Annals of Applied Probability*, 32(3):1830–1872, 2004.

F. Jamshidian. LIBOR and swap market models and measures. *Finance and Stochastics*, 1:293–330, 1997.

M. Jeanblanc-Picqué and M. Pontier. Optimal portfolio for a small investor in a market model with discontinuous prices. *Applied Mathematics and Optimization*, 22(3):287–310, 1990.

M. Johannes and N. Polson. MCMC methods for continuous-time financial econometrics. In Y. Ait-Sahalia and L. Hansen, editors, *Handboook of Financial Econometrics*, Handbooks in Finance, 2nd volume, Elsevier, Amsterdam, Netherlands, 2010.

M. Joshi and A. Stacey. New and robust drift approximations for the LIBOR market model. *Quantitative Finance*, 8(4):427–434, 2008.

C. Kahl and P. Jäckel. Fast strong approximation Monte-Carlo schemes for stochastic volatility models. *Journal of Quantitative Finance*, 6(6):513–536, 2006.

I. Karatzas and S. E. Shreve. *Brownian Motion and Stochastic Calculus.* Springer, Berlin, Germany, 2nd edition, 1991.

I. Karatzas and S. E. Shreve. *Methods of Mathematical Finance.* Springer, Berlin, Germany, 1998.

A. Kebaier. Statistical Romberg extrapolation: A new variance reduction method and applications to option pricing. *Annals of Applied Probability*, 15(4):2681–2705, 2005.

A. Kemna and A. Vorst. A pricing method for options based on average asset values. *Journal of Banking and Finance*, 14:113–129, 1990.

P. E. Kloeden and E. Platen. *Numerical Solution of Stochastic Differential Equations.* Springer, Berlin, Germany, 1999.

C. Klüppelberg. Subexponential distributions and integrated tails. *Journal of Applied Probability*, 25(1):132–141, 1988.

C. Klüppelberg and T. Mikosch. Explosive Poisson shot noise processes with applications to risk reserves. *Bernoulli*, 1:125–147, 1995.

D. E. Knuth. *The Art of Computer Programming, Volume 2 (Seminumerical Algorithms).* Addison-Wesley, Reading, Massachusetts, USA, 3rd edition, 1998.

R. Korn and E. Korn. *Option Pricing and Portfolio Optimization.* Graduate Studies in Mathematics. American Mathematical Society, Providence, Rhode Island, USA, 2001.

R. Korn and L. Rogers. Stocks paying discrete dividends: Modelling and option pricing. *Journal of Derivatives*, 13(2):44–49, 2005.

R. Korn, K. Natcheva, and J. Zipperer. Longevity bonds: Pricing, modelling and application for German data. *Blätter der DGVFM*, XXVII(3), 2006.

S. G. Kou. A jump diffusion model for option pricing. *Management Science*, 48:1086–1101, 2002.

M. Krekel, J. de Kock, R. Korn, and T.-K. Man. An analysis of pricing methods for basket options. *Wilmott Magazine*, May:82–89, 2004.

U. Küchler, K. Neumann, M. Sørensen, and A. Streller. Stock returns and hyperbolic distributions. *Mathematical and Computer Modelling*, 29:1–15, 1999.

R. J. Laeven and M. J. Goovaerts. Premium calculation and insurance pricing. In E. L. Melnick and B. S. Everitt, editors, *Encyclopedia of Quantitative Risk Analysis and Assessment*, 1302–1314, Wiley, New York, USA, 2008.

P. L'Ecuyer. Quasi-Monte Carlo methods in finance. In R. Ingalls, M. Rossetti, J. Smith, and B. Peters, editors, *Proceedings of the 2004 Winter Simulation Conference*, 1645–1655, ACM, New York, USA, 2004.

P. L'Ecuyer. Uniform random number generation. *Annals of Operations Research*, 53:77–120, 1994.

P. L'Ecuyer. Combined multiple recursive random number generators. *Operations Research*, 44(5):816–822, 1996a.

P. L'Ecuyer. Maximally equidistributed combined Tausworthe generators. *Mathematics of Computation*, 65(213):203–213, 1996b.

P. L'Ecuyer. Bad lattice structure for vectors of non-successive values produced by some linear recurrences. *INFORMS Journal on Computing*, 9(1): 57–60, 1997.

P. L'Ecuyer. Good parameters and implementations for combined multiple recursive random number generators. *Operations Research*, 47(1):159–164, 1999a.

P. L'Ecuyer. Tables of maximally-equidistributed combined LFSR generators. *Mathematics of Computation*, 68(225):261–269, 1999b.

P. L'Ecuyer and J. Granger-Piché. Combined generators with components from different families. *Mathematics and Computers in Simulation*, 62: 395–404, 2003.

P. L'Ecuyer and F. Panneton. Fast random number generators based on linear recurrences modulo 2: Overview and comparison. In M. E. Kuhl, N. M. Steiger, F. B. Armstrong, and J. A. Joines, editors, *Proceedings of the 2005*

Winter Simulation Conference, 110–119, ACM, New York USA, 2005.

P. L'Ecuyer and F. Panneton. $\mathbb{F}_2$-linear random number generators. In *Advancing the Frontiers of Simulation: A Festschrift in Honor of George S. Fishman*, Springer, Berlin, Germany, 2007.

P. L'Ecuyer and R. Simard. TestU01: A software library in ANSI C for empirical testing of random number generators, 2002. URL `http://www.iro.umontreal.ca/~lecuyer`.

P. L'Ecuyer and R. Simard. Inverting the symmetrical beta distribution. *ACM Transactions on Mathematical Software*, 32(4):509–520, 2006.

P. L'Ecuyer, F. Panneton, and M. Matsumoto. Improved long-period generators based on linear recurrences modulo 2. *ACM Transactions on Mathematical Software*, 32(1):1–16, 2006.

P. Lee. *Bayesian Statistics: An Introduction.* Arnold Publishing, New York, USA, 1997.

R. Lee and L. Carter. Modeling and forecasting U.S. mortality. *Journal of the American Statistical Association*, 87(14):659–671, 1992.

D. Lehmer. Mathematical methods in large-scale computing units. In *Proceedings of the 2nd Symposium on Large-Scale Digital Calculating Machinery*, 141–146, Harvard University Press, Cambridge, Massachusetts, USA, 1949.

E. Lévy. Pricing European average rate currency options. *Journal of International Money and Finance*, 11(5):474–491, 1992.

J. S. Liu. *Monte Carlo Strategies in Scientific Computing.* Springer, Berlin, Germany, 2001.

W. Loh. On latin hypercube sampling. *The Annals of Statistics*, 24(5):2058–2080, 1996.

F. A. Longstaff and E. S. Schwartz. Valuing American options by simulation: A simple least-squares approach. *The Review of Financial Studies*, 14(1): 113–147, 2001.

R. Lord, R. Koekkoek, and D. van Dijk. A comparison of biased simulation schemes for stochastic volatility models, 2008. URL `http://papers.ssrn.com/sol3/papers.cfm?abstract_id=903116`.

F. Lundberg. *I. Approximerad framställning af sannolikhetsfunctionen. II. Återförsäkring af kollektivrisken.* Almqvist & Wiksell, Uppsala, Sweden, 1903.

D. Lunn, A. Thomas, N. Best, and D. Spiegelhalter. WinBUGS – A Bayesian modelling framework: Concepts, structure, and extensibility. *Statistics and Computing*, 10:325–337, 2000.

M. Lüscher. RANLUX. *Computer Physics Communications*, 79:110, 1994.

D. B. Madan and F. Milne. Option pricing with v.g. martingale components. *Mathematical Finance*, 1(4):39–55, 1991.

D. B. Madan and E. Seneta. The variance gamma model for share market returns. *Journal of Business*, 63:511–524, 1990.

D. B. Madan, P. P. Carr, and E. C. Chang. The variance gamma process and option pricing. *European Finance Review*, 2(1):79–105, 1998.

G. Marsaglia. Xorshift RNGs. *Journal of Statistical Software*, 8(14):1–6, 2003.

G. Marsaglia. Evaluating the normal distribution. *Journal of Statistical Software*, 11(4):1–11, 2004.

G. Marsaglia. The Marsaglia random number CDROM including the Diehard battery of tests of randomness, 1996. URL `http://stat.fsu.edu/pub/diehard`.

A. Marshall and I. Olkin. Families of multivariate distributions. *Journal of the American Statistical Association*, 83(403):834–841, 1988.

M. Matsumoto and Y. Kurita. Twisted GFSR generators. *ACM Transactions on Modeling and Computer Simulation*, 2:179–1940, 1992.

M. Matsumoto and T. Nishimura. Dynamic creation of pseudorandom number generators. In *Monte Carlo and Quasi-Monte Carlo Methods 1998*, 56–69, Springer, Berlin, Germany, 2000.

M. Matsumoto and T. Nishimura. Mersenne Twister: A 623-dimensionally equidistributed uniform pseudo-random number generator. *ACM Transactions on Modeling and Computer Simulation*, 8:3–30, 1998.

M. Matsumoto and M. Saito. SIMD-oriented fast Mersenne Twister: A 128-bit pseudorandom number generator. In *Monte Carlo and Quasi-Monte Carlo Methods 2006*, 607–622, Springer, Berlin, Germany, 2008.

M. D. McKay, R. Beckman, and W. J. Conover. A comparison of three methods for selecting values of input variables in the analysis of output from a computer code. *Technometrics*, 21:239–245, 1979.

A. McNeil, R. Frey, and P. Embrechts. *Quantitative Risk Management: Concepts, Techniques, and Tools.* Princeton Series in Finance. Princeton University Press, Princeton, New Jersey, USA, 2005.

R. C. Merton. Theory of rational option pricing. *Bell Journal of Economics and Management Science*, 4(1):141–183, 1973.

R. C. Merton. Option pricing when underlying stock returns are discontinuous. *Journal of Financial Economics*, 3:125–144, 1976.

N. Metropolis and S. Ulam. The Monte Carlo method. *Journal of the American Statistical Association*, 44:335–341, 1949.

N. Metropolis, A. Rosenbluth, M. Rosenbluth, A. Teller, and E. Teller. Equations of state calculations by fast computing machines. *Journal of Chemical Physics*, 21:1087–1091, 1953.

T. Mikosch. *Non-Life Insurance Mathematics: An Introduction with Stochastic Processes.* Springer, Berlin, Germany, 2004.

G. Milstein. A method of second-order accuracy integration of stochastic differential equations. *Theory of Probability and its Applications*, 19:557–562, 1978.

K. Miltersen, K. Sandmann, and D. Sondermann. Closed form solutions for term structure derivatives with log-normal interest rates. *Journal of Finance*, 52(1):409–430, 1997.

T. Møller and M. Steffensen. *Market-Valuation Methods in Life and Pension Insurance.* Cambridge University Press, Cambridge, UK, 2007.

K.-S. Moon. Efficient Monte Carlo algorithm for pricing barrier options. *Communications of the Korean Mathematical Society*, 23(2):285–294, 2008.

A. Müller and D. Stoyan. *Comparison Methods for Stochastic Models and Risks.* Wiley, New York, USA, 2002.

R. Myneni. The pricing of the American option. *Annals of Applied Probability*, 2(1):1–23, 1992.

V. Naik and M. Lee. General equilibrium pricing of options on the market portfolio with discontinuous returns. *Review of Financial Studies*, 3(4): 493–521, 1990.

H. Niederreiter. The multiple-recursive matrix method for pseudorandom number generation. *Finite Fields and their Applications*, 1:3–30, 1995.

R. Norberg. Credibility theory, 2002. URL `http://stats.lse.ac.uk/norberg/links/papers/CRED-eas.pdf`.

A. Owen. A central limit theorem for Latin hypercube sampling. *Journal of the Royal Statistical Society, Series B*, 54:541–551, 1992.

S. Park and K. Miller. Random number generators: Good ones are hard to find. *Communications of the ACM*, 31:1192–1201, 1988.

P. Pellizzari. Efficient Monte Carlo pricing of European options using mean value control variates. *Decisions in Economics and Finance*, 24(2):107–126, 2001.

E. Pitacco. Survival models in actuarial mathematics: From Halley to longevity risk. Technical report, University of Trieste, 2003. URL `http:`

//www.univ.trieste.it/~matappl/PDF%20file/155.ps.

V. V. Piterbarg. A practitioner's guide to pricing and hedging callable LIBOR exotics in forward LIBOR models, 2003. URL http://papers.ssrn.com/sol3/papers.cfm?abstract_id=427084.

V. V. Piterbarg. Computing deltas of callable LIBOR exotics in forward LIBOR models. *Journal of Computational Finance*, 7(2):107–143, 2004.

W. Press, S. Teukolsky, W. Vettering, and B. Flannery. *Numerical Recipes in C++*. Cambridge University Press, Cambridge, UK, 2nd edition, 2002.

P. Protter. *Stochastic Integration and Differential Equations.* Springer, Heidelberg, Germany, 2nd edition, 2004.

T. D. Protter P. Convergence rate of the Euler scheme for stochastic differential equations driven by Lévy processes. *Annals of Probability*, 25(1): 393–423, 1997.

N. S. Rasmussen. Control variates for Monte Carlo valuation of American options. *Journal of Computational Finance*, 9(1):84–102, 2005.

E. Reiner and M. Rubinstein. Breaking down the barriers. *Risk*, September: 28–35, 1991.

C. Ribeiro and N. Webber. A Monte Carlo method for the normal inverse Gaussian option valuation model using an inverse Gaussian bridge, 2003. URL http://www.mbs.ac.uk/research/accountingfinance/documents/Webberseminarpaper.pdf.

B. Ripley. *Stochastic Simulation.* Wiley, New York, USA, 1987.

P. Ritchken and L. Sankarasubramanian. Volatility structure of forward rates and the dynamics of the term structure. *Mathematical Finance*, 5(1):55–72, 1995.

L. Rogers. Monte Carlo valuation of American options. *Mathematical Finance*, 12(3):271–286, 2002.

L. Rogers. The potential approach to the term-structure of interest rates and foreign exchange rates. *Mathematical Finance*, 7:157–164, 1997.

L. Rogers and Z. Shi. The value of an Asian option. *Journal of Applied Probability*, 32(4):1077–1088, 1995.

J. Rosiński. Series representations of Lévy processes from the perspective of point processes. In O. Barndorff-Nielsen, T. Mikosch, and S. Resnick, editors, *Lévy Processes – Theory and Applications*, 410–415, Birkhäuser, Basel, Switzerland, 2001.

R. Y. Rubinstein. *Simulation and the Monte Carlo Method.* Wiley, New York, USA, 1981.

D. Rudolf. Explicit error bounds for lazy reversible Markov Chain Monte Carlo. *Journal of Complexity*, 25:11–24, 2009.

A. Rukhin, J. Soto, J. Nechvatal, M. Smid, E. Barker, S. Leigh, M. Levenson, M. Vangel, D. Banks, A. Heckert, J. Dray, and S. Vo. A statistical test suite for random and pseudorandom number generators for cryptographic applications. *National Institute of Standards and Technology (NIST) Special Publication 800-22*, 2001.

T. H. Rydberg. The normal inverse Gaussian Lévy process: Simulation and approximation. *Communications in Statistics: Stochastic Models*, 13(4): 887–910, 1997.

P. A. Samuelson. Proof that properly anticipated prices fluctuate randomly. *Industrial Management Review*, 6(2):41–49, 1965.

K. Sato. Semi-stable processes and their extensions. In N. Kono and N.-R. Shieh, editors, *Trends in Probability and Related Analysis Communications in Statistics – Stochastic Models, Proc. SAP 1998*, 129-145, World Scientific Publishing Company, Singapore, 1999.

J. G. Schoenmakers. *Robust LIBOR Modelling and Pricing of Derivative Products.* CRC Press, Boca Raton, Florida, USA, 2007.

P. J. Schönbucher. *Credit Derivatives Pricing Models.* Wiley, New York, USA, 2003.

W. Schoutens. *Stochastic Processes and Orthogonal Polynomials.* Springer, Berlin, Germany, 2000.

W. Schoutens. *Lévy Processes in Finance: Pricing Financial Derivatives.* Wiley, New York, USA, 2003.

W. Schoutens and J. Teugels. Lévy processes, polynomials and martingales. *Communications in Statistics – Stochastic Models*, 14:335–349, 1998.

M. Schroder. Computing the constant elasticity of variance option pricing formula. *Journal of Finance*, 44:211–219, 1989.

M. Schweizer. Option hedging for semimartingales. *Stochastic Processes and their Applications*, 37:339–363, 1991.

D. Scollnik. Actuarial modelling with MCMC and BUGS. *North American Actuarial Journal*, 5:96–124, 2001.

M. Sklar. Fonctions de répartition à n dimensions et leur marges. *Publications de l'Institut de Statistique de l'Université de Paris*, 8:229–231, 1960.

M. Steffensen. A no arbitrage approach to Thiele's differential equation. *Insurance Mathematics & Economics*, 27(2):201–214, 2000.

E. M. Stein and J. Stein. Stock price distributions with stochastic volatility:

An analytic approach. *Review of Financial Studies*, 4(4):727–752, 1991.

M. Stein. Large sample properties of simulations using latin hypercube sampling. *Technometrics*, 29:141–151, 1987.

J. Stoer and R. Bulirsch. *Introduction to Numerical Analysis*. Springer, Berlin, Germany, 2nd edition, 1993.

B. Sundt. *An Introduction to Non-Life Insurance Mathematics*. VVW Karlsruhe, Germany, 3rd edition, 1993.

D. Talay and L. Tubaro. Expansion of the global error for numerical schemes solving stochastic differential equations. *Stochastic Analysis and Applications*, 8(4):483–509, 1990.

J. N. Tsitsiklis and B. van Roy. Regression methods for pricing complex American-style options. *IEEE Transactions on Neural Networks*, 12(4): 694–703, 2001.

J. N. Tsitsiklis and B. van Roy. Optimal stopping of Markov processes: Hilbert space theory, approximation algorithms, and an application to pricing high-dimensional financial derivatives. *IEEE Transactions on Automatic Control*, 44(10):1840–1851, 1999.

S. Turnbull and L. Wakeman. A quick algorithm for pricing European average options. *Journal of Financial and Quantitative Analysis*, 26:377–389, 1991.

O. Ugur. *An Introduction to Computational Finance*. Series in Quantitative Finance. Imperial College Press, London, UK, 2009.

O. A. Vasicek. An equilibrium characterization of the term structure. *Journal of Financial Economics*, 5:177–188, 1977.

J.-Y. Wang. Variance reduction for multivariate Monte Carlo simulation. *The Journal of Derivatives*, 16:7–28, 2008.

X. Wang. Constructing robust good lattice rules for computational finance. *SIAM Journal on Scientific Computing*, 29(2):598–621, 2007.

S. Wendel. The Longstaff-Schwartz algorithm for pricing American options. Master's thesis, University of Kaiserslautern, Germany, 2009.

Index